D0142659

Introduction to Probability

Mark Daniel Ward
Associate Professor of Statistics
Purdue University

Ellen Gundlach
Education Specialist and Continuing Lecturer in Statistics
Purdue University

W. H. Freeman and Company
A Macmillan Education Imprint

Publisher: Terri Ward
Senior Acquisitions Editor: Karen Carson
Marketing Manager: Cara LeClair
Developmental Editor: Katrina Mangold
Editorial Assistant: Victoria Garvey
Photo Editor: Cecilia Varas
Cover Designer: Diana Blume
Text Designer: Mark Daniel Ward
Project Editor: Liz Geller
Copyeditor: Elizabeth Marraffino
Illustrations: Mark Daniel Ward; Eli Ensor
Illustration Coordinator: Matthew McAdams
Production Supervisor: Susan Wein
Printing and Binding: RR Donnelley
Cover Photo: Matthias Pahl/Shutterstock

Library of Congress Control Number: 2015935799

ISBN-13: 978-0-7167-7109-8
ISBN-10: 0-7167-7109-8

Printed in the United States of America

Second printing

W. H. Freeman and Company
41 Madison Avenue
New York, NY 10010
www.macmillanhighered.com

To my wife, Laura,
and to our children:
Bruce, Audrey, Mary, Luke, and Dean.
—MDW

To my boys, Callum and Philip;
to my parents, John and Dianne;
and to Judd.
—EG

Contents

Preface

We want to briefly justify why there should be another probability book, when so many others are available.

Motivation: Students from majors in the mathematical sciences and in other areas will be more engaged with the material if they are studying problems that are relevant to them. While testing drafts of the book in the classroom, the students who used this book were asked to contribute questions. As a result, many of the exercises in this text began with questions motivated by the students' own interests.

Example- and exercise-oriented approach: Our book serves as a student's first introduction to probability theory, so we devote significant attention to a wealth of exercises and examples. We encourage students to practice their skills by solving lots of questions. Our exercises are split into practice, extensions, and advanced types of questions. We recommend assigning a small number of questions to students on a daily basis. This promotes more interactive discussion in class between the students and the instructor. It consistently empowers the students to try their hand at some problems of their own. It reduces stress and "cramming" at exam time, as the students consistently develop their understanding during the course. It also provides a firm foundation for the students' long term understanding of probability. The exercises, theorems, definitions, and remarks are all numbered in one list (instead of numbered separately) because we believe that they all should be used in tandem to understand the chapter material.

Relationship between events and random variables: The jump from events to random variables is often a "leap" in other texts. In the present book, we devote significant attention to outcomes, events, sample spaces, and probabilities, before moving on to random variables. Random variables are introduced explicitly as real-valued functions on the sample space, i.e., as functions that depend on the outcome. The notation $X(\omega)$ is used to introduce a random variable at first (where ω is an outcome), so that students can more easily make a transition from studying outcomes to studying random variables that depend on such outcomes.

Jointly distributed random variables: Many probability texts first emphasize properties of one discrete random variable, followed by properties of one continuous random variable, and finally return to jointly distributed random variables. We believe, in contrast, that a firm understanding of jointly distributed random variables, *from the very beginning*, is most helpful for the students' comprehensive understanding of the material. Using jointly distributed random variables at an early stage of the book also allows for more intuitive definitions of some of the concepts. For instance, many probability texts introduce Binomial random variables by explaining the mass, which requires a good grasp of Binomial coefficients, i.e., of $\binom{n}{k}$. We believe, however, that Binomial random variables are introduced more intuitively by $X = X_1 + \cdots + X_n$, where the X_j's are indicator random variables. This requires the students to have familiarity with more than one random variable at a time (i.e., to understand joint distributions), but it allows the students to be more versatile in their thinking. It promotes the understanding of "big picture" kinds of insights. Students begin to think, from the very start, about the ways in which random variables are related, and the ways in which collections of random variables can give rise to new random variables.

Counting: Many probability books start with counting, which means this material is taught during the first two or three weeks of a course. Unfortunately, this means that a student who is weighing her/his interest in a course will not even begin to grasp the concepts of randomness until the registration period is over. This leads to attrition. Moreover, some questions in counting are best understood from a probabilistic point of view, using (for instance) the linearity of expectation, which is not available to the students at the start of the course. As an example, consider how many couples are expected to sit together, when people sit uniformly at random in a circle. This question can be answered very succinctly, using indicator functions and the linearity of expectation. (The probability mass function, in contrast, is cumbersome to compute.) In general, we believe that our approach to counting is significantly enhanced by the use of sums of indicator random variables. Therefore, we focus on counting after we finish a thorough treatment of discrete random variables, but before moving onwards to continuous random variables. This allows the students to feel confident in their understanding of the discrete world before they tackle difficult counting questions. We emphasize to students that combinatorics is a deep and beautiful subject (much of the Ward's research is motivated by problems in applied discrete mathematics). We also try to emphasize the connections between discrete random variables and counting. We firmly believe that this is best accomplished when the students already understand discrete random variables.

Comparison/summary chapters: A first course in probability theory can feel like a whirlwind tour. Thus, we have checkpoints throughout the book, where material is summarized and reviewed. This helps to ground the reader and build confidence. It also helps the students discriminate between the commonly confused distributions and counting techniques, mass functions vs. CDFs, etc. These summaries are useful while reviewing for examinations, e.g., the Actuarial P/1 exam given by the Society of Actuaries and the Casualty Actuarial Society. Passing the P/1 exam requires knowledge of all the material in this text. We also guide the students through ways to tell which kind of distribution they are working with. We give suggestions about how to grasp nuances in a problem that make it a Binomial or Geometric or Negative Binomial situation. These guides help students home in on what separates these distributions. We summarize each distribution for quick reference, but we also go into details to explain each discrete distribution's mass, expected value, variance, etc.

Our students have had an excellent experience using this probability book. Several of our own students have already passed the SOA/CAS P/1 exam after having learned probability using only the early drafts of this book. Our students seem to enjoy the many examples and friendly tone. We hope that you and your students also find our book approachable and thorough. We are delighted by the kind reception that our students and colleagues have given to the book during its pilot tests.

We have divided our book into seven main parts:

Part I: Randomness. We introduce outcomes, events, sample spaces, basic probability rules, independence, conditional probabilities, and Bayes' Theorem.

Part II: Discrete Random Variables. We discuss the difference between discrete and continuous random variables and introduce probability mass function, cumulative distribution function, expected value, variance, and joint distributions for discrete random variables.

Part III: Named Discrete Random Variables. We consider ways to distinguish between—and perform calculations with—the most common discrete random variables: Bernoulli, Binomial, Geometric, Negative Binomial, Poisson, Hypergeometric, and Discrete Uniform. We include a review chapter to help students see the similarities and differences between all of these distributions.

Part IV: Counting. We use indicator variables and the linearity of expectation as tools to help tackle several different types of counting problems: sampling with and without replacement; when order matters and doesn't matter; and rearrangement problems. We have case studies on poker and Yahtzee, two popular games many students will recognize.

Part V: Continuous Random Variables. We reinforce the difference between discrete and continuous random variables. Then we introduce the probability density function, cumulative distribution function, expected value, variance, and joint and conditional distributions for continuous random variables.

Part VI: Named Continuous Random Variables. We show ways to utilize (and quickly make distinctions between) the most common continuous random variables: Continuous Uniform, Exponential, Gamma, Beta and Normal. We show how the Central Limit Theorems and Laws of Large Numbers work. We include a review chapter to help students see the similarities and differences between all of these continuous distributions and between some of the continuous and discrete distributions.

Part VII: Additional Topics. Here we cover more advanced topics that could be optional, depending on how much time an instructor has in a semester or quarter. We treat the distribution of a function of one continuous random variable, the variance of sums of random variables, correlation, conditional expectation, Markov and Chebyshev Inequalities, order statistics, moment generating functions, and the joint density of two random variables that are functions of another pair of random variables.

Acknowledgements

We thank our students for their feedback, as well as their contributions to the exercises in this text. These students are:

Probability: The Science Of Uncertainty (HONR 399, Fall 2010) J. Blair, J. Covalt, S. Fancher, C. Fleming, B. Goosman, W. Hess, E. Hoffman, C. Holcomb, A. Hurlock, E. Jenkins, J. Ling, D. Mo, B. Morgan, C. Mullen, S. Mussmann, D. Rouleau, C. Sanor, V. Savikhin, S. Sheafer, S. Spence, R. Stevick, J. Wu, S. Yap, M. Zachman, B. Zhou.

Probability (MA/STAT 416, Fall 2011) K. Amstutz, C. Ben, M. Boing, K. Breneman, P. Coghlan, B. Copeland, O. El-Ghirani, M. Fronek, Z. Gao, A. Gerardi, T. He, X. He, K. Hopkins, J. Jaagosild, N. Johnson, K. Kemmerling, J. Kwong, N. Lawrence, K. Leow, B. Lewis, K. Li, S. Li, Y. Liang, E. Luo Cao, M. Mazlan, J. Milligan, E. Myers, Y. Nazarbekov, C. Ng, R. Schwartz, S. Scott, S. Seffrin, J. Sheng, D. Snyder, T. Sun, B. Wilson, E. Winkowska, R. Xue, Y. Yang, G. Zhao.

We heartily thank our colleague Dr. Frederi Viens (Purdue University) for piloting the book in his STAT/MA 416 course in Fall 2011 and again in Spring 2013. We extend our thanks also to Dr. David Galvin (University of Notre Dame), Dr. Patricia Humphrey (Georgia Southern University), Dr. Hosam Mahmoud (The George Washington University), and Dr. Meike Neiderhausen (University of Portland), for their willingness to test our book in Fall 2012 in several different types of courses at their own universities. These colleagues and their students provided very valuable feedback about early drafts of the book. We benefited a great deal from their guidance, insight, and suggestions.

This book was genuinely a team effort. We are very grateful to Deborah Sutton for preparing the answers to the odd-numbered exercises. Dr. Patricia Humphrey provided numerous insights about all aspects of the content and checked the accuracy of the entire book. Pat also produced the solutions in the student and instructor manuals. We appreciate Jackie Miller's insights and ideas at the beginning of this project. We are thankful for the support of Terri Ward, our publisher at W. H. Freeman, and for the contributions of the entire W. H. Freeman team. In particular, Liz Geller and Elizabeth Marraffino provided hundreds of helpful suggestions for improvement. We are also especially thankful for the marketing expertise of Karen Carson, Cara LeClair, and Katrina Mangold. We are grateful to Diana Blume and Matt McAdams for their help with the design, layout, and art, and to Susan Wein for making sure our files were technically sound and printer-ready.

We are also thankful for the feedback of the following reviewers:

David Anderson, *University of Wisconsin*
Alireza Arasteh, *Western New Mexico University*
G. Jogesh Babu, *Pennsylvania State University*
Christian Beneš, *Brooklyn College of the City University of New York*
Kiran R. Bhutani, *The Catholic University of America*
Ken Bosworth, *Idaho State University*
Daniel Conus, *Lehigh University*
Michelle Cook, *Stephen F. Austin State University*
Dana Draghicescu, *City University of New York, Hunter College*
Randy Eubank, *Arizona State University*
Marian Frazier, *Gustavus Adolphus College*
Rohitha Goonatilake, *Texas A&M University*
Patrick Gorman, *Kutztown University*
Ross Gosky, *Appalachian State University*
Marc Goulet, *University of Wisconsin–Eau Claire*
Susan Herring, *Sonoma State University*
Christopher Hoffman, *University of Washington*
Patricia Humphrey, *Georgia Southern University*
Ahmad Kamalvand, *Huston-Tillotson University*
Judy Kasabian, *El Camino College*
Syed Kirmani, *University of Northern Iowa*
Charles Lindsey, *Florida Gulf Coast University*
Susan Martonosi, *Harvey Mudd College*
Ronald Mellado Miller, *Brigham Young University–Hawaii*
Joseph Mitchell, *Stony Brook University*
Sumona Mondal, *Clarkson University*
Cindy Moss, *Skyline College*
Meike Niederhausen, *University of Portland*
Will Perkins, *Georgia Institute of Technology*
Rebecca L. Pierce, *Ball State University*
David J. Rader, Jr., *Rose-Hulman Institute of Technology*
Aaron Robertson, *Colgate University*
Charles A. Rohde, *Johns Hopkins University*
Seyed Roosta, *Albany State University*
Juana Sanchez, *University of California, Los Angeles*
Yiyuan She, *Florida State University*
Therese Shelton, *Southwestern University*
A. Robert Sinn, *University of North Georgia*
Clifton D. Sutton, *George Mason University*
Mahbobeh Vezvaei, *Kent State University*

Frederi Viens, *Purdue University*
P. Patrick Wang, *University of Alabama*
Daniel Weiner, *Boston University*
Alison Weir, *University of Toronto, Mississauga*
Reem Yassawi, *Trent University*
Norbert Youmbi, *Saint Francis University*

M. D. Ward's work is supported by the National Science Foundation grants #0939370, #1140489, and #1246818.

We welcome readers' feedback. Please contact us at wgbook@purdue.edu. We will maintain a list of all the corrections and suggestions sent to us, at http://www.stat.purdue.edu/~mdw/book.html.

Notation Review

Notation for Named Sets of Numbers:

$\mathbb{Z}^{\geq 0}$	the set of nonnegative integers $\mathbb{Z}^{\geq 0} = \{0, 1, 2, \ldots\}$
$\mathbb{Z}$	the set of integers, $\mathbb{Z} = \{\ldots, -2, -1, 0, 1, 2, \ldots\}$
$\mathbb{N}$	the natural (a.k.a. counting) numbers, $\mathbb{N} = \{1, 2, \ldots\}$
$\mathbb{R}^{>0}$	the set of positive real numbers
$\mathbb{R}^{\geq 0}$	the set of nonnegative real numbers
$\mathbb{R}$	the set of real numbers, including positive and negative numbers, i.e., whole numbers, fractions, decimals, roots (although not imaginary), and transcendental numbers like π and e.

Notation for Events:

sets:	{things in the set | conditions on those things}; e.g., $S = \{(x, y) \mid x + y = 2\}$.
$\emptyset$	the empty set, i.e., event with no outcomes e.g., $A \cap B = \emptyset$ if A, B have no outcomes in common; see the definition of $\cap$ below to clarify further
S	the sample space, i.e., event with all outcomes
ω	an outcome in the sample space
$\in$	inclusion in an event (i.e., $x \in A$ if outcome x is in event A)
$\subset$	subset (i.e., $A \subset B$ if every outcome of A is also in B)
$\|A\|$	the number of outcomes (also called the size) of event A

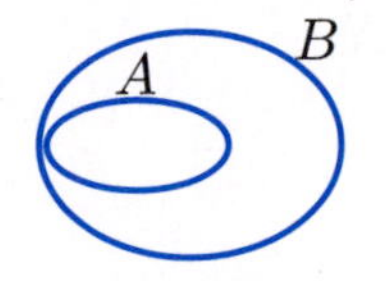

Notation for Building New Events Using Known Ones:

^{c} complement of an event, which corresponds with the word "not"
i.e., x is in A^c exactly when x is not in A

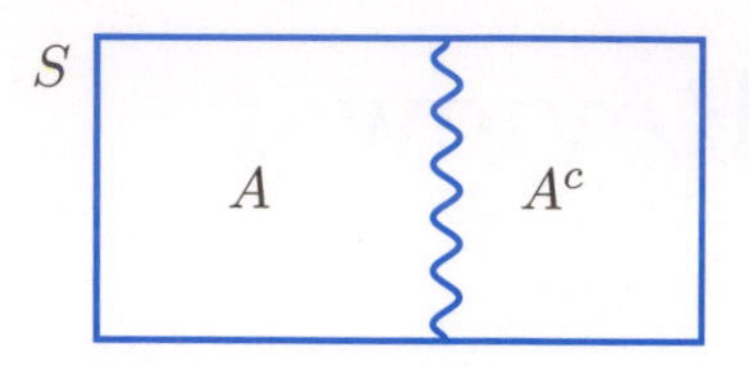

$\setminus$ setminus (i.e., $B \setminus A = B \cap A^c$;
the event containing outcomes in B that are not in A)

$\cup$ union of events, which corresponds with the word "or,"
i.e., $A \cup B$ is the event containing outcomes in A or B or both.
The set $A \cup B$ corresponds to the region containing shading, lines, or both, in the figure below.

$\cap$ intersection of events, which corresponds with the word "and,"
i.e., $A \cap B$ is the event containing outcomes in A *and* B.
The set $A \cap B$ is the region that contains lines overlapping the shading in the figure below.

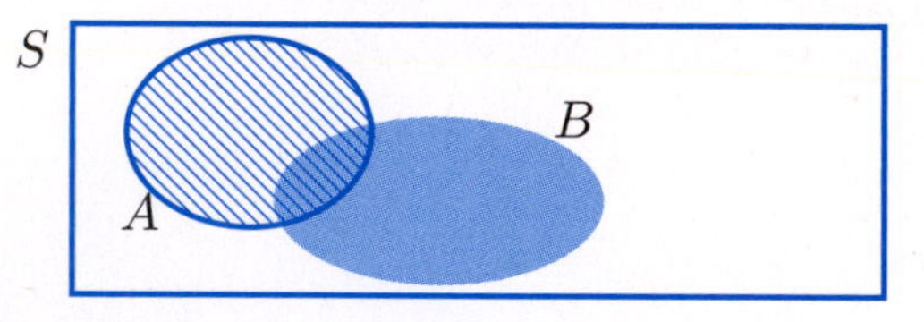

Notation for Random Variables:

X	a random variable (always written in capital letters)
x	a value that a random variable might take on
$P(A)$	probability that event A occurs
$P(X = x)$	is a shorthand notation for $P(\{\omega \mid X(\omega) = x\})$
$\mathbb{E}(X) = \mu$	expected value of X
$\text{Var}(X) = \sigma^2$	variance of X

Math Review

Geometric Sums

For $-1 < a < 1$, recall these two finite summations of geometric terms,

$$1 + a + a^2 + a^3 + \cdots + a^r = \sum_{j=0}^{r} a^j = \frac{1 - a^{r+1}}{1 - a}$$

and

$$a + a^2 + a^3 + a^4 + \cdots + a^r = \sum_{j=1}^{r} a^j = \frac{a - a^{r+1}}{1 - a}.$$

For $-1 < a < 1$, these yield two infinite summations of geometric terms,

$$1 + a + a^2 + a^3 + \cdots = \sum_{j=0}^{\infty} a^j = \frac{1}{1 - a}$$

and

$$a + a^2 + a^3 + a^4 + \cdots = \sum_{j=1}^{\infty} a^j = \frac{a}{1 - a}.$$

Exponential Function

For any real-valued x, the power series definition of the exponential function evaluated at x is

$$e^x = \sum_{n=0}^{\infty} \frac{x^n}{n!},$$

where $n! := (1)(2) \cdots (n)$.

Sum of Integers, Sum of Squares

It is also helpful to know that, for positive integers n,

$$1 + 2 + \cdots + n = \frac{(n)(n+1)}{2}$$

and

$$1^2 + 2^2 + \cdots + n^2 = \frac{(n)(n+1)(2n+1)}{6}$$

Floor and Ceiling Functions

$\lfloor x \rfloor$ round x down to the closest, lower integer, e.g., $\lfloor 14.37 \rfloor = 14$
$\lceil x \rceil$ round x up to the closest, higher integer, e.g., $\lfloor 14.37 \rfloor = 15$

Binomial Coefficient

$\binom{n}{k} = \frac{n!}{k!(n-k)!}$ number of ways to choose k out of n objects, when the order of selection does not matter, e.g., $\binom{5}{3} = 10$ since there are 10 ways to choose 3 out of 5 objects:

1 **2** **3** 4 5 | **1** 2 **3** **4** 5 | 1 **2** **3** **4** 5
1 **2** 3 **4** 5 | **1** 2 **3** 4 **5** | 1 **2** **3** 4 **5**
1 **2** 3 4 **5** | **1** 2 3 **4** **5** | 1 **2** 3 **4** **5**
| | 1 2 **3** **4** **5**

Gamma Function

$\Gamma(n) = (n-1)!$ The gamma function extends the notion of factorials beyond the set of nonnegative integers. We will only use this one fact about the gamma function in this text.

Double Integration

$\int_a^b \int_c^d f(x, y)\, dy\, dx$ This denotes the double integral of $f(x, y)$ over the range where $a \leq x \leq b$ and $c \leq y \leq d$. Under most conditions used in this book, the order of integration can often be switched. (In more advanced courses, this must be done with caution, and suitable convergence theorems must hold—but we do not cover such topics in this text.) If the order of integration is switched, we have instead $\int_c^d \int_a^b f(x, y)\, dx\, dy$. We urge students to make sure that the outer integral corresponds to the outer variable, and the inner integral corresponds to the inner variable.

Dice

We assume that all dice in this text are six-sided, numbered 1, ..., 6, unless stated otherwise.

Cards

We assume that all decks of playing cards have 52 cards, consisting of four suits (spades ♠, hearts ♡, diamonds ◇, clubs ♣), 13 values each: A, 2, 3, ..., 10, J, Q, K.

Part I

Randomness

At the beginning of this course in probability, we consider the basic aspects of randomness. We first discuss the outcomes that are possible when something random occurs, with an emphasis on how these outcomes can be collected into events. Then we give the basic, fundamental notions of probability theory. These are very simple to state, but they constitute the groundwork on which the rest of our study of probability theory is based. We also consider independence of events, as well as the way that the occurrence of one event will affect the probability of occurrence of another event. The first part of the book concludes with the introduction of Bayes' Theorem, which allows us to manipulate probabilities and conditional probabilities.

Even small children learn basic probability ideas simply from observing the world around them. We will formalize these probability ideas and introduce mathematical calculations to go with them. Drawing pictures to help visualize the information in the stories is strongly recommended.

By the end of this part of the book, you should be able to:

1. Define basic terms related to probability and events.
2. Use proper set notation for events.
3. Characterize the possible outcomes, when something random occurs.
4. Describe the events into which outcomes can be grouped.
5. Assign probabilities to events, and perform calculations using probability rules.
6. Calculate whether two or more events are independent.
7. Calculate the probability of an event occurring, given that another event occurred.
8. Calculate the conditional probability of an event using Bayes' Theorem.

Math skills you will need: basic understanding of set notation, unions, intersections, and summation $\sum$ notation.

Additional resources: Calculators may be used to assist in the calculations. Colored pencils may be helpful for drawing Venn diagrams clearly.

Chapter 1

Outcomes, Events, and Sample Spaces

On Monday in math class, Mrs. Fibonacci says, "You know, you can think of almost everything as a math problem." On Tuesday I start having problems.

—*Math Curse* by Jon Scieszka and Lane Smith (Viking, 1995)

In a National Public Radio story from November 30, 2012, "That's So Random: The Evolution of an Odd Word," Neda Ulaby writes about the many misuses of the word "random" in our modern culture, including snippets from the comedian Spencer Thompson's routine, "I Hate When People Misuse the Word Random." For example, Thompson explains that if your friends talk about a "random party" they went to, it probably wasn't as random is they think since it was likely to be held within a reasonably small community and planned with some people that your friends already knew. What do mathematicians and statisticians mean by the word "random"?

1.1 Introduction

Probability theory is the study of randomness and all things associated with randomness. Examples abound everywhere. From the time that we are children, we play guessing games, roll dice, and flip coins. We frequently encounter the unknown and the uncertain. We turn on an mp3 player in a "shuffle" mode, or listen to the radio, eagerly waiting to see what song will come on next. The time until an something happens is often random, e.g., until a traffic light turns green, an email arrives, the telephone rings, or a text message buzzes. The sex of a baby remains unknown until birth (or an ultrasound). An athlete runs a race, but the exact finishing time is unknown beforehand. Millions of people play lotteries and other games of chance, often wagering large amounts of money. Throughout this book, we study probability using examples that will be familiar to the reader.

Definition 1.1. When something happens at **random** there are several potential **outcomes**. *Exactly one* of these outcomes occurs. An **event** is defined to be a collection of some outcomes.

Even though the empty set never happens, we will need it to understand **disjoint** events, i.e., events that have no outcome in common.

Two extreme events have names: The **empty set** $\emptyset$ consists of no outcomes (so the empty set never happens). The **sample space** S consists of all outcomes (so the sample space always happens).

Example 1.2. You roll a 6-sided die.

The sample space is $S = \{1, 2, 3, 4, 5, 6\}$. Only one of these six outcomes actually occurs; for instance, 2 is a possible outcome, or 5 is a possible outcome, etc. We cannot "solve" for which outcome occurs because, as we know from practical experience, we do not know (in advance) which outcome will occur. The outcome is random.

(Q: How many events are there altogether? Hint: It's a power of 2.)

One event is $\{1, 3, 5\}$, i.e., the event that the outcome is odd. The event that the roll is 2 or higher is $\{2, 3, 4, 5, 6\}$. The event that 4 does not appear is $\{1, 2, 3, 5, 6\}$. The event $\{3\}$ consists of only one outcome. Event $\{1, 6\}$ has the smallest and largest possible outcomes.

One event is a **subset** of another if every outcome from the first event is contained in the second event too. Subsets are denoted with the "$\subset$" symbol. For instance, an event with one outcome (such as 5) is a subset of a larger event (such as $\{1, 2, 5\}$), which is a subset of the sample space:

$$\{5\} \subset \{1, 2, 5\} \subset \{1, 2, 3, 4, 5, 6\}.$$

Definition 1.3. Event A is a **subset** of event B, written $A \subset B$, if every outcome in A is also an outcome in B.

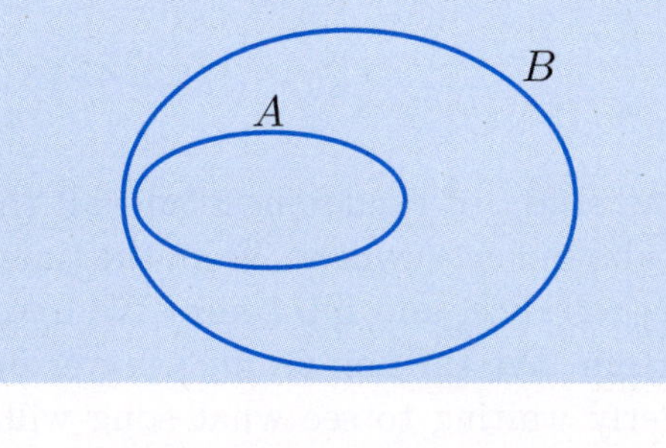

Example 1.4. A student buys a book and opens it to a random page. He notes the number of typographical errors on the page.

The sample space is $S = \mathbb{Z}^{\geq 0}$, i.e., the set of nonnegative integers.

The event that the page contains at most 2 errors is $\{0, 1, 2\}$.

Example 1.5. A new mother delivers one baby.

The sample space is $S = \{\text{boy}, \text{girl}\}$. Although there is just one baby, we can describe four events:

$$\emptyset, \qquad \{\text{boy}\}, \qquad \{\text{girl}\}, \qquad \{\text{boy}, \text{girl}\} = S.$$

Example 1.6. A new mother delivers at least one baby.

One possible outcome is that the the mother has triplets, which are all girls; we denote this outcome as (g, g, g). If she delivers a boy and then a girl, the outcome is (b, g). So the sample space is

$$\begin{aligned} S = \{&(b), (g), \\ &(b, b), (b, g), (g, b), (g, g), \\ &(b, b, b), (b, b, g), (b, g, b), (b, g, g), (g, b, b), (g, b, g), (g, g, b), (g, g, g), \ldots\}. \end{aligned}$$

Note: A new mother may have a single baby, twins, triplets, octuplets or any other (relatively small) number of babies at one time. We only listed the possibilities up to triplets explicitly, but the other possibilities are included in S too; hence, the "..." at the end of S.

A set of octuplets (8 babies) was born in 1998 and also in 2009 in the United States.

Let A be the event that the mother has at least one boy and at least one girl. So A does not contain the outcomes (b) or (b, b) or (b, b, b) etc., and does not contain the outcomes (g) or (g, g) or (g, g, g) etc. Thus

$$A = \{(b, g), (g, b), (b, b, g), (b, g, b), (b, g, g), (g, b, b), (g, b, g), (g, g, b), \ldots\}.$$

Example 1.7. You wait at a red traffic light and record the time (in seconds) until the light turns green.

The sample space is the set of all positive real numbers, $\mathbb{R}^{>0}$. One event is $[5, 10]$, the event consisting of all outcomes between 5 and 10 seconds (inclusive). Another event is $(12.7, \infty)$, i.e., the waiting time is strictly more than 12.7 seconds. Another event is $\{32.7\}$ seconds, the event consisting of only the outcome 32.7 seconds. Events can be built using unions and intersections, e.g., $(0, 60) \cup (120, 180)$ is the event consisting of all outcomes less than 1 minute and also consisting of all outcomes of 2 to 3 minutes.

Example 1.8. You notice the color of the next car to pass on the street.

The sample space is the set of all possible colors in the scheme used to classify this car's color, for instance, perhaps it is classified according to the sample space

$$S = \{\text{red}, \text{yellow}, \text{green}, \text{blue}, \text{orange}, \text{silver}, \text{brown}, \text{black}, \text{white}, \text{other}\}.$$

As we see in the examples with the baby's sex or car's color, outcomes do not have to be numbers.

At the most fundamental level, it is essential to consider how we classify the outcomes. There are often several valid viewpoints. As an example:

Example 1.9. We hit or miss the bullseye with a dart (two possible outcomes).

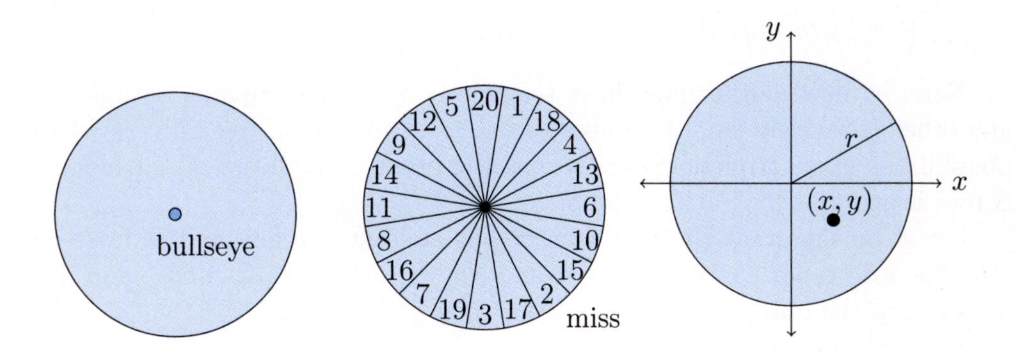

FIGURE 1.1: Different sample spaces for a dart throw. Left: Two outcomes in the sample space. Middle: Twenty-one outcomes in the sample space (the 21st outcome denotes missing the board altogether). Right: Sample space consists of the outcomes, according to location, given as coordinates.

The sample space is $S = \{\text{hit}, \text{miss}\}$. This is depicted on the left side of Figure 1.1. There are four events:

$$\emptyset, \qquad \{\text{hit}\}, \qquad \{\text{miss}\}, \qquad \{\text{hit}, \text{miss}\} = S.$$

(The empty set never happens because it has no outcomes. Sometimes event $\{\text{hit}\}$ happens; sometimes event $\{\text{miss}\}$ happens. Event $\{\text{hit}, \text{miss}\} = S$ always happens.)

Example 1.9 (continued) When throwing a dart, we hit one of twenty regions, or we miss the entire board (twenty-one possible outcomes). Notice: this classification of the outcomes is very different than the "hit" or "miss" setup.

The sample space is $S = \{\text{miss}, 1, 2, 3, \ldots, 20\}$, consisting of the twenty-one possible outcomes: either we "miss" the board altogether or we hit one of the 20 specified regions. This is depicted in the middle of Figure 1.1. (The board has metal ridges between the regions, so that a dart cannot land exactly on the boundary of two regions.)

Example 1.9 (continued) When throwing a dart, we note the exact location where the dart lands.

The sample space is

$$S = \{(x, y) \mid x, y \in \mathbb{R}\},$$

consisting of the outcome listed according to the x (horizontal) and y (vertical) locations where it landed (using the origin at the center of the dartboard as a reference point). This is depicted on the right side of Figure 1.1.

The set notation we used for S is nice, because we are unable to list *all* of the possible points, in this last version of Example 1.9; in fact, there are infinitely many points in the sample space $S = \{(x, y) \mid x, y \in \mathbb{R}\}$. Set notation is also nice because we can add other conditions to the allowable points. For instance, if we want to consider only the situation in which the dart hits the dartboard, then the sample space could be narrowed to

$$\{(x, y) \mid x^2 + y^2 \leq r^2\},$$

where r is the radius of the dartboard. (In this case, we have not handled darts that miss the board entirely.) For instance, if $r = 9$ inches, then the sample space includes outcomes such as $(x, y) = (3.6, -1.35)$, etc.

Set notation is a way to describe a collection of things, sometimes using an annotation about conditions on these things (maybe it should be called "set annotation"). The things that are inside the set are written on the left, and any conditions about these things go on the right.

Notation 1.10. The notation for a set uses braces, with the contents of the set, often followed by a line and then any conditions on the contents of the set.

$$\left\{ \begin{array}{c|c} \text{things} & \text{conditions on} \\ \text{in the set} & \text{these things} \end{array} \right\}$$

The dartboard example should illustrate that it is really important to understand what outcomes are possible when something random happens. This takes some practice at the outset. Sometimes it is helpful to write a list—even an incomplete list—of the different outcomes that are possible from a random phenomenon. (As a rule of thumb, we often encourage students to write five different possible outcomes, if the problem is complicated, just to develop some intuition.) With the darts in Example 1.9, we can certainly write down both outcomes in the first scenario, i.e., "bullseye" or "not bullseye." In the second scenario, the list of all possible outcomes would be "miss," 1, 2, 3, ..., 20. In the third scenario, as soon as we begin to try to write down all of the possible locations on the board by their (x, y) coordinates, we quickly realize that this is a hopeless task. It will not be possible for us to write down every potential outcome, so the concise set notation is crucial to use.

Definition 1.11. We use the **union** notation "$\cup$" when a new set is formed that contains each outcome found in any of the component events. E.g., $A \cup B$ contains each outcome that is found in A, or in B, or in both.

Definition 1.12. We use the **intersection** notation "$\cap$" to construct a new event that contains only the outcomes found in all of the components. E.g., $A \cap B$ contains each outcome that is found in both A and B; it is insufficient to be in just one of these sets.

Example 1.13. A student shuffles a deck of cards thoroughly (one time) and then selects cards from the deck *without replacement* until the ace of spades appears.

"**Without replacement**" means that the cards are not put back into the deck after they are drawn. So on the first draw there are 52 cards available, but on the second draw there are only 51 cards available, and 50 cards available on the third draw, etc. So the ace of spades is certain to appear sometime during the 52 draws. Also, because they are selected without replacement, the chosen cards will be distinct.

The event that exactly three draws are needed to see the ace of spades is

$$\{(x_1, x_2, x_3) \mid x_3 = \mathbf{A}\spadesuit, \text{ and the } x_j\text{'s are distinct}\}.$$

The sample space S consists of all possible draws of distinct cards that end with the ace of spades:

$$\begin{aligned} S = \{(\mathbf{A}\spadesuit)\} &\cup \{(x_1, x_2) \mid x_2 = \mathbf{A}\spadesuit, \text{ and the } x_j\text{'s are distinct}\} \\ &\cup \{(x_1, x_2, x_3) \mid x_3 = \mathbf{A}\spadesuit, \text{ and the } x_j\text{'s are distinct}\} \\ &\cup \{(x_1, x_2, x_3, x_4) \mid x_4 = \mathbf{A}\spadesuit, \text{ and the } x_j\text{'s are distinct}\} \\ &\vdots \\ &\cup \{(x_1, x_2, \ldots, x_{52}) \mid x_{52} = \mathbf{A}\spadesuit, \text{ and the } x_j\text{'s are distinct}\}. \end{aligned}$$

Equivalently, if $B_k = \{(x_1, x_2, \ldots, x_k) \mid x_k = \mathbf{A}\spadesuit, \text{ for distinct } x_j\text{'s}\}$, then the sample space is $S = \bigcup_{k=1}^{52} B_k$.

Example 1.14. A student draws cards from a standard deck of playing cards until the ace of spades appears. After every unsuccessful draw, the student replaces the card and shuffles the deck thoroughly before selecting a new card.

The set of outcomes in which the ace of spades *first appears* on the kth draw is

$$B_k = \{(x_1, \ldots, x_k) \mid \text{only } x_k \text{ is } \mathbf{A}\spadesuit\}$$

Notice that we dropped the condition about the cards being distinct.

The set of all possibilities in which the student actually finds the ace of spades is $\bigcup_{j=1}^{\infty} B_k$. The astute reader will notice that we did not yet mention the possibility that the aces of spades never appears. We write this event as

$$C = \{(x_1, x_2, x_3, \ldots) \mid \text{none of the } x_k\text{'s is } \mathbf{A}\spadesuit\}.$$

So the entire sample space is

$$S = \Big(\bigcup_{k \geq 1} B_k\Big) \cup C.$$

Since the cards are replaced after each draw, this scenario is quite different from Example 1.13.

Example 1.15. A traffic engineer records times (in seconds) between the next six cars that pass.

For example, consider when the next six cars arrive:

car 1 car 2 car 3 car 4 car 5 car 6

0.62 1.31 0.58 1.77 1.97 1.10

0 0.62 1.93 2.51 4.28 6.25 7.35 x

The sample space is

$$S = \{(x_1, \ldots, x_6) \mid x_j \in \mathbb{R}^{>0} \text{ for each } j\}.$$

The engineer uses x_1 for the time until the first car passes, and x_2 is the time between the first and second cars, and in general, x_j is the time between $(j-1)$st and jth cars.

The event where there are at least 3 seconds between all pairs of consecutive cars is

$$\{(x_1, \ldots, x_6) \mid x_j \geq 3 \text{ for each } j\}.$$

The event in which the cars have consecutively longer and longer inter-arrival times (i.e., the distance between cars 1 and 2 is shorter than the distance between cars 2 and 3, which is shorter than the distance between cars 3 and 4, etc.), is

$$\{(x_1, \ldots, x_6) \mid x_j < x_{j+1} \text{ for each } j\}.$$

Many other possible events can be written. The possibilities are endless.

Example 1.16. A meteorologist records the quantity of rain (in centimeters) in a city on a certain day.

The sample space is $S = \mathbb{R}^{\geq 0}$. The event in which it rains between 3 to 5 centimeters (inclusive) is $[3, 5] = \{x \mid 3 \leq x \leq 5\}$. The event in which it rains more than 2 centimeters is $(2, \infty) = \{x \mid x > 2\}$.

Example 1.17. A student hears ten songs (in a random shuffle mode) on her music player, noting how many of these songs belong to her favorite type of music.

If she uses F to denote when a song belongs to her favorite type of music, and N for not-favorite, then the sample space consists of all ten-tuples of F's and N's. In other words, the sample space is

$$S = \{(x_1, \ldots, x_{10}) \mid x_j \in \{F, N\}\}.$$

The event that none of the first three songs is her favorite type of music is

$$A = \{(N, N, N, x_4, \ldots, x_{10}) \mid x_j \in \{F, N\}\}.$$

Songs 1, 2, 3 must be of type "N," but each of songs 4, 5, 6, 7, 8, 9, 10 have two possible assignments of types, either N or "F." So A contains $1 \times 1 \times 1 \times 2 \times 2 \times 2 \times 2 \times 2 \times 2 \times 2 = 2^7$ outcomes. (In Part IV of the book, and in the chapters on discrete random variables, we will investigate more thoroughly the ways that counting is used in probability theory.)

The event that the even-numbered songs are from her favorite type of music is

$$B = \{(x_1, F, x_3, F, x_5, F, x_7, F, x_9, F) \mid x_j \in \{F, N\}\}.$$

Event B contains $2^5 = 32$ outcomes.

No selection of songs would be an outcome in both A and B, so $A \cap B = \emptyset$; in particular, the second song is of type N if the outcome is in event A, but the second song must be of type F if the outcome is in event B.

The event that the last five songs are from type F is

$$C = \{(x_1, x_2, x_3, x_4, x_5, F, F, F, F, F) \mid x_j \in \{F, N\}\};$$

there are $2^5 = 32$ outcomes in event C.

An outcome is in $B \cap C$ if and only if the 2nd, 4th, 6th, 7th, 8th, 9th, and 10th songs are of type "F." So the event $B \cap C$ can be written as

$$B \cap C = \{(x_1, F, x_3, F, x_5, F, F, F, F, F) \mid x_j \in \{F, N\}\}.$$

So event $B \cap C$ has $2^3 = 8$ outcomes.

1.2 Complements and DeMorgan's Laws

In Chapter 2, we introduce probabilities of events. If events are disjoint (have no overlap), we will see that we can calculate the probability of each, and then sum the probabilities. If the events are overlapping, however, we need to use unions and intersections to handle the overlaps. *DeMorgan's Laws give us the dexterity of moving our view between unions and intersections.*

Before introducing DeMorgan's Laws, however, we need to introduce the idea of the complement of an event.

Example 1.18. Consider two events: A is the event that the amount of rainfall on a given day is strictly less than 2.8 inches, and B is the event that the amount of rainfall is 2.8 inches or more. Thus

$$A = [0, 2.8) \qquad\qquad B = [2.8, \infty).$$

Then any possible outcome (amount of rainfall) is in either A or B but not both. So A and B are complements, i.e., B contains exactly the outcomes not found in A, and vice versa.

Whenever the entire sample space is split into two events without overlap, the two events are **complements** of each other. Thus B is the complement of A; this is written as $B = A^c$. Similarly, $A = B^c$.

Some students call "$\setminus$" the "throwaway" operator, i.e., $S \setminus A$ consists of all of S, "throwing away" any outcomes in A.

Definition 1.19. For an event A, the **complement** is the set of all outcomes in the sample space S that are not in A. The complement of A is written as A^c or as $S \setminus A$, using the "setminus" notation given below.

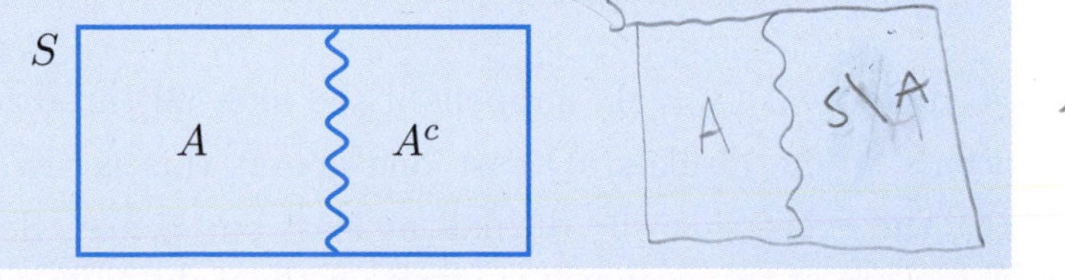

Notation 1.20. The **setminus** "$\setminus$" notation can be used for any pair of events. The event $B \setminus A$ contains all outcomes found in B but not found in A.

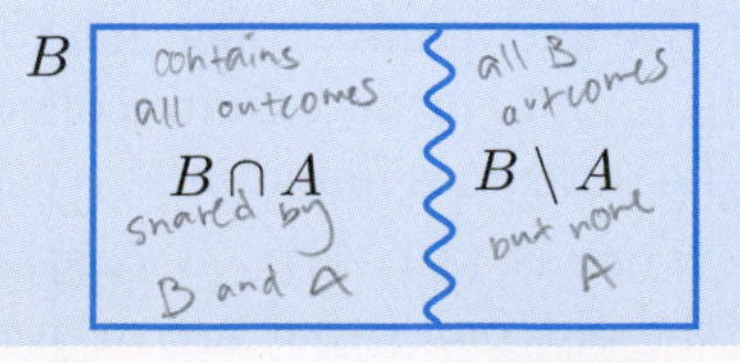

Example 1.21. If $S = \{0, 1, 2, 3, \ldots\}$ is the set of all nonnegative integers, and $A = \{1, 2, 3, \ldots\}$ is the event containing all strictly positive integers, then $A^c = \{0\}$, i.e., A^c contains only 1 outcome, namely, the integer 0.

Now we are prepared to discuss DeMorgan's Laws, as seen in the following example.

Example 1.22. A student is randomly selected, and she is asked about her movie preferences. Let A_1, A_2, A_3 be the event that she enjoys adventure, comedy, or romance movies, respectively. She might enjoy more than one genre.

Event $\bigcup_{j=1}^3 A_j$ occurs if she likes at least one of these genres. Thus, $\left(\bigcup_{j=1}^3 A_j\right)^c$ occurs if she dislikes all three genres; this is the same event as $\bigcap_{j=1}^3 A_j^c$, i.e., the event that she dislikes each of the three genres. So $\left(\bigcup_{j=1}^3 A_j\right)^c = \bigcap_{j=1}^3 A_j^c$. A picture of this scenario is given on the left side of Figure 1.2.

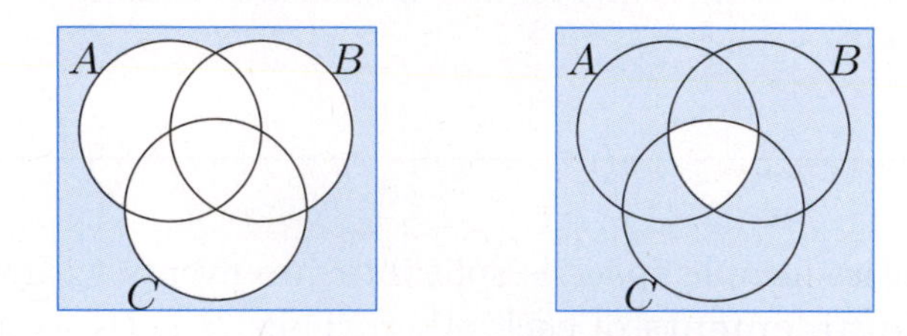

FIGURE 1.2: Left: Visualization of DeMorgan's first law, with three events. The blue shaded area shows $\left(\bigcup_{j=1}^3 A_j\right)^c = \bigcap_{j=1}^3 A_j^c$. Right: Visualization of DeMorgan's second law, with three events. The blue shaded area shows $\left(\bigcap_{j=1}^3 A_j\right)^c = \bigcup_{j=1}^3 A_j^c$.

The event $\bigcap_{j=1}^3 A_j$ happens if she likes all three genres. Thus, $\left(\bigcap_{j=1}^3 A_j\right)^c$ occurs if she dislikes at least one genre; this is the same event as $\bigcup_{j=1}^3 A_j^c$, i.e., the event that she dislikes at least one genre. So $\left(\bigcap_{j=1}^3 A_j\right)^c = \bigcup_{j=1}^3 A_j^c$. A picture of this scenario is given on the right side of Figure 1.2.

These ideas about complements of unions and intersections of events hold much more generally, for both finite and infinite collections of events. Consider a finite collection of events $A_1, A_2, \ldots, A_n$ or an infinite collection of events $A_1, A_2, \ldots$. The event that contains the outcomes found in at least one of the A_j's is $\bigcup_j A_j$. The complement of this event is $\left(\bigcup_j A_j\right)^c$; it contains the outcomes missing from all of the A_j's, i.e., the outcomes that are in every A_j^c and thus in $\bigcap_j A_j^c$.

Theorem 1.23. DeMorgan's first law (complement of the union equals the intersection of the complements)
For a finite or infinite collection of events $A_1, A_2, \ldots$,

$$\Big(\bigcup_j A_j\Big)^c = \bigcap_j A_j^c.$$

Similarly, the event that contains the outcomes found in all of the A_j's is $\bigcap_j A_j$. The complement of this event is $\left(\bigcap_j A_j\right)^c$; it contains the outcomes missing from at least one of the A_j's, i.e., the outcomes that are in at least one A_j^c and thus in $\bigcup_j A_j^c$.

Theorem 1.24. DeMorgan's second law (complement of the intersection equals the union of the complements)
For a finite or infinite collection of events $A_1, A_2, \ldots$,

$$\Big(\bigcap_j A_j\Big)^c = \bigcup_j A_j^c,$$

1.3 Exercises

In some of these scenarios, several different interpretations are possible. These early exercises are intended to inspire discussion. A key goal is to effectively communicate your understanding of the sample space and of the various attributes of the scenario that the outcomes exhibit.

1.3.1 Practice

In each of Exercises 1.1 to 1.6, find:
(1) one specific outcome; (2) one specific event; and (3) the sample space.

Exercise 1.1. **Skydiver.** A skydiver jumps out of a plane and lands somewhere at random inside a circle with radius one mile. What is his landing location?

Exercise 1.2. **Q library books.** A library worker named Jim is going through the returned books. Books are constantly arriving, and Jim's quirky boss, Quinten Quirrell, forces his workers to sort books until they sort a book with a title beginning with the letter Q. How many books will Jim have to sort until he gets a break?

Exercise 1.3. **Gatorade.** Chris has an 18-pack of Gatorade sports drink: 6 orange, 6 lemon-lime, and 6 fruit punch. He blindly grabs one out of the pack over and over if necessary—without replacement—until he finds an orange one. How many bottles will he have to pull out of the pack?

Exercise 1.4. A random hand. You are dealt a hand of five cards (without replacement) from a standard deck of fifty-two playing cards. You note the suits and values of the cards (the order does not matter). Which cards are you dealt?

Exercise 1.5. Cell phone minutes. Your parents restricted your cell phone minutes to 400 minutes this month. You call your boyfriend 75 times during the month. What are the lengths of your calls, if you don't exceed your allotted 400 minutes?

Exercise 1.6. Crayons. A little girl picks out crayons (without replacement) from her 24-pack of Crayolas until she gets to the pink crayon. How many crayons are needed?

1.3.2 Extensions

Exercise 1.7. Moving chairs. Four chairs are placed in a row; two of them are red and look identical; the other two are blue and look identical.

a. How many outcomes are in the sample space? What are they?

b. How many different events are there?

Exercise 1.8. Abstract art. A painter has four different jars of paint colors available, exactly one of which is purple. She wants to paint something abstract, so she blindfolds herself, randomly dips her brush, and paints on the canvas. She continues trying paint jars until she finally gets some purple onto the canvas (her assistant will tell her when this happens). Assume that she does not repeat any of the jars because her assistant removes a jar once it has been used.

a. How many outcomes are in the sample space? What are they?

b. How many different events are there?

c. Another painter borrows the four jars of paint and performs the same experiment; i.e., selects paint at random; but she allows the jars to be reused, perhaps over and over many times (assume each contains an unlimited amount of paint). List a few of the outcomes in the sample space, when repetitions are allowed.

d. In the scenario from part c, write an expression for the sample space.

Exercise 1.9. Waiting for a text message. On Wednesday evening at 5 PM, a student waits for a text message to arrive (hoping to arrange plans for the weekend). Each time a text arrives, the student quickly answers and then begins to wait for the next message. The student quits waiting after receiving three messages.

a. What is the sample space that describes the set of waiting times between the messages that the student receives?

b. Assume the time required to type a text response also takes a random amount of time. What is the sample space that describes the set of all waiting times and also the lengths of typing the responses as well?

c. Write an expression for the event that the waiting times get longer and longer, but the times used to type responses get shorter and shorter.

Exercise 1.10. Double die rolling. Two friends are playing a board game that requires each of them to roll a die. Each player uses her/his own die.

a. What is the sample space for a single roll if their dice are painted two different colors?

b. What if both dice are white—does this change anything?

c. What if one person rolls both dice—does this change anything?

1.3.3 Advanced

Exercise 1.11. Choose a point in a triangle. A point is chosen at random inside the triangle in Figure 1.3.

What is the sample space? (Use set notation for the constraints on x and y.)

S : {

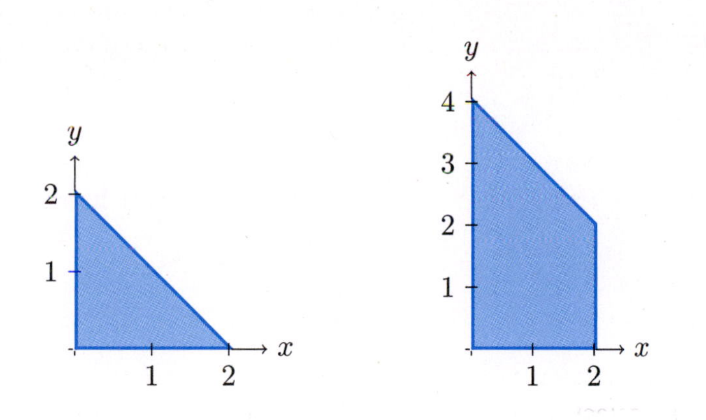

FIGURE 1.3: Left: A triangle. Right: A quadrilateral.

Exercise 1.12. Choose a point in a quadrilateral. A point is chosen at random inside the quadrilateral in Figure 1.3.

What is the sample space? (Hint: Give bounds on x and then on y.)

Exercise 1.13. Building a loft. You are assembling a loft. One piece of wood has 8 screw holes in a straight row, but you can only find 6 screws (which look identical). In a hurry, you put the 6 screws in the 8 holes. How many outcomes are in the sample space, if exactly 6 holes are selected (and the order of selection does not matter)?

Exercise 1.14. Seating arrangements. Alice, Bob, Catherine, Doug, and Edna are randomly assigned seats at a circular table in a perfectly circular room. Assume that rotations of the table do not matter, so there are exactly 24 possible outcomes in the sample space.

Bob and Catherine are married. Doug and Edna are married. In how many of these 24 outcomes are both married couples sitting together and therefore happy?

Exercise 1.15. Sum of three dice. Roll three distinguishable dice (e.g., assume that there is a way to tell them apart, for instance, that the dice are three different colors). There are $6 \times 6 \times 6 = 216$ possible outcomes.

For $3 \leq j \leq 18$, define A_j as the event that the sum of the dice equals j. Find $|A_j|$, i.e., the number of outcomes in A_j? (For instance, $|A_3| = 1$ since A_3 contains only the one outcome $(1, 1, 1)$. Similarly, $|A_{18}| = 1$ since A_{18} contains only one outcome, $(6, 6, 6)$.)

Chapter 2

Probability

In the fields of observation chance favors only the prepared mind.
—Louis Pasteur (Lecture, University of Lille, December 7, 1854)

You have 22 songs on your mp3 player's playlist, and you set the player on shuffle mode, where songs are allowed to repeat, while you study. Country music is your favorite, but only 12 of the songs on this playlist are country with the rest being rock. What is the probability the first song will be country music? What is the probability the first song will not be country music? If 3 songs play, what is the probability that all 3 are country music? Or the probability that exactly 2 of them are country music? Or the probability that none of them are country music? If 3 songs play, what is the probability that only the first song is country, and is this different from the probability that exactly one song will be country? Why or why not?

2.1 Introduction

The probability of an event is a number between 0 and 1 (inclusive).[1] In Chapter 37, we further explore the idea of a probability of an event as the percentage of time that an event occurs, in the long run. For now, however, we begin with only three basic assumptions. We build the framework for our understanding from these 3 intuitive ideas:

[1]In more advanced courses, some events are so complicated that there is not a reasonable way to assign them a probability. We will not study such complications here. See, e.g., Billingsley [1] or Durrett [2].

Remark 2.1. Three Probability Axioms (Intuitive Statements)

1. Any event occurs a certain percentage of the time, so probabilities are always between 0 and 1.

2. With probability 1, some outcome in the sample space occurs.

3. If events have no outcomes in common, then the probability of their union is the sum of the probabilities of the individual events.

Definition 2.2. A pair of events A, B is **disjoint** (also called **mutually exclusive**) if they have no outcome in common, i.e., if their intersection is empty, $A \cap B = \emptyset$. A collection of events is **disjoint** if every pair of events is disjoint.

Definition 2.3. We always use "P" to denote a probability that is defined on the events associated with some random phenomenon.

Now we can state the fundamental ideas (mentioned earlier) in a precise way.

Axioms 2.4. Three Probability Axioms (Mathematical Statements)

1. For each event A,
$$0 \leq P(A) \leq 1.$$

2. For the sample space S,
$$P(S) = 1.$$

3. If $A_1, A_2, \ldots$ is a collection of *disjoint* events, then
$$P\Big(\bigcup_{j=1}^{\infty} A_j\Big) = \sum_{j=1}^{\infty} P(A_j).$$

Theorem 2.5. The probability of the empty set $\emptyset$ is always 0.

This theorem makes sense intuitively, but does it fit with our basic assumptions? Yes! Here is the reasoning:

$$\begin{aligned} 1 &= P(S) && \text{by Axiom 2.4.2} \\ &= P(S \cup \emptyset \cup \emptyset \cup \cdots) && \text{since } S = S \cup \emptyset \cup \emptyset \cdots \\ &= P(S) + P(\emptyset) + P(\emptyset) + \cdots && \text{by Axiom 2.4.3 } (A_1 = S;\ A_j = \emptyset \text{ for } j \geq 2) \end{aligned}$$

By Axiom 2.4.2, $P(S) = 1$. The rest of the right hand side consists of nonnegative terms $P(\emptyset)$, which must therefore each be 0. So $P(\emptyset) = 0$.

Theorem 2.6. If $A_1, A_2, \ldots, A_n$ is a collection of finitely many *disjoint* events, then the probability of the union of the events equals the sum of the probabilities of the events:

$$P\Big(\bigcup_{j=1}^{n} A_j\Big) = \sum_{j=1}^{n} P(A_j).$$

Again, this makes intuitive sense. To prove it, using our basic assumptions, define $A_j = \emptyset$ for all $j > n$. Then we use the probability theory axioms, as follows:

$$\begin{aligned} P\Big(\bigcup_{j=1}^{n} A_j\Big) &= P\Big(\bigcup_{j=1}^{\infty} A_j\Big) && \text{since } A_j = \emptyset \text{ for } j > n \\ &= \sum_{j=1}^{\infty} P(A_j) && \text{by Axiom 2.4.3} \\ &= \sum_{j=1}^{n} P(A_j) && \text{since } P(A_j) = 0 \text{ for } j > n \end{aligned}$$

2.2 Equally Likely Events

We begin by considering the probabilities assigned to some of the events that were discussed at the start of Chapter 1.

Example 2.7. When rolling a die, each of the six outcomes should be equally likely. This means that each single-outcome event should have the same probability.

The probability of each seems (intuitively) to be 1/6. Using our simple assumptions,

$$\begin{aligned} 1 &= P(S) && \text{by Axiom 2.4.2} \\ &= P(\{1,2,3,4,5,6\}) \\ &= P(\{1\} \cup \{2\} \cup \{3\} \cup \{4\} \cup \{5\} \cup \{6\}) \\ &= P(\{1\}) + P(\{2\}) + P(\{3\}) + P(\{4\}) + P(\{5\}) + P(\{6\}) && \text{by Theorem 2.6} \end{aligned}$$

If all $P(\{j\})$'s are the same, then $1 = 6P(\{j\})$, so $P(\{j\}) = 1/6$ for each j. So our intuition is correct. Now we can compute any kind of probability associated with one roll of a die. For instance, the probability a die roll is odd is:

$$P(\{1,3,5\}) = P(\{1\}) + P(\{3\}) + P(\{5\}) = \frac{1}{6} + \frac{1}{6} + \frac{1}{6} = \frac{1}{2}.$$

Example 2.8. A pregnancy that yields exactly one baby would yield an outcome of either a boy or a girl, which are equally likely (as in the die example above).

www.cdc.gov/nchs/data/nvsr/nvsr61/nvsr61_01.pdf suggests that the odds are really closer to 51.17% for boys vs 48.83% for girls, but we assume a 50/50 ratio.

The four relevant probabilities are

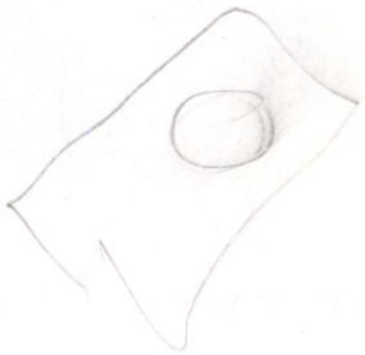

1. $P(\emptyset) = 0$,
2. $P(\{\text{boy}\}) = 1/2$,
3. $P(\{\text{girl}\}) = 1/2$,
4. $P(\{\text{boy, girl}\}) = P(S) = 1$.

In the last case, $S = \{\text{boy, girl}\}$, so we are really just emphasizing the fact that $P(S) = 1$.

These observations about equally likely outcomes are handy and very general:

Theorem 2.9. If a sample space S has n **equally likely** outcomes, then each outcome has probability $1/n$ of occurring.

This is true for just the same reasons as in the die example. Let $x_1, x_2, \ldots, x_n$ be the n outcomes. Then

$$\begin{aligned} 1 &= P(S) \qquad \text{by Axiom 2.4.2} \\ &= P(\{x_1, x_2, \ldots, x_n\}) \\ &= P(\{x_1\} \cup \{x_2\} \cup \cdots \cup \{x_n\}) \\ &= P(\{x_1\}) + P(\{x_2\}) + \cdots + P(\{x_n\}) \qquad \text{by Axiom 2.4.3} \end{aligned}$$

If all $P(\{x_j\})$'s are the same, then $1 = nP(\{x_j\})$, so $P(\{x_j\}) = 1/n$ for each j.

Corollary 2.10. If sample space S has n **equally likely** outcomes, and A is an event with j outcomes, then event A has probability j/n of occurring, i.e., $P(A) = j/n$.

To prove this corollary, write $y_1, \ldots, y_j$ as the j outcomes in A. Then

$$\begin{aligned} P(A) &= P(\{y_1, y_2, \ldots, y_j\}) \\ &= P(\{y_1\} \cup \{y_2\} \cup \cdots \cup \{y_j\}) \\ &= P(\{y_1\}) + P(\{y_2\}) + \cdots + P(\{y_j\}) \qquad \text{by Axiom 2.4.3} \\ &= \frac{1}{n} + \frac{1}{n} + \cdots + \frac{1}{n} \qquad \text{by Theorem 2.9} \\ &= j/n \end{aligned}$$

Definition 2.11. The number of outcomes in an event A, also called the size of A, is denoted as $|A|$.

Using the notation of $|S|$ and in $|A|$ as the number of items in S and A, respectively, Corollary 2.10 can be rewritten as follows:

Corollary 2.12. If sample space S has a finite number of equally likely outcomes, then event A has probability

$$P(A) = |A|/|S|,$$

where $|S|$ and $|A|$ denote the number of items in S and A, respectively.

We will study equally likely outcomes to a much greater extent, in Chapters 20 and 22. For now, however, we give a few examples.

Example 2.13. As in Example 1.9, consider a dartboard split into 20 regions. We treat each of the 20 regions as an outcome, and the possibility of a "miss" as a 21st potential outcome. For simplicity, suppose that the probability of a miss is 0. (Just allow the player to try again, if initially missing the board.) Also suppose the player is equally likely to hit any of the 20 regions on the board.

The sample space is $S = \{1, 2, 3, \ldots, 20\}$. We have 20 disjoint events, each with one outcome: $A_1 = \{1\}$, $A_2 = \{2\}$, and in general,

$$A_j = \{j\}, \qquad \text{for } 1 \leq j \leq 20.$$

(This is a gentle introduction to enumerating events. We will not always use $A_j = \{j\}$.) We assumed $P(\{\text{miss}\}) = 0$. The other 20 outcomes are each equally likely, and the jth event has just 1 outcome, so

$$P(A_j) = 1/20 \qquad \text{for each } j.$$

Example 2.13 (continued) We can now split the dartboard into four regions.

If a new event R_1 is constructed as the union of several of the A_j's, then the probability of R_1 is just the sum of the probabilities. E.g., if

$$R_1 = \{1, 18, 4, 13, 6\} = A_1 \cup A_{18} \cup A_4 \cup A_{13} \cup A_6,$$

(i.e., R_1 is the event that the dart lands in the northeast portion), then

$$P(R_1) = P(A_1) + P(A_{18}) + P(A_4) + P(A_{13}) + P(A_6) = 5/20 = 1/4.$$

This could also seen by using Corollary 2.10, since R_1 contains 5 of the 20 equally likely outcomes, so $P(R_1) = 5/20$.

As in Figure 2.1, define northeast, southeast, southwest, and northwest regions as

$$R_1 = \{1, 18, 4, 13, 6\}, \qquad R_2 = \{10, 15, 2, 17, 3\},$$
$$R_3 = \{19, 7, 16, 8, 11\}, \qquad R_4 = \{14, 9, 12, 5, 20\}.$$

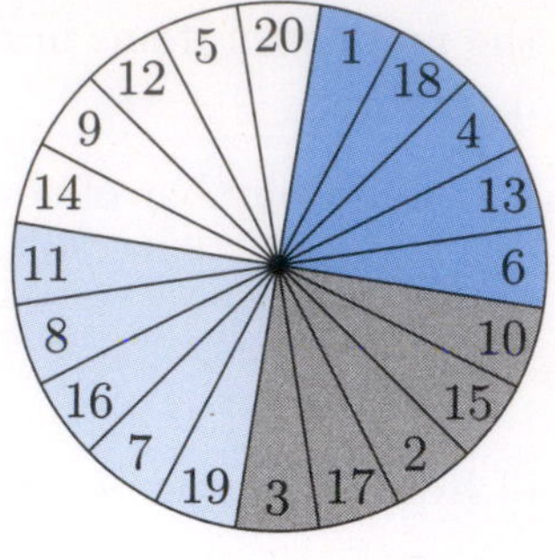

(or could miss the dartboard altogether)

FIGURE 2.1: Twenty-one possible outcomes. Four colors for the northeast, southeast, southwest, and northwest regions.

A classification of the sample space, as we made in Example 2.13, when splitting the sample space into four regions, is called a **partition** of the outcomes in the sample space. More generally, in a partition, every outcome belongs in exactly one of the events, and the events in the partition are disjoint. Partitions will be very helpful in Chapter 5, on Bayes' Theorem.

Definition 2.14. If a collection of nonempty events is disjoint, and their union is the entire sample space, then the collection is called a **partition**. If $\bigcup_j B_j = S$ and the B_j's are disjoint events, then the collection of B_j's is called a **partition**.

In a partition, not all of the regions need to have the same size.

Example 2.15. For instance, a different partition of the dartboard could consist of three disjoint events:

$$\text{the upper portion}, T_1 = \{12, 5, 20, 1, 18\}$$
$$\text{the lower left portion}, T_2 = \{3, 19, 7, 16, 8, 11, 14, 9\}$$
$$\text{the lower right portion}, T_3 = \{4, 13, 6, 10, 15, 2, 17\}$$

See Figure 2.2. In this partition, $P(T_1) = 5/20$, $P(T_2) = 8/20$, and $P(T_3) = 7/20$. Notice $P(T_1) + P(T_2) + P(T_3) = 1$.

Remark 2.16. The probabilities of events in a partition always sum to 1.

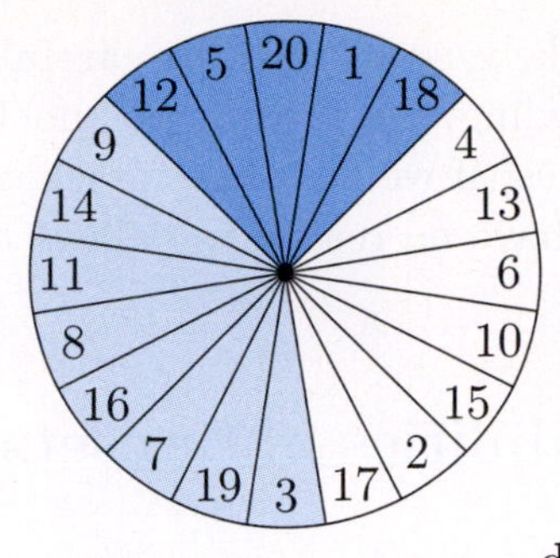

(or could miss the dartboard altogether)

FIGURE 2.2: Twenty-one possible outcomes. Three colors for Example 2.15 regions.

To see this, in a partition consisting B_j's, every outcome is in one of the events, so $S = \bigcup_j B_j$. Also, each outcome is in exactly one of these events in the partition, so the B_j's are disjoint. Thus $P(\bigcup_j B_j) = \sum_j P(B_j)$. Putting these together, we get

$$1 = P(S) = P\Big(\bigcup_j B_j\Big) = \sum_j P(B_j).$$

So the sum of the probabilities of the events in a partition is always 1.

Example 2.17. As in Example 1.13, a student shuffles a deck of cards thoroughly (one time) and then selects cards from the deck *without replacement* until the ace of spades appears.

Let B_k denote the event that the ace of spades is drawn on exactly the kth draw:

$$B_k = \{(x_1, x_2, \ldots, x_k) \mid x_k = \mathbf{A}\spadesuit, \text{ and the } x_j\text{'s are distinct}\}.$$

We emphasize that $P(B_k) = 1/52$ for each k, since the initial placement of the ace of spades (i.e., during the initial shuffle) completely determines when the ace of spades will appear. Since the ace of spades is equally likely to be in any of the 52 places in the deck, then the ace of spades is equally likely to appear on any of the 52 draws.

The B_k's are disjoint events, since it is impossible for an outcome to simultaneously be in more than one of the B_k's. Also, every outcome is in exactly one of the events. So $B_1, B_2, \ldots, B_{52}$ form a partition of the sample space.

Not every set of outcomes is equally likely, so we must be careful when applying Corollary 2.10. For instance, when bowling, it is possible to knock down between 0 and 10 pins, so there are 11 outcomes (if we only keep track of the score, not the specific pins that fall down). We have no reason to believe that all of these 11 outcomes are equally likely.

2.3 Complements; Probabilities of Subsets

Example 2.18. As in Example 1.18, consider two events: A is the event that the amount of rainfall on a given day is strictly less than 2.8 inches, and B is the event that the amount of rainfall is 2.8 inches or more. Thus

$$A = [0, 2.8) \qquad\qquad B = [2.8, \infty).$$

The sample space consisting of all possible amounts of rain is $S = [0, \infty)$ so $S = A \cup B$. Also, A and B are disjoint. So

$$1 = P(S) = P(A \cup B) = P(A) + P(B),$$

so $P(B) = 1 - P(A)$. E.g., if the probability of "rainfall less than 2.8 inches" is 83%, then the probability of "rainfall 2.8 inches or more" must be $1-0.83 = 0.17$, i.e., 17%.

Theorem 2.19. The complement A^c of event A has probability $P(A^c) = 1 - P(A)$.

The argument is straightforward: An event and its complement are disjoint, and their union is the whole sample space:

$$A \cup A^c = S$$

So

$$1 = P(S) = P(A \cup A^c) = P(A) + P(A^c),$$

and the theorem follows, $P(A^c) = 1 - P(A)$.

Now we consider a bound for the probability of an event that is the subset of another event. Recall that A is a subset of B if every outcome of event A is contained in B too (denoted by $A \subset B$). Intuitively, B seems more likely to occur than A in this case. This intuition is correct:

Theorem 2.20. If $A \subset B$ then $P(A) \leq P(B)$.

The event B can be expressed as a disjoint union $B = A \cup (B \setminus A)$. So

$$P(B) = P(A) + P(B \setminus A),$$

but probability is always positive, so $P(B \setminus A) \geq 0$, and thus $P(B) \geq P(A)$.

2.4 Inclusion-Exclusion

The method of inclusion-exclusion allows us to relate overlaps among subsets to unions and intersections. This decomposition enables us to calculate probabilities for events that are overlapping.

Example 2.21. Consider an observer who randomly chooses a car without knowing its color. There are exactly 10 cars available, one from each of these colors: red, blue, yellow, green, lime, teal, orange, silver, brown, or black. Let A denote the event that the car is red, blue, yellow, orange, or silver; let B denote the event that the car is orange, silver, brown, or black. So $A \cap B$ is the event that the car is orange or silver. Thus

$$P(A) = 5/10, \qquad P(B) = 4/10, \qquad P(A \cap B) = 2/10, \qquad P(A \cup B) = 7/10.$$

These probabilities correspond to the scenario in Figure 2.3.

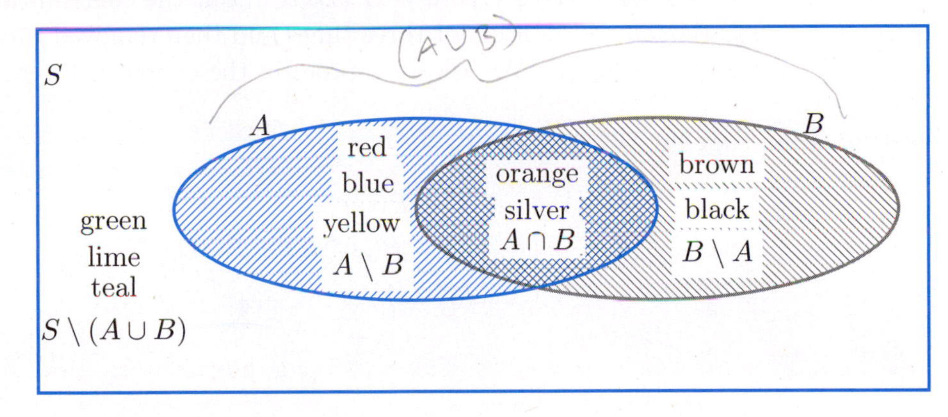

FIGURE 2.3: Observing ten colors of cars.

Notice

$$P(A \cup B) = P(A) + P(B) - P(A \cap B)$$

This might seem intuitively clear because $A \cup B$ accounts for each of the colors red, blue, yellow, orange, silver one time, while

1. A accounts for red, blue, yellow, orange, silver
2. B accounts for orange, silver, brown, black
3. $A \cap B$ accounts for orange, silver

So we are just removing the duplication, i.e., correcting for the fact that the orange and silver were accounted for twice.

The very same argument works much more generally:

Theorem 2.22. For any two events A and B,

$$P(A \cup B) = P(A) + P(B) - P(A \cap B).$$

To prove this, write

$$\begin{aligned} P(A \cup B) &= P(A \setminus B) + P(A \cap B) + P(B \setminus A) \\ &= P(A \setminus B) + P(A \cap B) + P(B \setminus A) + P(A \cap B) - P(A \cap B) \\ &= P(A) + P(B) - P(A \cap B) \end{aligned}$$

Similar nice things happen if we try to characterize $P(A \cup B \cup C)$ by taking each set into account the proper amount of times. Intuitively, we might first guess that $P(A \cup B \cup C)$ and $P(A) + P(B) + P(C)$ are close in value, but we have double-counted all of the contributions from the outcomes in $A \cap B$ and $A \cap C$ and $B \cap C$, so we remove those. Then $P(A \cup B \cup C)$ is close to $P(A) + P(B) + P(C) - P(A \cap B) - P(A \cap C) - P(B \cap C)$, but the contribution from $A \cap B \cap C$ was originally accounted for three times and then removed three times, so we must add it back on. (We ask for a proof in the exercises.)

Theorem 2.23. For any three events A, B, C,

$$\begin{aligned} P(A \cup B \cup C) = P(A) &+ P(B) + P(C) \\ &- P(A \cap B) - P(A \cap C) - P(B \cap C) \\ &+ P(A \cap B \cap C). \end{aligned}$$

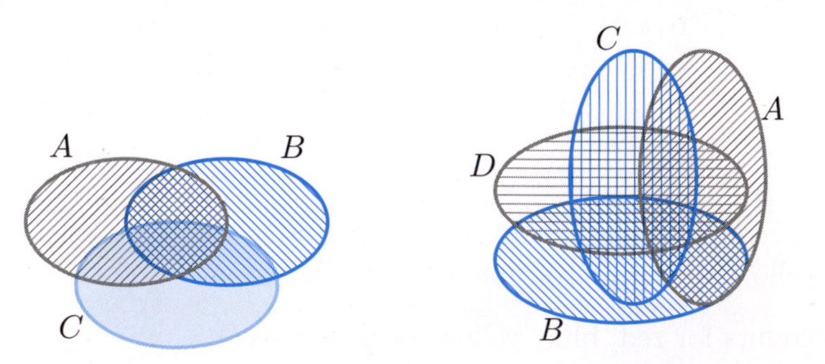

FIGURE 2.4: Left: Overlaps among three events A, B, C. Right: Overlaps among four events A, B, C, D.

The overlaps among A, B, C can be visualized on the left side of Figure 2.4. The overlaps among four events, A, B, C, D can be visualized on the right side of the same figure. It is true, furthermore, by similar reasoning, that

Theorem 2.24. For any four events A, B, C, D,

$$\begin{aligned} P(A \cup B \cup C \cup D) = &P(A) + P(B) + P(C) + P(D) \\ &- P(A \cap B) - P(A \cap C) - P(A \cap D) \\ &- P(B \cap C) - P(B \cap D) - P(C \cap D) \\ &+ P(A \cap B \cap C) + P(A \cap B \cap D) \\ &+ P(A \cap C \cap D) + P(B \cap C \cap D) \\ &- P(A \cap B \cap C \cap D). \end{aligned}$$

This same kind of reasoning can continue, and we get a general inclusion-exclusion formula.

Theorem 2.25. Inclusion-Exclusion Rule
For any finite sequence of events $A_1, A_2, \ldots, A_n$,

$$\begin{aligned} P\left(\bigcup_{j=1}^{n} A_j\right) = &\sum_{j=1}^{n} P(A_j) - \sum_{i<j} P(A_i \cap A_j) + \sum_{i<j<k} P(A_i \cap A_j \cap A_k) \\ &- \sum_{i<j<k<l} P(A_i \cap A_j \cap A_k \cap A_l) \\ &\pm \cdots + (-1)^{n+1} P(A_1 \cap A_2 \cap \cdots \cap A_n) \end{aligned}$$

2.5 More Examples of Probabilities of Events

Example 2.26. As in Example 1.14, consider a student who draws cards from a deck but this time he always replaces the card after each selection and then reshuffles the deck. He only stops if he reaches the ace of spades.

Let B_k be the set of outcomes in which the ace of spades is discovered for the first time on the kth draw:

$$B_k = \{(x_1, \ldots, x_k) \mid \text{only } x_k \text{ is } \mathbf{A}\spadesuit\}$$

Then the set of all possibilities in which the student actually finds the ace of spades is $\bigcup_{k=1}^{\infty} B_k$. Also let

$$C = \{(x_1, x_2, x_3, \ldots) \mid \text{none of the } x_k\text{'s is } \mathbf{A}\spadesuit\}$$

So the entire sample space can be partitioned using the B_k's and C, i.e.,

$$S = \Big(\bigcup_{k \geq 1} B_k\Big) \cup C.$$

The probability that a single draw does not contain the ace of spades is always 51/52. Since the draws do not affect each other (a phenomenon that we will explore much further in Chapter 3 on independence), it follows that

$$P(B_k) = \overbrace{\left(\frac{51}{52}\right)\left(\frac{51}{52}\right)\cdots\left(\frac{51}{52}\right)}^{k-1}\left(\frac{1}{52}\right) = \left(\frac{51}{52}\right)^{k-1}\left(\frac{1}{52}\right).$$

$\sum_{j=0}^{\infty}\left(\frac{51}{52}\right)^j = \frac{1}{1-\frac{51}{52}}$ is a geometric sum; see the Math Review.

We know that the B_k's and C are all disjoint, and also

$$S = \left(\bigcup_{k=1}^{\infty} B_k\right) \cup C.$$

So

$$1 = P(S) = P(C \cup B_1 \cup B_2 \cup B_3 \cup \cdots) = P(C) + \sum_{k=1}^{\infty} P(B_k).$$

We already computed $P(B_k) = (51/52)^{k-1}(1/52)$, so

$$\sum_{k=1}^{\infty} P(B_k) = \sum_{k=1}^{\infty}\left(\frac{51}{52}\right)^{k-1}\left(\frac{1}{52}\right) = \left(\frac{1}{1-\frac{51}{52}}\right)\left(\frac{1}{52}\right) = 1.$$

In summary, $1 = P(C) + 1$, so $P(C)$ must be 0. In other words, the probability that the ace of spades never appears is 0.

Example 2.27. **(continued from Example 1.17)** A student hears ten songs (in a random shuffle mode) on her music player, paying special attention to how many of these songs belong to her favorite type of music. We assume that the songs are picked independently of each other and that each song has probability p of being a song of the student's favorite type.

As in Chapter 1, use F and N to denote when a song belongs to her favorite, or not to her favorite, type of music. So, for each song, the probability that the song is one of her favorite type can be called p, and the probability that the song is not one of her favorite type is $1 - p$.

Now consider the event A that none of the first three songs is her favorite type of music. This event is

$$A = \{(N, N, N, x_4, \ldots, x_{10}) \mid x_j \in \{F, N\}\},$$

and the probability of the event is $P(A) = (1 - p)^3$, because all that the event requires is that none of the first three songs is from her favorite type of music. We do not impose any restrictions on songs 4, 5, ..., 10, so these do not affect the probability of event A occurring.

Now consider the event that the even-numbered songs are from her favorite type of music. We write this event as

$$B = \{(x_1, F, x_3, F, x_5, F, x_7, F, x_9, F) \mid x_j \in \{F, N\}\},$$

and the probability of B is $P(B) = p^5$, because we only require that five specific songs are from her favorite type of music.

Similarly, if

$$C = \{(x_1, x_2, x_3, x_4, x_5, F, F, F, F, F) \mid x_j \in \{F, N\}\},$$

then $P(C) = p^5$ too.

An outcome is in $B \cap C$ if and only if the 2nd, 4th, 6th, 7th, 8th, 9th, and 10th songs are favorites, i.e., of type "F." So the event $B \cap C$ is

$$B \cap C = \{(x_1, F, x_3, F, x_5, F, F, F, F, F) \mid x_j \in \{F, N\}\},$$

which has probability p^7.

As one final event, let A_j denote the event that exactly j of the 10 songs are from her favorite type of music. The probability of A_j is

$$P(A_j) = \binom{10}{j} p^j (1-p)^{10-j},$$

where $\binom{10}{j} = \frac{10!}{j!(10-j)!}$ is the number of ways to pick exactly j out of 10 songs. We will explore this idea in greater depth in Chapter 15, on Binomial random variables.

2.6 Exercises

2.6.1 Practice

Exercise 2.1. Songs by genre. A song is chosen at random from a person's mp3 player. The student makes a partition of the sample space, according to genre of music. The table below gives the number of outcomes in each part of the partition. There are 27,333 songs altogether.

1032	Alternative	83	Electronic	56	Metal
330	Blues	508	Folk	2718	Other
275	Books & Spoken	183	Gospel	1786	Pop
1468	Children's Music	82	Hip-Hop	403	R&B
921	Classical	564	Holiday	8286	Rock
6169	Country	537	Jazz	1432	Soundtrack
178	Easy Listening	106	Latin	216	World

Let A be the event that a song is either blues, jazz, or rock, i.e.,

$$A = \{x \mid x \text{ is a blues, jazz, or rock song}\},$$

when one song is chosen at random. Assume that all songs are equally likely to appear.

a. What is the probability of randomly selecting a blues song?

b. What is $P(A)$?

c. Did you need to use inclusion-exclusion for part b? (Explain very briefly; one sentence will do.)

d. What is $P(A^c)$?

Exercise 2.2. Rock climbing. I am out rock climbing, and the rock face has 4 easy, 7 challenging, and 3 extreme routes to get to the top. The routes are poorly marked, so I just choose one at random, with all routes equally likely. What is the probability that I do not choose an extreme route?

Exercise 2.3. Student interests. A student is chosen at random. Let A, B, C be the events that the student is an Aeronautics major, a Basketball player, or a Co-op student. The events are not disjoint; we are told

$$P(A) = P(B) = P(C) = 0.38,$$

and

$$P(A \cap B) = P(A \cap C) = P(B \cap C) = 0.12,$$

and

$$P(A \cap B \cap C) = 0.05.$$

Find the probability that the student participates in at least one of these three programs, i.e., find $P(A \cup B \cup C)$.

Exercise 2.4. Dining with Dad. At a random meal during a parent weekend in the dining hall, a student notices the food chosen by her father. Let A, B, C be the events that his meal include Artichokes, Broccoli, or Cauliflower. These events have the property that

$$\begin{aligned}
P(B) &= 0.39 \\
P(C) &= 0.44 \\
P(A \cap B) &= 0.13 \\
P(A \cap C) &= 0.12 \\
P(B \cap C) &= 0.13 \\
P(A \cap B \cap C) &= 0.09 \\
P(A \cup B \cup C) &= 0.89
\end{aligned}$$

What is the probability that the father includes Artichokes in his meal, i.e., what is $P(A)$?

Exercise 2.5. Selecting a pair of shoes. A woman has 3 pairs of sneakers, 8 pairs of flip flops, 6 pairs of flats, 4 pairs of wedges, and 9 pairs of high heels. (Hint for those of you unfamiliar with women's footwear: The wedges and high heels will make her look taller.)

a. What is the probability that she selects a pair of shoes that makes her taller if she pulls a pair from her closet without looking?

b. What is the probability that she selects a pair of shoes that does not make her look taller?

c. Create another type of partition for this woman's collection of shoes.

Exercise 2.6. Lollipops and licorice. In a kindergarten class, there are 30 children. Altogether, 19 of them like lollipops, and 10 of them like licorice (some like both). There are 8 students who don't like either of these. A child is chosen at random. What is the probability that the child likes both lollipops and licorice?

Exercise 2.7. Application to weather. Measure the amount of rainfall that occurs in your city for a year.

a. List 5 possible outcomes.

b. What is the sample space?

c. Devise a useful way to partition the sample space so that the partition sheds some insight for the general public about the annual rainfall.

d. Explain how the answer to part c meets the definition of a partition; see Definition 2.14.

Exercise 2.8. Coin flips. You flip a coin 5 times. What is the probability the first 4 are heads and the last one is a tail?

Exercise 2.9. Pizza meat. The guys on one floor of a college dorm all decide to get pizzas to share. They get 3 pepperoni pizzas, 2 bacon pizzas, 1 cheese pizza, 3 sausage pepperoni pizzas, and 3 meat lovers pizzas with sausage, pepperoni, and bacon. What is the probability of a randomly selected slice of pizza containing:

a. Bacon?

b. Pepperoni?

c. Sausage?

Exercise 2.10. Math and physics. In a class of 100 people, 60 of them are math majors, and 75 of them are physics majors. There are no students majoring in anything else in this class. This means that some of the students are double-majoring in both math and physics. A student is picked at random. What is the probability the student is double-majoring?

2.6.2 Extensions

Exercise 2.11. Monkey keystrokes. A monkey is let loose in a computer lab and starts playing with a keyboard. What is the probability that the monkey, without any comprehension or intention, types out the word "bananas" if he types exactly 7 keys? The typical keyboard has 101 keys, and the monkey only presses one key at a time.

Exercise 2.12. Apples. There are 6 apples in a basket. Two of them are red, and four are green.

a. What is the probability of selecting a red apple when choosing at random?

b. What is the probability that, if one apple is randomly chosen per day (and then eaten, not replaced), red apples are chosen on the first two days and green apples are chosen on the last four days?

Exercise 2.13. Shuffling and star ratings. I have 20 five-star songs and 200 four-star songs on my iPod, which has 2000 songs total.

a. What is the probability of shuffling to a five-star song?

b. What is the probability of shuffling to a four-star song?

c. What is the probability of shuffling to either a five- or four-star song?

d. What is the probability of shuffling to a song with fewer than four stars?

Exercise 2.14. Pizza toppings. Consider the following preferences: 35% of people like olive pizza, 54% like sausage pizza, and 12% like both olive and sausage pizza. What is the probability that a randomly chosen person likes neither olive pizza nor sausage pizza?

Exercise 2.15. Roll a die. If you roll a die, event A contains outcomes 1, 3, and 6; event B contains outcomes 1 and 6, and event C contains outcomes 4 and 6.

a. What is the probability of having $A \cup B \cup C$ occur?

b. What is the probability of having $A \cap B \cap C$ occur?

Exercise 2.16. DeMorgan's first law. For three events, A, B, C, consider the following probabilities: A, B, C have probability 20% of occurring, the probability of any two of these events occurring is 3% for each pair, and the probability of all three occurring is 1%. Using DeMorgan's first law, find the probability of $(A \cup B \cup C)^c$.

Exercise 2.17. Abstract art. A painter has three different jars of paint colors available, in colors green, yellow, and purple. She wants to paint something abstract, so she blindfolds herself, randomly dips her brush, and paints on the canvas. She continues trying paint jars until she finally gets some purple onto the canvas (her assistant will tell her when this happens) and then she stops.

Assume that she does not repeat any of the jars because her assistant removes a jar once it has been used. So the sample space is

$$S = \{(P), (G, P), (Y, P), (Y, G, P), (G, Y, P)\}.$$

Find the probabilities of each of the following events:

$$\{(P)\}, \ \{(G, P), (Y, P)\}, \ \{(G, P), (G, Y, P)\},$$

$$\{(Y, G, P), (G, Y, P)\}, \ \{(P), (Y, P)\}$$

Exercise 2.18. Yahtzee. In the game Yahtzee, there are 5 dice with 6 possible numbers on each. What is the probability for a Yahtzee on a player's first roll? (In other words, what is the probability that all 5 dice show the same number the first time that they are rolled)?

Exercise 2.19. Locker combinations. You just forgot your locker combination and are too embarrassed to ask for it. You know for sure that the first number is 22, or was it 32? It's one of those. You're certain that the middle number is a one-digit number (0–9), and the last number could be anything between 0 and 45. If the lock is a 3-number lock with numbers 0 through 45, what is the maximum number of tries needed to open it, assuming you don't repeat any combinations?

Exercise 2.20. Mathematical Science majors. There are 89 students in a volunteer group for majors in math, statistics, or actuarial science. Thirteen students have a double-major in math and statistics. Twelve students have a double major in math and actuarial science. Thirteen students have a double major in statistics and actuarial science. There are 9 students doing a triple major. Thirty-nine students are majoring in at least statistics, and forty-four are majoring in at least actuarial science. How many students are majoring in at least mathematics?

Exercise 2.21. Postal customers. A sequence of seven people walk into a post office and only their sexes are noted (in sequence) as they enter.

a. How many outcomes are in the sample space?

b. Let A_2 be the event that exactly two of the people are females. How many outcomes are in A_2?

c. Let A_j be the event that exactly j of the people are females. How many outcomes are in A_j?

d. If each of the seven customers is equally likely to be male or female, what probability should be associated with event A_j?

Exercise 2.22. Deducing a probability. Events A, B, C are to be considered with the following properties:

$$P(A) = 0.17$$
$$P(B) = 0.37$$
$$P(C) = 0.19$$
$$P(A \cap B) = 0.07$$
$$P(B \cap C) = 0.11$$
$$P(A \cap B \cap C) = 0.03$$
$$P(A \cup B \cup C) = 0.48$$

Find the probability of $A \cap C$.

Exercise 2.23. Mystery probability. Suppose there are 3 events such that

$$P(A) = 0.20$$
$$P(B) = 0.10$$
$$P(C) = 0.40$$
$$P(A \cap B) = 0.05$$
$$P(A \cap C) = 0.10$$
$$P(B \cap C) = 0.03$$
$$P(A \cap B \cap C) = 0.01$$

What is the probability that none of the events happens?

Exercise 2.24. Prove Theorem 2.23.

Exercise 2.25. Prove Theorem 2.24.

2.6.3 Advanced

Exercise 2.26. Prove Theorem 2.25.

Exercise 2.27. Is the whole smaller than the sum of the parts?

a. It is always true, for any events A, B, that $P(A \cup B) \leq P(A) + P(B)$. Why? Explain briefly with words or a very clear picture.

b. Is it always true that $P(A \cup B \cup C) \leq P(A) + P(B) + P(C)$? If so, explain why, either using words or a very clear picture. If not, please give a counterexample.

Exercise 2.28. Grabbing a pen. You find a container of 27 old pens in your school supplies and continue to test them (without replacement), until you find one that works. If each individual pen works 25% of the time (regardless of the other pens), what is the probability that you find one that works within the first four tries?

Exercise 2.29. Die rolls. You roll a die three times. What is the probability the sum of the first two rolls is equal to the third roll?

Exercise 2.30. Cookies. Consider a jar of 9 chocolate chip and 11 peanut butter cookies. You randomly select 2 cookies to eat. All possible choices are equally likely.

a. What is the probability that the 2 you select will both will be chocolate chip?

b. What is the probability that at least one of your cookies will be peanut butter?

c. What is the probability that last 2 cookies left in the jar (after 18 have been eaten) will be chocolate chip? (Is this answer the same or different than part a? Why or why not?)

Exercise 2.31. Seating arrangements. Alice, Bob, Catherine, Doug, and Edna are randomly assigned seats at a circular table in a perfectly circular room. Assume that rotations of the table do not matter, so there are exactly 24 possible outcomes in the sample space.

Bob and Catherine are married. Doug and Edna are married.

Let A_j denote the event that exactly j of the married couples are happy because they are sitting together. Find $P(A_0)$ and $P(A_1)$ and $P(A_2)$.

Exercise 2.32. Socks. In your drawer you have 10 white socks, 6 black socks, 4 red socks, and 2 purple socks. Your roommate is still asleep, and you can't turn the light on while you're getting dressed. You reach in blindly and grab two socks. What is the probability of pulling out a matching pair of purple socks?

Exercise 2.33. Maximum of three dice. Roll three distinguishable dice (e.g., assume that there is a way to tell them apart, for instance, that the dice are three different colors). There are $6 \times 6 \times 6 = 216$ possible outcomes.

Let B_k be the event that the maximum value that appears on all three dice when they are rolled is less than or equal to k. Find $P(B_1)$, $P(B_2)$, $P(B_3)$, $P(B_4)$, $P(B_5)$, and $P(B_6)$. If you prefer, you are welcome to just give a general formula that covers all six of these cases, i.e., you are welcome to just give a formula for $P(B_k)$ itself.

Exercise 2.34. If $A_1, A_2, \ldots, A_n$ is a collection of events, is it always true that

$$P\Big(\bigcup_{k=1}^{n} A_k\Big) \leq \sum_{k=1}^{n} P(A_k)?$$

Prove this inequality, or give a counterexample.

Exercise 2.35. Tickets. Consider n people who drop a ticket into a box. The box of tickets is thoroughly shaken and randomized. Each person then draws one ticket, without replacement. A person is a winner if she selects her own ticket.

a. What is the probability that nobody is a winner?

b. Find the limit of the probability in part a, as $n \to \infty$.

Chapter 3

Independent Events

The word *probability*, in its mathematical acceptation, has reference to the state of our knowledge of the circumstances under which an event may happen or fail.
—*Collected Logical Works, Volume 2: The Laws Of Thought* by George Boole (Walton and Maberly, 1854)

A mother has 2 different pregnancies, each producing a single baby. What is the chance both babies are girls? What is the chance the older child is a girl? What is the chance exactly one of the babies is a girl? What is the chance neither baby is a girl? Are these probabilities the same? Why or why not?

3.1 Introduction

We have an intuitive understanding of the word "independence": Two events A and B are independent if the occurrence of one of the events does not affect the probability of occurrence of the other event. This is exactly right, but we need the concept of conditional probabilities (to be covered in Chapter 4), to use this viewpoint. In the present chapter, we define events A and B as **independent** if the probability that A and B both occur equals the probability that A occurs times the probability that B occurs. We will also discuss the notion of independence among more than two events. Afterwards, we will give examples of dependent events, as well as a very general fact about sequences of independent attempts, in which we are waiting for the first "good" result to occur.

Definition 3.1. Independence
Events A and B are called **independent** if

$$P(A \cap B) = P(A)P(B).$$

Definition 3.2. Dependence
Events A and B are called **dependent** if they are not independent. In other words, A and B are dependent if

$$P(A \cap B) \neq P(A)P(B).$$

We give a multitude of examples to clarify the concept of independence.

Example 3.3. Consider the birth of two children from two separate pregnancies (in particular, we are not considering the birth of twins, in which one baby's sex might affect the other).

If A is the event that the first baby is a girl, and B is the event that the second baby is a girl, then $P(A) = 1/2$, and $P(B) = 1/2$, and $P(A \cap B) = 1/4$, so $P(A \cap B) = P(A)P(B)$. Thus, events A and B are independent. This matches our traditional understanding of the word "independent," because the sex of the first baby does not affect the sex of the second baby.

Let C denote the event that both children are girls. Then $P(A \cap C) = 1/4$ but $P(A)P(C) = 1/8$, so A and C are dependent (intuitively, if C happens, then A must happen).

A question immediately arises:

Remark 3.4. Is "independent" the same thing as "disjoint"? Answer: "No"!

E.g., consider the outcome "girl, girl" in Example 3.3, which is found in both events A and B, and thus in $A \cap B$ too. So A and B are independent but are not disjoint.

More generally, consider the picture in Figure 3.1, for two different situations.

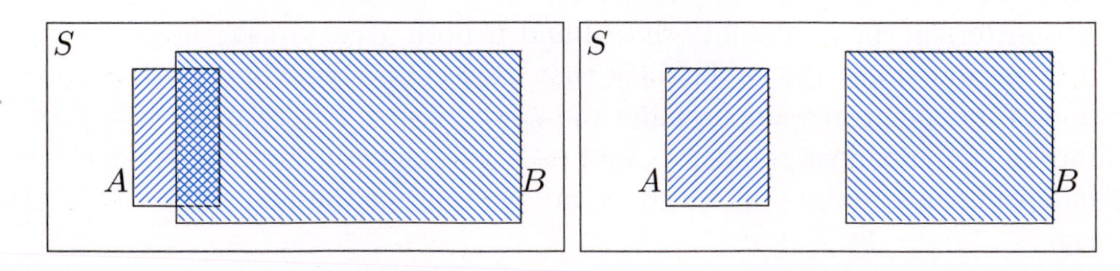

FIGURE 3.1: Left: Independent events A and B. Right: Disjoint events A and B.

Remark 3.5. Can A and B be both independent and disjoint? Only in a very special case: if $P(A) = 0$ or $P(B) = 0$.

Why? If A and B are disjoint, then they have no overlap, i.e., $A \cap B = \emptyset$. So $P(A \cap B) = P(\emptyset) = 0$. On the other hand, since A and B are independent, then $P(A)P(B) = P(A \cap B)$. So

$$P(A)P(B) = 0.$$

So either $P(A) = 0$ or $P(B) = 0$ (or both). In summary:

Remark 3.6. Independent vs Disjoint
If A and B both have positive probabilities, they cannot be both independent and disjoint.

Remark 3.7. When $P(A) = 0$, then A is independent from any event B, because, in such a case,

$$P(A \cap B) = 0 = P(A)P(B).$$

Example 3.8. When rolling a die, let A denote the event consisting of outcomes $\{1, 2, 3\}$, and let B denote the event consisting of outcomes $\{3, 4\}$, so $P(A) = 1/2$ and $P(B) = 1/3$. Also $A \cap B = \{3\}$, so $P(A \cap B) = 1/6$. So $P(A \cap B) = P(A)P(B)$, and this means that A and B are independent.

We will return to this scenario in Example 4.14.

Continuing the scenario, let $C = \{1, 2, 3, 5, 6\}$. So $P(C) = 5/6$. Also $P(B \cap C) = P(\{3\}) = 1/6$, but $P(B)P(C) = (1/3)(5/6) = 5/18 \neq 1/6$. So B, C are dependent.

Theorem 3.9. Subsets are dependent. If $A \subset B$ and neither $P(A) = 0$ nor $P(B) = 1$, then A, B are dependent.

To see this, consider such events A and B. We have $P(A \cap B) = P(A)$. Also $P(B) < 1$, so multiplying both sides by $P(A)$ (which is strictly positive) preserves the strict inequality. Thus $P(A)P(B) < P(A) = P(A \cap B)$, so A, B are dependent.

Theorem 3.10. Complements are dependent. If neither $P(A) = 0$ nor $P(A) = 1$, then A, A^c are dependent.

Note $P(A \cap A^c) = P(\emptyset) = 0$, but $P(A) \neq 0$ and $P(A^c) \neq 0$, so $P(A)P(A^c) \neq 0 = P(A \cap A^c)$, so A, A^c are dependent.

Example 3.11. Consider the songs from Exercise 2.1. Suppose that songs are chosen in such a way that each song is chosen at random, and repetitions are allowed, and every outcome is equally likely (an "outcome" is a particular song, not a genre).

1032	Alternative	83	Electronic	56	Metal
330	Blues	508	Folk	2718	Other
275	Books & Spoken	183	Gospel	1786	Pop
1468	Children's Music	82	Hip-Hop	403	R&B
921	Classical	564	Holiday	8286	Rock
6169	Country	537	Jazz	1432	Soundtrack
178	Easy Listening	106	Latin	216	World

Let A be the event that the first song is either blues or jazz. Let B be the event that the second song is jazz. Let C be the event that the third song is blues or rock.

Notice A and B are independent. Also, A and C are independent. Also, B and C are independent. In the scenario when song repetitions are allowed, the type of one song does not affect the types of other songs.

So far, we have only characterized what it means for a pair of events to be "independent." Now, we see the need to discuss the conditions that are necessary for a collection of three or more events to be collectively called "independent."

Definition 3.12. Independence of three events
A collection three events A, B, C is called (mutually) *independent* if all four of the following are satisfied:

$$\begin{aligned} P(A \cap B) &= P(A)P(B) \\ P(A \cap C) &= P(A)P(C) \\ P(B \cap C) &= P(B)P(C) \\ P(A \cap B \cap C) &= P(A)P(B)P(C) \end{aligned}$$

Example 3.11 (continued)

For instance, in the example above, about various combinations of blues and jazz and rock songs, the collection of events A, B, C is independent.

$$P(A \cap B) = \left(\frac{330 + 537}{27{,}333}\right)\left(\frac{537}{27{,}333}\right) = P(A)P(B)$$

$$P(A \cap C) = \left(\frac{330 + 537}{27{,}333}\right)\left(\frac{330 + 8286}{27{,}333}\right) = P(A)P(C)$$

$$P(B \cap C) = \left(\frac{537}{27{,}333}\right)\left(\frac{330 + 8286}{27{,}333}\right) = P(B)P(C)$$

$$P(A \cap B \cap C) = \left(\frac{330 + 537}{27{,}333}\right)\left(\frac{537}{27{,}333}\right)\left(\frac{330 + 8286}{27{,}333}\right) = P(A)P(B)P(C)$$

Definition 3.13. Independence for a finite collection of events
A finite collection of events $A_1, A_2, \ldots, A_n$ are called (mutually) *independent* if, for *every subcollection* of the events, the probability of the intersection is equal to the product of the probabilities of the individual events in the collection.

Independence for an infinite collection of events
An infinite collection of events $A_1, A_2, \ldots$ is called (mutually) *independent* if every finite collection of the events is independent.

Example 3.14. Consider a student who flips twenty coins in a row. Let A_j denote the event that the jth coin shows a head. Then the events $A_1, \ldots, A_{20}$ are independent.

Example 3.15. Consider a student who flips coins for an arbitrarily long amount of time. As before, let A_j denote the event that the jth coin shows a head. Again, the individual coin flips do not impact each other, so the collection of all of the A_j's is independent.

Example 3.16. If A_j represents the event that there are two or more errors on the jth page of a book, then the collection of A_j's is perhaps independent, because the errors on the individual pages of a book should not affect the errors that occur on other pages of the book.

Example 3.17. The lifetimes of 100 randomly selected light bulbs are measured. Let A_j denote the event that the jth bulb lasts for at least 60 days. Then the collection of 100 events, $A_1, \ldots, A_{100}$, is independent.

Example 3.18. Two hundred customers' purchases are randomly inspected at the grocery store. The event A_j denotes the event that the jth customer purchased at least 3 dairy items. Then the collection of two hundred events, $A_1, \ldots, A_{200}$, are again (perhaps) independent.

When a finite or infinite sequence of random phenomenon are observed, and the results are recorded, we often use the word **trials** to denote such a collection of events. Trials are usually independent from each other, although they do not need to be independent. For instance, if we repeatedly draw cards from a deck, looking for the ace of spades, we might call each draw a "trial," regardless of whether the cards are replaced before the next trial. If the cards are replaced (and reshuffled!), then the trials are independent. If the cards are not replaced, then the trials are dependent.

In Examples 3.14 and 3.15, each coin flip is a trial. In Example 3.16, each examination of a page of a book is a trial. In Example 3.17, each test of the lifetime of a bulb is a trial. In Example 3.18, the examination of a customer's purchase is a trial.

Example 3.19. A student flips a coin until the tenth head appears. See Figure 3.2. Let A denote the event that at least 3 flips are needed between the 7th and 8th heads; let B denote the event that at least 3 flips are needed between the 8th and 9th heads. Then A and B are independent. The coin flips are trials.

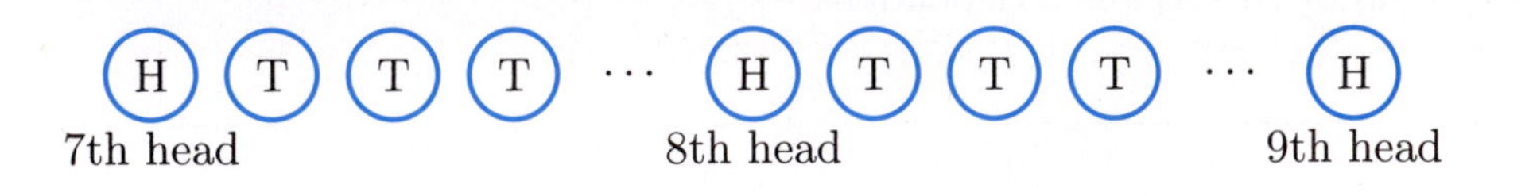

FIGURE 3.2: The number of flips between the 7th and 8th heads do not affect the number of flips between the 8th and 9th heads.

3.2 Some Nice Facts about Independence

Theorem 3.20. Independence among complements
When events are independent, their complements are too, i.e., if A, B are independent, then A^c, B are independent, A, B^c are independent, and A^c, B^c are independent too.

To see this, if A, B are independent, then

$$P(A)P(B) = P(A \cap B).$$

Subtracting both sides from $P(B)$ yields

$$P(B) - P(A)P(B) = P(B) - P(A \cap B),$$

i.e., $P(B)(1 - P(A)) = P(B \setminus A)$. So

$$P(B)P(A^c) = P(B \cap A^c),$$

i.e., A^c, B are independent.

Switching A and B, it follows too that if A, B are independent then A, B^c are independent too. Replacing B by B^c in the previous paragraph, we see that if A, B^c are independent then A^c, B^c are independent too.

Remark 3.21. These nice rules about independence between complements of events also extend to collections of three or more events.

Example 3.22. Two randomly chosen people are selected from a large college campus, and their heights are measured.

Let A denote the event that the height of the first person is 70 inches or greater; let B denote the event that the height of the second person is less than 68.5 inches. Then A and B are independent.

Let C denote the event that the first student's height is less than 68.5 inches. Then A and C are disjoint, so by Remark 3.5, A and C are dependent too.

3.3 Probability of Good Occurring before Bad

Example 3.23. Consider a very large container of carbonated beverages, of which 37% are apple flavor, 20% are orange flavor, and the other 43% are other flavors (cherry, lemon, etc.). If we repeatedly reach for a beverage until we get an apple or orange flavor, what is the probability that an apple flavored drink appears first?

Let A_n denote the event that the nth trial is apple and none of the earlier trials are apple or orange. Then we are looking for $P\left(\bigcup_{n=1}^{\infty} A_n\right)$.

Notice that the A_n's are disjoint. If A_3 occurs, then apple first appears on the 3rd trial, so apple cannot appear for the first time on the 1st trial, or 2nd trial, so neither A_1 nor A_2 can occur. Since the A_n's are disjoint, then

$$P\Big(\bigcup_{n=1}^{\infty} A_n\Big) = \sum_{n=1}^{\infty} P(A_n)$$

Also $P(A_n) = (1 - 0.37 - 0.20)^{n-1}(0.37)$, since we need $n-1$ choices that are not apple or orange, followed by a choice that is apple.

So the desired probability is

$$\begin{aligned}\sum_{n=1}^{\infty} P(A_n) &= \sum_{n=1}^{\infty} (1 - 0.37 - 0.20)^{n-1}(0.37)\\ &= \frac{0.37}{1-(1-0.37-0.20)}\\ &= \frac{0.37}{0.37+0.20}\\ &= 0.65.\end{aligned}$$

This idea works much more generally and can be helpful in many situations in probability theory.

Theorem 3.24. Consider a sequence of independent trials, each of which can be classified as good, bad, or neutral, which happen (on any given trial) with probabilities p, q, and $1-p-q$, respectively. (We do not necessarily have $q = 1-p$ here, although that case is allowed.) Then the probability that something good happens before something bad happens is $p/(p+q)$.

To see this, consider a sequence of *independent* events $A_1, A_2, A_3, \ldots$, where the nth event indicates that something good happens on the nth trial, and neither good nor bad things happen on the previous trials. The same kind of reasoning from the apples and oranges argument works here.

We compute $P\big(\bigcup_{n=1}^{\infty} A_n\big)$. The A_n's are disjoint, since something good cannot happen *for the first time* on two different trials! Since the A_n's are disjoint, this gives

$$P\Big(\bigcup_{n=1}^{\infty} A_n\Big) = \sum_{n=1}^{\infty} P(A_n).$$

Also, A_n occurs if $n-1$ neutral trials are followed by a good one, so

$$P(A_n) = (1-p-q)^{n-1}p.$$

So the desired probability is

$$P\Big(\bigcup_{n=1}^{\infty} A_n\Big) = \sum_{n=1}^{\infty} (1-p-q)^{n-1}p = \frac{1}{1-(1-p-q)}p = \frac{p}{p+q}.$$

So the probability something good happens before something bad happens is $p/(p+q)$.

3.4 Exercises

3.4.1 Practice

Exercise 3.1. Graduation. Jack and Jill are independently struggling to pass their last (one) class required for graduation. Jack needs to pass Calculus III, but he only has probability 0.30 of passing. Jill needs to pass Advanced Pharmaceuticals, but she only has probability 0.46 of passing. They work independently. What is the probability that at least one of them gets a diploma?

Exercise 3.2. Japanese pan noodles. Ten students order noodles at a certain local restaurant. Their orders are placed independently. Each student is known to prefer Japanese pan noodles 40% of the time (it is a very popular and tasty dish!).

a. What is the probability that all ten of the students order Japanese pan noodles?

b. What is the probability that none of the students order Japanese pan noodles?

c. What is the probability that at least one of the students orders Japanese pan noodles?

Exercise 3.3. Off to the races. Suppose Mike places three separate bets on three separate horse races at three separate tracks. Each bet is for a specific horse to win. His horse in race 1 wins with probability $1/5$. His horse in race 2 wins with probability $2/5$. His horse in race 3 wins with probability $3/5$. What is the probability that he made the correct bet in exactly one of these three races?

Exercise 3.4. Early class. Consider these 3 independent trials: On Monday you wake up 45 minutes before class, and the probability that you get to class on time is 0.98. On Tuesday you wake up 32 minutes before class, and your chance of being on time is 0.71. On Wednesday you wake up very, very late, and your probability of being on time is only 0.16.

a. What is the probability that you were on time to class all 3 days?

b. What is the probability that you were never on time?

c. What is the probability that you were on time at least 1 day?

Exercise 3.5. Home for the holidays. A holiday flight from New York to Indianapolis has a probability of 0.75 each time it flies (independently) of taking less than 4 hours.

a. What is the probability that at least one of 3 flights arrives in less than 4 hours?

b. What is the probability that exactly 2 of the 3 flights arrive in less than 4 hours?

3.4.2 Extensions

Exercise 3.6. Hoops. Your sister is playing basketball. She makes 4 tosses to a lowered basketball hoop, and whether the ball goes in each time is independent of the other trials. Her chance of making the basket on a trial is 60%.

For each j with $0 \le j \le 4$, what is the probability that she makes exactly j baskets?

Exercise 3.7. Abstract art. A painter has three different jars of paint colors available, in colors green, yellow, and purple. She wants to paint something abstract, so she blindfolds herself, randomly dips her brush, and paints on the canvas. She continues trying paint jars, without replacement, until all three have been used. (Her assistant helps with this blindfolded process!) So sample space S is

$$S = \{(G, P, Y), (G, Y, P), (P, G, Y), (P, Y, G), (Y, G, P), (Y, P, G)\}.$$

Let A be the event that purple is found in the second jar tested by the painter. Let B be the event that green is found before yellow. Are events A and B independent?

Exercise 3.8. Even versus four or less. Roll a die. Let A be the event that the outcome on the die is an even number. Let B be the event that the outcome on the die is 4 or smaller. Let C be the event that the outcome on the die is 3 or larger.

a. Are A and B independent?

b. Are B and C independent?

Exercise 3.9. Vegetarian dilemma. In a very large collection of sandwiches, 40% are cheese, 45% have steak, and 15% have tofu. A person is vegetarian and therefore samples the sandwiches randomly until finding a cheese or tofu sandwich. What is the probability that they find a cheese sandwich before finding a tofu sandwich?

Exercise 3.10. Guessing on an exam. While taking a probability exam, you come to three questions that you have no clue how to answer. You would have known the answers if you had taken the time to study the night before instead of going to a party, but you did not make a good life choice, and you vow to never party on a school night again if you fail this exam. Each question on the exam is multiple choice with the correct answer being either a, b, c, d, or e. (Your guesses are independent.)

What is the probability that:

a. you randomly guess the right answer to all three questions?

b. you randomly guess the right answer to none of the three questions?

c. you randomly guess the right answer to exactly one of the three questions?

d. you randomly guess the right answer to exactly two of the three questions?

e. Do the probabilities in parts a–d sum to 1?

3.4.3 Advanced

Exercise 3.11. Can the sum be greater than 1? Is it possible to have two *independent* events A and B with the property that

$$P(A) + P(B) > 1\ ?$$

If your answer is "no," give a brief justification of why this is impossible.

If your answer is "yes," please give a brief example, in which you list the numbers $P(A)$ and $P(B)$ and also $P(A \cup B)$ and $P(A \cap B)$.

Exercise 3.12. Seating arrangements. Alice, Bob, Catherine, Doug, and Edna are randomly assigned seats at a circular table in a perfectly circular room. Assume that rotations of the table do not matter, so there are exactly 24 possible outcomes in the sample space.

Bob and Catherine are married. Doug and Edna are married.

Let T denote the event that Bob and Catherine are sitting next to each other. Let U be the event that Alice and Bob are sitting next to each other. Are events T and U independent?

Exercise 3.13. Political survey. On a large campus, 53% of the students are Democrats, and 47% are Republicans. A political survey is conducted. Assume that the students respond independently. How many students are needed, so that the probability of at least 1 Democratic participant exceeds 99%?

Chapter 4

Conditional Probability

Probability is the very guide of life.
—*The Analogy of Religion* by Joseph Butler (Knapton, 1736)

Your dad is visiting you at college, and you have taken him to lunch in your dorm's dining hall to give him the full college student experience. Your dad is a bit of a health nut and loves vegetables, but he thinks some vegetables pair better than others. For example, he's more likely to put broccoli and cauliflower together than broccoli and green beans. Your cafeteria has a fairly extensive selection of vegetables available in the lunch buffet. If you know your dad already picked up cauliflower, what's the chance he will also pick up broccoli? How does knowing that your dad already picked up cauliflower change the probability that he will pick up broccoli compared to when you walked in the door before he had selected any vegetables?

4.1 Introduction

When we have some additional information about a random phenomenon, we can take advantage of the concept of **conditional probability**. When we know (or assume) something about a random phenomenon in advance, it allows us to essentially shrink the sample space to a smaller set of possible outcomes. This fundamentally alters the probabilities. Consider the following example:

Example 4.1. A student on a large college campus is chosen at random. We want to know the probability that the student gets a job next summer, given that the student is pursuing a major in French.

An advantage of conditional probability is that, given some information, we can effectively narrow the sample space. This often makes it easier to compute a probability. E.g., we are not interested in the entire sample space of students at the university. We only care about the probability that the student gets a job, given that they are studying French. There could be (say) 40,000 possible outcomes when choosing a general student from the whole campus, i.e., the original sample space is large. The (much smaller) set of students who study French is more feasible to analyze. So we change our view from the original sample space (all of the college's students) to just the French majors (a much smaller sample space), and the problem gets easier.

If A is the event that the randomly selected student gets a job, and B is the event that the randomly selected student is a French major, the conditional probability of A given B is written as $P(A \mid B)$.

In $P(A|B)$, the bar is read as "given."

The reason for conditional probability is that we sometimes have additional information that we want (or need) to incorporate into the problem. In general, we denote conditional probability as follows:

Definition 4.2. The **conditional probability** of event $\boldsymbol{A}$, given event $\boldsymbol{B}$ (i.e., B has occurred, or will occur, or is assumed to occur, etc.) is written as $P(A \mid B)$.

In other words, conditional probability is written as

$$P(\text{event under consideration} \mid \text{a given event}).$$

In general, if event B has nonzero probability (i.e., $P(B) > 0$), then the conditional probability $P(A \mid B)$ of A given B is defined as

$$P(A \mid B) = \frac{P(A \cap B)}{P(B)}.$$

Equivalently,

$$P(A \cap B) = P(B)P(A \mid B).$$

The statement $P(A \cap B) = P(B)P(A \mid B)$ provides another very natural interpretation of conditional probability. What is the probability that A and B simultaneously occur? Event B needs to occur (hence, the presence of $P(B)$) and—once B is given—the likelihood that A occurs too is $P(A \mid B)$. Thus $P(A \cap B)$ is the product of these two percentages, $P(B)P(A \mid B)$.

Conditional probabilities reduce our sample space to just the part where B has occurred. This shrinks our world and often makes calculations easier to perform, i.e., having the knowledge that some event B occurred can often be helpful for computing the probability that another event occurred too.

Remark 4.3. Consider event B with $P(B) > 0$. Recall A and B are independent exactly when $P(A)P(B) = P(A \cap B)$, i.e., exactly when

$$P(A) = \frac{P(A \cap B)}{P(B)},$$

but the right-hand side is always equal to the conditional probability $P(A \mid B)$. So A and B are independent exactly when

$$P(A) = P(A \mid B),$$

i.e., when B's occurrence does not affect the probability of A occurring.

Theorem 4.4. If $P(B) > 0$, then A and B are independent if and only if $P(A) = P(A \mid B)$, i.e., when B's occurrence does not affect the probability of A occurring.

Example 4.5. When a die is rolled, let B be the event that the outcome is "odd." Then, for example,

$$P(\{3,5\} \mid B) = 2/3, \qquad P(\{1,3,5\} \mid B) = 1, \qquad P(\{4,6\} \mid B) = 0.$$

Since we know that B occurs, the sample space has essentially been reduced from the original sample space, with 6 outcomes,

1	2	3	4	5	6

to a smaller, conditional sample space, with only 3 outcomes,

1	2	3	4	5	6

All of the probabilities from the original model are scaled by a factor of $\frac{1}{P(B)}$ to get the conditional probabilities.

Example 4.6. Suppose that a friend will call one time during the next 60 minutes. We measure (in minutes) the waiting time until she calls. Let $A = \{x \mid x \leq 30\}$, and let $B = \{x \mid x \leq 10\}$.

Given that A occurs, the probability of B occurring is $1/3$; in other words,

$$P(B \mid A) = 1/3.$$

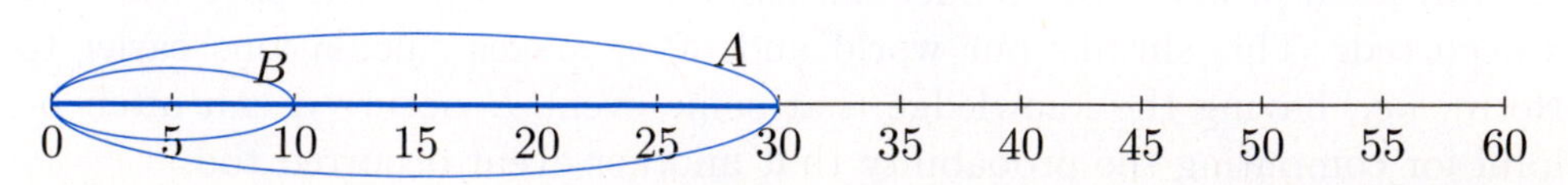

Once we know that A occurs, we can ignore any of the outcomes that are bigger than 30. Also, conditional probabilities satisfy all of the requirements of probabilities, as we will see at the end of this chapter. So, given that A occurs, if we know B occurs 1/3 of the time, it must be the case that B^c occurs the other 2/3 of the time. Thus, the conditional probability of B not occurring (given A occurred) must be 2/3, i.e.,

$$P(B^c \mid A) = 2/3.$$

Example 4.7. On a dartboard let $C = \{9, 12, 5, 20, 1, 18, 4\}$ be the event that the dart lands in the upper portion of the dartboard. If $A = \{9, 12, 5\}$, we have $P(A \mid C) = 3/7$, as in Figure 4.1.

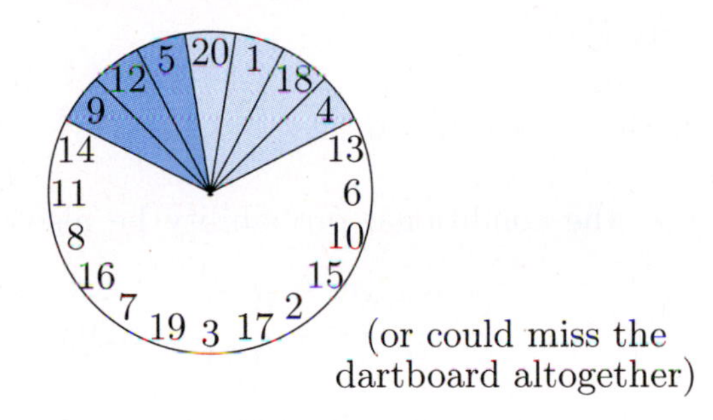

Figure 4.1: Event C is shaded in blue. More specifically, the event $A \cap C$ is shaded in dark blue, and the event $A^c \cap C$ is shaded in light blue.

We can also consider some events that are not completely within C. For instance, say $B = \{1, 2, 3, 4, 5, 6, 7, 8, 9, 10\}$. Some of B overlaps with C and some does not; this is depicted in Figure 4.2.

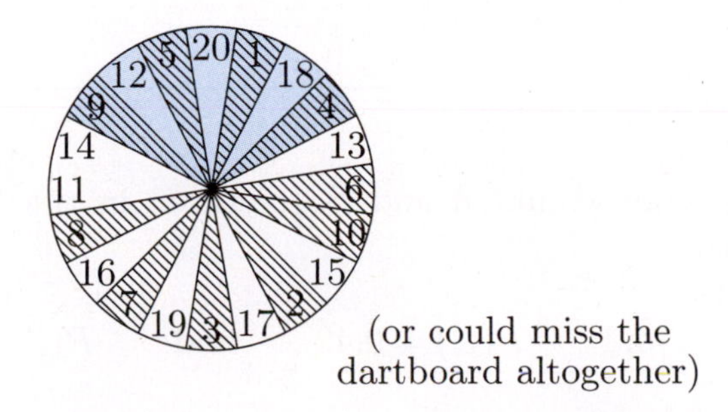

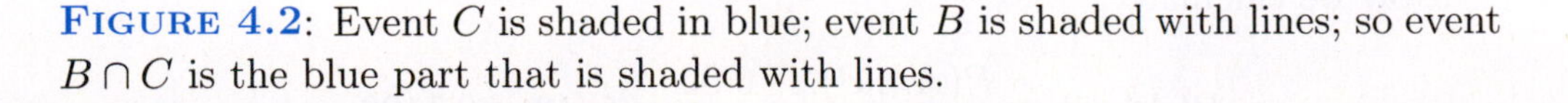
Figure 4.2: Event C is shaded in blue; event B is shaded with lines; so event $B \cap C$ is the blue part that is shaded with lines.

If we compute $P(B \mid C)$, we will end up basically ignoring that part of B outside of C, because these will not be relevant. We compute

$$P(B \mid C) = \frac{P(B \cap C)}{P(C)} = \frac{P(\{1,4,5,9\})}{P(C)} = \frac{4/20}{7/20} = 4/7.$$

So the outcomes 8, 7, 3, 2, 10, 6 from B were essentially not used in this solution. These are not possible outcomes, because they are outside C. Event C can be viewed as the (new) sample space, when calculating probabilities conditioned on C.

Example 4.8. In Exercise 2.4, at a random meal during a parent weekend in the dining hall, a student notices the food chosen by her father. Let A, B, C be the events that his meal include Artichokes, Broccoli, or Cauliflower. These events have the property that

$$P(B) = 0.39, \qquad P(C) = 0.44, \qquad P(B \cap C) = 0.13.$$

If he orders cauliflower, the conditional probability he also orders broccoli is 0.295:

$$P(B \mid C) = \frac{P(B \cap C)}{P(C)} = \frac{0.13}{0.44} = 0.295.$$

If he orders broccoli, the conditional probability he also orders cauliflower is 0.333:

$$P(C \mid B) = \frac{P(B \cap C)}{P(B)} = \frac{0.13}{0.39} = 0.333.$$

Notice $P(B \mid C)$ and $P(C \mid B)$ are not usually the same.

Example 4.8 demonstrates that we cannot swap the events A and B in the conditional probability $P(A \mid B)$.

Example 4.9. Consider events A and B (pictured in the Venn diagram in Figure 4.3) such that

$$P(A) = 0.10 \qquad \text{and} \qquad P(B) = 0.42 \qquad \text{and} \qquad P(A \cap B) = 0.05$$

Now we calculate

$$P(A \mid B) = \frac{P(A \cap B)}{P(B)} = \frac{0.05}{0.42} = 5/42 = 0.1190.$$

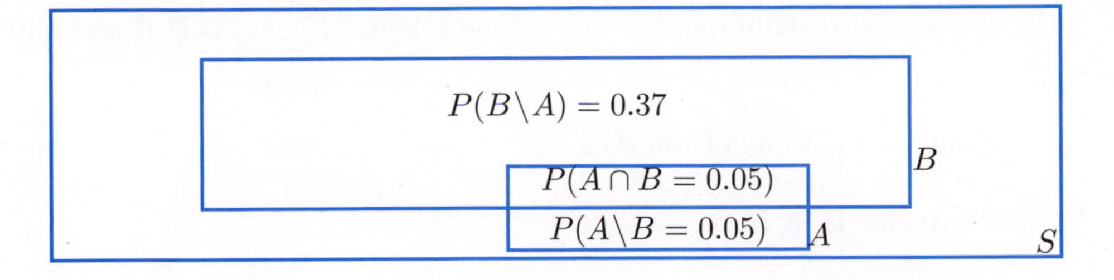

FIGURE 4.3: Venn diagram for events A and B.

This is a very small probability. Intuitively: The event B is very large, compared to $A \cap B$. So if we know that B has occurred, the probability of $A \cap B$ is very small. On the other hand,

$$P(B \mid A) = \frac{P(A \cap B)}{P(A)} = \frac{0.05}{0.10} = 1/2.$$

Intuitively: The event A is twice as large as $A \cap B$. The overlapping part of A with B is half of the size of A itself. Once we know that A has occurred, the outcome will be in $A \cap B$ half the time (and of course in $A \setminus B$ the other half of the time).

4.2 Distributive Laws

The distributive laws for events are helpful to mention at this point because they will be used when showing that conditional probabilities satisfy all of the properties of probability.

Theorem 4.10. Distributive Laws For any events $A_1, A_2, \ldots$,

$$\left(\bigcup_j A_j\right) \cap B = \bigcup_j (A_j \cap B);$$

and

$$\left(\bigcap_j A_j\right) \cup B = \bigcap_j (A_j \cup B);$$

The unions and intersections over j can be finite, written as $\bigcup_{j=1}^{n} A_j$, or can be infinite, $\bigcup_{j=1}^{\infty} A_j$.

The first of the distributive laws is perhaps easier to understand: An outcome is found in $\left(\bigcup_j A_j\right) \cap B$ exactly when the outcome is found in B and in at least one of the A_j's, or equivalently, found in $A_j \cap B$ for at least one j, or equivalently, in $\bigcup_j (A_j \cap B)$.

For the second distributive law, an outcome is in $\left(\bigcap_j A_j\right) \cup B$ if and only if:

1. the outcome is in all of the A_j's, or

2. the outcome is in B, or

3. both are true, i.e., the outcome is in all of the A_j's and also in B.

This is the same as the requirement for an outcome to be in $\bigcap_j (A_j \cup B)$; the outcome must be in B, or if not in B, it must be in all of the A_j's.

Example 4.11. A student is chosen at random. Let A_1, A_2, A_3 be the events that the student is majoring in biology, chemistry, or mathematics, respectively. (Students can pursue more than one major.) Let B be the event that the student is on the volleyball team.

Event $\left(\bigcup_{j=1}^{3} A_j\right) \cap B$ occurs if and only if the student pursues at least one of these three majors and is also on the volleyball team; this is the same event as $\bigcup_{j=1}^{3}(A_j \cap B)$. Event $\left(\bigcap_{j=1}^{3} A_j\right) \cup B$ occurs if and only if the student pursues all three of these majors or is on the volleyball team (or both); this is the same event as $\bigcap_{j=1}^{3}(A_j \cup B)$.

4.3 Conditional Probabilities Satisfy the Probability Axioms

In this section, we show that, when conditioning on any event B, the **conditional probabilities satisfy all of the three axioms of probabilities**, discussed in Section 2.1.

Remark 4.12. The probabilities of the form $P(A)$ that we have studied in Chapters 2 and 3 were unconditional. They did not assume that any event occurred. Such probability could be treated as "conditional" if we just make S the condition, i.e., $P(A)$ is the same as $P(A \mid S)$.

The intuitive way to think about this is that, when B is a given event throughout a problem or scenario, it is just the same as if B is written into the fabric of what is known. So everything is conditioned on B. The event B can just be viewed as replacing the original sample space S.

Theorem 4.13. Consider an event B with $P(B) > 0$.

1. For any event A,
$$0 \le P(A \mid B) \le 1;$$
2. For the sample space S,
$$P(S \mid B) = 1;$$
3. For any disjoint events $A_1, A_2, \ldots,$
$$P\Big(\bigcup_{j=1}^{\infty} A_j \Bigm| B\Big) = \sum_{j=1}^{\infty} P(A_j \mid B).$$

To see (1), notice $0 \le P(A \cap B) \le P(B)$ since $A \cap B$ is a subset of B. Dividing throughout by $P(B)$ gives $0 \le \frac{P(A \cap B)}{P(B)} \le 1$.

For (2), we observe $P(S \mid B) = \frac{P(S \cap B)}{P(B)}$, but $S \cap B = B$, so $P(S \mid B) = \frac{P(B)}{P(B)} = 1$.

For (3), consider disjoint events $A_1, A_2, \ldots$. By the definition of conditional events,
$$P\Big(\bigcup_{j=1}^{\infty} A_j \Bigm| B\Big) = \frac{P\big((\bigcup_{j=1}^{\infty} A_j) \cap B\big)}{P(B)}. \tag{4.1}$$

Now we observe that
$$\Big(\bigcup_{j=1}^{\infty} A_j\Big) \cap B = \bigcup_{j=1}^{\infty} (A_j \cap B)$$

by the first distributive law, explained above. So Equation (4.1) becomes
$$P\Big(\bigcup_{j=1}^{\infty} A_j \Bigm| B\Big) = \frac{P\big(\bigcup_{j=1}^{\infty} (A_j \cap B)\big)}{P(B)}.$$

The $A_j \cap B$ are disjoint since the A_j's are disjoint. So $P\big(\bigcup_{j=1}^{\infty}(A_j \cap B)\big) = \sum_{j=1}^{\infty} P(A_j \cap B)$. So we conclude that
$$P\Big(\bigcup_{j=1}^{\infty} A_j \Bigm| B\Big) = \frac{\sum_{j=1}^{\infty} P(A_j \cap B)}{P(B)} = \sum_{j=1}^{\infty} \frac{P(A_j \cap B)}{P(B)} = \sum_{j=1}^{\infty} P(A_j \mid B).$$

Example 4.14. When rolling a die, let A denote the event consisting of outcomes $\{3, 5\}$, and let B denote the event consisting of outcomes $\{1, 2, 3\}$. Find $P(A \mid B)$ and $P(A^c \mid B)$.

We have $P(A \mid B) = \frac{P(A \cap B)}{P(B)} = \frac{1/6}{1/2} = 1/3$. Since conditional probabilities satisfy the rules of probability, then the conditional probability of the complement must be $2/3$, i.e., $P(A^c \mid B) = 2/3$. (We can check this directly: $P(A^c \mid B) = \frac{P(A^c \cap B)}{P(B)} = \frac{2/6}{1/2} = 2/3$.)

Example 4.15. When rolling a die, let A denote the event consisting of outcomes $\{1, 2, 3\}$, and let B denote the event consisting of outcomes $\{3, 4\}$.

Then $P(A \mid B) = \frac{P(A \cap B)}{P(B)} = \frac{1/6}{1/3} = 1/2 = P(A)$. Since $P(A \mid B) = P(A)$, then A and B are independent (as we also verified in Example 3.8).

Example 4.16. Let A denote the event that the amount of rain on July 1, 2011, in Lafayette, Indiana, is 0.10 inches or more. Let B denote the event that the amount of rain on July 1, 2012 (i.e., one year later), in Lafayette, is 0.10 inches or more. The occurrence of A should not affect the likelihood of occurrence of B, so $P(B \mid A) = P(B)$, and thus A and B are independent.

4.4 Exercises

4.4.1 Practice

Exercise 4.1. Pick ten songs. As in Example 2.27, a student hears ten songs (in a random shuffle mode) and uses F to denote when a song belongs to her favorite type of music, and N for not-favorite, so the sample space consists of all ten-tuples of F's and N's. For each song, the probability that the song is one of her favorite type can be called p, and the probability that the song is not one of her favorite type is $1 - p$. Define

$$A = \{(N, N, N, x_4, \ldots, x_{10}) \mid x_j \in \{F, N\}\},$$
$$B = \{(x_1, F, x_3, F, x_5, F, x_7, F, x_9, F) \mid x_j \in \{F, N\}\},$$
$$C = \{(x_1, x_2, x_3, x_4, x_5, F, F, F, F, F) \mid x_j \in \{F, N\}\},$$

so that, for instance, $P(A) = (1 - p)^3$, and $P(B) = P(C) = p^5$. Find the following conditional probabilities:

$$P(B \mid C), \quad P(C \mid B), \quad P(A \mid B), \quad P(B \mid A), \quad P(A \mid C), \quad P(C \mid A).$$

Exercise 4.2. Dining with Dad. Consider the events from Exercise 2.4, i.e., at a random meal during a parent weekend in the dining hall, a student notices the food chosen by her father. Let A, B, C be the events that his meal include Artichokes, Broccoli, or Cauliflower. These events have the property that: $P(A) = 0.35; P(B) = 0.39; P(C) = 0.44; P(A \cap B) = 0.13; P(A \cap C) = 0.12; P(B \cap C) = 0.13.$

Find the following conditional probabilities:
$P(B \mid C)$, $P(C \mid B)$, $P(A \mid B)$, $P(B \mid A)$, $P(A \mid C)$, $P(C \mid A)$.

Exercise 4.3. Songs by genre. As in Exercise 2.1 and Example 3.11, a song is chosen at random from a person's mp3 player. The student makes a partition of the sample space, according to genre of music. The table below gives the number of outcomes in each part of the partition. There are 27,333 songs altogether.

1032	Alternative	83	Electronic	56	Metal
330	Blues	508	Folk	2718	Other
275	Books & Spoken	183	Gospel	1786	Pop
1468	Children's Music	82	Hip-Hop	403	R&B
921	Classical	564	Holiday	8286	Rock
6169	Country	537	Jazz	1432	Soundtrack
178	Easy Listening	106	Latin	216	World

Let A be the event that a song is either blues, jazz, or rock, i.e.,

$$A = \{x \mid x \text{ is a blues, jazz, or rock song}\},$$

when one song is chosen at random. Let B, J, R denote the events that the song is a blues, jazz, or rock song, respectively.

Find $P(B \mid A)$ and $P(J \mid A)$ and $P(R \mid A)$. Hint: Since B, J, R are disjoint and $A = B \cup J \cup R$, these three answers should sum to 1.

Exercise 4.4. Golf. In the PGA, on par 3 holes, golfers hit the green in one shot 80% of the time. In fact, 20% of the time, they hit the green in one shot and then need only one putt to complete the hole; so 60% of the time, they hit the green in one shot but are unsuccessful on their putt. What is the probability that a PGA golfer only needs one putt, given that he hits the green in one shot?

Exercise 4.5. Parity of spinning. A spinner has the left side (numbers 1, 2, 3, 4, and 5) colored red and the right side colored white (numbers 6, 7, 8, and 9), with all numbers equally likely.

a. What is the probability the spinner lands on an odd number?

b. Given that the spinner landed on an odd number, what is the chance the spinner landed on a white number?

c. Given that the spinner landed on a white number, what is the chance it landed on an odd number?

Exercise 4.6. Conditioning on cards. Draw one card from a shuffled deck of 52 cards.

a. What is the probability that the card is a spade if you know the card is a 7?

b. What is the probability that the card is a spade if you know the card is black?

c. What is the probability that the card is a 7 if you know the card is a spade?

d. What is the probability that the card is a 7 if you know the card is black?

e. What is the probability that the card is black if you know the card is a spade?

f. What is the probability the card is black if you know the card is a 7?

Exercise 4.7. Puppets. You are previewing movies for your young nephew. You have 1,284 movies available to view, 272 of which are G-rated. Your nephew enjoys movies with puppets, which make up 94 of your G-rated movies. There are only 30 movies with puppets that are not G-rated. If you happen to pick a movie that has puppets, what is the probability that it is G-rated?

4.4.2 Extensions

Exercise 4.8. Dice. You roll two dice. Let A be the event that the sum of the dice is an even number. Let B be the event that the two results are different. If B has occurred, what is the probability A has also occurred?

Exercise 4.9. Pair of dice. Roll a pair of dice. Given that the two dice have different values, find the probability that the sum of the dice is an even number.

Exercise 4.10. Pair of dice. Roll a pair of dice. Given that the sum of the pair of dice is 9 or larger, find the probability that the sum of the pair of dice is exactly 10.

Exercise 4.11. Random sexes. A couple has two children. At least one is a boy. What is the probability that the couple has one child of each sex?

Exercise 4.12. More random sexes. A couple has three children. They are not all girls.

a. What is the probability of exactly one boy?

b. What is the probability of exactly two boys?

c. What is the probability of exactly three boys?

4.4.3 Advanced

Exercise 4.13. Seating arrangements. Alice, Bob, Catherine, Doug, and Edna are randomly assigned seats at a circular table in a perfectly circular room. Assume that rotations of the table do not matter, so there are exactly 24 possible outcomes in the sample space. Bob and Catherine are married. Doug and Edna are married. Given that Bob and Catherine are sitting next to each other, find the conditional probability that Doug and Edna are sitting next to each other.

Exercise 4.14. Randomly choose a page. Randomly open a 300-page book, and mark a page.

a. Given that at least one of the digits on the chosen page is a 5, find the probability that the page is 255.

b. Given that at least two of the digits on the chosen page are 5's, find the probability that the page is 255.

Exercise 4.15. Even more random sexes. A couple has n children, where $n \geq 1$ is fixed. They are not all girls. What is the probability of exactly j boys (where $j \geq 1$)?

Chapter 5

Bayes' Theorem

Are no probabilities to be accepted, merely because they are not certainties?
—*Sense and Sensibility* by Jane Austen (Egerton, 1811)

We all know that using cell phones while driving is dangerous, but many people do it anyway. If an accident occurs, the first question an insurance company asks the driver is, "Was anybody using a cell phone when the accident occurred?" What is the probability of a randomly selected driver having an accident in the month of September? How does that probability change if we know that driver is a "regular" cell phone user? If a driver has an accident, what is the chance the driver is a "regular" cell phone user? Why are these last two probabilities not the same?

5.1 Introduction to Versions of Bayes' Theorem

Consider two events that we are interested in, say A and B. Suppose that we know how the occurrence of event A affects the occurrence of event B. How can we use this information to evaluate the way that the occurrence of event B affects the occurrence of event A? Bayes' Theorem tells us exactly how to do this.

For "**b**ackward," think "**B**ayes."

Bayes' Theorem lets us flip conditional probabilities. If the conditional probability we know is backward from the conditional probability we want, use Bayes' Theorem.

The explanation for Bayes' Theorem is straightforward; no memorizing is necessary. We first write $P(A \cap B)$ in two ways, as $P(B)P(A \mid B)$ and also $P(A)P(B \mid A)$:

$$P(B)P(A \mid B) = P(A)P(B \mid A).$$

Then we divide by $P(B)$ throughout, and we get Bayes' Theorem:

Theorem 5.1. Bayes' Theorem
For any two events A and B with nonzero probabilities,

$$P(A \mid B) = \frac{P(A)P(B \mid A)}{P(B)}.$$

Example 5.2. In a certain household, 20% of the milk has two-percent milkfat, and the other 80% of the milk is whole milk. The whole milk is spoiled 5% of the time; overall, the milk is spoiled 4.7% of the time. Find the conditional probability that, if you just poured a spoiled cup of milk from a random jug, it is whole milk.

Let A denote the event that the milk came from a whole milk carton, and let B denote the event that the milk was spoiled. Then we want to find $P(A \mid B)$. We are given $P(B \mid A) = 0.05$, i.e., the probability that a whole milk carton will be spoiled is 5%. We are also told that $P(A) = 0.80$ and that $P(B) = 0.047$. So we get

$$P(A \mid B) = \frac{P(A)P(B \mid A)}{P(B)} = \frac{(0.80)(0.05)}{0.047} = 0.85.$$

So if the milk in a cup is spoiled, it is whole milk with probability 0.85.

Often we are not given $P(B)$ directly, and we have to decompose $P(B)$ by writing:

$$\begin{aligned} P(B) &= P(A \cap B) + P(A^c \cap B) \\ &= P(A)P(B \mid A) + P(A^c)P(B \mid A^c) \end{aligned}$$

We consider several examples like this.

Example 5.3. In a recent study about driving safety, drivers were classified as either "regularly" or "rarely" using a cell phone while driving. Each person who regularly talks on a cell phone while driving has a probability of 1/250 of causing an accident in September. Each person who rarely talks on a cell phone while driving has a probability of 1/2000 of causing an accident in September. Also, 35% of people were classified as regularly using a cell phone while driving.

Suppose that a randomly selected person causes an accident in September. What is the probability that the person is a "regular" cell phone user while driving?

We write A for the event that the person who caused the accident was a "regular" user, and B for the event that the person selected causes an accident in September. We want to know $P(A \mid B)$. We are given:

A is the event that the person is a "regular" cell phone user	B is the event that the person selected causes an accident in September
$P(A) = 0.35$	$P(B \mid A) = 1/250$
so $P(A^c) = 0.65$	$P(B \mid A^c) = 1/2000$

We want $P(A \mid B)$, which we get from

$$P(A \mid B) = \frac{P(A)P(B \mid A)}{P(B)},$$

and we now have all of the pieces except $P(B)$. There are 2 ways in which an accident can happen: with a driver who regularly uses a cell phone or with a driver who rarely uses a cell phone. We need to combine these possibilities for the overall probability that there will be an accident.

We can compute $P(B)$ by writing

$$\begin{aligned} P(B) &= P(A \cap B) + P(A^c \cap B) \\ &= P(A)P(B \mid A) + P(A^c)P(B \mid A^c) \\ &= (0.35)(1/250) + (0.65)(1/2000) \\ &= 0.001725 \end{aligned}$$

Thus,

$$P(A \mid B) = \frac{P(A)P(B \mid A)}{P(B)} = \frac{(0.35)(1/250)}{0.001725} = 0.81.$$

So, given that a randomly selected person causes an accident in September, that person is a "regular" user with probability 0.81.

There are several alternative formulations which can be derived directly from Bayes' Theorem. The first one is a generalization of the example about driver safety.

Remark 5.4. We can decompose B into two parts: $A \cap B$ and $B \setminus A = A^c \cap B$. Since $B = (A \cap B) \cup (A^c \cap B)$ is a disjoint union, then if $P(A)$ is not 0 or 1,

$$P(B) = P(A \cap B) + P(A^c \cap B) = P(A)P(B \mid A) + P(A^c)P(B \mid A^c).$$

Substituting this into the denominator of the original statement of Bayes' Theorem, we get the following alternative formulation:

Theorem 5.5. Bayes' Theorem (Decomposition of Sample Space into Two Parts)
If $P(B) \neq 0$ and $0 < P(A) < 1$, then,

$$P(A \mid B) = \frac{P(A)P(B \mid A)}{P(A)P(B \mid A) + P(A^c)P(B \mid A^c)}.$$

This is a really nice version because we can now compute $P(A \mid B)$ using only three numbers: $P(A)$, $P(B \mid A)$, and $P(B \mid A^c)$. The last piece of information, $P(A^c)$, we can get for free, because we always have $P(A^c) = 1 - P(A)$.

Example 5.6. If a student chooses a playlist created by his roommate on his mp3 player, he finds a song from his favorite type of music 62% of the time. If he chooses a playlist created by his girlfriend, he finds a song from his favorite type of music 88% of the time. These are his only two playlists. He chooses his roommate's playlist 40% of the time. If the student is listening his favorite type of music, what is the probability he is using his roommate's playlist?

Let A and A^c be the events that the song was chosen from his roommate's or girlfriend's playlist, respectively. Let B be the event a song is from the student's favorite type of music. We want to find $P(A \mid B)$. We are given:

A is the event that the song is from the roommate's playlist	B is the event that the song is one of the favorite type
$P(A) = 0.4$	$P(B \mid A) = 0.62$
thus $P(A^c) = 0.6$	$P(B \mid A^c) = 0.88$

We can compute $P(A \mid B)$ directly. Here are the steps:,

$$\begin{aligned} P(A \mid B) &= \frac{P(A \cap B)}{P(B)} \\ &= \frac{P(A \cap B)}{P(A \cap B) + P(A^c \cap B)} \\ &= \frac{P(A)P(B \mid A)}{P(A)P(B \mid A) + P(A^c)P(B \mid A^c)} \\ &= \frac{(0.40)(0.62)}{(0.40)(0.62) + (0.60)(0.88)} \\ &= 0.32. \end{aligned}$$

Thus, given that his favorite type of music is playing, he chose his roommate's playlist 32% of the time.

In Example 5.6, we split the sample space into two pieces, e.g., whether a song was chosen from the playlist of a person's roommate or girlfriend, or whether the randomly chosen person regularly or rarely uses the cell phone while driving. Sometimes it is helpful to be able to split the sample space into more than two groups. Here is an example:

Example 5.7. Consider three types of computer monitors. The probabilities that each type lasts more than five years are, respectively, $0.8, 0.7, 0.3$. It is known that 10% of monitors are type 1; 30% of monitors are type 2; and 60% of monitors are type 3. What is the probability a randomly chosen monitor lasts over 5 years? Given that a monitor lasts over 5 years, what is the probability that it was a monitor of type 1? Type 2? Type 3?

Let B be the probability a randomly chosen monitor lasts over 5 years. Let A_j be the event that a monitor is of type j. We want $P(A_j \mid B)$. We are given:

A_j is the event that a monitor type is j	B is the event that a monitor lasts 5 years
$P(A_1) = 0.1$	$P(B \mid A_1) = 0.8$
$P(A_2) = 0.3$	$P(B \mid A_2) = 0.7$
$P(A_3) = 0.6$	$P(B \mid A_3) = 0.3$

The probability that a randomly chosen person's monitor lasts over 5 years is

$$\begin{aligned}
P(B) &= P(A_1 \cap B) + P(A_2 \cap B) + P(A_3 \cap B) \\
&= P(A_1)P(B \mid A_1) + P(A_2)P(B \mid A_2) + P(A_3)P(B \mid A_3) \\
&= (0.1)(0.8) + (0.3)(0.7) + (0.6)(0.3) \\
&= 0.47
\end{aligned}$$

Given that a randomly chosen person's monitor lasts over 5 years, the probabilities that it is of type 1, or type 2, or type 3 (respectively) are:

$$\begin{aligned}
P(A_1 \mid B) &= \frac{P(A_1)P(B \mid A_1)}{P(B)} = \frac{(0.1)(0.8)}{0.47} = 0.17 \\
P(A_2 \mid B) &= \frac{P(A_2)P(B \mid A_2)}{P(B)} = \frac{(0.3)(0.7)}{0.47} = 0.45 \\
P(A_3 \mid B) &= \frac{P(A_3)P(B \mid A_3)}{P(B)} = \frac{(0.6)(0.3)}{0.47} = 0.38
\end{aligned}$$

The probabilities 0.17, 0.45, and 0.38 must add to 1 because the three possible events $A_1 \cap B$, $A_2 \cap B$, $A_3 \cap B$ are disjoint and their union is all of B.

The following version of Bayes' Theorem allows us to split the sample space into a finite number of different pieces:

Remark 5.8. Let $A_1, A_2, \ldots, A_n$ be any finite partition of the sample space (i.e., $\bigcup_{k=1}^{n} A_k = S$ and the A_k's are disjoint). Then a natural partition of B into n parts is $A_1 \cap B$, $A_2 \cap B$, ..., $A_n \cap B$, i.e., $B = (A_1 \cap B) \cup (A_2 \cap B) \cup \cdots \cup (A_n \cap B)$ is a union of disjoint events, so

$$P(B) = P(A_1 \cap B) + \cdots + P(A_n \cap B) = P(A_1)P(B \mid A_1) + \cdots + P(A_n)P(B \mid A_n).$$

Substituting into the denominator of the first version of Bayes' Theorem, we get:

Theorem 5.9. Bayes' Theorem (Decomposition of Sample Space into Finitely Many Parts)
If $P(B) > 0$, and if $A_1, A_2, \ldots, A_n$ form a partition of S, with all $P(A_j) > 0$, then

$$P(A_k \mid B) = \frac{P(A_k)P(B \mid A_k)}{P(A_1)P(B \mid A_1) + P(A_2)P(B \mid A_2) + \cdots + P(A_n)P(B \mid A_n)}.$$

Thus, to compute $P(A_k \mid B)$ we only need the $P(A_k)$'s and $P(B \mid A_k)$'s.

Finally, sometimes we need to split the sample space into infinitely many pieces. A version of Bayes' Theorem is possible for this situation too:

Remark 5.10. Let $A_1, A_2, \ldots$ be a partition of the sample space (i.e., $\bigcup_{j=1}^{\infty} A_j = S$ and the A_j's are disjoint). Then we have the decomposition of B into infinitely many parts, $A_1 \cap B$, $A_2 \cap B$, etc. So $B = (A_1 \cap B) \cup (A_2 \cap B) \cup \cdots$ is a union of disjoint events. Thus

$$P(B) = P(A_1 \cap B) + P(A_2 \cap B) + \cdots = \sum_{j=1}^{\infty} P(A_j \cap B) = \sum_{j=1}^{\infty} P(A_j)P(B \mid A_j).$$

Substituting this into the denominator of the original statement of Bayes' Theorem, we get the following alternative formulation:

Theorem 5.11. Bayes' Theorem (Decomposition of Sample Space into Infinitely Many Parts)
If $P(B) > 0$, and if $A_1, A_2, \ldots$ form a partition of S, with all $P(A_j) > 0$, then

$$P(A_k \mid B) = \frac{P(A_k)P(B \mid A_k)}{\sum_{j=1}^{\infty} P(A_j)P(B \mid A_j)}.$$

Example 5.12. Consider a game with two stages. In the first stage, a player flips a fair coin until a head appears (usually this only requires a small number of flips). Say that it takes j flips to get this first head. Then, in the second stage, he random draws an integer between 1 and 2^j to get his winnings. E.g., if the player takes 3 flips to get a head, he wins an amount between 1 and 8.

Given that the player wins 1 dollar, what is the probability that the coin was flipped exactly 6 times, i.e., that $j = 6$?

Let A_j be the event that exactly j coin flips are needed to get a head on the coin. Let B_k be the event that the player wins k dollars. We want $P(A_6 \mid B_1)$. We have

$$P(A_6 \mid B_1) = \frac{P(B_1 \mid A_6)P(A_6)}{\sum_{j=1}^{\infty} P(B_1 \mid A_j)P(A_j)}.$$

Also, $P(A_j) = 1/2^j$. Given that A_j occurred, the player will win one of the amounts between 1 and 2^j inclusive, and these 2^j outcomes are equally likely, so the player wins 1 with probability $1/2^j$. So we have

$$P(A_6 \mid B_1) = \frac{\frac{1}{2^6}\frac{1}{2^6}}{\sum_{j=1}^{\infty} \frac{1}{2^j}\frac{1}{2^j}} = \frac{\frac{1}{4^6}}{\sum_{j=1}^{\infty} \left(\frac{1}{4}\right)^j}$$

As a side computation for the series above, we compute

$$\sum_{j=1}^{\infty} \left(\frac{1}{4}\right)^j = \frac{1}{4}\sum_{j=0}^{\infty} \left(\frac{1}{4}\right)^j = \frac{1}{4}\left(\frac{1}{1-\frac{1}{4}}\right) = \frac{1}{4}\left(\frac{1}{3/4}\right) = \frac{1}{3}.$$

So we conclude that

$$P(A_6 \mid B_1) = \frac{1/4^6}{1/3} = \frac{3}{4^6} = 0.000732.$$

5.2 Multiplication with Conditional Probabilities

We close this chapter with one more idea that is useful in working with conditional problems. We assume throughout this section that all of the intersections of events here have positive probability, so that we are not dividing by zero.

Example 5.13. We already know that

$$P(A_2 \mid A_1) = \frac{P(A_1 \cap A_2)}{P(A_1)},$$

or equivalently

$$P(A_1 \cap A_2) = P(A_1)P(A_2 \mid A_1).$$

Informally, if we want A_1 and A_2 to occur:

we just need A_1 to occur, and then,
given that A_1 occurs, we need A_2 to occur too.

Example 5.14. Next we note that

$$\frac{P(A_1 \cap A_2 \cap A_3)}{P(A_1 \cap A_2)} = P(A_3 \mid A_1 \cap A_2).$$

Multiplying on the left- and right-hand sides of the equation from Example 5.13, yields

$$P(A_1 \cap A_2 \cap A_3) = P(A_1)P(A_2 \mid A_1)P(A_3 \mid A_1 \cap A_2).$$

Informally, if we want A_1 and A_2 and A_3 to occur:

> we just need A_1 to occur, and then,
> given that A_1 occurs, we need A_2 to occur too, and then,
> given that A_1 and A_2 occur, we need A_3 to occur too.

Example 5.15. Similarly, we note that $\frac{P(A_1 \cap A_2 \cap A_3 \cap A_4)}{P(A_1 \cap A_2 \cap A_3)} = P(A_4 \mid A_1 \cap A_2 \cap A_3)$. Multiplying on the left- and right-hand sides of the equation from Example 5.14, yields

$$P(A_1 \cap A_2 \cap A_3 \cap A_4) = P(A_1)P(A_2 \mid A_1)P(A_3 \mid A_1 \cap A_2)P(A_4 \mid A_1 \cap A_2 \cap A_3).$$

Informally, if we want A_1 and A_2 and A_3 and A_4 to occur:

> we just need A_1 to occur, and then,
> given that A_1 occurs, we need A_2 to occur too, and then,
> given that A_1 and A_2 occur, we need A_3 to occur too, and then,
> given that A_1 and A_2 and A_3 occur, we need A_4 to occur too.

Similar reasoning continues, and we make the following general assumption:

Remark 5.16. Multiplication with Conditional Probabilities
For any events $A_1, A_2, \ldots, A_n$ with $A_1 \cap A_2 \cap \cdots \cap A_n \neq \emptyset$,

$$\begin{aligned} P(A_1 \cap A_2 \cap \cdots \cap A_n) = P(A_1)P(A_2 \mid A_1)&P(A_3 \mid A_1 \cap A_2) \\ &\times P(A_4 \mid A_1 \cap A_2 \cap A_3) \\ &\cdots \times P(A_n \mid A_1 \cap A_2 \cap \cdots \cap A_{n-1}). \end{aligned}$$

Example 5.17. Consider a student's playlist that has 10 rock songs and 12 country songs. The student chooses at random—with all selections equally likely, and no repeats—three songs from the playlist. What is the probability that all three are country songs?

Let A_1, A_2, A_3 be the event that the first, second and third songs, respectively, are country songs. Then the desired probability is

$$P(A_1 \cap A_2 \cap A_3) = P(A_1)P(A_2 \mid A_1)P(A_3 \mid A_1 \cap A_2)$$

The probability of A_1 is $12/22$ since there are initially 22 songs, 12 of which are country, and all of the selections are equally likely.

Next, given that A_1 occurs, there are 21 songs remaining—10 rock songs and 11 country songs—so $P(A_2 \mid A_1) = 11/21$.

Finally, given that $A_1 \cap A_2$ occurs, there are 20 songs remaining—10 rock songs and 10 country songs—so $P(A_3 \mid A_1 \cap A_2) = 10/20$.

So the desired probability is

$$\begin{aligned} P(A_1 \cap A_2 \cap A_3) &= P(A_1)P(A_2 \mid A_1)P(A_3 \mid A_1 \cap A_2) \\ &= (12/22)(11/21)(10/20) \\ &= 1/7. \end{aligned}$$

This kind of example works much more generally:

Example 5.18. Consider any scenario in which there are N items of one type and M items of another type, and we need to choose n items altogether. Suppose all choices are equally likely. If we want to calculate the probability that all n of the chosen items are of the second type, we can do this systematically.

We write $A_1, A_2, \ldots, A_n$ for the events that the first, second, third, ..., nth items are of the second type. Then we use the fact that

$$\begin{aligned} P(A_1 \cap A_2 \cap A_3 \cap \cdots \cap A_n) = P(A_1)P(A_2 \mid A_1)&P(A_3 \mid A_1 \cap A_2) \\ &\cdots \times P(A_n \mid A_1 \cap A_2 \cap \cdots \cap A_{n-1}). \end{aligned}$$

As above, we have

$$\begin{aligned}
P(A_1) &= \frac{M}{N+M} \\
P(A_2 \mid A_1) &= \frac{M-1}{N+M-1} \\
P(A_3 \mid A_1 \cap A_2) &= \frac{M-2}{N+M-2} \\
&\vdots \\
P(A_n \mid A_1 \cap A_2 \cap \cdots \cap A_{n-1}) &= \frac{M-(n-1)}{N+M-(n-1)}
\end{aligned}$$

Multiplying this system of equations yields:

$$P(A_1 \cap A_2 \cap \cdots \cap A_n) = \frac{(M)(M-1)\cdots(M-n+1)}{(N+M)(N+M-1)\cdots(N+M-n+1)}.$$

5.3 Exercises

5.3.1 Practice

Exercise 5.1. **Car safety.** A car safety institute observes that 12% of cars on the road are manufactured by Honda. They also observe that 98% of Honda vehicles are classified as "safe" at the time of inspection. The percentage of all cars (regardless of brand) classified as "safe" is 72%. A car is randomly chosen on the road and inspected. It is classified as "safe." What is the probability that it is a Honda?

Exercise 5.2. **Happy courses.** Twenty percent of students will participate in an Honors course this semester. Students who have an Honors course on their schedule are known to have a 99% chance of fully enjoying their semester. Students who are not in an Honors course during the semester have only a 30% chance of fully enjoying their semester. With these assumptions, if a randomly chosen student is fully enjoying his semester, what is the probability that he has an Honors course on his schedule?

Exercise 5.3. **Stubborn dogs.** I am at a dog training class. Twenty percent of the dogs are hounds. The other eighty percent of the dogs are of the toy variety. Ninety percent of the hounds are stubborn and difficult to train. Thirty percent of the dogs in general are stubborn. What is the probability that a stubborn dog is a hound?

Exercise 5.4. mp3 players. Eighty percent of all mp3 players are iPods. Five percent of iPods are defective. Seven percent of all mp3 players are defective. A randomly chosen mp3 player is taken to a repair shop because it is defective. What is the probability it is an iPod?

Exercise 5.5. Chalkboards and dry-erase boards. In the Department of Mathematical and Computational Science, 83% of lecture halls have chalkboards, and the other 17% have dry-erase boards. Of the classes taught in rooms with chalkboards, 75% are mathematics courses, 15% are computer science courses, and 10% are statistics courses. Of the classes taught in rooms with dry-erase boards, 65% are computer science courses, 8% are mathematics courses, and 27% are statistics courses. What percentage of courses at the college are mathematics or statistics?

Exercise 5.6. Math and art. In the theory of general intelligence, it is stated that being good in one intelligence, like math, increases the chance of one being good in another intelligence. Suppose 10% of the people are good at art, and that 40% of the people who are good at art are also good at math. If a person is not good at art, they have only a 30% chance of being good at math. What is the probability that a person who is good at math will also be good at art?

Exercise 5.7. Students with music players. A student is chosen at random. You want to know the probability that the student has an iPod. This could be hard to determine since there are over 40,000 students on the given campus. Fortunately, a current survey recorded that 47% of first-year students have iPods, and you know that 32% of the students are first-year students. The survey indicates that 56.2% of upperclass students (i.e., non first-years) have iPods.

a. What is the probability that the student you randomly select has an iPod?

b. Given that the selected student has an iPod, what is the probability that he/she is an upperclass student?

c. Given that the selected student does not have an iPod, what is the probability that he/she is an upperclass student?

Exercise 5.8. Smoke detectors. It is estimated that 82% of homes have working smoke detectors. On average, 22% of fires result in fatalities, but the presence of a working smoke detector cuts the risk to just 7%.

a. If a random fire resulted in a fatality, what is the probability that the house had a working smoke detector?

b. In homes without a working smoke detector, what is the risk of fatalities?

5.3.2 Extensions

Exercise 5.9. Coins and dice. Consider the following game: The player flips a fair coin. If it shows a head, he gets to roll a 4-sided die. If it shows a tail, he gets to roll a 6-sided die. In either case, let A denote the event that he gets 1 on the die roll. Let H denote the event that his coin flip shows a head, i.e., that he uses the 4-sided die. Let T denote the event that his coin flip shows a tail, i.e., that he uses the 6-sided die.

a. Find $P(H \mid A)$.

b. Find $P(T \mid A)$.

Hint: To verify your answers, note $P(H \mid A) + P(T \mid A) = 1$.

Exercise 5.10. French classes. In a certain school, 4 levels of French are taught with 40% of the students being enrolled in level 1, 30% enrolled in level 2, 20% enrolled in level 3, and 10% enrolled in level 4. The percentage of people who enjoy their French class is 70% in level 1, 80% in level 2, 85% in level 3, and 90% in level 4.

Given that a person enjoys his/her French class, let p_j be the probability that the student is enrolled in level j. Find p_j for $j = 1, 2, 3, 4$. (Check: Your four answers should sum to 1 altogether.)

Exercise 5.11. Sex and switching majors. At a certain university, 60% of undergraduate students are male, and 40% are female. Ten percent of females change their majors at least once. Overall, 30% of students change their majors at least once.

a. Find the probability that, given a randomly selected student who changes majors, the student is a female.

b. What percentage of males change their majors at least once?

Exercise 5.12. Engineering majors. At First Street Towers, an equal number of students live on each floor. On floors 1, 2, 3, 4, 5, the percentage of students who study engineering are, respectively, 80%, 52%, 74%, 67%, and 29%. Upon meeting an Engineering student who lives in First Street Towers, what is the probability that the student lives on the 4th floor?

Exercise 5.13. Babies. Allison delivers one baby, and a year later she delivers a second baby. Let C be the event that at least one of the babies is a girl (either the first, or the second, or both). Let D be the event that both of the babies are girls. Find $P(D \mid C)$.

5.3.3 Advanced

Exercise 5.14. Pair of dice. Roll a blue die and a red die. Given that the blue die has an odd value, what is the probability that the sum of the two dice is exactly 4?

Exercise 5.15. Pair of dice. Roll a blue die and a red die. Given that the blue die has a value of 4 or smaller, what is the probability that the sum of the two dice is 7 or larger?

Exercise 5.16. Fuses. Two fuses in series are built to shut down if an overload occurs. If the first fuse shuts down properly 90% of the time, there is no need for the second fuse to do anything. If the first fuse fails to shut down properly, the second fuse shuts down properly 95% of the time. What is the probability the whole system operates correctly during an overload (with one fuse or the other shutting down properly)?

Exercise 5.17. Weather. The weather on any given day can either be sunny, cloudy, or partially cloudy. Each day is also classified as dry or rainy. The probability of a sunny day is 0.48, and the probability of a cloudy day is 0.39.

The probability of having a sunny and dry day is 0.48. The probability of a cloudy and dry day is 0.14. The probability of a partially cloudy and dry day is 0.09.

a. Find the probability of a dry day.

b. Find the probability of a rainy day.

c. Find the probability of a sunny day given that it is a dry day.

d. Find the probability of a dry day given that it is a sunny day.

e. Find the probability of a cloudy day given that it is a rainy day.

f. Find the probability that it is a rainy day given that it is a cloudy day.

Exercise 5.18. Coin flips and then dice. Claire flips a coin until she gets a head for the first time. Say it takes her n times. Then (afterwards) she rolls exactly n dice. What is the probability that none of the dice show the value 1?

(For instance, if it takes her 7 flips to get a head for the first time, then she rolls 7 dice. Hint: It is enough to use Remark 5.8. You do not need Bayes' Theorem itself for this problem.)

Exercise 5.19. Randomly chosen beverage. Lakisha has two identical-looking beverage containers. She randomly picks one (each is equally likely to be chosen). Then she reaches inside the chosen container and picks a drink for herself and a drink for her boyfriend. Suppose that one cooler has 3 sodas and 2 fruit juices; the other cooler has 2 sodas and 2 fruit juices. If her drink is a fruit juice, find the probability that her boyfriend's drink is also a fruit juice.

Chapter 6

Review of Randomness

6.1 Summary of Randomness

How do we classify random things?

- When something random occurs, exactly one outcome happens.
- An event is a collection of outcomes.
- Empty set $\emptyset$ contains no outcomes. Sample space S contains all outcomes.
- An event A and its complement A^c are disjoint (have no overlap) and contain all of the events from the sample space S, i.e., $A \cup A^c = S$.
- DeMorgan's laws $\left(\bigcup_j A_j\right)^c = \bigcap_j A_j^c$ and $\left(\bigcap_j A_j\right)^c = \bigcup_j A_j^c$ allow us to work with the complement of a union or an intersection.

What are the basic probability axioms?

- Any event occurs a certain percentage of the time, so probabilities are always between 0 and 1.
- With probability 1, some outcome in the sample space occurs.
- If events have no outcomes in common, then the probability of their union is the sum of the probabilities of the individual events.

Can we add the probabilities of events?

- If events are disjoint, we can sum their probabilities.
- If $A_1, A_2, \ldots$ is a collection of *disjoint* events, then

$$P\Big(\bigcup_{j=1}^{\infty} A_j\Big) = \sum_{j=1}^{\infty} P(A_j).$$

Can we add the probabilities of events if the events are overlapping?

If A, B, C are overlapping events, then $P(A \cup B \cup C) = P(A) + P(B) + P(C) - P(A \cap B) - P(A \cap C) - P(B \cap C) + P(A \cap B \cap C)$. More general versions of the inclusion-exclusion rule are available as well.

How do we know if two events are independent?

Events A and B are independent if $P(A \cap B) = P(A)P(B)$.

Can two events be independent and disjoint?

No, if A and B both have positive probabilities, they cannot be both independent and disjoint.

Can subsets be independent?

If $A \subset B$ and neither $P(A) = 0$ nor $P(B) = 1$, then A, B are dependent.

Can complements be independent?

If neither $P(A) = 0$ nor $P(A) = 1$, then A, A^c are dependent.

How do we calculate conditional probabilities?

If event B has nonzero probability (i.e., $P(B) > 0$), then the conditional probability $P(A \mid B)$ of A given B is defined as

$$P(A \mid B) = \frac{P(A \cap B)}{P(B)}.$$

Equivalently,

$$P(A \cap B) = P(B)P(A \mid B).$$

How can we decompose an event in parts?

If A is an event, then we can decompose B into two parts: $A \cap B$ and $B \setminus A = A^c \cap B$. Since $B = (A \cap B) \cup (A^c \cap B)$ is a disjoint union, then if $P(A)$ is not 0 or 1,

$$P(B) = P(A \cap B) + P(A^c \cap B) = P(A)P(B \mid A) + P(A^c)P(B \mid A^c).$$

Or we can decompose B into finitely many parts: Let $A_1, A_2, \ldots, A_n$ be any finite partition of the sample space. Then we can partition B into n parts as $A_1 \cap B$, $A_2 \cap B$, $\ldots$, $A_n \cap B$, i.e., $B = (A_1 \cap B) \cup (A_2 \cap B) \cup \cdots \cup (A_n \cap B)$ is a union of disjoint events, so

$$P(B) = P(A_1 \cap B) + \cdots + P(A_n \cap B) = P(A_1)P(B \mid A_1) + \cdots + P(A_n)P(B \mid A_n).$$

When do we want to use Bayes' Theorem? How do we use it?

If you want $P(A \mid B)$ but you instead know (or are told) $P(B \mid A)$, Bayes' Theorem lets you turn this around and compute $P(A \mid B)$.

If $P(B) \neq 0$ and $0 < P(A) < 1$, then,

$$P(A \mid B) = \frac{P(A)P(B \mid A)}{P(A)P(B \mid A) + P(A^c)P(B \mid A^c)}.$$

If $P(B) > 0$, and if $A_1, A_2, \ldots, A_n$ form a partition of S, with all $P(A_j) > 0$, then

$$P(A_k \mid B) = \frac{P(A_k)P(B \mid A_k)}{P(A_1)P(B \mid A_1) + P(A_2)P(B \mid A_2) + \cdots + P(A_n)P(B \mid A_n)}.$$

6.2 Exercises

Exercise 6.1. Probabilities in a chain of subsets. Four events A, B, C, D could be compared in the following way:

$$A \subset B \subset C \subset D.$$

The probability of A is 0.03, and the probability of C is 0.27. With the given information, what are potential probabilities of B and D?

Exercise 6.2. Waiting for the bus. A customer waits for a bus to appear.

a. List 5 possible outcomes.

b. What is the sample space?

c. Write a partition for the customer's waiting time, using five-minute intervals for the partition. (Assume that the customer can wait as long as needed for the bus.)

d. Explain how the answer to part c meets the definition of a partition; see Definition 2.14.

e. Modify your answer to part c, with the assumption that the customer has a maximum time that they can afford to wait before giving up.

Exercise 6.3. Randomly choose a page. Randomly open a 300-page book, and mark a page. What is the probability that the number of the page contains the lucky number 5?

Exercise 6.4. Cell phone. A student loses her first cell phone. Her new phone number is randomly chosen by the store, but the area code is fixed, so there are exactly 7 randomly selected digits. (Assume all 10^7 possibilities are equally likely.) Give the probabilities of the following events:

a. Her phone number ends in a 2.

b. Her phone number ends in an odd number.

c. Her phone number ends in a 5 or a 7.

Exercise 6.5. Egg-citing! There are 45 egg boxes in a store. Twenty are Brand A, fifteen are brand B, and ten are Brand C. Brand A and C each have half green boxes and half yellow boxes. Brand B is all yellow.

a. If you have Brand B or C, what is the chance that you have a green box?

b. If you have a yellow box, what is the chance the brand is B?

Exercise 6.6. Skittles. Chris has 32 Skittles candies. Nine are red, three are blue, seven are yellow, five are orange, and eight are purple. Exactly four are sour, and all of these sour Skittles are purple. Chris picks one Skittle at random.

a. What is the probability he selected a red or blue one?

b. What is the probability he selected a sour one?

c. Given that he selected one that is not sour, what is the probability it was purple?

Exercise 6.7. Homework. Joe works on homework 25% of the time. He is on the computer 40% of the time. He likes to do homework on the computer, so 30% of the time that he is on the computer is spent doing homework. What is the probability that he is on his computer, given that he is doing his homework?

Exercise 6.8. Depression and anxiety. According to the National Institute of Mental Health (from `www.nimh.nih.gov/statistics/1ANYANX_ADULT.shtml`), in any year, 18.1% of all U.S. adults suffer from anxiety disorders. Among people who have anxiety disorders, 30.2% are between 18 and 29 years old. According to the 2010 U.S. Census, approximately 10% of the U.S. adults are between 18 and 29 years old.

a. Calculate the chance that a randomly selected 18–29 year old American has an anxiety disorder.

b. Calculate the chance that a randomly selected American is 18–29 years old AND has an anxiety disorder.

Part II

Discrete Random Variables

In this part of the book, we study discrete random variables and their distributions. We will discuss the differences between discrete and continuous random variables and introduce some useful formulas for calculating other information about discrete random variables. (We will return to continuous random variables starting in Chapter 24.) The rules and formulas in the next few chapters will work for all discrete random variables. This will build a base of understanding about common themes of discrete random variables that will be useful in the subsequent part of the book, on named discrete random variables.

By the end of this part of the book, you should be able to:

1. Distinguish between discrete and continuous random variables when reading a story.
2. Calculate probabilities for discrete random variables.
3. Calculate and graph a probability mass function (PMF).
4. Calculate and graph a cumulative distribution function (CDF).
5. Calculate the mean, variance, and standard deviation of discrete random variables.
6. Calculate the mean and variance of sums of discrete random variables.
7. Calculate the mean and variance of functions of discrete random variables.

Math skills you will need: Binomial coefficients ("choose" parentheses), factorials, summation $\sum$ notation, step functions, and the floor function $\lfloor x \rfloor$.

Additional resources: We will be calculating probabilities by hand, but there are mathematical and statistical software programs (such as Excel, Maple, Mathematica, Matlab, Minitab, R, SPSS, etc.), which could help you to perform these calculations.

Chapter 7

Discrete Versus Continuous Random Variables

While writing my book I had an argument with Feller. He asserted that everyone said "random variable" and I asserted that everyone said "chance variable." We obviously had to use the same name in our books, so we decided the issue by a stochastic procedure. That is, we tossed for it and he won.

—Joe Doob, from "A Conversation with Joe Doob," by J. Laurie Snell, from *Statistical Science*, volume 12, number 4, November 1997

We do not measure everything with the same types of variables. Can what we are measuring be counted? Or does it fall within a certain range without specific levels? For instance, consider the difference in how we measure the outcome of a die roll versus the distance a dart lands from the bullseye. If our types of variables are different, we will need different ways of calculating probabilities.

7.1 Introduction

We have already seen that, in a random phenomenon, there are many possible outcomes; exactly one of these outcomes occurs. The set of all possible outcomes is the sample space. Now we introduce random variables, which are the main topic of study throughout the rest of this book:

Definition 7.1. A **random variable** assigns a real number to each outcome in the sample space. The random variable's value is completely determined by the outcome. A random variable is a function from the sample space to the set of real numbers.

Often an outcome contains much more information than a random variable with which it is associated. A random variable is very simple; it is a one-number way to summarize an outcome.

Example 7.2a If we roll three dice, there are $6^3 = 216$ possible outcomes. One possible random variable, which we call X, is the sum of the three dice.

1 die has 6 possible outcomes; 2 dice have $6 \times 6 = 36$ possible outcomes; 3 dice have $6 \times 6 \times 6 = 216$ possible outcomes.

If the outcome is $\omega = (6, 4, 3)$, then sum of the dice is $X(\omega) = 13$. If the outcome is $\omega = (5, 2, 2)$, the sum is $X(\omega) = 9$. In general, if the outcome is $\omega = (a, b, c)$, then the sum is $X(\omega) = a + b + c$. In this example, X is always an integer from 3 and 18, depending on the outcome. *The outcome completely and uniquely determines the value of the random variable.*

Example 7.2.b *As before, roll three dice. Let the random variable Y be the maximum of the three dice.*

If the outcome is $\omega = (a, b, c)$, then $Y(\omega) = \max(a, b, c)$. For instance, if the outcome is $\omega = (4, 5, 2)$, then $Y(\omega) = 5$.

Example 7.2.c *Let the random variable Z denote the value of the first die rolled.*

If the outcome is $\omega = (a, b, c)$, then $Z(\omega) = a$. E.g., if the outcome is $\omega = (6, 1, 2)$, then $Z(\omega) = 6$.

When working with random variables, we must resist the urge to "solve" for the value of the random variable. Unlike in algebra, we will not spend time studying equations such as $x^2 - x - 6 = 0$ and trying to solve for x. As a constant reminder that we are not solving such equations, we always use a capital letters (such as X) to denote a random variable.

The outcome of a random phenomenon is, well, *random*! Different outcomes happen, so random variables can have different values. Every random variable depends on the underlying outcome.

Also, random variables must assign a real number to each outcome. E.g., if a randomly chosen person ω (the outcome) in a classroom is selected, we could define a random variable $X(\omega)$ as the person's age. So $X(\omega)$, the person's age, is a random variable that completely depends on ω, the person selected.

Remark 7.3. Suppressing Outcomes in Random Variable Notation
We have been writing ω as the outcome and $X(\omega)$ as the random variable, but often the outcome is well-understood and the random variable is the thing we want to study. Thus, we just write X (instead of $X(\omega)$), to simplify our notation. We understand, however, that the there is actually an underlying outcome ω, and that X actually means $X(\omega)$.

Since a random variable must assign a real number to each outcome, then a random variable cannot be a person's name, the color of a car, a country, a suit in a deck of cards, heads or tails, etc. These types of things might be the underlying outcomes, but they cannot be the random variables themselves.

Random variables can be either **discrete** or **continuous**. Parts II and III of the book are all about discrete random variables. Parts V and VI are dedicated to continuous random variables.

The list of all values taken on by a discrete random variable is either finite:

$$\text{e.g., } \{1, 2, 3, 4, 5, 6\} \text{ or } \{-1, 0, 1\} \text{ or } \{2, 3, 4, \ldots, 12\} \text{ or } \{-5.62, 2.33, 7.7\}, \text{ etc.}$$

or countably infinite:

$$\text{e.g., } \{1, 1/2, 1/4, 1/8, \ldots\} \text{ or } \{1, 2, 3, 4, \ldots\} \text{ or } \{\ldots, -2, -1, 0, 1, 2, \ldots\}, \text{ etc.}$$

Discrete random variables can assume any kind of real numbers; e.g., negative numbers are allowed; decimals, fractions, transcendental numbers, etc., are all allowed too. All that matters for a random variable to be discrete is that the set of possible values is either finite or can be put into a countably infinite list.

Continuous random variables, on the other hand, to be covered in Part V, take values on continuous intervals (or on the union of continuous intervals):

$$\text{e.g., } [0, \infty) \text{ or } (-\infty, \infty) \text{ or } [0, 1] \text{ or } [-3.7, 10.2), \text{ etc.}$$

7.2 Examples

Some examples of random variables that are discrete are:

- let X be the number of male puppies in a litter of Golden Retrievers;
- let X be "1" if a newborn baby is a girl, or "0" if it is a boy;
- let X be "1" if the next car to pass is blue, or "2" if red, or "3" if silver, or "-1" otherwise;
- let X be the number of the region, 1 through 20, on which a dart lands.

Remember, random variables can be negative too!

Examples of random variables that are continuous are:

- let X be the length of your left foot (in inches);

- let X be the time, in seconds, for a traffic light to turn green.

Here are some more examples of random variables.

Example 7.4. Flip a coin three times, and let X denote the total number of heads that appear.

The set of values X can assume is $\{0, 1, 2, 3\}$, so X is a discrete random variable. We can make a chart of all of the possible outcomes and the associated values of X:

outcome	prob. of outcome	value of X	prob. of each X value
(H, H, H)	$1/8$	3	$1/8 = P(X = 3)$
(H, H, T)	$1/8$	2	
(H, T, H)	$1/8$	2	$3/8 = P(X = 2)$
(T, H, H)	$1/8$	2	
(T, T, H)	$1/8$	1	
(T, H, T)	$1/8$	1	$3/8 = P(X = 1)$
(H, T, T)	$1/8$	1	
(T, T, T)	$1/8$	0	$1/8 = P(X = 0)$

Each of the outcomes (H, H, T) and (H, T, H) and (T, H, H) will cause X to be 2. Each of these outcomes has probability 1/8, so the event containing all three of these outcomes has probability 3/8, i.e.,

$$P(\{(H, H, T), (H, T, H), (T, H, H)\}) = 3/8.$$

We abbreviate this by writing

$$P(X = 2) = 3/8.$$

So the probability of each possible value of X can be written succinctly as

x	0	1	2	3
$P(X = x)$	1/8	3/8	3/8	1/8

Example 7.5. a Let $\omega = (x, y)$ be the Cartesian coordinates where a dart lands on a dart board. We could let $X(\omega) = x$ be a random variable that denotes the first coordinate and let $Y(\omega) = y$ be a random variable that denotes the second coordinate.

Both X and Y are continuous random variables that each assume values (for instance) on the interval $(-9, 9)$ if the dartboard has radius 9 inches. If $\omega = (3.6, -1.35)$ is the location where the dart lands, then $X(\omega) = 3.6$ and $Y(\omega) = -1.35$. Again, we often write (more simply) just $X = 3.6$ and $Y = -1.35$, in such a case.

Example 7.5.b We could also consider other random variables, for instance, if $\omega = (x, y)$ is the location of the dart's landing, then $Z(\omega) = \sqrt{x^2 + y^2}$ is the distance of the dart from the center of the board. More simply, we can drop the notation for ω and just write Z as the distance to the center of the dartboard. Notice Z is a continuous random variable.

Example 7.6. A traffic engineer observes the next three cars that pass. He uses X as the time until the arrival of the third car, Y as the time between the arrivals of the second and third cars, and Z as the speed of the third car. Since X, Y, Z each take values on the interval $(0, \infty)$, then X, Y, Z are each continuous random variables.

Example 7.7. A student flips a coin until the 10th head appears. Each outcome is a string of heads and tails. For instance, an outcome ω might be

$$\omega = (H, H, H, T, H, T, T, T, H, T, T, T, H, H, T, H, T, H, T, H),$$

He writes:

$$\begin{aligned}
&X_1 \text{ as the number of flips until the 1st head,}\\
&X_2 \text{ as the number of flips after the 1st head until the 2nd head,}\\
&\vdots \qquad\quad \vdots\\
&X_j \text{ as the number of flips after the } (j-1)\text{st head until the } j\text{th head,}\\
&\vdots \qquad\quad \vdots
\end{aligned}$$

In this case, outcome ω causes these random variables to have the following values:

$$X_1 = 1, \quad X_2 = 1, \quad X_3 = 1, \quad X_4 = 2, \quad X_5 = 4,$$

$$X_6 = 4, \quad X_7 = 1, \quad X_8 = 2, \quad X_9 = 2, \quad X_{10} = 2.$$

Each of the X_j's takes on a positive integer value and is therefore a discrete random variable. Finally, $X_1 + \cdots + X_{10}$ is the total number of flips needed until the 10th head appears.

Example 7.8. A student is selected at random and her mp3 player is examined. Let X denote the number of songs on her music player and let Y denote the number of songs on the first playlist on the music player, or $Y = 0$ if there are no playlists on the mp3 player. (Let $X = 0$ and $Y = 0$ if she doesn't have an mp3 player at all.) Notice that X and Y are random variables. Also, we know $Y \leq X$, since the number of songs on the playlist is limited by the number of songs on the mp3 player altogether.

We cannot assign an artist's name, or a genre of song, as a random variable, because these are not real numbers. Nonetheless, we could (say) assign a numeric scheme, such as $Z = 1$ if the majority of the songs are blues songs, $Z = 2$ if the majority of the songs are rock songs, $Z = 3$ if the majority of the songs are jazz songs, or $Z = 4$ otherwise (including if there is a "tie" for the majority).

Both X and Y assume values that are nonnegative integers and thus are discrete random variables. Since Z takes values in the range $\{1, 2, 3, 4\}$, then Z is a discrete random variable too.

At this point, it should be clear that there are ample, real-world possibilities for assigning random variables according to all kinds of random phenomenon.

Example 7.9. A student is selected at random. Let W be her height (in inches); let X denote her body temperature; let Y be her age; and let Z denote the zip code of her mailing address.

Her height is a continuous random variable that is in the range $[22.5, 107.1]$ inches. Her body temperature is a continuous random variable number that is close to $98.6°F$ a large percentage of the time and is almost always in the range (say) 92–104$°F$, even in the most extreme situations. Her age is a discrete random variable in the set $\{0, 1, 2, 3, \ldots, 122\}$, if rounded (down) to the nearest integer, or (on the other hand) her age is a continuous random variable in the range $[0, 122.3]$ if the age is not rounded to the nearest integer. Her zip code is a discrete random variable with values in the set $\{00000, 00001, \ldots, 99999\}$.

Guinness World Records (2005) claim adult humans range in height from 22.5 inches to 107.1 inches.

Sometimes it is helpful to make a list of all of the possible outcomes for a random phenomenon, and then to list the associated random variable that goes along with each outcome.

Example 7.10. Roll a pair of dice, and let X denote the sum of the two values that appear. So if the outcome is (i, j), then $X = i + j$.

The sum X is a discrete random variable that has values in the set $\{2, 3, \ldots, 12\}$. There are several outcomes that cause X to be equal to 5. These outcomes are those in the event

$$\{(1,4), (2,3), (3,2), (4,1)\}.$$

Each event has a probability associated with it. In this case,

$$P(\{(1,4), (2,3), (3,2), (4,1)\}) = 4/36,$$

but we usually use a more succinct notation and write

$$P(X = 5) = 4/36$$

as a shorthand for the equation above.

Next we compute all of the probabilities associated with the possible values of X:

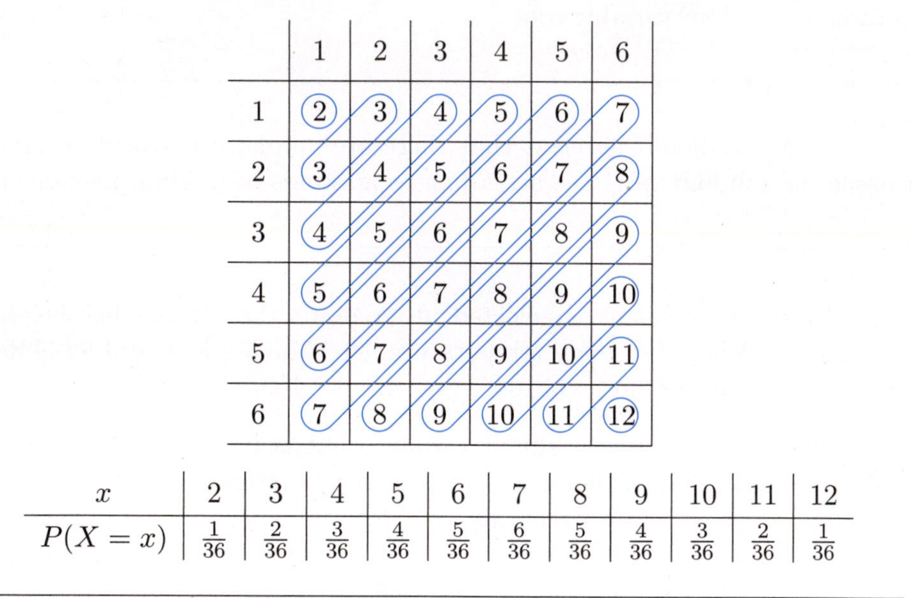

x	2	3	4	5	6	7	8	9	10	11	12
$P(X = x)$	$\frac{1}{36}$	$\frac{2}{36}$	$\frac{3}{36}$	$\frac{4}{36}$	$\frac{5}{36}$	$\frac{6}{36}$	$\frac{5}{36}$	$\frac{4}{36}$	$\frac{3}{36}$	$\frac{2}{36}$	$\frac{1}{36}$

Example 7.11. Select a random song on a student's music player. Let X be equal to 1 if the song is a blues song, or otherwise let X be 0; let Y be equal to 1 if the song is jazz, or otherwise Y is 0; let $Z = 1$ if the song is rock, or otherwise $Z = 0$.

Random variables that are 1 when an event occurs or 0 when the event does not occur are called **indicator random variables**, sometimes abbreviated as **indicators**; they are also called Bernoulli random variables. They will be studied in Chapter 14. Indicator random variables are used extensively when working with sums of random variables, to be covered in Chapter 11.

7.3 Exercises

7.3.1 Practice

In Exercises 7.1 to 7.15, identify whether the random variables are discrete or continuous (an exercise might contain one or more of both types). Also identify the possible values that each random variable could assume. In some cases, answers will differ according to interpretation of the problems by the students.

Exercise 7.1. **Pair of dice.** Roll two dice. Let X be the absolute value of the difference of the two values that appear. (If D_1, D_2 are the two die values, then $X = |D_1 - D_2|$.)

Exercise 7.2. **Onions.** Pick a basket of onions. Let X be the weight of the onions (in pounds). Let Y be the number of onions in the basket.

Exercise 7.3. **Grades.** Let X be the cumulative grade point average of a randomly chosen student. Let Y be the score on the student's most recent exam. Let Z be the number of exams that the student takes during the current semester.

Exercise 7.4. **Harmonicas.** When ordering a new box of harmonicas, let X denote the time (in hours) until the box arrives, and let Y denote the number of harmonicas that work properly.

Exercise 7.5. **Music genres.** Select a random student's music player. Let X denote the number of blues songs on the music player; let Y denote the number of jazz songs; let Z denote the number of rock songs.

Exercise 7.6. **Average distance.** Let X be the average distance (in miles) that a student drives in a week.

Exercise 7.7. **Sexes of babies.** Consider the births of ten consecutive babies. Let X be the number of babies that are girls. Let Y be the number of the first baby that is a boy (if none of the babies are boys, then just let $Y = 0$, or suggest your own, alternative value for such a case).

Exercise 7.8. **Calling boyfriend/girlfriend.** Consider the time X (in minutes) from now until your boyfriend/girlfriend calls on the telephone. Let Y be the length of the call. Let Z be the number of times that he/she calls in one evening.

Exercise 7.9. **Lightbulbs.** Consider a box of two lightbulbs. Let X be the lifetime of the first bulb removed from the box, and let Y denote the lifetime of the second bulb removed from the box.

Exercise 7.10. **More lightbulbs.** Consider a collection of 100 lightbulbs. Let X_j denote the lifetime of the jth bulb. Let Y denote the *sum* of the lifetimes of all 100 bulbs.

Exercise 7.11. Social networking. Place "friend" requests on a social networking site until the first person accepts your request. Let X be the number of people needed until this first acceptance. Let Y be the time until the first request is accepted.

Exercise 7.12. Puzzle pieces. A pile of puzzle pieces falls onto the floor. Let X denote the number of pieces that are edge pieces, and let Y denote the number of pieces that are interior (i.e., not edge) pieces.

Exercise 7.13. Milk. A half-gallon container of milk is tested by a quality control company. Let X be a rating of the milk quality, given as an integer from 1 to 10; let Y denote the exact volume of milk in the selected half-gallon container; let Z be "1" if the half-gallon of milk is 1% milkfat, or "2" if it is 2% milkfat, or "3" if it is fat-free, or "4" otherwise.

Exercise 7.14. Snake eyes. Roll a pair of dice until "snake eyes" (i.e., a pair of 1's) appear. Let X denote the total number of rolls required. Let Y denote the sum of all of the dice rolled during this process.

Exercise 7.15. Student age and name. Select a random student from your course; let X denote the age, in days, of the selected student; let Y denote the length of the student's name.

7.3.2 Extensions

Exercise 7.16. Toy ballet dancers. When opening a container of 32 toy ballet dancers performing pirouettes, let X denote the number of broken dancers, and let Y denote the number of whole (unbroken) dancers.

Write $Z = X/Y$, i.e., Z is the ratio of broken dancers to whole dancers. What complication arises in this definition of Z as a random variable?

Exercise 7.17. Pick two cards. Pick two cards at random from a well-shuffled deck of 52 cards (pick them simultaneously, i.e., grab two cards at once—so they are not the same card!). There are 12 cards which are considered face cards (4 Jacks, 4 Queens, 4 Kings). Let X be the number of face cards that you get.

a. Find $P(X = 0)$.

b. Find $P(X = 1)$.

c. Find $P(X = 2)$.

7.3.3 Advanced

Exercise 7.18. Mixed random variables. Do you think that there are random variables which are neither discrete nor continuous? If yes, try to construct a simple example. If no, then discuss why not.

Chapter 8

Probability Mass Functions and CDFs

The 50-50-90 rule: Anytime you have a 50-50 chance of getting something right, there's a 90% probability you'll get it wrong.
—Andy Rooney

If we toss a single coin, we have a 1/2 chance of getting a head. How do the probabilities change if we toss a coin 3 times and total up the number of heads? Is the chance of getting 0, 1, 2, or 3 heads exactly the same? Why or why not? Is the probability of getting exactly 1 head the same as the probability of getting head, tail, tail?

8.1 Introduction

In many situations, it is very helpful to consider the probability that a random variable assumes a specific value or is found in an entire range of numbers. For instance, it might be helpful to know whether a pregnant mother will have 2 or more babies during her delivery; if X denotes the number of babies to be born, the probability of this event is written as $P(X \geq 2)$. If X is the number of guitar strings in a package to be shipped, and it is supposed to have 6 strings, the manufacturing company will be very interested in $P(X = 6)$, i.e., if the package contains the right number of strings. When baking cookies, X could be the number of cookies that your roommates would like to eat. If there are 8 cookies in a container, then $P(X \leq 8)$ is the probability that you have enough cookies to feed them.

We use the concepts $P(X = x)$ and $P(X \leq x)$ so often that we give them each a special name:

Definition 8.1. Probability Mass Function (a.k.a. PMF or mass)
If X is a random variable, the probability that X is exactly equal to x is

$$p_X(x) = P(X = x).$$

This is called the **probability mass function (PMF)**, or just the **mass**, of X.

Definition 8.2. Cumulative Distribution Function (a.k.a. CDF)
If X is a random variable, the probability that X does not exceed x is written as

$$F_X(x) = P(X \leq x).$$

This is called the **cumulative distribution function (CDF)** of X.

8.2 Examples

Example 8.3. Roll a die, and let X denote the value that appears.

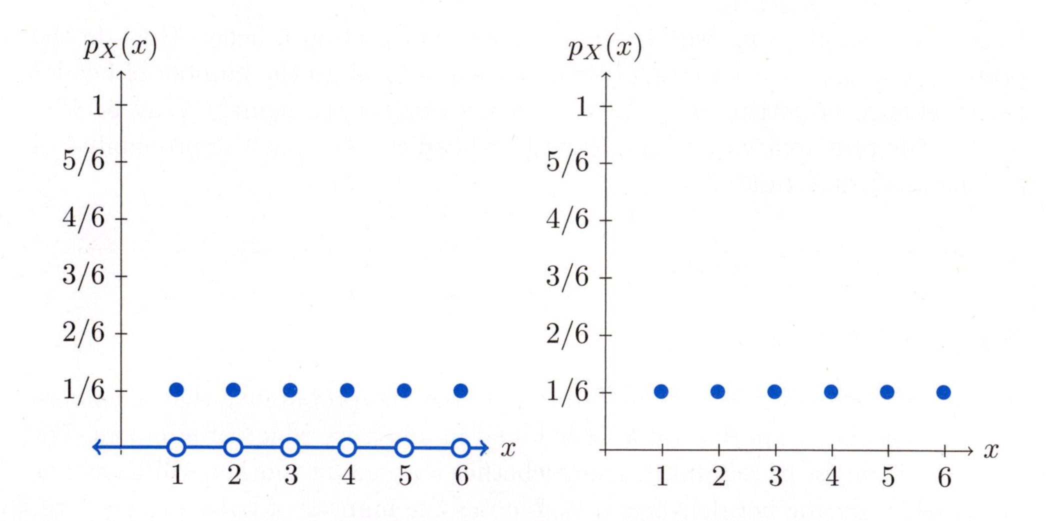

FIGURE 8.1: Left: Mass $p_X(x) = P(X = x)$ of the value on a die roll. Right: Same plot but with the values of 0 not shown in the plot.

In the mass of a die roll X, the left hand side of Figure 8.1 illustrates the fact that

$$p_X(x) = P(X = x) = 1/6$$

for all x in the set $\{1, 2, 3, 4, 5, 6\}$, and $p_X(x) = P(X = x) = 0$ otherwise. Since the "otherwise" encompasses most values of x, we usually suppress the values for which the mass is 0. Thus, the right hand side of Figure 8.1 is the way that we will usually show such a mass (with the 0 values omitted from the plot).

To compute the cumulative distribution function of X, we compute

$$\begin{aligned}
F_X(1) &= P(X \le 1) = P(X = 1) = 1/6;\\
F_X(2) &= P(X \le 2) = P(X = 1) + P(X = 2) = 2/6;\\
F_X(3) &= P(X \le 3) = P(X = 1) + P(X = 2) + P(X = 3) = 3/6;\\
F_X(4) &= P(X \le 4) = P(X = 1) + \cdots + P(X = 4) = 4/6;\\
F_X(5) &= P(X \le 5) = P(X = 1) + \cdots + P(X = 5) = 5/6;\\
F_X(6) &= P(X \le 6) = P(X = 1) + \cdots + P(X = 6) = 1.
\end{aligned}$$

So if we begin to draw the cumulative distribution function in this case, we know 6 of the values of $F_X(x) = P(X \le x)$; see the left side of Figure 8.2.

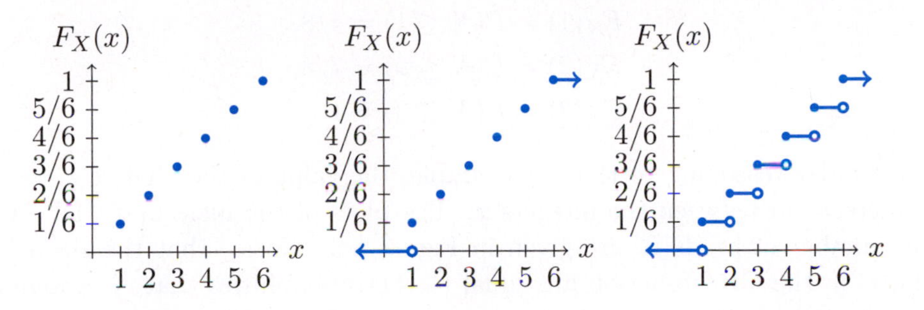

FIGURE 8.2: Left: Starting to construct the CDF $F_X(x) = P(X \le x)$ for the value of a die roll, for x in $\{1, 2, 3, 4, 5, 6\}$. Middle: The CDF $F_X(x)$ including $x < 0$ and $x > 6$. Right: The CDF $F_X(x)$ for all values of x.

Of course, we know that the value on a die is 6 or smaller. So $P(X \le 100) = 1$ and $P(X \le 12) = 1$ and $P(X \le 7) = 1$ and $P(X \le 6.2) = 1$, etc., etc. Thus, $P(X \le x) = 1$ for any $x \ge 6$.

Similarly, we know that the value on a die is 1 or larger. So $P(X \le -10) = 0$ and $P(X \le -3) = 0$ and $P(X \le -1) = 0$ and $P(X \le 0) = 0$ and $P(X \le 0.7) = 0$, etc., etc. In fact, $P(X \le x) = 0$ for any $x < 1$.

All that remains is to fill in the picture in between the integers from 1 to 6. Here is an example. We determine $P(X \le 3.1)$. If $X \le 3.1$, then $X = 1$ or $X = 2$ or $X = 3$, so

$$P(X \le 3.1) = P(X = 1) + P(X = 2) + P(X = 3) = 3/6.$$

This is the same value as $P(X \le 3)$. The reason is that X cannot be between 3 and 3.1, so the CDF does not increase in this interval. The CDF $F_X(x) = P(X \le x)$ is constant for $3 \le x < 3.1$. This reasoning similarly shows that $F_X(x) = P(X \le x) = 3$ for any x with $3 \le x < 4$. We use similar reasoning to finish the CDF. See the right side of Figure 8.2.

Example 8.4. Flip a coin three times; let X denote the total number of heads.

The mass of X is

$$\begin{aligned} p_X(0) &= P(X = 0) = 1/8, \\ p_X(1) &= P(X = 1) = 3/8, \\ p_X(2) &= P(X = 2) = 3/8, \\ p_X(3) &= P(X = 3) = 1/8. \end{aligned}$$

Thus, the CDF of X is

$$\begin{aligned} F_X(0) &= P(X \le 0) = 1/8, \\ F_X(1) &= P(X \le 1) = 4/8, \\ F_X(2) &= P(X \le 2) = 7/8, \\ F_X(3) &= P(X \le 3) = 1. \end{aligned}$$

With similar reasoning to the last example, the value of the CDF $F_X(x)$ does not increase in between the integers x. The plots of the mass and CDF of the total number of heads X are given in Figure 8.3. Notice that the size of the "jumps" in the CDF function are equal to the probabilities that X is found at a specific value:

1. The jump in $F_X(x)$ at $x = 0$ of size $1/8$ corresponds to $P(X = 0) = 1/8$;
2. The jump in $F_X(x)$ at $x = 1$ of size $3/8$ corresponds to $P(X = 1) = 3/8$;
3. The jump in $F_X(x)$ at $x = 2$ of size $3/8$ corresponds to $P(X = 2) = 3/8$;
4. The jump in $F_X(x)$ at $x = 3$ of size $1/8$ corresponds to $P(X = 3) = 1/8$.

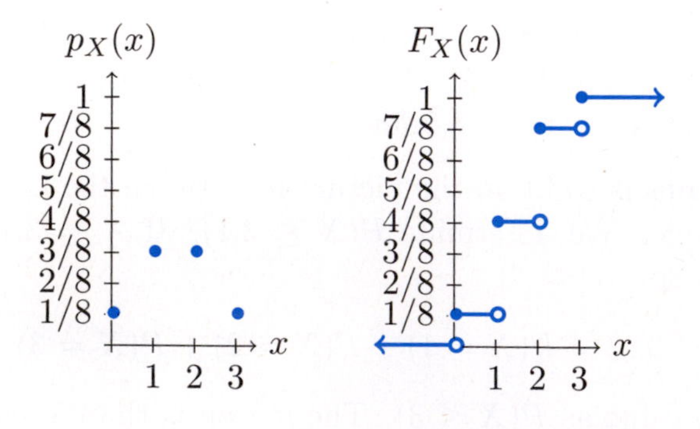

FIGURE 8.3: Left: The mass $p_X(x)$ of X, the number of heads in three tosses of a fair coin. Right: The CDF $F_X(x)$ of X.

The reasoning from Example 8.4 works in general. As we move from left to right across the mass, we start with probability 0 on the extreme left-hand side of the CDF. After sweeping all the way across the mass, we eventually accumulate all of the probability from the mass, so that we get probability 1 on the extreme right-hand side of the CDF.

8.3 Properties of the Mass and CDF

Remark 8.5. The mass $p_X(x) = P(X = x)$ and the CDF $F_X(x) = P(X \le x)$ are both probabilities, so they both take on values between 0 and 1, i.e.,

$$0 \le p_X(x) \le 1, \qquad \text{and} \qquad 0 \le F_X(x) \le 1, \qquad \text{for all } x.$$

Remark 8.6. Since discrete random variables assume only a finite or countable number of values, we can sum over all the nonzero masses, and we must get sum 1:

$$\sum_{x\,:\,p_X(x)\neq 0} p_X(x) = 1.$$

We usually drop the notation about restricting to x's for which $p_X(x) = 0$:

$$\sum_x p_X(x) = 1.$$

Alternatively, if X takes on only the finite values $x_1, x_2, \ldots, x_n$, then we have

$$\sum_{j=1}^{n} P(X = x_j) = \sum_{j=1}^{n} p_X(x_j) = 1,$$

or if X takes on infinitely many values, say $x_1, x_2, \ldots$, then we have

$$\sum_{j=1}^{\infty} P(X = x_j) = \sum_{j=1}^{\infty} p_X(x_j) = 1.$$

For example, in Example 8.4, the random variable X can only take on values $0, 1, 2, 3$, so $\sum_{j=0}^{3} p_X(j) = 1$.

Remark 8.7. The CDF is a non-decreasing function. From a visual perspective, this means that as we look left to right across the plot, the CDF is always increasing or flat. So if $a \le b$, then $F_X(a) \le F_X(b)$.

To see this, consider any two real numbers $a \leq b$. If an outcome ω causes X to be less than or equal to a, then X is automatically smaller than b too (since $a \leq b$). So the event $X \leq a$ is contained in the event $X \leq b$, i.e.,

$$\{\omega \mid X(\omega) \leq a\} \subset \{\omega \mid X(\omega) \leq b\}.$$

(Subsets were discussed in Section 1.1.) Our shorthand for this is

$$\{X \leq a\} \subset \{X \leq b\}.$$

Whenever one event is contained in another, then the probability of the first event is smaller than the probability of the second event. Thus

$$P(\{X \leq a\}) \leq P(\{X \leq b\}),$$

i.e., $F_X(a) \leq F_X(b)$, as claimed.

Remark 8.8. Recall from Example 8.4 that, as we move from left to right across the mass, we pick up the probabilities that build the CDF.

On the extreme left-hand side of the CDF, we would have probabilities near (or at) 0, and on the extreme right-hand side of the CDF, we would have probabilities near (or at) 1. Unfortunately, for instance, the mass might be nonzero for arbitrarily large x, e.g., consider the example of the number of coin tosses to see the first head, given below in Example 8.10. So the best thing we can say with precision about the extreme behavior of the CDF is:

Remark 8.9. Concerning the right-hand behavior of the CDF,

$$\lim_{x \to \infty} F_X(x) = 1,$$

and similarly for the left-hand behavior of the CDF,

$$\lim_{x \to -\infty} F_X(x) = 0.$$

Intuitively, for instance, $F_X(100) = P(X \leq 100)$ is less than $F_X(10{,}000) = P(X \leq 10{,}000)$, which is less than $F_X(1{,}000{,}000) = P(X \leq 1{,}000{,}000)$, etc., etc., with the limit always equal to 1. Consider what would go wrong if $\lim_{x\to\infty} F_X(x) < 1$. E.g., if $\lim_{x\to\infty} F_X(x) = 0.95$, then X would be equal to $+\infty$ five percent of the time, but that is not allowed, because random variables only take on real (finite) values.

8.4 More Examples

Example 8.10. Flip a coin until the first head appears; let X denote the total number of flips until the first head appears.

The mass of X is depicted on the left of Figure 8.4; the values are:

$$\begin{aligned} p_X(1) &= P(X = 1) = P(\{H\}) = 1/2, \\ p_X(2) &= P(X = 2) = P(\{T, H\}) = 1/4, \\ p_X(3) &= P(X = 3) = P(\{T, T, H\}) = 1/8, \\ p_X(4) &= P(X = 4) = P(\{T, T, T, H\}) = 1/16, \end{aligned}$$

and in general

$$p_X(j) = P(X = j) = P(\{\overbrace{T, T, \ldots, T}^{j-1}, H\}) = 1/2^j.$$

Thus, the CDF of X (shown on the right of Figure 8.4) has the following values at the integers:

$$\begin{aligned} F_X(1) &= P(X \le 1) = 1/2, \\ F_X(2) &= P(X \le 2) = 3/4, \\ F_X(3) &= P(X \le 3) = 7/8, \\ F_X(4) &= P(X \le 4) = 15/16, \end{aligned}$$

and in general, for each positive integer x,

$$F_X(x) = P(X \le x) = P(\{\text{first head within } x \text{ tosses}\}) = 1 - 1/2^x.$$

Another viewpoint is that, for positive integers x, we have $F_X(x) = P(X \le x) = 1 - P(X > x)$, but $X > x$ only if the first x tosses are T, which has probability $1/2^x$. Thus $F_X(x) = 1 - 1/2^x$.

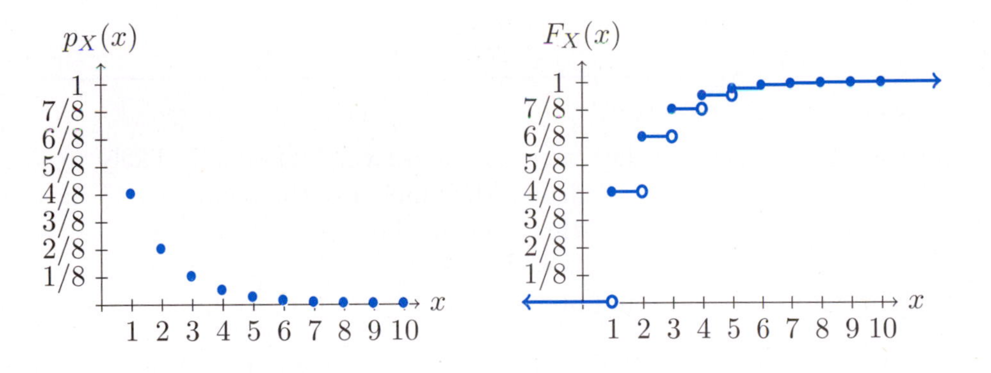

FIGURE 8.4: Left: The mass $p_X(x)$ of X, the number of tosses until the first head. Right: The CDF $F_X(x)$ of X.

Remark 8.11. In Example 8.10, the observant reader will notice that, for the outcome $\omega = (T, T, T, T, T, T, \ldots)$, we did not define the value of X. This outcome requires an infinite number of tosses until the first head is reached. This outcome has probability 0, so it does not affect any of our calculations.

Example 8.12. As in Example 7.10, roll a pair of dice, and let X denote the sum of the two values that appear. In other words, if the outcome is (i, j), then we let $X = i + j$.

The sum X is a discrete random variable with values in $\{2, 3, 4, \ldots, 12\}$. We computed the mass back in Example 7.10; we reproduce it here:

x	2	3	4	5	6	7	8	9	10	11	12
$P(X = x)$	$\frac{1}{36}$	$\frac{2}{36}$	$\frac{3}{36}$	$\frac{4}{36}$	$\frac{5}{36}$	$\frac{6}{36}$	$\frac{5}{36}$	$\frac{4}{36}$	$\frac{3}{36}$	$\frac{2}{36}$	$\frac{1}{36}$

We plot the mass and CDF of the sum of two dice in Figure 8.5.

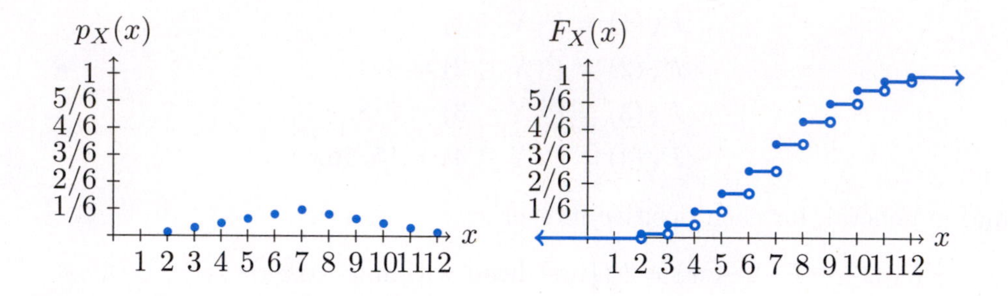

FIGURE 8.5: Left: The mass $p_X(x)$ of X, the sum of values on two dice. Right: The CDF $F_X(x)$ of X.

Example 8.13. Throw a dart at a dartboard with 20 equally likely regions, numbered 1 through 20. Let X denote the number of the region where the dart lands. Assume that the person throwing the dart can always land the dart on the board (otherwise, they are allowed to try again until they hit the board; the probability of missing over and over—forever—will be 0, so it will not matter).

The mass of X is $p_X(x) = 1/20$ for $x = 1, 2, 3, \ldots, 20$, and $p_X(x) = 0$ otherwise. The CDF of X is $F_X(x) = x/20$ for each integer x with $1 \le x \le 20$. Thus

$$F_X(x) = \begin{cases} 0 & \text{for } x < 1, \\ \lfloor x \rfloor / 20 & \text{for } 1 \le x < 20, \\ 1 & \text{for } x \ge 20. \end{cases}$$

Here $\lfloor x \rfloor$ is obtained by rounding x down to an integer, e.g., $\lfloor 14.37 \rfloor = 14$.

Example 8.14. Suppose that 20.2% of cars are blue, 32.7% of cars are red, 3.06% of cars are silver, and the rest of the cars on the road are other colors. Let X be "1" if the next car to pass is blue, or "2" if the next car is red, or "3" if the next car is silver, or "0" otherwise.

The mass of X at the values 1, 2, 3 is given to be (respectively)

$$p_X(1) = 0.202; \qquad p_X(2) = 0.327; \qquad p_X(3) = 0.0306.$$

Since the rest of the mass is at $x = 0$, and the mass adds to 1, the mass at 0 must be

$$p_X(0) = 1 - 0.202 - 0.327 - 0.0306 = 0.4404.$$

For any values of x not in $\{0, 1, 2, 3\}$, we must have $p_X(x) = 0$. The CDF is

$$F_X(x) = \begin{cases} 0 & \text{for } x < 0, \\ 0.4404 & \text{for } 0 \le x < 1, \\ 0.6424 & \text{for } 1 \le x < 2, \\ 0.9694 & \text{for } 2 \le x < 3, \\ 1 & \text{for } x \ge 3. \end{cases}$$

Example 8.15. Draw a card from a well-shuffled deck until the ace of spades appears. If a draw is unsuccessful, then replace and reshuffle the deck before making the next selection. Let X be the number of draws needed until the ace of spades appears for the first time.

(The event that the ace of spades never appears has probability 0. In such a case, we could, for instance, make $X = -1$, or any other suitable solution. It will not matter, because such an event has probability 0 anyway.)

The mass of X is $p_X(j) = (51/52)^{j-1}(1/52)$ for positive integers j, and $p_X(x) = 0$ otherwise.

Example 8.16. Suppose that a basketball player makes 80% of her free throws successfully. She shoots as many times as necessary, until scoring the first basket. Let X denote the number of necessary attempts.

(The probability that she never scores is 0, so we can safely ignore this outcome; the other probabilities will not be affected.)

The mass of X is $p_X(j) = (0.20)^{j-1}(0.80)$ for positive integers j, and $p_X(x) = 0$ otherwise.

Example 8.17. Draw a card from a well-shuffled deck until the ace of spades appears. If a draw is unsuccessful, do not replace the card—just continue to draw. Let X be the number of draws needed until the ace of spades appears for the first time.

Here, X will be one of the integers between 1 and 52, inclusive. As discussed in Example 2.17, the placement of the ace of spades is equally likely to be anywhere in the deck, so any of the first 52 draws are equally likely to be the ace of spades. So the mass of X is $p_X(j) = 1/52$ for $j = 1, 2, 3, \ldots, 52$, and $p_X(x) = 0$ otherwise. The CDF of X is $F_X(x) = x/52$ for each integer x with $1 \leq x \leq 52$. Thus

$$F_X(x) = \begin{cases} 0 & \text{for } x < 1, \\ \lfloor x \rfloor / 52 & \text{for } 1 \leq x < 52, \\ 1 & \text{for } x \geq 52. \end{cases}$$

Again $\lfloor x \rfloor$ is the integer obtained by rounding x down to an integer.

Example 8.18. A con artist has a trick die which is weighted so that a 1 will come up half of the time. The other numbers are equally likely to appear. Let X denote the number that appears when you roll the die once.

The possible values X assumes are still $\{1, 2, 3, 4, 5, 6\}$. We are given

$$p_X(1) = P(X = 1) = 1/2.$$

Since the other values are equally likely to appear, we must have

$$p_X(2) + \cdots + p_X(6) = P(X = 2) + \cdots + P(X = 6) = 1/2,$$

so $p_X(j) = P(X = j) = 1/10$ for each j in $\{2, 3, 4, 5, 6\}$.

8.5 Exercises

8.5.1 Practice

Exercise 8.1. **Snacks.** A student makes a trip once per day to the store, and he always buys a snack. The student eats the snack 70% of the time, but the other 30% of the time, his roommate eats it first.

a. During a period of four days, the student keeps track of whether he gets to eat his snack. What is the sample space of possible outcomes?

b. Let X be the number of times, within the four day period, that he gets to eat his snack. What is the mass of X?

c. What is the CDF of X?

Exercise 8.2. Post office. As in Exercise 2.21, a sequence of seven people walk into a post office and only their sexes are noted. Assume that each of the seven customers is equally likely to be a man or a woman. Let X denote the number of customers that are female.

a. Find the mass of X.

b. Find the CDF of X.

Exercise 8.3. Songs by genre. As in Exercises 2.1 and 4.3, and in Example 3.11, a randomly chosen song is from the blues genre with probability 330/27333; from the jazz genre with probability 537/27333; from the rock genre with probability 8286/27333; or from some other genre with probability 18180/27333. Let X be "1" if a randomly selected song is from the "blues" genre, or "2" if "jazz," or "3" if "rock," or "-1" otherwise. Find the mass and CDF of X.

Exercise 8.4. Milkfat. Suppose that 13% of all milk cartons have 1% milkfat, and 28% of all milk cartons have 2% milkfat, and 18% of all milk cartons are fat-free, and 41% of all milk cartons are of some other type. Randomly choose a carton of milk, and let X be "1" if the carton of milk is 1% milkfat, or "2" if the carton of milk is 2% milkfat, or "3" if the carton of milk is fat-free, or "4" otherwise.

a. What is the mass of X?

b. Make a plot of the probability mass function.

c. What is the CDF of X?

d. Make a plot of the CDF.

Exercise 8.5. Super Breakfast Challenge. The Super Breakfast Challenge (SBC) consists of bacon, eggs, oatmeal, orange juice, milk, and several other foods, and it costs \$12.99 per person to order at a local restaurant. It is known to be very difficult to consume the entire SBC. Only 10% of people are able to eat all of the SBC. The other 90% of people will be unable to eat the whole SBC (it is too much food!).

A probability student hears about the SBC and goes to the local restaurant. He observes the number of customers, X, that attempt to eat the SBC, until the first success. So if there are 4 failures and then 1 success (i.e., the outcome is (F, F, F, F, T)), then $X = 5$.

a. What is the mass of X?

b. Make a plot of the probability mass function.

c. What is the CDF of X?

d. Make a plot of the CDF.

Exercise 8.6. Cereal box prizes. A cereal company puts a Star Wars toy watch in each of its boxes as a sales promotion. Twenty percent of the cereal boxes contain a watch with Obi Wan Kenobi on it. You are a huge Obi Wan fan, so you decide to buy 8 boxes of the cereal in hopes that you will find an Obi Wan watch. Let X denote the number of Obi Wan watches you will have from these boxes.

a. What are the possible values of X?

b. What is the mass of X?

c. Make a plot of the probability mass function.

d. What is the CDF of X?

e. Make a plot of the CDF.

Exercise 8.7. Number of hearts. Recall that a standard deck of 52 playing cards has 4 suits (hearts, spades, clubs, and diamonds) and 13 cards in each suit (labeled 2, 3, 4, 5, 6, 7, 8, 9, 10, J, Q, K, A). The cards are shuffled. You are dealt the first 5 cards off the top of the deck. Let X be the number of hearts you get.

a. What are the possible values of X?

b. What is the mass of X?

c. Make a plot of the probability mass function.

d. What is the CDF of X?

e. Make a plot of the CDF.

Exercise 8.8. Mystery constant. Suppose X is a discrete random variable with a probability mass function $p_X(x) = c(4 - x)$ for x in $\{-1, 2, 3\}$, and $p_X(x) = 0$ otherwise.

a. What is the value of c so that $p_X(x)$ is a mass?

b. Make a plot of the probability mass function.

c. What is the CDF of X?

d. Make a plot of the CDF.

Exercise 8.9. Discrete Uniform distribution. Suppose X is a discrete random variable with mass

$$p_X(x) = 1/4 \qquad \text{for } x = 1, 2, 3, 4,$$

and $p_X(x) = 0$ otherwise. For future reference, this is a called a Discrete Uniform distribution. We will study these random variables more in Chapter 20.

a. Make a plot of the probability mass function.

b. What is the CDF of X?

c. Make a plot of the CDF.

Exercise 8.10. Mystery mass. Consider a random variable X that has CDF

$$F_X(x) = \begin{cases} 0 & \text{if } x < 2, \\ 0.3 & \text{if } 2 \leq x < 4, \\ 0.8 & \text{if } 4 \leq x < 6, \\ 0.95 & \text{if } 6 \leq x < 8, \\ 1 & \text{if } 8 \leq x. \end{cases}$$

What is the mass of X?

Exercise 8.11. Mystery mass. Consider a random variable X that has CDF

$$F_X(x) = \begin{cases} 0 & \text{if } x < -10, \\ 0.1 & \text{if } -10 \leq x < -5, \\ 0.8 & \text{if } -5 \leq x < 0, \\ 1 & \text{if } 0 \leq x. \end{cases}$$

What is the mass of X?

Exercise 8.12. Butterflies. Alice, Bob, and Charlotte are looking for butterflies. They look in three separate parts of a field, so that their probabilities of success are independent.

- Alice finds 1 butterfly with probability 17%, and otherwise does not find one.
- Bob finds 1 butterfly with probability 25%, and otherwise does not find one.
- Charlotte finds 1 butterfly with probability 45%, and otherwise does not find one.

Let X be the total number of butterflies that they find. Find the probability mass function of X.

Exercise 8.13. Appetizers. At a restaurant that sells appetizers:

- 8% of the appetizers cost $1 each,
- 20% of the appetizers cost $2 each,

- 32% of the appetizers cost \$3 each,
- 40% of the appetizers cost \$4 each.

An appetizer is chosen at random, and X is its price. Draw the CDF of X.

Exercise 8.14. Two 4-sided dice. Roll two 4-sided dice and let X denote the sum.

a. Draw the mass of X.

b. Draw the CDF of X.

8.5.2 Extensions

Exercise 8.15. Poisson distribution. Suppose X has mass

$$p_X(x) = \frac{\lambda^x e^{-\lambda}}{x!} \qquad \text{for } x \in \mathbb{Z}^{\geq 0},$$

and $p_X(x) = 0$ otherwise.

a. For $\lambda = 2$, make a plot of the probability mass function.

b. What is the CDF of X, when $\lambda = 2$?

c. Make a plot of the CDF, when $\lambda = 2$.

For future reference, this is called a Poisson random variable. You may want to refer to the Math Review for help with the summation of the terms of the form $\frac{\lambda^x}{x!}$. We will study these random variables more in Chapter 18.

Exercise 8.16. Coin flips. Flip a coin three times. Let X denote the number of heads minus the number of tails. So, for instance, if (H, T, T) is the outcome, then $X = 1 - 2 = -1$.

a. What are the possible values of X?

b. What is the mass of X?

c. Make a plot of the probability mass function.

d. What is the CDF of X?

e. Make a plot of the CDF.

8.5.3 Advanced

Exercise 8.17. Magic. Using a shuffled standard deck of 52 playing cards, a magician wants to do a trick where he tries to guess which card an audience member has selected if the audience member chooses a card at random and doesn't show the magician. Let X be the number of cards the magician will guess correctly if he tries the trick with 6 audience members. Each audience member starts with a complete and well-shuffled deck. The magician decides to call a trial a success if he guesses the number on the card correctly even if he doesn't get the right suit.

a. What are the possible values of X?

b. What is the mass of X?

c. Make a plot of the probability mass function.

d. What is the CDF of X?

e. Make a plot of the CDF.

Exercise 8.18. Wastebasket basketball. Chris tries to throw a ball of paper in the wastebasket behind his back (without looking). He estimates that his chance of success each time, regardless of the outcome of the other attempts, is 1/3. Let X be the number of attempts required. If he is not successful within the first 5 attempts, then he quits, and he lets $X = 6$ in such a case.

a. Draw the mass of X.

b. Draw the CDF of X.

Exercise 8.19. Pick two cards. (See also Exercise 7.17) Pick two cards at random from a well-shuffled deck of 52 cards (pick them simultaneously, i.e., grab two cards at once—so they are not the same card!). There are 12 cards which are considered face cards (4 Jacks, 4 Queens, 4 Kings). Let X be the number of face cards that you get. Draw the CDF $F_X(x)$ of X.

Chapter 9

Independence and Conditioning

The most misleading assumptions are the ones you don't even know you're making.

—"Meeting a Gorilla," by Douglas Adams and Mark Carwardine, from Chapter 2 of *The Great Ape Project*, edited by Paola Cavalieri and Peter Singer (St. Martin's Griffin, 1993)

Seven people stand in a post office lobby. We don't know who came in first. What is the probability the first person from that group who walked in the door was female? How does that probability change if we know the total number of females in the group of 7? If all 7 people are female, what do we know about the probability that the first person was a female? If all 7 people are male, what do we know about the probability that the first person was a female? What happens to the probabilities for the gender of the first person for total numbers of women in between 0 and 7?

9.1 Joint Probability Mass Functions

When dealing with two or more random variables at a time, it is usually helpful to utilize a joint probability mass function (also called a joint PMF, or simply a joint mass) or a joint cumulative distribution function (also called a joint CDF). A joint mass of two random variables X, Y specifies, for each pair x, y, the probability of X, Y taking on these specific values. Similarly, a joint CDF of X and Y gives, for each x, y, the probability of $X \leq x$ and $Y \leq y$.

Definition 9.1. Joint probability mass function (a.k.a. joint PMF or joint mass)
The joint probability mass function of a pair of discrete random variables X and Y is

$$p_{X,Y}(x,y) = P(\{\omega \mid X(\omega) = x \text{ and } Y(\omega) = y\}),$$

i.e.,

$$p_{X,Y}(x,y) = P(X = x \text{ and } Y = y).$$

Definition 9.2. Joint cumulative distribution function (a.k.a. joint CDF)
The joint CDF of a pair of discrete random variable X and Y is

$$F_{X,Y}(x,y) = P(\{\omega \mid X(\omega) \le x \text{ and } Y(\omega) \le y\}),$$

or equivalently,

$$F_{X,Y}(x,y) = P(X \le x \text{ and } Y \le y).$$

Example 9.3. Roll two dice. Let X denote the minimum of the two values that appear, and let Y denote the maximum of the two values that appear.

For instance,

$$\begin{aligned} p_{X,Y}(3,5) &= P(\{\text{min} = 3 \text{ and max} = 5\}) \\ &= P(\{\text{die values } 3{,}5\}) + P(\{\text{die values } 5{,}3\}) \\ &= 2/36. \end{aligned}$$

Also,

$$\begin{aligned} p_{X,Y}(3,3) &= P(\{\text{min} = 3 \text{ and max} = 3\}) \\ &= P(\{\text{die values } 3{,}3\}) \\ &= 1/36. \end{aligned}$$

In general, for $1 \le x < y \le 6$,

$$\begin{aligned} p_{X,Y}(x,y) &= P(\{\text{min} = x \text{ and max} = y\}) \\ &= P(\{\text{die values } x,y\}) + P(\{\text{die values } y,x\}) \\ &= 2/36, \end{aligned}$$

and for $1 \le x = y \le 6$,

$$\begin{aligned} p_{X,Y}(x,y) &= P(\{\text{min} = y = x = \text{max}\}) \\ &= P(\{\text{die values } x,x\}) \qquad \text{since } x \text{ and } y \text{ are the same in this case} \\ &= 1/36, \end{aligned}$$

and

$$p_{X,Y}(x,y) = 0 \qquad \text{otherwise.}$$

So the case $p_{X,Y}(x,y) = 2/36$ with $1 \le x < y \le 6$ corresponds to the two possible outcomes (x,y) or (y,x), and the case $p_{X,Y}(x,y) = 1/36$ with $1 \le x = y \le 6$ corresponds to the one possible outcome (x,x) (here, we emphasize that $y = x$, i.e., the maximum and minimum are exactly the same because the die rolls are the same).

The cumulative distribution function is calculated similarly, e.g.,

$$\begin{aligned} F_{X,Y}(2,4) &= P(\{\min \le 2 \text{ and } \max \le 4\}) \\ &= P(\{(1,1),(1,2),(1,3),(1,4),(2,1),(2,2), \\ &\qquad (2,3),(2,4),(3,1),(3,2),(4,1),(4,2)\}) \\ &= 12/36. \end{aligned}$$

As with CDFs for one variable, nothing changes if we calculate, for instance,

$$F_{X,Y}(2.9,4.1) = P(X \le 2.9 \text{ and } Y \le 4.1) = P(X \le 2 \text{ and } Y \le 4) = 12/36.$$

Using the chart below, we can easily obtain all of the values of $F_{X,Y}(x,y)$. As another example,

$$\begin{aligned} F_{X,Y}(5,2) &= P(\{\min \le 5 \text{ and } \max \le 2\}) \\ &= P(\{(1,1),(1,2),(2,1),(2,2)\}) \\ &= 4/36. \end{aligned}$$

In general, we have

$F_{X,Y}(x,y)$	$y=1$	$y=2$	$y=3$	$y=4$	$y=5$	$y=6$
$x=1$	1/36	3/36	5/36	7/36	9/36	11/36
$x=2$	1/36	4/36	8/36	12/36	16/36	20/36
$x=3$	1/36	4/36	9/36	15/36	21/36	27/36
$x=4$	1/36	4/36	9/36	16/36	24/36	32/36
$x=5$	1/36	4/36	9/36	16/36	25/36	35/36
$x=6$	1/36	4/36	9/36	16/36	25/36	36/36

Example 9.4. Flip a fair coin three times. Let X be the total number of heads that appear, and let Y be the total number of tails that appear.

First of all, we notice that $X + Y = 3$ always. So we are certain to have $p_{X,Y}(x,y) = 0$ if $x + y \ne 3$. We can easily make a chart of all of the possible outcomes—organized strategically into events according to the values of X and Y that they induce—and the associated values of X:

Event	Probability	Joint Mass of X and Y
$\{(H,H,H)\}$	1/8	$p_{X,Y}(3,0) = 1/8$
$\{(H,H,T),(H,T,H),(T,H,H)\}$	3/8	$p_{X,Y}(2,1) = 3/8$
$\{(T,T,H),(T,H,T),(H,T,T)\}$	3/8	$p_{X,Y}(1,2) = 3/8$
$\{(T,T,T)\}$	1/8	$p_{X,Y}(0,3) = 1/8$

Otherwise, $p_{X,Y}(x,y) = 0$.

Using the chart below, we can easily obtain all of the values of $F_{X,Y}(x,y)$. For example,

$$\begin{aligned} F_{X,Y}(1,2) &= P(\{X \leq 1 \text{ and } Y \leq 2\}) \\ &= P(\{(T,T,H),(T,H,T),(H,T,T)\}) \\ &= 3/8. \end{aligned}$$

As another example, $F_{X,Y}(2,2) = F_{X,Y}(1,2) + F_{X,Y}(2,1) = 3/8 + 3/8 = 6/8$. In general, we have

$F_{X,Y}(x,y)$	$y=0$	$y=1$	$y=2$	$y=3$
$x=0$	0	0	0	1/8
$x=1$	0	0	3/8	4/8
$x=2$	0	3/8	6/8	7/8
$x=3$	1/8	4/8	7/8	8/8

Example 9.5. Roll a 6-sided fair die and let X denote the outcome. Also flip a spinner that shows "1" with probability 0.30, or shows "2" with probability 0.32, or shows "3" with probability 0.38, and let Y denote the outcome.

Then for each integer j from 1 to 6 inclusive,

$$\begin{aligned} p_{X,Y}(j,1) &= (1/6)(0.30); \\ p_{X,Y}(j,2) &= (1/6)(0.32); \\ p_{X,Y}(j,3) &= (1/6)(0.38); \end{aligned}$$

and $p_{X,Y}(x,y) = 0$ otherwise.

Remark 9.6. Since the joint mass and joint CDF are probabilities, then

$$0 \leq p_{X,Y}(x,y) \leq 1 \qquad \text{and} \qquad 0 \leq F_{X,Y}(x,y) \leq 1 \qquad \text{for all } x, y.$$

The joint mass, summed over all x's and y's, takes the probabilities from the whole sample space into account, so

$$\sum_x \sum_y p_{X,Y}(x,y) = 1.$$

Example 9.7. As in Example 9.3, roll two dice. Let X denote the minimum of the two values that appear, and let Y denote the maximum of the two values that appear.

We can calculate the mass of X directly, e.g.,

$$p_X(3) = P(\{(3,3),(3,4),(3,5),(3,6),(4,3),(5,3),(6,3)\}) = 7/36.$$

The mass of X is:

x	1	2	3	4	5	6
$p_X(x)$	11/36	9/36	7/36	5/36	3/36	1/36.

Similarly, the mass of Y is

y	1	2	3	4	5	6
$p_Y(y)$	1/36	3/36	5/36	7/36	9/36	11/36.

Alternatively, we can compute the mass of X directly from the joint mass of X and Y, by letting Y take on any value. For instance,

$$p_X(3) = \sum_{y=1}^{6} p_{X,Y}(x,y) = 0 + 0 + \frac{1}{36} + \frac{2}{36} + \frac{2}{36} + \frac{2}{36} = 7/36.$$

A similar concept works much more generally:

Remark 9.8. Calculating the mass of one variable from the joint mass. The mass of X can be calculated by summing the joint mass over all possible values of Y:

$$p_X(x) = \sum_y p_{X,Y}(x,y).$$

Similarly, the mass of Y is the sum of the joint mass over all possible values of X:

$$p_Y(y) = \sum_x p_{X,Y}(x,y).$$

Example 9.9. Again, as in Example 9.3, roll two dice. Let X denote the minimum of the two values that appear, and let Y denote the maximum of the two values that appear.

We can calculate the CDF of X directly, e.g.,

$$\begin{aligned} F_X(2) = P(\{&(1,1),(1,2),(1,3),(1,4),(1,5),(1,6), \\ &(2,1),(2,2),(2,3),(2,4),(2,5),(2,6) \\ &(3,1),(3,2),(4,1),(4,2),(5,1),(5,2),(6,1),(6,2)\}) \\ = 20/36&. \end{aligned}$$

The CDF of X is:

x	1	2	3	4	5	6
$F_X(x)$	11/36	20/36	27/36	32/36	35/36	36/36

Similarly, the CDF of Y is

x	1	2	3	4	5	6
$F_Y(y)$	1/36	4/36	9/36	16/36	25/36	36/36

As an alternative, we can compute the CDF of X directly from the joint CDF of X and Y, by letting $Y \to \infty$, e.g.,

$$F_X(3) = \lim_{y \to \infty} p_{X,Y}(3, y) = 27/36.$$

Again, we can generalize:

Remark 9.10. Calculating the CDF of one variable from the joint CDF. The CDF of X can be calculated by taking the limit as $y \to \infty$ in the joint CDF:

$$F_X(x) = \lim_{y \to \infty} F_{X,Y}(x, y).$$

Intuitively, when taking $y \to \infty$ in the joint CDF, we are letting $Y < \infty$, i.e., we let Y take on *any* value, since we only want to extract information about X.

Similarly, the CDF of Y is the limit as $x \to \infty$ in the joint CDF:

$$F_Y(y) = \lim_{x \to \infty} F_{X,Y}(x, y).$$

9.2 Independent Random Variables

Just as we discussed the notion of independence of events in Chapter 3, we have a notion of independent random variables as well.

Example 9.11. As in Example 9.5, the mass of X is $p_X(j) = 1/6$ for $j = 1, 2, 3, 4, 5, 6$, and $p_X(x) = 0$ otherwise. The mass of Y is $p_Y(1) = 0.30$ and $p_Y(2) = 0.32$ and $p_Y(3) = 0.38$, and $p_Y(y) = 0$ otherwise. So in every possible case, we have the very nice condition

$$p_{X,Y}(x, y) = p_X(x)p_Y(y).$$

This is one way to characterize two random variables as being independent. There are several other equivalent formulations of this idea. It should make *intuitive* sense that we might call X and Y independent, because the value of X has no bearing whatsoever on the value of Y, and vice versa. If we think about each underlying outcome as containing two pieces of information, one about the die and one about the spinner, then only the piece of information about the die has any impact on X, and only the piece of information about the spinner has any impact on Y. As another example:

Example 9.12. Roll a die and flip a fair coin. Let X be the result of the die roll. Let Y be 0 if the coin shows a "tail" or 1 if the coin shows a "head." Notice

$$p_{X,Y}(x, y) = 1/12$$

for all $1 \leq x \leq 6$ and $0 \leq y \leq 1$, since all twelve of these outcomes are equally likely. Also

$$p_X(x) = 1/6 \qquad \text{for } 1 \leq x \leq 6,$$
$$p_Y(y) = 1/2 \qquad \text{for } 0 \leq y \leq 1.$$

So $p_{X,Y}(x, y) = p_X(x)p_Y(y)$, so X and Y are independent.

We introduce several equivalent ways to state that two discrete random variables are independent, and all of these ways are equivalent to each other. (Exercise 9.11 asks how to prove that these statements are equivalent.)

Definition 9.13. Independent discrete random variables: Several equivalent formulations

1. Joint mass factors into a function of x times a function of y. These functions of x and y can also be normalized so they are the masses of X and Y, respectively:

$$p_{X,Y}(x, y) = p_X(x)p_Y(y) \qquad \text{for all } x \text{ and } y.$$

2. Joint CDF factors into a function of x times a function of y. These functions of x and y can also be normalized so they are the CDFs of X and Y, respectively:

$$F_{X,Y}(x, y) = F_X(x)F_Y(y) \qquad \text{for all } x \text{ and } y.$$

3. We will define the conditional mass later in this chapter, but we go ahead and state a way to use conditional mass for independence.

(a) Mass of X is conditional mass $p_{X|Y}(x \mid y)$ of X given $Y = y$,

or (b) Mass of Y is conditional mass $p_{Y|X}(y \mid x)$ of Y given $X = x$.

An informal way to summarize is to say that X and Y are independent if their joint behavior can be completely separated, so that the probability of X and Y jointly behaving in some way is equal to the probability of X behaving some way times the probability of Y behaving some way.

In practice, to show that two discrete random variables X and Y are independent, we usually either show that

$$p_{X,Y}(x, y) = p_X(x)p_Y(y),$$

or (to be discussed in Section 9.4) that

$$p_{X|Y}(x \mid y) = p_X(x),$$

or

$$p_{Y|X}(y \mid x) = p_Y(y).$$

Once we get more familiar with these concepts, we will not go through such tedious calculations. Sometimes we will just go ahead and observe that "X and Y are independent," but for now it is good to practice a little bit.

Example 9.14. Let X indicate whether the first baby born to a certain mother is a girl, i.e., $X = 1$ if the first baby born is a girl; otherwise, $X = 0$. Let Y indicate whether the second baby born to a certain mother is a girl, i.e., $Y = 1$ if the second baby born is a girl; otherwise, $Y = 0$. (We are not considering the birth of twins, in which one baby's sex might affect the other.) Since

$$p_{X,Y}(x, y) = 1/4 \qquad \text{for } 0 \le x, y \le 1,$$

and since X and Y each take values in the set $\{0, 1\}$, we can factor the joint mass $p_{X,Y}(x, y) = 1/4$ into $1/4 = (1/2) \cdot (1/2)$, and we must have $p_X(x) = 1/2$ for $x = 0, 1$ and $p_Y(y) = 1/2$ for $y = 0, 1$. So X and Y are independent.

Example 9.15. Roll a die. Let X be 1 if the outcome is 1, 3, or 5; let X be 0 otherwise. Let Y be 1 if the outcome is 5 or 6; let Y be 0 otherwise. Since

$$\begin{aligned}
p_{X,Y}(0, 0) &= P(\{\text{outcome is 2, 4}\}) = 2/6 \\
p_{X,Y}(0, 1) &= P(\{\text{outcome is 6}\}) = 1/6 \\
p_{X,Y}(1, 0) &= P(\{\text{outcome is 1, 3}\}) = 2/6 \\
p_{X,Y}(1, 1) &= P(\{\text{outcome is 5}\}) = 1/6
\end{aligned}$$

Thus $p_{X,Y}(x,y)$ can be factored as $p_{X,Y}(x,y) = p_X(x)p_Y(y)$, by writing

$$p_X(x) = 1/2 \qquad \text{for } x = 0 \text{ or } x = 1,$$

and

$$p_Y(1) = 1/3 \qquad \text{and} \qquad p_Y(0) = 2/3.$$

So X, Y are independent random variables.

Theorem 9.16. Indicators of independent events are independent random variables

Consider events A and B. Let X be an indicator for event A, i.e., $X = 1$ if A occurs, and $X = 0$ otherwise. Let Y be an indicator for event B, i.e., $Y = 1$ if B occurs, and $Y = 0$ otherwise. Then, A and B are independent events, if and only if X and Y are independent random variables.

To see that the above box is true, note that

$$p_{X,Y}(1,1) = P(A \cap B), \qquad P(A)P(B) = p_X(1)p_Y(1);$$

$$p_{X,Y}(1,0) = P(A \cap B^c), \qquad P(A)P(B^c) = p_X(1)p_Y(0);$$

$$p_{X,Y}(0,1) = P(A^c \cap B), \qquad P(A^c)P(B) = p_X(0)p_Y(1);$$

$$p_{X,Y}(0,0) = P(A^c \cap B^c), \qquad P(A^c)P(B^c) = p_X(0)p_Y(0).$$

A moment's thought shows that the commas (in the four lines above) can be replaced with equalities if and only if A and B are independent events or equivalently if X and Y are independent random variables.

Example 9.17. Flip a coin until a head appears. Let A denote the event that an even number of flips are required. (The case that a head never appears can be included in A; it does not matter, since the probability that a head never appears is 0.) Let B denote the event that 11 or more flips are necessary. Let X indicate whether A occurs (i.e., $X = 1$ if A occurs, and $X = 0$ otherwise). Let Y indicate whether B occurs (i.e., $Y = 1$ if B occurs, and $Y = 0$ otherwise).

In this example, X and Y are independent random variables. To see this, it is enough to show that A and B are independent events. We notice that B occurs if the first ten flips are tails, so $P(B) = (1/2)^{10}$. Also, A occurs if there are an even number of flips until a head, which happens with probability

$$\begin{aligned}
P(A) &= P(TH) + P(TTTH) + P(TTTTTH) + P(TTTTTTTH) + \cdots \\
&= (1/2)^2 + (1/2)^4 + (1/2)^6 + (1/2)^8 + \cdots \\
&= 1/4 + (1/4)^2 + (1/4)^3 + (1/4)^4 + \cdots \\
&= (1/4)(1 + (1/4) + (1/4)^2 + (1/4)^3 + \cdots) \\
&= \left(\frac{1}{4}\right)\left(\frac{1}{1-1/4}\right) \\
&= \left(\frac{1}{4}\right)\left(\frac{1}{3/4}\right) \\
&= 1/3
\end{aligned}$$

To get an intuitive idea why $P(A) = 1/3$, look at the first two flips:

We are using geometric sums.

1. If the first two flips are HH, then A^c occurs, i.e., an odd number of flips was needed to see the first head.

2. If the first two flips are HT, then A^c occurs, i.e., an odd number of flips was needed to see the first head.

3. If the first two flips are TH, then A occurs, i.e., an even number of flips was needed to see the first head.

4. If the first two flips are TT, then we essentially start over; just look at the next pair of flips.

Thus A occurs in exactly 1 out of the 3 deciding cases (and the 3 deciding cases are all equally likely).

Finally, we calculate the probability of A and B occurring:

$$\begin{aligned}
P(A \cap B) &= P(\overbrace{TTTTTTTTTTT}^{11} H) + P(\overbrace{TTTTTTTTTTTTT}^{13} H) \\
&\quad + P(\overbrace{TTTTTTTTTTTTTTT}^{15} H) + \cdots \\
&= (1/2)^{12} + (1/2)^{14} + (1/2)^{16} + (1/2)^{18} + \cdots \\
&= (1/4)^6 + (1/4)^7 + (1/4)^8 + (1/4)^9 + \cdots \\
&= (1/4)^6(1 + (1/4) + (1/4)^2 + (1/4)^3 + \cdots) \\
&= \left(\frac{1}{4}\right)^6\left(\frac{1}{1-1/4}\right) \\
&= \left(\frac{1}{4}\right)^6\left(\frac{1}{3/4}\right) \\
&= (1/4)^5(1/3)
\end{aligned}$$

So we conclude

$$P(A \cap B) = (1/4)^5(1/3) = (1/2)^{10}(1/3) = P(A)P(B).$$

So events A and B are independent. Since X and Y are, respectively, indicator random variables for A and B, this means that X and Y are independent random variables too.

9.3 Three or More Independent Random Variables

To be independent, as a collection, three or more random variables must satisfy rules similar to those for collections of three or more events. Possibilities include the following:

Definition 9.18. Joint probability mass function
If $X_1, X_2, \ldots, X_n$ are random variables, then the joint probability mass function (the joint mass) of $X_1, X_2, \ldots, X_n$ is the probability that $X_j = x_j$ for each j, i.e.,

$$p_{X_1,X_2,\ldots,X_n}(x_1, x_2, \ldots, x_n) = P(X_j = x_j \text{ for all } j).$$

Remark 9.19. Independent discrete random variables: Joint mass equals the product of the masses
A finite number of discrete random variables $X_1, X_2, \ldots, X_n$ are independent if and only if their joint mass is equal to the product of their masses, i.e., if

$$p_{X_1,X_2,\ldots,X_n}(x_1, x_2, \ldots, x_n) = p_{X_1}(x_1)p_{X_2}(x_2)\cdots p_{X_n}(x_n)$$

for all $x_1, x_2, \ldots, x_n$.

Another equivalent way to say that $X_1, X_2, \ldots, X_n$ are independent can be given, after we introduce the concept of a joint cumulative distribution function for n variables.

Definition 9.20. Joint cumulative distribution function
If $X_1, X_2, \ldots, X_n$ are random variables, then the joint cumulative distribution function (also called joint CDF) of $X_1, X_2, \ldots, X_n$ is the probability that $X_j \leq x_j$ for every j, i.e.,

$$F_{X_1,X_2,\ldots,X_n}(x_1, x_2, \ldots, x_n) = P(X_j \leq x_j \text{ for all } j).$$

Remark 9.21. Independent discrete random variables: Joint CDF equals the product of the CDFs

A collection of random variables $X_1, X_2, \ldots, X_n$ are independent if and only if their joint CDF is equal to the product of their CDFs, i.e., if

$$F_{X_1,X_2,\ldots,X_n}(x_1, x_2, \ldots, x_n) = F_{X_1}(x_1)F_{X_2}(x_2)\cdots F_{X_n}(x_n)$$

for all $x_1, x_2, \ldots, X_n$.

Remark 9.22. Indicators of independent events are independent random variables

Consider a collection of events $A_1, A_2, \ldots, A_n$. For each j, let X_j be an indicator for event A_j, i.e., $X_j = 1$ if A_j occurs, and $X_j = 0$ otherwise. Then, the A_j's are independent events if and only if $X_1, X_2, \ldots, X_n$ are independent random variables.

9.4 Conditional Probability Mass Functions

When two random variables are under consideration at the same time, sometimes it is helpful to know how one of them will impact the other one. In order to characterize such an effect, we introduce the conditional probability mass function (also called a conditional PMF, or simply a conditional mass). A conditional mass completely specifies the probability of one random variable, given that we know the value of the other random variable. Such a conditional mass of X, given the value of Y, is specified as follows:

Definition 9.23. Conditional probability mass function

The conditional probability mass function of a discrete random variable X, given the value of another discrete random variable Y, is

$$p_{X|Y}(x \mid y) = P(X = x \mid Y = y).$$

The conditional probability mass function of X, given the value of Y, is also referred to as "the conditional PMF of X given Y" or as "the conditional mass of X given Y."

The conditional mass is defined as the ratio of the joint mass divided by the mass of Y:

$$p_{X|Y}(x \mid y) = \frac{P(X = x \text{ and } Y = y)}{P(Y = y)} = \frac{p_{X,Y}(x, y)}{p_Y(y)}.$$

For the conditional mass of X given Y to make sense, we must use Y values such that $P(Y = y) > 0$.

Example 9.24. Roll two dice. Let X denote the value of the first die, and let Y denote the value of the sum of the two dice. If we are given $Y = 4$, then we know that the set of possible outcomes are

$$\{(1,3),(2,2),(3,1)\}.$$

Thus X is either 1, 2, or 3. So the condition mass of X, given $Y = 4$, is

$$p_{X|Y}(1 \mid 4) = \frac{p_{X,Y}(1,4)}{p_Y(4)} = \frac{P(\{(1,3)\})}{P(\{(1,3),(2,2),(3,1)\})} = \frac{1/36}{3/36} = \frac{1}{3},$$
$$p_{X|Y}(2 \mid 4) = \frac{p_{X,Y}(2,4)}{p_Y(4)} = \frac{P(\{(2,2)\})}{P(\{(1,3),(2,2),(3,1)\})} = \frac{1/36}{3/36} = \frac{1}{3},$$
$$p_{X|Y}(3 \mid 4) = \frac{p_{X,Y}(3,4)}{p_Y(4)} = \frac{P(\{(3,1)\})}{P(\{(1,3),(2,2),(3,1)\})} = \frac{1/36}{3/36} = \frac{1}{3},$$

and $p_{X|Y}(x \mid y) = 0$ otherwise.

Example 9.25. Flip the cards from a deck over, one at a time, until the whole deck has been flipped over. Let X denote the number of cards until the first ace appears. Let Y denote the number of cards until the first queen appears. Then X and Y are dependent.

To see that X and Y are dependent, first notice that $p_Y(1) > 0$ because it is possible (i.e., the probability is positive) that the first card is a queen. On the other hand, if $X = 1$, i.e., if the first card is an ace, then the first card is not a queen, so $p_{Y|X}(1 \mid 1) = 0$. Thus $p_Y(y) \neq p_{Y|X}(y \mid x)$ when x and y are both equal to 1. So X and Y are not independent. In fact, X and Y are dependent. (It is worthwhile to compare and contrast with Example 9.15, to see what is similar and different in these two examples.)

Example 9.26. Roll a die. Let X be 1 if the outcome is 1, 2, or 3; let X be 0 otherwise. Let Y be 1 if the outcome is even (2, 4, or 6); let Y be 0 otherwise.

To see that X and Y are dependent, first notice that $p_Y(1) = 1/2$ because the outcome is even with probability 1/2. On the other hand, if $X = 1$, i.e., if the outcome is 1, 2, or 3, then the outcome is even with probability 1/3. so $p_{Y|X}(1 \mid 1) = 1/3$. Thus $p_Y(y) = 1/2 \neq 1/3 = p_{Y|X}(y \mid x)$ when x and y are both equal to 1. So X and Y are not independent. In fact, X and Y are dependent.

Example 9.27. As in Example 5.2, in a certain household, 20% of the milk has two-percent milkfat, and the other 80% of the milk is whole milk. The whole milk is spoiled 5% of the time; overall, the milk is spoiled 4.7% of the time.

In that example, we let A denote the event that the milk is came from a whole milk carton, and we let B denote the event that the milk was spoiled. We were given $P(A) = 0.80$, and we calculated

$$P(\text{whole milk} \mid \text{spoiled}) = P(A \mid B) = 0.85.$$

In this example, let X indicate whether A occurred, i.e.,

$$X = \begin{cases} 1 & \text{if } A \text{ occurs,} \\ 0 & \text{otherwise.} \end{cases}$$

Let Y indicate whether B occurred, i.e.,

$$Y = \begin{cases} 1 & \text{if } B \text{ occurs (milk was spoiled),} \\ 0 & \text{otherwise (milk was not spoiled).} \end{cases}$$

Then X and Y are dependent. To see this, notice $P(A) = 0.80$ is the same as $p_X(1) = 0.80$, and $P(A \mid B) = 0.85$ is the same as $p_{X|Y}(1 \mid 1) = 0.85$. So $p_X(x) \neq p_{X|Y}(x \mid y)$ when $x = 1$ and $y = 1$.

Example 9.28. In Example 9.15, X and Y are independent. (In that example, we roll a die and let X be 1 if the outcome is 1, 3, or 5; let X be 0 otherwise. Let Y be 1 if the outcome is 5 or 6; let Y be 0 otherwise.) An alternative way to see that X and Y are independent is to show that $p_{Y|X}(y \mid x) = p_Y(y)$.

We have $p_Y(1) = 1/3$ and $p_Y(0) = 2/3$.

Also, given that $X = 1$, then the outcome is 1, 3, or 5, so $Y = 1$ exactly 1/3 of the time, thus $p_{Y|X}(1 \mid 1) = 1/3$ and $p_{Y|X}(0 \mid 1) = 2/3$; similarly, given that $X = 0$, then the outcome is 2, 4, or 6, so $Y = 1$ exactly 1/3 of the time, so $p_{Y|X}(1 \mid 0) = 1/3$ and $p_{Y|X}(0 \mid 0) = 2/3$. So, regardless of the value of x, we have $p_{Y|X}(1 \mid x) = 1/3$ and $p_{Y|X}(0 \mid x) = 2/3$.

Thus $p_{Y|X}(y \mid x) = p_Y(y)$ in all cases. So X and Y are independent.

Example 9.29. As in Exercises 2.21 and 8.2, a sequence of seven people walk into a post office (one at a time) and only their sexes are noted. Assume that each of the seven customers is equally likely to be a man or a woman. Let Y denote the number of customers that are female. Let $X = 0$ if the first person is a male, or $X = 1$ if the first person is a female.

Method #1 of computing the conditional mass. Before computing the conditional mass of X given Y, it is usually helpful to compute the mass of Y separately, and to compute the joint mass of X and Y. This is the method we follow here. The mass of Y is the following:

$$p_Y(0) = 1/128 \quad p_Y(1) = 7/128 \quad p_Y(2) = 21/128 \quad p_Y(3) = 35/128$$
$$p_Y(4) = 35/128 \quad p_Y(5) = 21/128 \quad p_Y(6) = 7/128 \quad p_Y(7) = 1/128$$

and $p_Y(y) = 0$ otherwise.

Now we compute the joint mass of X and Y. We have $X = 0$ only if the first person is a male, which happens 1/2 of the time. This leaves Y as an integer between 0 and 6, so we have

$$p_{X,Y}(0,0) = (1/2)(1/64) = 1/128 \quad p_{X,Y}(0,1) = (1/2)(6/64) = 6/128$$
$$p_{X,Y}(0,2) = (1/2)(15/64) = 15/128 \quad p_{X,Y}(0,3) = (1/2)(20/64) = 20/128$$
$$p_{X,Y}(0,4) = (1/2)(15/64) = 15/128 \quad p_{X,Y}(0,5) = (1/2)(6/64) = 6/128$$
$$p_{X,Y}(0,6) = (1/2)(1/64) = 1/128$$

We have $X = 1$ only if the first person is a female, which happens 1/2 of the time. This leaves Y as an integer between 1 and 7, so we have

$$p_{X,Y}(1,1) = (1/2)(1/64) = 1/128 \quad p_{X,Y}(1,2) = (1/2)(6/64) = 6/128$$
$$p_{X,Y}(1,3) = (1/2)(15/64) = 15/128 \quad p_{X,Y}(1,4) = (1/2)(20/64) = 20/128$$
$$p_{X,Y}(1,5) = (1/2)(15/64) = 15/128 \quad p_{X,Y}(1,6) = (1/2)(6/64) = 6/128$$
$$p_{X,Y}(1,7) = (1/2)(1/64) = 1/128$$

and $p_{X,Y}(x,y) = 0$ otherwise.

Now we are ready to compute $p_{X|Y}(x \mid y)$ in all cases:

$$p_{X|Y}(0 \mid 0) = \frac{p_{X,Y}(0,0)}{p_Y(0)} = \frac{1/128}{1/128} = 1$$
$$p_{X|Y}(0 \mid 1) = \frac{p_{X,Y}(0,1)}{p_Y(1)} = \frac{6/128}{7/128} = \frac{6}{7}$$
$$p_{X|Y}(0 \mid 2) = \frac{p_{X,Y}(0,2)}{p_Y(2)} = \frac{15/128}{21/128} = \frac{5}{7}$$
$$p_{X|Y}(0 \mid 3) = \frac{p_{X,Y}(0,3)}{p_Y(3)} = \frac{20/128}{35/128} = \frac{4}{7}$$
$$p_{X|Y}(0 \mid 4) = \frac{p_{X,Y}(0,4)}{p_Y(4)} = \frac{15/128}{35/128} = \frac{3}{7}$$
$$p_{X|Y}(0 \mid 5) = \frac{p_{X,Y}(0,5)}{p_Y(5)} = \frac{6/128}{21/128} = \frac{2}{7}$$
$$p_{X|Y}(0 \mid 6) = \frac{p_{X,Y}(0,6)}{p_Y(6)} = \frac{1/128}{7/128} = \frac{1}{7}$$
$$p_{X|Y}(0 \mid 7) = \frac{p_{X,Y}(0,7)}{p_Y(7)} = \frac{0/128}{1/128} = 0$$

and

$$p_{X|Y}(1 \mid 0) = \frac{p_{X,Y}(1,0)}{p_Y(0)} = \frac{0}{1/128} = 0$$
$$p_{X|Y}(1 \mid 1) = \frac{p_{X,Y}(1,1)}{p_Y(1)} = \frac{1/128}{7/128} = \frac{1}{7}$$
$$p_{X|Y}(1 \mid 2) = \frac{p_{X,Y}(1,2)}{p_Y(2)} = \frac{6/128}{21/128} = \frac{2}{7}$$
$$p_{X|Y}(1 \mid 3) = \frac{p_{X,Y}(1,3)}{p_Y(3)} = \frac{15/128}{35/128} = \frac{3}{7}$$
$$p_{X|Y}(1 \mid 4) = \frac{p_{X,Y}(1,4)}{p_Y(4)} = \frac{20/128}{35/128} = \frac{4}{7}$$
$$p_{X|Y}(1 \mid 5) = \frac{p_{X,Y}(1,5)}{p_Y(5)} = \frac{15/128}{21/128} = \frac{5}{7}$$
$$p_{X|Y}(1 \mid 6) = \frac{p_{X,Y}(1,6)}{p_Y(6)} = \frac{6/128}{7/128} = \frac{6}{7}$$
$$p_{X|Y}(1 \mid 7) = \frac{p_{X,Y}(1,7)}{p_Y(7)} = \frac{1/128}{1/128} = 1$$

Notice: The two terms in each row have a sum of 1, i.e., $p_{X|Y}(0 \mid y) + p_{X|Y}(1 \mid y) = 1$ for each y, because the sum over x's, i.e., $\sum_x p_{X|Y}(x \mid y)$ must be 1 for each fixed value of y, because the first person to enter is either male ($X = 0$) or female ($X = 1$).

Method #2 of computing the conditional mass. In hindsight, we could have computed $p_{X|Y}(x \mid y)$ directly, but it was perhaps helpful to go through such a rigorous exercise. Now we directly compute $p_{X|Y}(x \mid y)$; this is much, much shorter but requires some subtle insight. Suppose that it is known that $Y = y$. Then, in other words, it is known that exactly y out of the 7 people are female. The first person is equally likely to be any of these 7 people, of which y are female and $7-y$ are male. So the probability that the first person is male is exactly $(7 - y)/7$; the probability that the first person is female is exactly $y/7$. So we always have the following: Given that $Y = y$ is an integer between 1 and 7, the probability that the first person is a male is exactly

$$p_{X|Y}(0 \mid y) = (7 - y)/7;$$

the probability that the first person is a female is exactly

$$p_{X|Y}(1 \mid y) = y/7.$$

In closing, we note that the joint mass of two random variables X, Y is equal to the conditional mass of X given Y multiplied by the mass of Y.

Remark 9.30. For all random variables X and Y (regardless of independence), we have

$$p_{X,Y}(x, y) = p_{X|Y}(x \mid y)p_Y(y).$$

9.5 Exercises

9.5.1 Practice

Exercise 9.1. **Random employee hiring.** Ten students apply for a job opening, but only 1 of the students will be selected. The employer chooses randomly; all ten outcomes are equally likely. If person 3, 5, 7, or 9 gets the job, let $X = 1$; otherwise, $X = 0$. If person 1, 2, 3, 4, or 5 gets the job, let $Y = 1$; otherwise, $Y = 0$. Are X and Y independent random variables? Justify your answer.

Exercise 9.2. **Forgetful morning.** Each day, Maude has a 1% chance of losing her cell phone (her behavior on different days is independent). Each day, Maude has a 3% chance of forgetting to eat breakfast (again, her behavior on different days is independent). Her breakfast and cell phone habits are independent. Let X be the number of days until she first loses her cell phone. Let Y be the number of days until she first forgets to eat breakfast. (Here, X and Y are independent.) Find the joint mass of X and Y.

Exercise 9.3. **Dependence/independence among coin flips.** A student flips a fair coin until a head appears. Let X be the number of flips until (and including) this first head. Afterwards, he begins flipping again until he gets another head. Let Y be the number of flips, after the first head, up to (and including) the second head. E.g., if the sequence of flips is TTTTTTHTTH then $X = 7$ and $Y = 3$. Are X and Y independent? Justify your answer.

Exercise 9.4. **More dependence/independence among coin flips.** Same scenario as problem 9.3. Let Z be the total number of flips until (and including) the second head. So $Z = X + Y$; e.g., in the example given, $Z = 10$. Are X and Z independent? Justify your answer.

Exercise 9.5. **Butterflies.** Alice, Bob, and Charlotte are looking for butterflies. They look in three separate parts of a field, so that their probabilities of success do not affect each other.

- Alice finds 1 butterfly with probability 17%, and otherwise does not find one.
- Bob finds 1 butterfly with probability 25%, and otherwise does not find one.
- Charlotte finds 1 butterfly with probability 45%, and otherwise does not find one.

Let X be the number of butterflies that they find altogether. Let Y be the number of people who do not find a butterfly.

Find the joint mass $p_{X,Y}(x, y)$ of X and Y.

Exercise 9.6. Dependence/independence among dice rolls. A student rolls a die until the first "4" appears. Let X be the numbers of rolls required until (and including) this first "4." After this is completed, he begins rolling again until he gets a "3." Let Y be the number of rolls, after the first "4," up to (and including) the next "3." E.g., if the sequence of rolls is 213662341261613 then $X = 8$ and $Y = 7$. Are X and Y independent? Justify your answer.

Exercise 9.7. Pick two cards. Pick two cards at random from a well-shuffled deck of 52 cards (pick them simultaneously, i.e., grab two cards at once—so they are not the same card!). There are 12 cards which are considered face cards (4 Jacks, 4 Queens, 4 Kings). There are 4 cards with the value 10. Let X be the number of face cards in your hand; let Y be the number of 10's in your hand. Are X and Y dependent or independent?

9.5.2 Extensions

Exercise 9.8. Dice. Roll two dice, one colored red and one colored blue. Let Y denote the maximum value that appears on the two dice. Let X denote the value of the blue die. Find the conditional mass of X given Y.

Exercise 9.9. Wastebasket basketball. Chris tries to throw a ball of paper in the wastebasket behind his back (without looking). He estimates that his chance of success each time, regardless of the outcome of the other attempts, is $1/3$. Let X be the number of attempts required. If he is not successful within the first 5 attempts, then he quits, and he lets $X = 6$ in such a case.

Let Y indicate whether he makes the basket successfully within the first three attempts. Thus $Y = 1$ if his first, second, or third attempt is successful, and $Y = 0$ otherwise.

Find the conditional mass of X given Y. You will need to list 12 values altogether, i.e., you need to compute $p_{X|Y}(x \mid y)$ for $1 \le x \le 6$ and $0 \le y \le 1$.

Exercise 9.10. Two 4-sided dice. Consider some 4-sided dice. Roll two of these dice. Let X denote the minimum of the two values that appear, and let Y denote the maximum of the two values that appear.

a. Find the joint mass $p_{X,Y}(x, y)$ of X and Y.

b. Find the joint CDF $F_{X,Y}(x, y)$ of X and Y. It suffices to give the values $F_{X,Y}(x, y)$ when x and y are integers between 1 and 4. You do not have to list any other values; these 16 possibilities will suffice.

9.5.3 Advanced

Exercise 9.11. Prove that the statements of independence in Definition 9.13 are equivalent.

Chapter 10

Expected Values of Discrete Random Variables

Your teacher can open the door, but you must enter by yourself.
—Proverb

You are having your weekly poker game with some of the people from your dorm. The dealer shuffles the cards and then deals each person five cards. How many of the cards in your hand do you expect to be hearts?

10.1 Introduction

An *expected value* (also called a (weighted) *average* or *mean*) says something succinct about a random variable. The expected value is one way to measure the center of the distribution. It does not tell us everything about the distribution; many different kinds of random variables can have the same expected value. Nonetheless, an expected value of a random variable is helpful. It describes the sum of the values that a random variable takes, *in proportion to the mass the random variable has at each value.*

Definition 10.1. Expected value of a discrete random variable
A discrete random variable X that takes on values $x_1, x_2, \ldots, x_n$ has expected value

$$\mathbb{E}(X) = \sum_{j=1}^{n} x_j p_X(x_j).$$

If X takes on a countably infinite number of values, $x_1, x_2, \ldots$, then the expected value of X is

$$\mathbb{E}(X) = \sum_{j=1}^{\infty} x_j p_X(x_j).$$

10.2 Examples

Example 10.2. Flip a coin three times. Let X denote the number of heads.

Since $X = 0$ with probability $1/8$, $X = 1$ with probability $3/8$, $X = 2$ with probability $3/8$, and $X = 3$ with probability $1/8$, then the expected value of X is

$$\mathbb{E}(X) = (0)(1/8) + (1)(3/8) + (2)(3/8) + (3)(1/8) = 3/2.$$

For a second way to view the expected values, notice that we have eight possible outcomes:

$$\omega_1 = TTT,$$
$$\omega_2 = HTT,\ \omega_3 = THT,\ \omega_4 = TTH,$$
$$\omega_5 = HHT,\ \omega_6 = HTH,\ \omega_7 = THH,$$
$$\omega_8 = HHH,$$

and we get, with this new interpretation of expected value, the very same result at the end:

$$\begin{aligned}\mathbb{E}(X) &= (0)(1/8)\\ &\quad + (1)(1/8) + (1)(1/8) + (1)(1/8)\\ &\quad + (2)(1/8) + (2)(1/8) + (2)(1/8)\\ &\quad + (3)(1/8)\\ &= 3/2.\end{aligned}$$

The reason that the two different definitions are equivalent is that we could just group the 2nd, 3rd, and 4th terms above, which all have value "1" as the value of X, to get

$$(1)(1/8) + (1)(1/8) + (1)(1/8) = (1)(3/8),$$

which was also found in the first definition. Similarly, we could just group the 5th, 6th, and 7th terms above, which all have value "2" as the value of X, and we get

$$(2)(1/8) + (2)(1/8) + (2)(1/8) = (2)(3/8),$$

which was also found in the first definition. So these are just two different ways of grouping things.

Thus, we can use either of these equivalent formulations:

Remark 10.3. Two equivalent ways to get the expected value of a random variable

1. a sum over all of the possible values of X, each weighted by the probability of X taking on that value, or
2. a sum over all of the possible outcomes, taking the value of X from such an outcome, weighted by the probability of that outcome

In the second method, where we enumerate the possible outcomes, we have the following:

Remark 10.4. Expected value of a discrete random variable (always gives same result as the original definition)
Consider a random phenomenon with possible outcomes $\omega_1, \omega_2, \ldots, \omega_n$. Suppose that the outcome ω_j causes random variable X to take on value x_j. Then the discrete random variable X has expected value

$$\mathbb{E}(X) = \sum_{j=1}^{n} x_j P(\{\omega_j\}).$$

If the random phenomenon can take on one of infinitely many possible outcomes $\omega_1, \omega_2, \ldots$, and we again suppose that the outcome ω_j causes random variable X to take on value x_j, then the expected value of X is

$$\mathbb{E}(X) = \sum_{j=1}^{\infty} x_j P(\{\omega_j\}).$$

Example 10.5. Consider a class consisting of eight students, who earn the following scores on an exam: 75, 92, 88, 94, 89, 60, 83, 84. Let X be the exam score of a randomly chosen student. Each student is equally likely to be selected. What is the expected value of X?

There are eight possible outcomes in this random phenomenon, each of which has probability 1/8. So, following the second definition of expected value, we obtain

$$\begin{aligned}\mathbb{E}(X) = {} & (75)(1/8) + (92)(1/8) + (88)(1/8) + (94)(1/8) \\ & + (89)(1/8) + (60)(1/8) + (83)(1/8) + (84)(1/8) \\ = {} & 83.125.\end{aligned}$$

Example 10.6. Roll a die three times. Let X be the number of times that a 6 appears. Find the expected value of X.

We compute:

$$P(X = 0) = (5/6)^3;$$

$$P(X = 1) = (3)(1/6)(5/6)^2$$

(since there are 3 ways to have 1 occurrence of 6, and each way has probability $(1/6)(5/6)^2$);

$$P(X = 2) = (3)(1/6)^2(5/6)$$

(since there are 3 ways to have 2 occurrences of 6, and each way has probability $(1/6)^2(5/6)$);

$$P(X = 3) = (1/6)^3.$$

So the expected value of X is

$$\mathbb{E}(X) = (0)(5/6)^3 + (1)(3)(1/6)(5/6)^2 + (2)(3)(1/6)^2(5/6) + (3)(1/6)^3 = 1/2.$$

If we were to have 100 people each roll a die three times, and then we took the average of these results (i.e., added the 100 results and divided by 100), the average from these 100 experiments would likely be very close to 0.5. It is worthwhile to try it and see. One thousand people would likely give us an average even closer to 0.5. We will consider what actually happens in the long run with experiments later in the book, when discussing Normal approximations and the Central Limit Theorem.

Remark 10.7. Helpful fact for double-checking an answer:
When writing the expected value of a discrete random variable, we always have a sum of terms, where each term is a value times a probability. We must consider all possible cases, so the probabilities under consideration had better add up to 1.

For instance, in the example above, the probabilities are $(5/6)^3$ and $(3)(1/6)(5/6)^2$ and $(3)(1/6)^2(5/6)$ and $(1/6)^3$. We can easily check that these add up to 1:

$$(5/6)^3 + (3)(1/6)(5/6)^2 + (3)(1/6)^2(5/6) + (1/6)^3 = 1.$$

Example 10.8. As in Examples 1.13 and 2.17, a student shuffles a deck of cards thoroughly (one time) and then selects cards from the deck *without replacement* until the ace of spades appears. How many cards does the student expect to draw until the ace of spades appears?

Let X be the number of cards needed until the ace of spades appears. Then

$$P(X = j) = 1/52 \qquad \text{for } 1 \le j \le 52.$$

Thus,

$$\begin{aligned}\mathbb{E}(X) &= (1)P(X=1) + (2)P(X=2) + (3)P(X=3) + \cdots + (52)P(X=52)\\ &= (1)(1/52) + (2)(1/52) + (3)(1/52) + \cdots + (52)(1/52)\\ &= (1+2+\cdots+52)(1/52)\\ &= 53/2\end{aligned}$$

So the student expects to draw 53/2 cards (i.e., 26.5 cards) to see the ace of spades appear. Here we used the helpful fact that $1+2+\cdots+n = (n)(n+1)/2$.

(We will further study such random variables, which are equally like to be any one of the values $1, 2, \ldots, N$, in Chapter 18.)

Example 10.9. A student draws cards from a standard deck of playing cards until the ace of spades appears for the first time. After every unsuccessful draw, the student replaces the card and shuffles the deck thoroughly before selecting a new card. How many cards does the student expect to draw until the ace of spades appears?

Let X denote the number of cards needed until the ace of spades appears for the first time. Then $P(X = j) = (51/52)^{j-1}(1/52)$ for each positive integer j. So

$$\begin{aligned}\mathbb{E}(X) &= (1)P(X=1) + (2)P(X=2) + (3)P(X=3) + \cdots\\ &= \sum_{j=1}^{\infty} j(51/52)^{j-1}(1/52)\\ &= \frac{1}{52}\sum_{j=1}^{\infty} j(51/52)^{j-1}\end{aligned}$$

We notice that $j(51/52)^{j-1}$ is the derivative of x^j (with respect to x), evaluated at $x = 51/52$. So we rewrite the equation above as follows:

$$\mathbb{E}(X) = \frac{1}{52}\sum_{j=1}^{\infty} \frac{d}{dx}x^j\bigg|_{x=51/52} = \frac{1}{52}\frac{d}{dx}\sum_{j=1}^{\infty} x^j\Big|_{x=51/52}.$$

We use our scrap paper to compute:

$$\sum_{j=1}^{\infty} x^j = x\sum_{j=0}^{\infty} x^j = \frac{x}{1-x}.$$

So we get

$$\begin{aligned}
\mathbb{E}(X) &= \frac{1}{52}\frac{d}{dx}\frac{x}{1-x}\bigg|_{x=51/52} \\
&= \frac{1}{52}\frac{(1-x)(1)-(x)(-1)}{(1-x)^2}\bigg|_{x=51/52} \\
&= \frac{1}{52}\frac{1}{(1-x)^2}\bigg|_{x=51/52} \\
&= \frac{1}{52}\frac{1}{\left(1-\frac{51}{52}\right)^2} \\
&= \frac{1}{52}\frac{1}{(1/52)^2} \\
&= 52
\end{aligned}$$

So the student expects to draw 52 cards to get the ace of spades. We will study more examples like this in Chapter 16, on Geometric random variables.

Example 10.10. A standard deck of 52 cards has 13 hearts. The cards in such a deck are shuffled, and the top five cards are dealt to a player. What is the expected number of hearts that the player receives?

Let X denote the number of hearts that the player receives. Then

$$\begin{aligned}
P(X=0) &= \frac{(39)(38)(37)(36)(35)}{(52)(51)(50)(49)(48)} = \frac{2109}{9520} \\
P(X=1) &= 5\frac{(13)(39)(38)(37)(36)}{(52)(51)(50)(49)(48)} = \frac{27417}{66640} \\
P(X=2) &= 10\frac{(13)(12)(39)(38)(37)}{(52)(51)(50)(49)(48)} = \frac{9139}{33320} \\
P(X=3) &= 10\frac{(13)(12)(11)(39)(38)}{(52)(51)(50)(49)(48)} = \frac{2717}{33320} \\
P(X=4) &= 5\frac{(13)(12)(11)(10)(39)}{(52)(51)(50)(49)(48)} = \frac{143}{13328} \\
P(X=5) &= \frac{(13)(12)(11)(10)(9)}{(52)(51)(50)(49)(48)} = \frac{33}{66640}
\end{aligned}$$

(Double check: $P(X=0)+P(X=1)+\cdots+P(X=5)=1$; thus, we have correctly considered all of the possibilities.) So we get

$$\mathbb{E}(X) = (0)\frac{2109}{9520}+(1)\frac{27417}{66640}+(2)\frac{9139}{33320}+(3)\frac{2717}{33320}+(4)\frac{143}{13328}+(5)\frac{33}{66640} = 5/4.$$

Example 10.11. Jim and his brother both like chocolate chip cookies best. They have a jar of cookies with 5 chocolate chip cookies, 3 oatmeal cookies, and 4 peanut butter cookies. They are each allowed to have 3 cookies. To be fair, they agree to randomly select their cookies without peeking, and they each must keep the cookies that they select. How many chocolate chip cookies does Jim expect to get? (Notice that it does not matter whether Jim or his brother selects the cookies first—the answer will be the same, either way.)

Let X the number of chocolate chip cookies that Jim selects. Since there are 12 cookies, there are (12)(11)(10) equally likely outcomes for Jim. Exactly (7)(6)(5) of them have no chocolate chips. So

$$P(X = 0) = \frac{(7)(6)(5)}{(12)(11)(10)} = \frac{7}{44}.$$

There are exactly 3 ways that one cookie could be chocolate chip; 5 such cookies could be the chocolate one; the other cookies could be selected in (7)(6) ways. So

$$P(X = 1) = \frac{(3)(5)(7)(6)}{(12)(11)(10)} = \frac{21}{44}.$$

There are exactly 3 ways that two cookies could be chocolate chips; 7 cookies could be the non-chocolate chip one; the other cookies could be chocolate chips, selected in (5)(4) ways. So

$$P(X = 2) = \frac{(3)(7)(5)(4)}{(12)(11)(10)} = \frac{7}{22}.$$

Exactly (5)(4)(3) of the outcomes are all chocolate chips. So

$$P(X = 3) = \frac{(5)(4)(3)}{(12)(11)(10)} = \frac{1}{22}.$$

So the expected value of X is

$$\mathbb{E}(X) = (0)\frac{7}{44} + (1)\frac{21}{44} + (2)\frac{7}{22} + (3)\frac{1}{22} = 5/4.$$

Note: The total number of outcomes is $(12)(11)(10) = 1320$. As a check, $210 + 630 + 420 + 60 = 1320$, so all 1320 possible outcomes have been taken into account.

10.3 Exercises

10.3.1 Practice

Exercise 10.1. Graduation. As in Exercise 3.1, Jack and Jill are independently struggling to pass their last (one) class required for graduation. Jack needs to pass Calculus III, but he only has probability 0.30 of passing. Jill needs to pass Advanced Pharmaceuticals, but she only has probability 0.46 of passing. They work independently. Let $X = 0$ if neither of them graduates, or $X = 1$ if exactly one of them graduate, or $X = 2$ if both of them graduate. Find the expected value of X.

Exercise 10.2. **Japanese pan noodles.** As in Exercise 3.2, Four students order noodles at a certain local restaurant. Their orders are placed independently. Each student is known to prefer Japanese pan noodles 40% of the time. How many of them do we expect to order Japanese pan noodles?

Exercise 10.3. **Waiting for folk song.** Angelina's music player, in "shuffle" mode, will play songs *without any repetitions*, until every song has been played exactly once. The number of songs of each genre is the following: 330 blues songs; 537 jazz songs; and 1 folk song. So there are 868 songs altogether. (All possible "shuffles"—i.e., all possible orderings of the 868 songs—are equally likely.) How many songs does Angelina expect to listen to, until the folk song finally appears? Please justify your answer. Hint: Use the method in Example 10.8.

Exercise 10.4. **Bowling strikes.** In a bowling alley, 20% of the time when someone bowls, he or she gets a strike. If there are 3 people in the bowling alley, and X is the total number of people who get a strike on their current attempt, what is the expected value of X?

Exercise 10.5. **Career fair.** You go to a career fair and have some job interviews. Based on career fair data, you think you have a 30% chance of getting an offer for $40,000, a 40% chance of getting an offer of $44,000, a 25% chance of getting an offer for $51,000, and a 5% chance of an offer for $57,000. What is your expected offer salary?

Exercise 10.6. **Golden ticket.** A student was at work at the county amphitheater, and was given the task of cleaning 1500 seats. To make the job more interesting, his boss hid a golden ticket somewhere in the seats. The ticket is equally likely to be in any of the seats. Let X be the number of seats cleaned until the ticket is found. Calculate the expected value of X.

Exercise 10.7. **Football.** A man is looking for a football game to watch on a Sunday night. He has ten channels, and only one of them shows football, but he can't remember which one. Assume they are all equally likely to have it, and let X be the number of channels he tries until he finds it. What is the expected number of channels he tries until he finds it?

Exercise 10.8. **Cashews.** There is a bowl containing 30 cashews, 20 pecans, 25 almonds, and 25 walnuts. I am going to randomly pick and eat 3 nuts. What is the expected number of cashews I will eat?

Exercise 10.9. **Chess.** Let X be the number of games of chess I win against Stephen. Assume that I have a 30% chance of winning in any particular game against him, and we play 5 games, with outcomes assumed to be independent. Find the expected value of X.

Exercise 10.10. Sports fans. The All-Star Pigeons lose 10 fans every time they lose a game and gain 100 fans every time they win. Assume they have a 30% chance of losing each game and that their performance in each game is independent. What is the expected value of X, the number of fans they will gain or lose over the next 3 games?

Exercise 10.11. Slot machine. People who play a slot machine win a prize 7% of the time. Let X denote the number of times the slot machine is used until the next winner is found. What is the expected value of X?

10.3.2 Extensions

Exercise 10.12. Butterflies. Alice, Bob, and Charlotte are looking for butterflies. They look in three separate parts of a field, so that their probabilities of success do not affect each other.

- Alice finds 1 butterfly with probability 17%, and otherwise does not find one.
- Bob finds 1 butterfly with probability 25%, and otherwise does not find one.
- Charlotte finds 1 butterfly with probability 45%, and otherwise does not find one.

Let X be the number of butterflies that they find altogether. Find $\mathbb{E}(X)$.

Exercise 10.13. Dependence/independence among dice rolls. A student rolls a die until the first "4" appears. Let X be the numbers of rolls required until (and including) this first "4." After this is completed, he begins rolling again until he gets a "3." Let Y be the number of rolls, after the first "4," up to (and including) the next "3." E.g., if the sequence of rolls is 213662341261613 then $X = 8$ and $Y = 7$.

a. Find the expected value of X.

b. Find the expected value of Y.

Exercise 10.14. Two 4-sided dice. Consider some 4-sided dice. Roll two of these dice. Let X denote the minimum of the two values that appear, and let Y denote the maximum of the two values that appear.

a. Find the expected value of X.

b. Find the expected value of Y.

Exercise 10.15. Pick two cards. Pick two cards at random from a well-shuffled deck of 52 cards (pick them simultaneously, i.e., grab two cards at once—so they are not the same card!). There are 12 cards which are considered face cards (4 Jacks, 4 Queens, 4 Kings). There are 4 cards with the value 10. Let X be the number of face cards in your hand; let Y be the number of 10's in your hand.

a. Find the expected value of X.

b. Find the expected value of Y.

Exercise 10.16. Wastebasket basketball. Chris tries to throw a ball of paper in the wastebasket behind his back (without looking). He estimates that his chance of success each time, regardless of the outcome of the other attempts, is $1/3$. Let X be the number of attempts required. If he is not successful within the first 5 attempts, then he quits, and he lets $X = 6$ in such a case. Find the expected value of X.

Exercise 10.17. Claw. Jorge has three kids who spotted a Claw machine with toys they want. In stores the toys costs $10 each, but each play on the claw only costs $1. The probability of Jorge winning a game on the Claw is 0.12. Should he use the Claw to get the toys (he needs one toy per kid), or does he expect it to be cheaper to buy them in the stores?

Exercise 10.18. Super Breakfast Challenge. As in Exercise 8.5, the Super Breakfast Challenge (SBC) consists of bacon, eggs, oatmeal, orange juice, milk, and several other foods, and it costs $12.99 per person to order at a local restaurant. It is known to be very difficult to consume the entire SBC. Only 10% of people are able to eat all of the SBC. The other 90% of people will be unable to eat the whole SBC (it is too much food!).

A probability student hears about the SBC and goes to the local restaurant. He observes the number of customers, X, that attempt to eat the SBC, until the first success. So if there are 4 failures and then 1 success (i.e., the outcome is (F, F, F, F, T)), then $X = 5$.

Find the expected value of X, i.e., the number of customers *expected* to try the SBC until the first success.

Exercise 10.19. Sum of dice. Two fair dice are rolled. Let X be the sum of the dice. What is the expected value of X?

Chapter 11

Expected Values of Sums of Random Variables

I, at any rate, am convinced that [God] does not throw dice.

—Albert Einstein, from a letter to Max Born (December 4, 1926), as quoted in *The Born-Einstein Letters*, translated by Irene Born (Walker, 1971)

While procrastinating before starting his probability homework, a student shuffles a deck of cards and deals a randomly selected card to himself once a minute, putting the used card back in the deck and reshuffling each time. He tells himself that when he gets the ace of spades, he will stop procrastinating and get to work. How many minutes does he expect to procrastinate?

11.1 Introduction

One useful property of expected values of random variables is *linearity*, i.e.:

1. the expected value of a sum of random variables is equal to the sum of the expected values, and

2. constants can be factored out of expected values.

Theorem 11.1. Expected value of the sum of discrete random variables

If $X_1, X_2, \ldots, X_n$ are discrete random variables with finite expected values, and $a_1, a_2, \ldots, a_n$ are constant numbers, then

$$\mathbb{E}(a_1X_1 + a_2X_2 + \cdots + a_nX_n) = a_1\mathbb{E}(X_1) + a_2\mathbb{E}(X_2) + \cdots + a_n\mathbb{E}(X_n).$$

The same property holds for a countably infinite collection of random variables. **This holds even when the X_j's are dependent.**

Justification of Theorem 11.1:

Consider any finite collection of discrete random variables $X_1, \ldots, X_n$ with finite expected values. Let $a_1, \ldots, a_n$ be any constants. Let S denote the underlying sample space. Let $X = a_1X_1 + \cdots + a_nX_n$. Then $X(\omega) = a_1X_1(\omega) + \cdots + a_nX_n(\omega)$ for each outcome $\omega \in S$. So

$$\begin{aligned}\mathbb{E}(a_1X_1 + a_2X_2 + \cdots + a_nX_n) &= \sum_{\omega \in S} (a_1X_1(\omega) + \cdots + a_nX_n(\omega))P(\{\omega\}) \\ &= \sum_{j=1}^{n} a_j \sum_{\omega \in S} X_j(\omega)P(\{\omega\}) \\ &= \sum_{j=1}^{n} a_j\mathbb{E}(X_j)\end{aligned}$$

The same type of justification works if there are a countably infinite number of random variables and constants.

We often use the property above in just the case where $n = 2$ and $a_2 = 1$ and X_2 is a constant, say "b." So Theorem 11.1 simplifies to

Corollary 11.2. For any random variable X and any constants a and b, we have

$$\mathbb{E}(aX + b) = a\mathbb{E}(X) + b.$$

This handy rule holds for any constants a and b.

Another popular use of the property in the box above is when the random variables are indicators for some events. In the case where X is an indicator random variable for event A, then $\mathbb{E}(X)$ is equal to $P(A)$:

$$\mathbb{E}(X) = 0P(X = 0) + 1P(X = 1) = 0P(A^c) + 1P(A) = P(A).$$

This rule is so helpful that we highlight it.

Theorem 11.3. Expected value of an indicator random variable
If X is an indicator random variable for event A, i.e., $X = 1$ if event A occurs and $X = 0$ otherwise, then the expected value of X is equal to the probability that event A occurs:

$$\mathbb{E}(X) = P(A).$$

Using Theorem 11.1 and Theorem 11.3 together, we have a very powerful probability tool that allows us to quickly resolve most of the problems from Chapter 10. We write a random variable as a sum of indicator random variables, and we compute the expected value of each indicator random variable by computing the probability of each corresponding event.

11.2 Examples

Example 11.4. Flip a coin three times. Let X denote the number of heads.

Let A_1, A_2, A_3 be the events that the first, second, third flips (respectively) are heads. Let X_1, X_2, X_3 be indicators for A_1, A_2, A_3, i.e., X_1, X_2, X_3 indicate whether the first, second, third flips (respectively) are heads. So $\mathbb{E}(X_j) = P(A_j) = 1/2$ for each j. Also, $X = X_1 + X_2 + X_3$. Thus,

$$\mathbb{E}(X) = \mathbb{E}(X_1 + X_2 + X_3) = \mathbb{E}(X_1) + \mathbb{E}(X_2) + \mathbb{E}(X_3) = 1/2 + 1/2 + 1/2 = 3/2.$$

Example 11.5. Consider a class consisting of eight students, who earn the following scores on an exam: 75, 92, 88, 94, 89, 60, 83, 84. Let X be the exam score of a randomly chosen student. Each student is equally likely to be selected. What is the expected value of X?

Let X_j be the jth student's score if the jth student is selected, and $X_j = 0$ otherwise. So, e.g., $\mathbb{E}(X_1) = (75)(1/8) + (0)(7/8) = (75)(1/8)$. Then always $X = X_1 + X_2 + \cdots + X_8$, because one of the X_j's will be the selected student's score, and the other X_j's will be 0. Therefore,

$$\begin{aligned}\mathbb{E}(X) &= \mathbb{E}(X_1 + \cdots + X_8)\\ &= \mathbb{E}(X_1) + \cdots + \mathbb{E}(X_8)\\ &= (75)(1/8) + (92)(1/8) + (88)(1/8) + (94)(1/8)\\ &\quad + (89)(1/8) + (60)(1/8) + (83)(1/8) + (84)(1/8)\\ &= 83.125.\end{aligned}$$

Example 11.6. Roll a die three times. Let X be the number of times that a 6 appears. Find the expected value of X.

Let A_j be the event that the jth roll is a 6, so $P(A_j) = 1/6$ for each j. Let X_j be the indicator for A_j, i.e., X_j indicates whether the jth roll is a 6, so $X_j = 1$ if the jth roll is a 6, and $X_j = 0$ otherwise. Thus, $\mathbb{E}(X_j) = P(A_j) = 1/6$ for each j. Also, $X = X_1 + X_2 + X_3$. So

$$\mathbb{E}(X) = \mathbb{E}(X_1 + X_2 + X_3) = \mathbb{E}(X_1) + \mathbb{E}(X_2) + \mathbb{E}(X_3) = 1/6 + 1/6 + 1/6 = 1/2.$$

More generally:

Theorem 11.7. Expected value of the sum of identically distributed discrete random variables

If $X_1, X_2, \ldots, X_n$ are identically distributed discrete random variables, then the expected value of $X_1 + X_2 + \cdots + X_n$ is equal to one of the expected values multiplied by n, i.e.,

$$\mathbb{E}(X_1 + X_2 + \cdots + X_n) = n\mathbb{E}(X_1).$$

This is true regardless of whether the X_j's are dependent or independent.

Example 11.8. A standard deck of 52 cards has 13 hearts. The cards in such a deck are shuffled, and the top five cards are dealt to a player. What is the expected number of hearts that the player receives?

Let X denote the number of hearts that the player receives. Let A_j be the event that the jth card is a heart, so $P(A_j) = 1/4$ for each j. Let X_j be the indicator for A_j, i.e., X_j indicates whether the jth card is a heart, so $X_j = 1$ if the jth card is a heart, and $X_j = 0$ otherwise. This yields $\mathbb{E}(X_j) = P(A_j) = 1/4$ for each j. (This is true because we are just focusing momentarily on the jth value; we are not considering the values of the other cards or conditioning on them.) Also, $X = X_1 + X_2 + X_3 + X_4 + X_5$. So

$$\mathbb{E}(X) = (5)(1/4) = 5/4.$$

Example 11.9. Jim and his brother both like chocolate chip cookies best. They have a jar of cookies with 5 chocolate chip cookies, 3 oatmeal cookies, and 4 peanut butter cookies. They are each allowed to have 3 cookies. To be fair, they agree to randomly select their cookies without peeking, and they each must keep the cookies that they select. How many chocolate chip cookies does Jim expect to get? (Notice that it does not matter whether Jim or his brother selects the cookies first—the answer will be the same, either way.)

Let X be the number of chocolate chip cookies that Jim selects. Let A_j be the event that the jth cookie that Jim selects is chocolate chip, so $P(A_j) = 5/12$ for each j. Let X_j be the indicator for A_j, i.e., X_j indicates whether the jth cookie that Jim selects is chocolate chip, so $X_j = 1$ if the jth cookie that Jim selects is chocolate chip, and $X_j = 0$ otherwise. So $\mathbb{E}(X_j) = P(A_j) = 5/12$ for each j. Also, $X = X_1 + X_2 + X_3$, so

$$\mathbb{E}(X) = (3)(5/12) = 5/4.$$

Indicator random variables are useful even when they do not necessary have the same distribution. Some creativity is required for seeing how to apply them. This creativity can be developed with experience and with persistence. It is often worthwhile to see if indicator random variables are relevant to apply in a problem. We present some different ways that they can be used in the two examples below, to demonstrate that there are often several different approaches—using different kinds of indicator random variables—to solve the same problem.

Example 11.10. A student shuffles a deck of cards thoroughly (one time) and then selects cards from the deck *without replacement* until the ace of spades appears. How many cards does the student expect to draw?

Method #1. Let A_j be the event that j or more draws are required, so $P(A_j) = 1 - \frac{j-1}{52}$, because A_j occurs if and only if the first $j-1$ draws are failures. Let X_j be the indicator for A_j, i.e., X_j indicates whether j or more draws are needed, so $X_j = 1$ if j or more draws are needed, and $X_j = 0$ otherwise. Therefore, we have $\mathbb{E}(X_j) = P(A_j) = 1 - \frac{j-1}{52}$ for each j. Also, $X = X_1 + X_2 + \cdots + X_{52}$ because (for instance) if exactly 3 draws are needed, then X_1, X_2, X_3 are each equal to 1, and the other X_j's are 0, so $X = X_1 + X_2 + X_3$, as desired. So

$$\begin{aligned}
\mathbb{E}(X) &= \mathbb{E}(X_1 + X_2 + \cdots + X_{52}) \\
&= \mathbb{E}(X_1) + \mathbb{E}(X_2) + \cdots + \mathbb{E}(X_{52}) \\
&= \sum_{j=1}^{52} \left(1 - \frac{j-1}{52}\right) \\
&= 52 - \frac{\sum_{j=1}^{52}(j-1)}{52}
\end{aligned}$$

but

$$\sum_{j=1}^{52}(j-1) = 0 + 1 + 2 + \cdots + 51 = (52)(51)/2,$$

since

$$1 + 2 + \cdots + n = (n)(n+1)/2.$$

So

$$\mathbb{E}(X) = 52 - \frac{(52)(51)/2}{52} = 52 - \frac{51}{2} = \frac{104 - 51}{2} = 53/2.$$

Method #2. Assign one indicator random variable to each card in the deck (except the ace of spades). The indictor is 1 if the corresponding card is drawn before the ace of spades, or 0 otherwise. The number of cards drawn is $X_1 +$

$\cdots + X_{51} + 1$, where the X_j's are the indicators. So the expected number of cards to draw until the ace of spades appears is

$$\begin{aligned}\mathbb{E}(X_1 + \cdots + X_{51} + 1) &= \mathbb{E}(X_1) + \cdots + \mathbb{E}(X_{51}) + \mathbb{E}(1)\\ &= \frac{1}{2} + \cdots + \frac{1}{2} + 1\\ &= \frac{51}{2} + 1\\ &= \frac{53}{2}.\end{aligned}$$

Method #3. (This method is only included here for readers who know and enjoy induction; if the reader does not know induction, it is OK to skip this method.) Let X_j denote the number of flips needed to find a particular card (e.g., the ace of spades) in a deck of j cards. Then we claim that $\mathbb{E}(X_j) = (j+1)/2$ for all j, and we prove it by induction. If $j = 1$, then the first card must be the desired card, so $\mathbb{E}(X_1) = (1+1)/2 = 1$; this shows the base case. Now we handle the inductive step: If $\mathbb{E}(X_{j-1}) = j/2$, we show that $\mathbb{E}(X_j) = (j+1)/2$. Flip the first card (from the deck of j cards); the first card is the desired card with probability $1/j$. The first card is not the desired card with probability $(j-1)/j$, and in such a case, this flip (that just occurred) was used, plus an additional $\mathbb{E}(X_{j-1})$ flips will be needed, because the problem is essentially starting again with $j-1$ cards. So

$$\mathbb{E}(X_j) = \frac{1}{j} + \frac{j-1}{j}(1 + \mathbb{E}(X_{j-1})) = \frac{1}{j} + \frac{j-1}{j}\left(1 + \frac{j}{2}\right) = \frac{j+1}{2}.$$

In particular, when starting with 52 cards, $\mathbb{E}(X_{52}) = 53/2$ flips are expected.

Example 11.11. A student draws cards from a standard deck of playing cards until the ace of spades appears for the first time. After every unsuccessful draw, the student replaces the card and shuffles the deck thoroughly before selecting a new card. How many cards does the student expect to draw until the ace of spades appears?

Method #1. Let A_j be the event that j or more draws are required, so $P(A_j) = (51/52)^{j-1}$, because A_j occurs if and only if the first $j-1$ draws are failures. Let X_j be the indicator for A_j, i.e., X_j indicates whether j or more draws are needed, so $X_j = 1$ if j or more draws are needed, and $X_j = 0$ otherwise. So $\mathbb{E}(X_j) = P(A_j) = (51/52)^{j-1}$ for each j. Also, $X = X_1 + X_2 + X_3 + \cdots$ because (for instance) if exactly 3 draws are needed, then X_1, X_2, X_3 are each equal to 1, and the other X_j's are 0, so $X = X_1 + X_2 + X_3$, as desired. Thus, we have

$$\mathbb{E}(X) = \mathbb{E}(X_1 + X_2 + \cdots) = \mathbb{E}(X_1) + \mathbb{E}(X_2) + \cdots = 1 + 51/52 + (51/52)^2 + \cdots.$$

It follows that

$$\mathbb{E}(X) = \sum_{j=0}^{\infty} (51/52)^j = \frac{1}{1 - \frac{51}{52}} = \frac{1}{1/52} = 52.$$

So the student expects to draw 52 cards to see the ace of spades for the first time.

Method #2. (Does not use indicators!) Let X be the number of flips that are necessary. With probability $1/52$, the ace of spades appears on the first draw. With probability $51/52$, the ace of spades does not appear on the first draw, and the problem essentially starts over again: in this case, the original flip was used, plus $\mathbb{E}(X)$ more flips will be needed. So we have

$$\mathbb{E}(X) = \frac{1}{52}(1) + \frac{51}{52}(1 + \mathbb{E}(X)) = \frac{1}{52} + \frac{51}{52} + \frac{51}{52}\mathbb{E}(X) = 1 + \frac{51}{52}\mathbb{E}(X).$$

Subtracting $\frac{51}{52}\mathbb{E}(X)$ from both sides, we get $\frac{1}{52}\mathbb{E}(X) = 1$, and we conclude that $\mathbb{E}(X) = 52$.

Example 11.12. As one last example, let X_1, X_2, X_3 be independent random variables that correspond to three consecutive flips of an unbiased coin. Then define $X = X_1 + X_2 + X_3$. We studied X in Example 11.4. In particular, we know that $\mathbb{E}(X) = 3/2$.

In contrast, consider the random variable $Y = 3X_1$. We also have $\mathbb{E}(Y) = \mathbb{E}(3X_1) = 3\mathbb{E}(X_1) = 3/2$. So X and Y have the same expected value.

On the other hand, X and Y are two very different types of random variables. For instance, X can take on four different values (0, 1, 2, or 3), and the mass of X is:

$$p_X(0) = 1/8, \qquad p_X(1) = 3/8, \qquad p_X(2) = 3/8, \qquad p_X(3) = 1/8.$$

On the other hand, Y can only take on the values 0 or 3, and the mass of Y is:

$$p_Y(0) = 1/2, \qquad p_Y(3) = 1/2.$$

Therefore, although $X_1 + X_2 + X_3$ and $3X_1$ have the same expected value, they have very different distributions. (More generally, $X_1 + X_2 + \cdots + X_n$ and nX_1 have different distributions in most similar examples.)

11.3 Exercises

11.3.1 Practice

Exercise 11.1. Graduation. As in Exercise 10.1, Jack and Jill are independently struggling to pass their last (one) class required for graduation. Jack needs to pass Calculus III, but he only has probability 0.30 of passing. Jill needs to pass Advanced Pharmaceuticals, but she only has probability 0.46 of passing. They work independently.

Following the notation of Exercise 10.1, let X be 0, 1, or 2, if (respectively) neither, one, or both of them graduate. Let Y indicate if Jack graduates, so $Y = 1$ if Jack graduates, and $Y = 0$ otherwise. Let Z indicate if Jill graduates, so $Z = 1$ if Jill graduates, and $Z = 0$ otherwise. Notice that $X = Y + Z$ always. Now find $\mathbb{E}(X)$ using $\mathbb{E}(Y)$ and $\mathbb{E}(Z)$.

Exercise 11.2. Japanese pan noodles. As in Exercise 10.2, four students order noodles at a certain local restaurant. Their orders are placed independently. Each student is known to prefer Japanese pan noodles 40% of the time. How many of them do we expect to order Japanese pan noodles?

Let A_1, A_2, A_3, A_4 be the events that (respectively) the first, second, third, or fourth person orders Japanese pan noodles. Let X_1, X_2, X_3, X_4 be indicator random variables for (respectively) A_1, A_2, A_3, A_4. Justify your answer using the values of the $\mathbb{E}(X_j)$'s.

Exercise 11.3. Yellow ducks. Three hundred little plastic yellow ducks are dumped in a pond; one of them contains a prize stamped on the bottom. Leonardo examines each duck until he discovers the prize. He discards each duck without a prize after he checks it, so that he never needs to check a duck more than one time. How many ducks does he expect to check until he discovers the prize?

Exercise 11.4. Super Breakfast Challenge. As in Exercises 8.5 and 10.18, the Super Breakfast Challenge (SBC) consists of bacon, eggs, oatmeal, orange juice, milk, and several other foods, and it costs \$12.99 per person to order at a local restaurant. It is known to be very difficult to consume the entire SBC. Only 10% of people are able to eat all of the SBC. The other 90% of people will be unable to eat the whole SBC (it is too much food!).

A probability student hears about the SBC and goes to the local restaurant. He observes the number of customers, X, that attempt to eat the SBC, until the first success. So if there are 4 failures and then 1 success (i.e., the outcome is (F, F, F, F, T)), then $X = 5$.

Find the expected value of X, i.e., the number of customers *expected* to try the SBC until the first success.

Exercise 11.5. Weather. A weather forecasting program gets the daily predictions right about 87% of the time. Assuming each day is independent, what is the expected number of days that will pass until the program gets the forecast wrong?

Exercise 11.6. Random movie picks. A man sits down to watch a movie. 40% of his options are action films, 35% are comedies, and 25% are drama. He wants to watch a comedy and starts picking movies at random from his box. If previous picks are put back in the box after marking the choice, how many movies is he expected to have to pick up until he finds a comedy?

Exercise 11.7. Dice. Robbie rolls a die until he gets a 6. What is the expected number of rolls?

Exercise 11.8. Free throws. While shooting free throws, you have a 37% chance of making a basket. How many shots do you expect to have to take until making a basket?

Exercise 11.9. Computer preferences. You randomly and independently ask students on campus whether they prefer Mac or PC. Each person prefers Mac 72% of the time. What is the expected number of students you ask until the first person prefers PC?

Exercise 11.10. Butterflies. Alice, Bob, and Charlotte are looking for butterflies. They look in three separate parts of a field, so that their probabilities of success do not affect each other.

- Alice finds 1 butterfly with probability 17%, and otherwise does not find one.
- Bob finds 1 butterfly with probability 25%, and otherwise does not find one.
- Charlotte finds 1 butterfly with probability 45%, and otherwise does not find one.

Let X be the number of butterflies that they find altogether.

Write X as the sum of three indicator random variables, X_1, X_2, X_3 that indicate whether Alice, Bob, Charlotte (respectively) found a butterfly. Then $X = X_1 + X_2 + X_3$. Find the expected value of X by finding the expected value of the sum of the indicator random variables.

Exercise 11.11. Dependence/independence among dice rolls. A student rolls a die until the first "4" appears. Let X be the numbers of rolls required until (and including) this first "4." After this is completed, he begins rolling again until he gets a "3." Let Y be the number of rolls, after the first "4," up to (and including) the next "3." E.g., if the sequence of rolls is 213662341261613 then $X = 8$ and $Y = 7$.

a. Let A_j be the event containing all outcomes in which "j or more rolls" are required to get the first "4." Let X_j indicate whether A_j occurs. Then

$X = X_1 + X_2 + X_3 + \cdots$. Find the expected value of X by finding the expected value of the sum of the indicator random variables.

b. Let B_j be the event containing all outcomes in which "j or more rolls" are required, after the first "4," until he gets a "3." Let Y_j indicate whether B_j occurs. Then $Y = Y_1 + Y_2 + Y_3 + \cdots$. Find the expected value of Y by finding the expected value of the sum of the indicator random variables.

Exercise 11.12. Wastebasket basketball. Chris tries to throw a ball of paper in the wastebasket behind his back (without looking). He estimates that his chance of success each time, regardless of the outcome of the other attempts, is 1/3. Let X be the number of attempts required. If he is not successful within the first 5 attempts, then he quits, and he lets $X = 6$ in such a case.

Let A_j be the event containing all outcomes in which "j or more attempts" are required to get the basket. Let X_j indicate whether A_j occurs. Then $X = X_1 + X_2 + X_3 + X_4 + X_5 + X_6$. Find the expected value of X by finding the expected value of the sum of the indicator random variables.

Exercise 11.13. Two 4-sided dice. Consider some 4-sided dice. Roll two of these dice. Let X denote the minimum of the two values that appear, and let Y denote the maximum of the two values that appear.

Let A_j be the event containing all outcomes in which the minimum of the two dice is "j or greater." Let X_j indicate whether A_j occurs. Then $X = X_1 + X_2 + X_3 + X_4$. Find the expected value of X by finding the expected value of the sum of the indicator random variables.

Exercise 11.14. Pick two cards. Pick two cards at random from a well-shuffled deck of 52 cards (pick them simultaneously, i.e., grab two cards at once—so they are not the same card!). There are 12 cards which are considered face cards (4 Jacks, 4 Queens, 4 Kings). There are 4 cards with the value 10. Let X be the number of face cards in your hand; let Y be the number of 10's in your hand.

a. Before looking at the cards, put one in your left hand and one in your right hand. Let X_1 and X_2 indicate, respectively, whether the cards in your left and right hands (respectively) are face cards. Then $X = X_1 + X_2$. Find the expected value of X by finding the expected value of the sum of the indicator random variables.

b. Before looking at the cards, put one in your left hand and one in your right hand. Let Y_1 and Y_2 indicate, respectively, whether the cards in your left and right hands (respectively) are 10's. Then $Y = Y_1 + Y_2$. Find the expected value of Y by finding the expected value of the sum of the indicator random variables.

11.3.2 Extensions

Exercise 11.15. Oreos. A box of Double Stuf Oreos has a defect. One of the Oreos is only single-stuffed. There are 6 Oreos in the pack. What is the expected number of Oreos you need to check in the pack until the defective one is found?

Exercise 11.16. **Flipping coins.** Flip a coin until the second head comes up. Let X be the number of flips needed to get the second head. What is the $\mathbb{E}(X)$?

Exercise 11.17. **Waiting for favorite song.** Michael puts his iTunes on shuffle mode where songs are not allowed to be replayed. He has 2,781 songs saved in iTunes, and exactly one of these is his favorite.

a. How many songs is he expected to have to listen to until his very favorite song comes up?

b. Now suppose that he allows songs to be repeated. How many songs does he expect to listen to until his very favorite song comes up?

Exercise 11.18. **Crayons.** A little girl has a 96-pack of crayons. She picks up crayons, at random, to check the color, and she leaves them in a separate pile after inspecting the color. She pulls crayons out of the pack until she gets the sea foam green crayon.

a. What is the expected number of crayons she will check until she finds sea foam green?

b. Assume instead that she puts each crayon back in the box before randomly drawing the next crayon. Now how many crayons does she expect to check until she finds sea foam green?

Exercise 11.19. **Lectures and labs.** Consider a group of 120 students. They are split into two lectures (60 students each). They are also split into six labs (20 students each). All assignments of students to lectures and labs are equally likely. How many classmates does Barry expect to be in both his lecture and his lab too?

Exercise 11.20. **Onions.** Eighty-seven percent of people tear up when cutting an onion. People are selected randomly and independently to cut up an onion. What is the expected number of people who cut up an onion until you find the third person who does it without crying?

Exercise 11.21. **Exiting the elevator.** Eight people enter an elevator in the parking garage below a building. Each person chooses her exit independently of the other people. The building has floors 1 through 10. What is the expected number of stops that the elevator makes?

Exercise 11.22. **Bears.** A total of 30 bears—consisting of 10 red bears, 10 yellow bears, and 10 blue bears—are randomly arranged into 10 groups of 3 bears each. Compute the expected number of groups in which all 3 bears are different colors.

Exercise 11.23. Rope tying. Consider n pieces of rope. Each piece is colored blue at one end and red at the other. The blue ends of the ropes are randomly paired with the red ends of the ropes and tied together, one-to-one, i.e., all $n!$ such possible methods of joining the ropes this way are equally likely. Let X be the number of loops that result from this method. Find $E(X)$.

Chapter 12

Variance of Discrete Random Variables

The probable is what usually happens.
—Aristotle

Two different teachers are teaching probability classes. Before registering, students want to know how past students' grades were distributed for each instructor. The students have a report of the mean and standard deviation for Exam 1 from last semester. The means are similar, but one instructor has a standard deviation that is much larger than the other instructor's. Is it good or bad for exam scores to be widely spread out from the mean? Why?

12.1 Introduction

The *expected value of a function of a random variable* is a sum over all values that a function of a random variable takes; as in Chapter 10, these the values are taken in proportion to the fraction of the time that the function of the random variable takes each value.

Definition 12.1. Expected value of a function of a discrete random variable
If g is any function, and X is a discrete random variable that takes on values $x_1, x_2, \ldots, x_n$, then the expected value of $g(X)$ is

$$\mathbb{E}(g(X)) = \sum_{j=1}^{n} g(x_j)P(X = x_j) = \sum_{j=1}^{n} g(x_j)p_X(x_j).$$

If the number of values X takes on is countably infinite, $x_1, x_2, \ldots$, then the same expressions hold, but the sums are taken over all $j \in \mathbb{N}$.

Example 12.2. Flip a coin three times. Let X denote the number of heads. Find $\mathbb{E}(X^2)$, $\mathbb{E}(X^3)$, $\mathbb{E}(e^X)$, $\mathbb{E}(2X-5)$, and $\mathbb{E}(17X^3+5X^2-7X+22)$.

Since $X=0$ with probability 1/8, $X=1$ with probability 3/8, $X=2$ with probability 3/8, and $X=3$ with probability 1/8, then the expected value of X^2 is

$$\mathbb{E}(X^2) = (0^2)(1/8) + (1^2)(3/8) + (2^2)(3/8) + (3^2)(1/8) = 3.$$

Warning: Note $\mathbb{E}(X^2) \neq (\mathbb{E}(X))^2$. This is one of the most common mistakes made when first learning these concepts.

The expected value of X^3 is

$$\mathbb{E}(X^3) = (0^3)(1/8) + (1^3)(3/8) + (2^3)(3/8) + (3^3)(1/8) = 27/4.$$

The expected value of e^X is

$$\mathbb{E}(e^X) = (e^0)(1/8) + (e^1)(3/8) + (e^2)(3/8) + (e^3)(1/8) = \frac{1}{8} + \frac{3}{8}e + \frac{3}{8}e^2 + \frac{1}{8}e^3.$$

The expected value of $2X-5$ is

$$\begin{aligned}\mathbb{E}(2X-5) &= ((2)(0)-5)\left(\frac{1}{8}\right) + ((2)(1)-5)\left(\frac{3}{8}\right) \\ &\quad + ((2)(2)-5)\left(\frac{3}{8}\right) + ((2)(3)-5)\left(\frac{1}{8}\right) \\ &= -2.\end{aligned}$$

As we learned in Chapter 11, the expected value is linear. Also, by treating the "5" as a random variable that is always equal to 5, we can use the fact from Chapter 10 that $\mathbb{E}(X)=3/2$, and we can simply calculate (as a shorter method than above),

$$\mathbb{E}(2X-5) = 2\mathbb{E}(X) - \mathbb{E}(5) = 2(3/2) - 5 = -2.$$

We can use the fact that $\mathbb{E}(X)=3/2$ and $\mathbb{E}(X^2)=3$ and $\mathbb{E}(X^3)=27/4$ to compute, for instance, $\mathbb{E}(17X^3+5X^2-7X+22)$, as follows:

$$\begin{aligned}\mathbb{E}(17X^3+5X^2-7X+22) &= 17\mathbb{E}(X^3) + 5\mathbb{E}(X^2) - 7\mathbb{E}(X) + 22 \\ &= 17(27/4) + 5(3) - 7(3/2) + 22 \\ &= 565/4 \\ &= 141.25.\end{aligned}$$

As in Chapter 10, we can also decompose things further, based upon the specific outcome of the underlying random phenomenon:

Definition 12.3. Expected value of a function of a discrete random variable (always gives same result as the original definition)
Consider a random phenomenon with possible outcomes $\omega_1, \omega_2, \ldots, \omega_n$. Suppose that the outcome ω_j causes random variable X to take on value x_j. Then $g(X)$ has expected value

$$\mathbb{E}(g(X)) = \sum_{j=1}^{n} g(x_j) P(\{\omega_j\}).$$

If there are a countably infinite number of such ω's, we just adjust to sums taken over all $j \in \mathbb{N}$.

In the setting of Example 12.2, this means we can just decompose the random phenomenon into all eight possible outcomes:

$$\begin{gathered}\omega_1 = TTT,\\ \omega_2 = HTT,\ \omega_3 = THT,\ \omega_4 = TTH,\\ \omega_5 = HHT,\ \omega_6 = HTH,\ \omega_7 = THH,\\ \omega_8 = HHH,\end{gathered}$$

and we get the same result:

$$\begin{aligned}\mathbb{E}(X^2) = {} & (0^2)(1/8) + (1^2)(1/8) + (1^2)(1/8) + (1^2)(1/8)\\ & + (2^2)(1/8) + (2^2)(1/8) + (2^2)(1/8) + (3^2)(1/8) = 3.\end{aligned}$$

(In an analogous way, we had two ways of computing expected values in Chapter 10, which basically amounted to whether we grouped the 2nd, 3rd, and 4th terms above, which all yield the same value of $g(X)$, and we could also group the 5th, 6th, and 7th terms above, which again all yield a common value of $g(X)$. So the two boxed methods above are, as in Chapter 10, just two different ways of grouping things.)

To summarize: When computing the expected value of a function g of a random variable X, we can take one of the two following approaches:

1. a sum over all of the possible values of $g(X)$, each weighted by the probability of X taking on value x, and thus $g(X)$ taking on value $g(x)$, or

2. a sum over all of the possible outcomes, taking the value of $g(X)$ from such an outcome, weighted by the probability of that outcome

---

Example 12.4. Consider a class consisting of eight students, who earn the following scores on an exam: 75, 92, 88, 94, 89, 60, 83, 84. Let X be the exam score of a randomly chosen student. Each student is equally likely to be selected. How much do we expect each student's score to exceed 60, i.e., what is the expected value of $X - 60$?

There are eight possible outcomes in this random phenomenon, each of which has probability 1/8. So, following the second definition of expected value, we obtain

$$\begin{aligned}\mathbb{E}(X-60) &= (75-60)(1/8)+(92-60)(1/8)+(88-60)(1/8)\\ &\quad +(94-60)(1/8)+(89-60)(1/8)+(60-60)(1/8)\\ &\quad +(83-60)(1/8)+(84-60)(1/8)\\ &= 23.125.\end{aligned}$$

Of course, since we already computed $\mathbb{E}(X) = 83.125$ in Chapter 10, we could have obtained the same answer using linearity of expected values:

$$\mathbb{E}(X-60) = \mathbb{E}(X) - 60 = 83.125 - 60 = 23.125.$$

12.2 Variance

One of the reasons that expected values of functions of random variables are so useful is the concept of variance. Since the variance of a random variable is positive or zero (as we will see below), we can always compute the square root of the variance; this quantity is referred to as the standard deviation of the random variable. The standard deviation measures the spread of a random variable around the mean, i.e., it describes how widely the mass of a random variable is spread out. A larger variance (or standard deviation) signifies that the values of the random variable tend to be spread relatively far from the expected value. A smaller variance (or standard deviation) signifies that the values of the random variable tend to be relatively closer to the expected value, i.e., the values of the random variable do not tend to be very spread out.

We use $\mu_X = \mathbb{E}(X)$ and $\sigma_X^2 = \mathrm{Var}(X)$ to simplify the notation.

Definition 12.5. Variance
The variance σ_X^2 of a random variable X, with expected value $\mu_X = \mathbb{E}(X)$, is

$$\mathrm{Var}(X) = \mathbb{E}((X-\mu_X)^2).$$

One property of the variance is that the variance of a random variable is always nonnegative, i.e., the variance is always 0 or a positive number. The reason for this is that $(X-\mu_X)^2 \geq 0$ always (i.e., the square of a number cannot be negative). Thus, the variance is the expected value of a nonnegative number, so the variance itself is nonnegative too.

Theorem 12.6. Variance is always nonnegative
For any random variable X, the variance of X is nonnegative:

$$\mathrm{Var}(X) = \mathbb{E}((X-\mu_X)^2) \geq 0.$$

We notice that $(X - \mu_X)^2 = X^2 - 2\mu_X X + \mu_X^2$. Using the linearity of expected values, this gives us

$$\mathrm{Var}(X) = \mathbb{E}((X - \mu_X)^2) = \mathbb{E}(X^2 - 2\mu_X X + \mu_X^2) = \mathbb{E}(X^2) - 2\mu_X \mathbb{E}(X) + \mu_X^2.$$

Since $\mu_X = \mathbb{E}(X)$, this simplifies to

$$\mathrm{Var}(X) = \mathbb{E}(X^2) - (\mathbb{E}(X))^2.$$

So we have an equivalent formulation of the variance, which is often used in practice to actually compute the variance.

Remark 12.7. Useful formulation for computing the variance of a random variable
The variance of a random variable X is

$$\mathrm{Var}(X) = \mathbb{E}(X^2) - (\mathbb{E}(X))^2.$$

Definition 12.8. Standard Deviation
For a random variable X with variance $\mathrm{Var}(X)$, the standard deviation of X is

$$\sigma_X = \sqrt{\mathrm{Var}(X)}.$$

In Example 12.9, we see random variables X and Y that have the same expected values but different variances.

Example 12.9. Recall Examples 10.5, 11.5, and 12.4, in which the expected value of a randomly chosen student's grade in a course is

$$\begin{aligned}\mathbb{E}(X) &= (75)(1/8) + (92)(1/8) + (88)(1/8) + (94)(1/8) \\ &\quad + (89)(1/8) + (60)(1/8) + (83)(1/8) + (84)(1/8) \\ &= 83.125.\end{aligned}$$

Consider eight other students in a different course, who earn the following scores on an exam: 82, 82, 81, 84, 89, 80, 83, 84. In this case, let Y be the exam score of a randomly chosen student. Find $\mathrm{Var}(X)$ and $\mathrm{Var}(Y)$.

The expected value of Y is

$$\begin{aligned}\mathbb{E}(Y) &= (82)(1/8) + (82)(1/8) + (81)(1/8) + (84)(1/8) \\ &\quad + (89)(1/8) + (80)(1/8) + (83)(1/8) + (84)(1/8) \\ &= 83.125,\end{aligned}$$

so that the average value of the students in the two scenarios is the same, i.e.,

$$\mathbb{E}(X) = \mathbb{E}(Y).$$

On the other hand, the variances of the grades from the two groups of students are different. We have

$$\begin{aligned}\mathbb{E}(X^2) &= (75^2)(1/8) + (92^2)(1/8) + (88^2)(1/8) + (94^2)(1/8)\\ &\quad + (89^2)(1/8) + (60^2)(1/8) + (83^2)(1/8) + (84^2)(1/8)\\ &= 7016.875.\end{aligned}$$

and

$$\begin{aligned}\mathbb{E}(Y^2) &= (82^2)(1/8) + (82^2)(1/8) + (81^2)(1/8) + (84^2)(1/8)\\ &\quad + (89^2)(1/8) + (80^2)(1/8) + (83^2)(1/8) + (84^2)(1/8)\\ &= 6916.375,\end{aligned}$$

So the variance of X is

$$\mathrm{Var}(X) = 7016.875 - (83.125)^2 = 107.109375,$$

and the variance of Y is

$$\mathrm{Var}(Y) = 6916.375 - (83.125)^2 = 6.609375.$$

The standard deviations of X and Y are, respectively,

$$\sigma_X = \sqrt{107.109375} = 10.349366 \qquad \text{and} \qquad \sigma_Y = \sqrt{6.609375} = 2.570870.$$

The variance of Y is much smaller than the variance of X. This signifies the fact that the values of Y tend to be much closer to the expected value of Y, as compared to the values of X, which tend to be very spread out far from the expected value of X. All of the values of Y are found in the 80's, while the values of X are spread from 60 all the way to 94.

Example 12.10. Flip a coin three times. Let X denote the number of heads. Find the variance of X.

We already computed $\mathbb{E}(X) = 3/2$ and $\mathbb{E}(X^2) = 3$. Thus

$$\mathrm{Var}(X) = \mathbb{E}(X^2) - (\mathbb{E}(X))^2 = 3 - (3/2)^2 = 3/4.$$

The standard deviation is $\sigma_X = \sqrt{3}/2$.

Example 12.11. Roll three dice. Let X denote the number of 6's that appear on the dice altogether. Find the variance of X.

In Example 10.6, we already computed the mass of X, which is the following:

$$\begin{aligned} P(X=0) &= (5/6)^3 \\ P(X=1) &= (3)(1/6)(5/6)^2 \\ P(X=2) &= (3)(1/6)^2(5/6) \\ P(X=3) &= (1/6)^3 \end{aligned}$$

So the expected value of X is

$$\mathbb{E}(X) = (0)(5/6)^3 + (1)(3)(1/6)(5/6)^2 + (2)(3)(1/6)^2(5/6) + (3)(1/6)^3 = 1/2.$$

The expected value of X^2 is

$$\mathbb{E}(X^2) = (0^2)(5/6)^3 + (1^2)(3)(1/6)(5/6)^2 + (2^2)(3)(1/6)^2(5/6) + (3^2)(1/6)^3 = 2/3.$$

So the variance of X is:

$$\mathrm{Var}(X) = \mathbb{E}(X^2) - (\mathbb{E}(X))^2 = 2/3 - (1/2)^2 = 5/12.$$

The standard deviation is $\sigma_X = \sqrt{5/12} = 0.645497$.

Example 12.12. Suppose that you are playing a game where you roll the die three times and count up the number of 6's you get in those rolls. You have to pay \$3 to play, but you win \$4 for every 6 that comes up. So your winnings are $4X - 3$ dollars, where X is the total number of 6's that appear.

Your expected winnings are

$$\mathbb{E}(4X-3) = 4\mathbb{E}(X) - 3 = 4(1/2) - 3 = -1 \text{ dollars.}$$

The variance of your winnings is

$$\begin{aligned} \mathrm{Var}(4X-3) &= \mathbb{E}((4X-3)^2) - (\mathbb{E}(4X-3))^2 \\ &= \mathbb{E}(16X^2 - 24X + 9) - (-1)^2 \\ &= 16\mathbb{E}(X^2) - 24\mathbb{E}(X) + 9 - 1 \\ &= 16(2/3) - 24(1/2) + 8 \\ &= 32/3 - 12 + 8 \\ &= 20/3 \\ &= 6.6667 \text{ dollars}^2. \end{aligned}$$

The standard deviation of your winnings is

$$\sigma_X = \sqrt{\mathrm{Var}(4X-3)} = \sqrt{20/3} = 2.581989 \text{ dollars.}$$

It is worthwhile to note that the dealer wins $-4X+3$ during this game. So dealer's expected winnings are

$$\mathbb{E}(-4X+3) = 1 \text{ dollar},$$

and the variance of the dealer's winnings is

$$\begin{aligned}\mathrm{Var}(-4X+3) &= \mathbb{E}((-4X+3)^2) - (\mathbb{E}(-4X+3))^2 \\ &= \mathbb{E}(16X^2 - 24X + 9) - (1)^2 \\ &= 16\mathbb{E}(X^2) - 24\mathbb{E}(X) + 9 - 1 \\ &= 16(2/3) - 24(1/2) + 8 \\ &= 32/3 - 12 + 8 \\ &= 20/3 \\ &= 6.6667 \text{ dollars}^2.\end{aligned}$$

The standard deviation of the dealer's winnings is

$$\sigma_X = \sqrt{\mathrm{Var}(-4X+3)} = \sqrt{20/3} = 2.581989 \text{ dollars}.$$

Importantly, we conclude that your expected winnings are the same as the dealer's expected winnings, but with the sign reversed (i.e., -1 and 1, respectively). We also conclude that your winnings and the dealer's winnings have the same variance and standard deviation. This is true more generally. The difference in the signs will disappear as the constant terms do not affect the variance, and the signs of the coefficients are lost during the squaring process in the variance.

Later in this chapter, we discuss the general remark:

Remark 12.13. For any random variable X and any constants a and b, we have

$$\mathrm{Var}(aX+b) = a^2\,\mathrm{Var}(X).$$

This rule will assist in computations like the previous example.

Example 12.14. A student shuffles a deck of cards thoroughly (one time) and then selects cards from the deck without replacement until the ace of spades appears. What is the variance of the number of cards that the student draws until the ace of spades appears?

Let X be the number of cards needed until the ace of spades appears. Then, as we observed in Example 10.8, we have $P(X = j) = 1/52$ for each j. So the expected value of X is:

$$\begin{aligned}\mathbb{E}(X) &= (1)P(X=1) + (2)P(X=2) + (3)P(X=3) + \cdots + (52)P(X=52)\\ &= (1)(1/52) + (2)(1/52) + (3)(1/52) + \cdots + (52)(1/52)\\ &= (1+2+\cdots+52)(1/52)\\ &= 53/2\end{aligned}$$

The expected value of X^2 is

$$\begin{aligned}\mathbb{E}(X^2) &= (1^2)P(X=1) + (2^2)P(X=2) + \cdots + (52^2)P(X=52)\\ &= (1^2)(1/52) + (2^2)(1/52) + \cdots + (52^2)(1/52)\\ &= (1^2 + 2^2 + \cdots + 52^2)(1/52)\\ &= \frac{(52)(52+1)((2)(52)+1)}{6}\left(\frac{1}{52}\right)\\ &= 1855/2\end{aligned}$$

We use the fact that $1^2 + 2^2 + \cdots + n^2 = \frac{(n)(n+1)(2n+1)}{6}$. So the variance of X is

$$\mathrm{Var}(X) = 1855/2 - (53/2)^2 = 901/4 = 225.25.$$

The standard deviation of X is $\sigma_X = \sqrt{225.25} = 15.01$.

Example 12.15. A standard deck of 52 cards has 13 hearts. The cards in such a deck are shuffled, and the top five cards are dealt to a player. What is the variance of the number of hearts that the player receives?

Let X denote the number of hearts that the player receives. In Example 10.10, we computed the mass of X:

$$\begin{aligned}P(X=0) &= \frac{(39)(38)(37)(36)(35)}{(52)(51)(50)(49)(48)} = \frac{2109}{9520}\\ P(X=1) &= 5\frac{(13)(39)(38)(37)(36)}{(52)(51)(50)(49)(48)} = \frac{27417}{66640}\\ P(X=2) &= 10\frac{(13)(12)(39)(38)(37)}{(52)(51)(50)(49)(48)} = \frac{9139}{33320}\\ P(X=3) &= 10\frac{(13)(12)(11)(39)(38)}{(52)(51)(50)(49)(48)} = \frac{2717}{33320}\\ P(X=4) &= 5\frac{(13)(12)(11)(10)(39)}{(52)(51)(50)(49)(48)} = \frac{143}{13328}\\ P(X=5) &= \frac{(13)(12)(11)(10)(9)}{(52)(51)(50)(49)(48)} = \frac{33}{66640}\end{aligned}$$

So we get

$$\mathbb{E}(X) = (0)\frac{2109}{9520} + (1)\frac{27417}{66640} + (2)\frac{9139}{33320} + (3)\frac{2717}{33320} + (4)\frac{143}{13328} + (5)\frac{33}{66640},$$

or more simply, $\mathbb{E}(X) = 5/4$; and

$$\mathbb{E}(X^2) = (0^2)\frac{2109}{9520} + (1^2)\frac{27417}{66640} + (2^2)\frac{9139}{33320} + (3^2)\frac{2717}{33320} + (4^2)\frac{143}{13328} + (5^2)\frac{33}{66640},$$

so $\mathbb{E}(X^2) = 165/68$. Thus, the variance of X is

$$\mathrm{Var}(X) = 165/68 - (5/4)^2 = 235/272.$$

Example 12.16. Jim and his brother both like chocolate chip cookies best. They have a jar of cookies with 5 chocolate chip cookies, 3 oatmeal cookies, and 4 peanut butter cookies. They are each allowed to have 3 cookies. To be fair, they agree to randomly select their cookies without peeking, and they each must keep the cookies that they select. What is the variance of the number of chocolate chip cookies that Jim gets? (Notice that it does not matter whether Jim or his brother selects the cookies first—the answer will be the same, either way.)

Let X denote the number of chocolate chip cookies that Jim selects. Notice that there are $(12)(11)(10) = 1320$ equally likely outcomes for Jim. We already computed the mass of X in Example 10.11:

$$P(X = 0) = \frac{(7)(6)(5)}{(12)(11)(10)} = \frac{7}{44}$$
$$P(X = 1) = \frac{(3)(5)(7)(6)}{(12)(11)(10)} = \frac{21}{44}$$
$$P(X = 2) = \frac{(3)(7)(5)(4)}{(12)(11)(10)} = \frac{7}{22}$$
$$P(X = 3) = \frac{(5)(4)(3)}{(12)(11)(10)} = \frac{1}{22}.$$

So, as we already computed in Example 10.11, the expected value of X is

$$\mathbb{E}(X) = (0)\frac{7}{44} + (1)\frac{21}{44} + (2)\frac{7}{22} + (3)\frac{1}{22} = 5/4.$$

The expected value of X^2 is

$$\mathbb{E}(X^2) = (0^2)\frac{7}{44} + (1^2)\frac{21}{44} + (2^2)\frac{7}{22} + (3^2)\frac{1}{22} = 95/44.$$

So the variance of X is

$$\mathrm{Var}(X) = 95/44 - (5/4)^2 = 105/176.$$

12.3 Five Friendly Facts with Independence

Caution: We strongly emphasize the fact that the random variables must be independent, in order to use the formulas in this section!!

Theorem 12.17. The expected value of the product of the functions of two independent random variables equals the product of the expected values of the functions of the two random variables
If X and Y are independent random variables, and g and h are any two functions, then

$$\mathbb{E}(g(X)h(Y)) = \mathbb{E}(g(X))\mathbb{E}(h(Y)).$$

To see this, we compute as follows (the second equality holds because X and Y are independent):

$$\begin{aligned}\mathbb{E}(g(X)h(Y)) &= \sum_x \sum_y g(x)h(y)P(X = x \text{ and } Y = y)\\ &= \sum_x \sum_y g(x)h(y)P(X = x)P(Y = y)\\ &= \sum_x g(x)P(X = x)\sum_y h(y)P(Y = y)\\ &= \mathbb{E}(g(X))\mathbb{E}(h(Y))\end{aligned}$$

Using identity functions $g(X) = X$ and $h(Y) = Y$, the following nice fact follows immediately:

Corollary 12.18. The expected value of the product of two independent random variables equals the product of the expected values of the two random variables
If X and Y are independent random variables, then

$$\mathbb{E}(XY) = \mathbb{E}(X)\mathbb{E}(Y).$$

When we are working with random variables, the variance of the sum equals the sum of the variances. When we pull constants outside of the variance, they get squared.

Theorem 12.19. If $X_1, \ldots, X_n$ are independent random variables, and $a_1, \ldots, a_n$ are constants, then

$$\text{Var}(a_1X_1 + \cdots + a_nX_n) = a_1^2 \text{Var}(X_1) + \cdots + a_n^2 \text{Var}(X_n).$$

To see this, we just compute directly:

$$\begin{aligned}
\operatorname{Var}(a_1X_1+\cdots+a_nX_n) &= \operatorname{Var}\left(\sum_{i=1}^{n} a_iX_i\right) \\
&= \mathbb{E}\left(\left(\sum_{i=1}^{n} a_iX_i\right)^2\right) - \left(\mathbb{E}\left(\sum_{i=1}^{n} a_iX_i\right)\right)^2 \\
&= \mathbb{E}\left(\sum_{i=1}^{n}\sum_{j=1}^{n} a_ia_jX_iX_j\right) - \left(\sum_{i=1}^{n} a_i\mathbb{E}(X_i)\right)^2 \\
&= \sum_{i=1}^{n}\sum_{j=1}^{n} a_ia_j\mathbb{E}(X_iX_j) - \sum_{i=1}^{n}\sum_{j=1}^{n} a_ia_j\mathbb{E}(X_i)\mathbb{E}(X_j),
\end{aligned}$$

but $\mathbb{E}(X_iX_j) = \mathbb{E}(X_i)\mathbb{E}(X_j)$ for $i \neq j$, since X_i and X_j are independent, so we get

$$\begin{aligned}
\operatorname{Var}(a_1X_1+\cdots+a_nX_n) = &\sum_{i=1}^{n} a_ia_i\mathbb{E}(X_i^2) \\
&+\sum_{i=1}^{n}\sum_{j\neq i} a_ia_j\mathbb{E}(X_i)\mathbb{E}(X_j) \\
&-\sum_{i=1}^{n}\sum_{j=1}^{n} a_ia_j\mathbb{E}(X_i)\mathbb{E}(X_j).
\end{aligned}$$

Now we can cancel all terms with $i \neq j$, and we get

$$\operatorname{Var}(a_1X_1+\cdots+a_nX_n) = \sum_{i=1}^{n} a_ia_i\mathbb{E}(X_i^2) - \sum_{i=1}^{n} a_ia_i\mathbb{E}(X_i)\mathbb{E}(X_i) = \sum_{i=1}^{n} a_i^2 \operatorname{Var}(X_i).$$

Now use $n = 2$, with $a_1 = a$ and $a_2 = 1$ and $X_1 = X$ and $X_2 = b$ (a constant). Then $\operatorname{Var}(X_2) = \mathbb{E}(X_2^2) - (\mathbb{E}(X_2))^2 = \mathbb{E}(b^2) - (\mathbb{E}(b))^2 = b^2 - b^2 = 0$, so this gives us immediately:

Corollary 12.20. If X is any random variable, and if a and b are any constants, then

$$\operatorname{Var}(aX+b) = a^2 \operatorname{Var}(X).$$

Thus $\sigma_{aX+b} = \sqrt{\operatorname{Var}(aX+b)} = \sqrt{a^2 \operatorname{Var}(X)} = |a|\sqrt{\operatorname{Var}(X)}$.

In the case where $a_i = 1$ for all i, Theorem 12.19 has a simpler form, given in Corollary 12.21. This corollary is very handy. (Just be sure that the random variables are independent when using these results; otherwise, these ideas cannot be applied.)

Corollary 12.21. The variance of the sum of independent random variables equals the sum of the variances
If $X_1, \ldots, X_n$ are independent random variables, then

$$\text{Var}(X_1 + \cdots + X_n) = \text{Var}(X_1) + \cdots + \text{Var}(X_n).$$

If the X_j's are independent and have the same distribution, then

$$\text{Var}(X_1 + \cdots + X_n) = n\,\text{Var}(X_1).$$

(Remember that the independence was not needed for the analogous method of computing the expected value of a sum of random variables.)

Corollary 12.21 allows us to recompute several of the variances from this chapter in a more direct way.

Example 12.22. As in Example 12.10, flip a coin three times. Let X denote the number of heads. Find the variance of X.

Let X_j indicate whether the jth flip is a head. Then $X = X_1 + X_2 + X_3$. Also

$$\text{Var}(X_j) = \mathbb{E}(X_j^2) - (\mathbb{E}(X_j))^2 = 1/2 - (1/2)^2 = 1/4.$$

Since the X_j's are independent, then

$$\text{Var}(X) = 3\,\text{Var}(X_1) = 3(1/4) = 3/4.$$

Example 12.23. As in Example 12.11, roll three dice. Let X denote the number of 6's that appear on the dice altogether. Find the variance of X.

Let X_j indicate whether the jth roll is a 6. Then $X = X_1 + X_2 + X_3$. Also

$$\text{Var}(X_j) = \mathbb{E}(X_j^2) - (\mathbb{E}(X_j))^2 = 1/6 - (1/6)^2 = 5/36.$$

Since the X_j's are independent, then

$$\text{Var}(X) = 3\,\text{Var}(X_1) = 3(5/36) = 5/12.$$

It should be evident from these revised versions of the examples that it is often possible to compute $\text{Var}(X)$ very quickly when X can be written as a sum of independent random variables, i.e., where $X = X_1 + \cdots + X_n$ and the X_j's are independent.

12.4 Exercises

12.4.1 Practice

Exercise 12.1. Variance of an indicator. Suppose event A occurs with probability p, and X is an indicator for A, i.e., $X = 1$ if A occurs, or $X = 0$ otherwise. We already know $\mathbb{E}(X) = p$. Find $\text{Var}(X)$.

Exercise 12.2. Concert tickets. At a concert on campus, 20% of people purchase Zone A tickets, for \$47 each. The other 80% of people purchase Zone B tickets, for \$38 each. If five people are selected at random, what is the variance of the revenue from these five ticket sales?

Exercise 12.3. Japanese pan noodles. As in Exercises 10.2 and 11.2, four students order noodles at a certain local restaurant. Their orders are placed independently. Each student is known to prefer Japanese pan noodles 40% of the time. Let X be the number of students who order Japanese pan noodles. What is the variance of X?

Exercise 12.4. Dice. Roll two dice; let X denote the maximum of the two values that appear.

a. Find $\mathbb{E}(X)$.

b. Find $\mathbb{E}(X^2)$.

c. Find $\text{Var}(X)$.

Exercise 12.5. Yellow ducks. Three hundred little plastic yellow ducks are dumped in a pond; one of them contains a prize stamped on the bottom. Leonardo examines each duck until he discovers the prize. He discards each duck without a prize after he checks it, so that he never needs to check a duck more than one time. Find the variance of the number of ducks he checks until he discovers the prize.

Exercise 12.6. Dice. Susan rolls a die 4 times. Let X be the number of 1s that appear.

a. Find $\mathbb{E}(X)$.

b. Find $\mathbb{E}(X^2)$.

c. Find $\text{Var}(X)$.

Exercise 12.7. Mystery mass. Suppose that the mass of X is

$$p_X(1) = 0.12$$
$$p_X(3) = 0.37$$
$$p_X(27) = 0.42$$
$$p_X(31) = 0.09$$

a. Find $\mathbb{E}(X)$.

b. Find $\mathbb{E}(X^2)$.

c. Find $\text{Var}(X)$.

d. Find $\mathbb{E}(X^2 - 4X)$.

Exercise 12.8. Smoke alarms. Let X be the number of smoke alarms in an apartment. From government data, you find the mass of X is

$$\begin{aligned} p_X(0) &= 0.05 \\ p_X(1) &= 0.15 \\ p_X(2) &= 0.35 \\ p_X(3) &= 0.30 \\ p_X(4) &= 0.10 \\ p_X(5) &= 0.05 \end{aligned}$$

Assume each alarm requires one battery per year, and batteries cost \$2.35.

a. What is the expected cost in batteries for smoke alarms for a randomly selected apartment?

b. What is the variance of the cost?

Exercise 12.9. Waiting for favorite song. Michael plays a random song on his iPod. He has 2,781 songs, but only one favorite song. Let X be the number of songs he has to play on shuffle (songs can be played more than once) in order to hear his favorite song.

a. Find $\mathbb{E}(X)$.

b. Find $\mathbb{E}(X^2)$.

c. Find $\text{Var}(X)$.

Exercise 12.10. Spinners. Two three-partitioned spinners are spun. Each of the three parts (labeled 1, 2, and 3) have an equally likelihood of occurring. Let X be the maximum of two spins.

a. Find $\mathbb{E}(X)$.

b. Find $\mathbb{E}(X^2)$.

c. Find $\text{Var}(X)$.

Exercise 12.11. Raffle tickets. You purchase 8 raffle tickets at the county fair. Each ticket costs \$5. A ticket is worth \$100 with probability 1/400, but is worthless otherwise. What is the expected value of your purchase? Be sure to take into account the original purchase price of the tickets.

Exercise 12.12. SAT scores. Suppose that, among a certain group of students, SAT scores have a mean value of 1026 and a standard deviation of 209. Let X denote a randomly chosen student's score. What is $\mathbb{E}(X^2)$?

Exercise 12.13. Butterflies. Alice, Bob, and Charlotte are looking for butterflies. They look in three separate parts of a field, so that their probabilities of success do not affect each other.

- Alice finds 1 butterfly with probability 17%, and otherwise does not find one.
- Bob finds 1 butterfly with probability 25%, and otherwise does not find one.
- Charlotte finds 1 butterfly with probability 45%, and otherwise does not find one.

Let X be the number of butterflies that they find altogether. Find the variance of X.

Exercise 12.14. Appetizers. At a restaurant that sells appetizers:

- 8% of the appetizers cost \$1 each,
- 20% of the appetizers cost \$2 each,
- 32% of the appetizers cost \$3 each,
- 40% of the appetizers cost \$4 each.

An appetizer is chosen at random, and X is its price. Each appetizer has 7% sales tax. So $Y = 1.07X$ is the amount paid on the bill (in dollars). Find the variance of Y.

Exercise 12.15. Two 4-sided dice. Consider some 4-sided dice. Roll two of these dice. Let X denote the minimum of the two values that appear, and let Y denote the maximum of the two values that appear.

Find the variance of X.

(Caution: If X_j is an indicator of whether the minimum of the two dice is "j or greater"—as in the previous homework—then $X = X_1 + X_2 + X_3 + X_4$, but the X_j's are dependent.)

Exercise 12.16. Pick two cards. Pick two cards at random from a well-shuffled deck of 52 cards (pick them simultaneously, i.e., grab two cards at once—so they are not the same card!). There are 12 cards which are considered face cards (4 Jacks, 4 Queens, 4 Kings). There are 4 cards with the value 10. Let X be the number of face cards in your hand; let Y be the number of 10's in your hand.

a. Find the variance of X.

b. Find the variance of Y.

Exercise 12.17. Die game. In each round of a game, you earn a dollar if a die shows 1 or 2, you lose a dollar if a die shows 5 or 6, and you neither earn nor lose anything if a die shows 3 or 4. Let X_j be $+1$, -1, or 0 according to your outcome on the jth round. Let $X = X_1 + X_2 + \cdots + X_{10}$. Find the variance of X.

Exercise 12.18. Sleeping schedule. A student gets at least 8 hours of sleep 45% of the nights; the sleeping schedule is independent from night to night. Let X_1, X_2, X_3, X_4 indicate whether the student gets at least 8 hours of sleep during the next four nights respectively. Let $X = X_1 + X_2 + X_3 + X_4$. Find the variance of X.

12.4.2 Extensions

Exercise 12.19. Magic. A magician wants to do a trick where he tries to guess the value of the card that an audience member has selected (the suit of the card is not taken into account). He has six decks of cards so he gives one deck to each of six audience members. Let X denote the number of cards that he guesses correctly, when each audience member selects one card at random.

a. Find $\mathbb{E}(X)$.

b. Find $\mathbb{E}(X^2)$.

c. Find $\text{Var}(X)$.

d. The magician gets paid \$500 for performing the show, plus \$20 for each card he guesses correctly. What is the average amount of money that he gets paid?

e. As in (d), what is the variance of the amount that he gets paid?

Exercise 12.20. Cookies. As in Example 12.16, Jim and his brother both like chocolate chip cookies best. They have a jar of cookies with 5 chocolate chip cookies, 3 oatmeal cookies, and 4 peanut butter cookies. They are each allowed to have 3 cookies. To be fair, they agree to randomly select their cookies without peeking, and they each must keep the cookies that they select. Suppose that the cookies are worth \$1.20 for each chocolate chip cookie, \$0.75 for each oatmeal cookie, and \$1.30 for each peanut butter cookie. find the expected value and variance of the total value of the cookies Jim gets.

Exercise 12.21. Mystery mass. Consider a random variable X with mass

x	1	2	3	4	5
$p_X(x)$	0.1	0.25	0.5	0.1	0.05

a. Find the expected value of $1/X$.

b. Find the standard deviation of $1/X$.

Exercise 12.22. Yahtzee. In a dice game, a "Yahtzee" is a result in which all five dice within a round have the same value. To simplify this problem, assume that the five dice are just rolled one time per round. (In the actual Yahtzee game, dice can be re-rolled.) Let X be the number of times that a player gets a Yahtzee in three separate rounds. Find the variance of X

Exercise 12.23. Die and coin. Roll a die and flip a coin. Let Y be the value of the die. Let $Z = 1$ if the coin shows a head, and $Z = 0$ otherwise. Let $X = Y + Z$. Find the variance of X.

Exercise 12.24. Golden ticket. A student was at work at the county amphitheater, and was given the task of cleaning 1500 seats. To make the job more interesting, his boss hid a golden ticket somewhere in the seats. The ticket is equally likely to be in any of the seats. Let X be the number of seats cleaned until the ticket is found. Calculate the variance of X.

Exercise 12.25. Dice. Two dice are rolled. Let X be the value of the first die minus the second. Find $\text{Var}(X)$.

12.4.3 Advanced

Exercise 12.26. Mystery mass. Consider the mass

$$p_X(x) = \begin{cases} 0.15 & \text{for } x = -1, \\ 0.35 & \text{for } x = 0, \\ 0.25 & \text{for } x = 1, \\ 0.25 & \text{for } x = 2. \end{cases}$$

a. Find the average value of X.

b. Find the average value of $4X$.

c. Find the average value of X^4.

d. Find the average value of 4^X.

e. Find the variance of X.

f. Find the variance of $4X$.

g. Find the variance of X^4.

h. Find the variance of 4^X.

Exercise 12.27. Mystery mass. Consider the mass

$$p_X(x) = \begin{cases} 0.8 & \text{for } x = -4, \\ 0.1 & \text{for } x = -2, \\ 0.07 & \text{for } x = 0, \\ 0.03 & \text{for } x = 2. \end{cases}$$

a. Find $\mathbb{E}(X)$.

b. Find $\mathbb{E}(2/X)$.

c. Find $2/\mathbb{E}(X)$.

d. Find $\mathbb{E}(|X|)$.

e. Find $\text{Var}(X)$.

f. Find $\text{Var}(|X|)$.

Exercise 12.28. If X and Y have joint density $f_{X,Y}(x, y) = 8xy$ on the triangle $0 \leq y \leq x \leq 1$, find $\mathbb{E}(XY)$.

Exercise 12.29. **Coins.** Consider a pile that has 9 coins altogether. Exactly 3 of the coins are dimes (worth 10 cents each), and the other 6 coins are pennies (worth 1 cent each). Emily picks up 4 of the coins blindly (all possibilities are equally likely) without replacement. Find the expected value (in cents) of the amount she picks up.

Exercise 12.30. Specify a discrete random variable that has the property $E(X_j) = 0$ for all odd integers j, and $E(X_j) = 1$ for all even integers j.

Chapter 13

Review of Discrete Random Variables

13.1 Summary of Discrete Random Variables

What can you do with a probability mass function?

- Find individual probabilities, $P(X = x)$.
- Graph the probability mass function.
- Calculate expected value (average), variance, and standard deviation for X and functions of X.
- Calculate the cumulative distribution function by adding up each individual probability step-wise.
- Remember that it is important to check that you have a valid probability mass function first.

How do you know if your probability mass function is valid?

- The probability mass function table must include every individual value that the random variable can take.
- Each individual probability must be between 0 and 1.

$$0 \leq P(X = x) \leq 1$$

- The probabilities—summed over all the possible values of the random variable—must sum to 1. (We're 100% sure that we will have an outcome of some sort!)

$$\sum_x P(X = x) = 1$$

What can you do with a cumulative distribution function?

- Find cumulative probabilities, $F_X(x) = P(X \le x)$.
- Graph the cumulative distribution function.
- Calculate the probability mass function or individual probabilities by subtracting two consecutive steps on the cumulative distribution function.

How do you find the cumulative distribution function (CDF)?

Just add up each probability mass function step one at a time so that you can make a list of the $F_X(x) = P(X \le x)$ accumulated probabilities.

How do you get the mean/(weighted) average/expected value?

$$\mathbb{E}(X) = \sum_x x\, p_X(x)$$
$$\mathbb{E}(g(X)) = \sum_x g(x)\, p_X(x)$$

How do you get the variance and standard deviation?

$$\text{Var}(X) = \mathbb{E}(X^2) - (\mathbb{E}(X))^2$$

$$\sigma_X = \sqrt{\text{Var}(X)}$$

What do you do if you have a linear combination of *X*? (Especially good for money problems.) Here a and b are constants.

$$\mathbb{E}(aX + b) = a\mathbb{E}(X) + b$$
$$\text{Var}(aX + b) = a^2\,\text{Var}(X)$$
$$\sigma_{aX+b} = \sqrt{a^2\,\text{Var}(X)} = |a|\sqrt{\text{Var}(X)}$$

What do you do if you have a sum of random variables and need an expected value? (This works whether the variables are independent or dependent.)

$$\mathbb{E}\left(\sum_{i=1}^{n} X_i\right) = \sum_{i=1}^{n} \mathbb{E}(X_i)$$

What if all of these random variables have the same distribution? (Again, this works whether the variables are independent or dependent.)

$$\mathbb{E}\left(\sum_{i=1}^{n} X_i\right) = n\mathbb{E}(X_1)$$

What do you do if you need the variance of n independent random variables?

$$\text{Var}\left(\sum_{i=1}^{n} X_i\right) = \sum_{i=1}^{n} \text{Var}(X)$$

$$\sigma_{\sum_{i=1}^{n} X_i} = \sqrt{\sum_{i=1}^{n} \text{Var}(X)}$$

What if all of these random variables are independent and also have the same distribution?

$$\text{Var}\left(\sum_{i=1}^{n} X_i\right) = n\,\text{Var}(X)$$

$$\sigma_{\sum_{i=1}^{n} X_i} = \sqrt{n\,\text{Var}(X)}$$

13.2 Exercises

Exercise 13.1. **Horse race.** There are 5 horses running in a race, and you own the one named Rosie. Let X be the order Rosie finishes the race (1st, 2nd, etc.). Below is a table for the probability for each x-value.

x	1	2	3	4	5
P(X = x)	0.4	0.2	0.1	?	0.02

a. Graph the probability mass function.

b. What place is Rosie expected to finish?

c. What is the standard deviation of the place Rosie will finish?

d. Find the cumulative distribution function.

e. Graph the cumulative distribution function.

Exercise 13.2. **Prizes for Rosie.** Using the horse race information from Exercise 13.1, if Rosie's owner pays \$500 for entering but wins \$1000 for first place, \$750 for second place, \$500 for third place, and nothing for placing any lower,

a. What are the expected winnings (or losses) for Rosie (i.e., including the entrance fee)?

b. What is the standard deviation of the winnings (or losses) for Rosie (i.e., including the entrance fee)?

Exercise 13.3. **Password.** A password is to be created at random by selecting 4 characters (with replacement) from the set $\{A, B, \ldots, Z, 0, 1, \ldots, 9\}$. Let X be the number of letters which are chosen for the password.

a. Find the probability mass function for the number of letters which are chosen.

b. Graph the probability mass function.

c. What is the mean number of letters which are chosen?

d. What is the standard deviation of the number of letters which are chosen?

e. Find the cumulative distribution function.

f. Graph the cumulative distribution function.

Exercise 13.4. **More passwords.** Assume that each password is created according to Exercise 13.3. If passwords are independent and 10 passwords are created:

a. What is the expected total number of letters which are used?

b. What is the standard deviation of the total number of letters which are used?

c. What is the probability that at least 2 of these passwords are identical?

d. If 100 passwords are created, what is the probability that at least 2 of the passwords are identical?

Exercise 13.5. **Math team.** There are 20 students competing to be on the Math Team. Thirteen of them specialize in algebra, 4 of them specialize in geometry, and 3 of them specialize in calculus. Only 5 spots on the team are available, and team members will be chosen at random, regardless of specialty. Let X be the number of calculus specialists who make the team.

a. Find the probability mass function for the number of math team members who are calculus specialists. (Hint: Can 5 calculus specialists make the team? Why or why not?)

b. Find the cumulative distribution function for the number of math team members who are calculus specialists.

c. What is the expected number of calculus specialists who make the team?

d. What is the variance of the number of calculus specialists who make the team?

Exercise 13.6. **Grading.** Joan will grade four statistics projects, selected at random from a large stack, this evening. From past experience, Joan thinks that 30% of the projects will be "A" quality projects. Let X be the number of "A" projects she grades tonight. (Assume that all the projects are independent and that Joan is completely fair and objective with her grading of each one.)

a. Find the probability mass function.

b. Is this a valid probability mass function? Justify your answer.

c. Find the cumulative distribution function.

d. What is the probability that Joan grades at least two "A" projects tonight?

e. What is the probability that Joan grades at most two "A" projects tonight?

f. What is the probability that Joan grades exactly four "A" projects given that she grades at least two "A" projects tonight?

g. What is the probability that Joan grades at most four "A" projects given that she grades strictly more than two "A" projects tonight?

Exercise 13.7. Mystery CDF. For the following cumulative distribution function for a discrete random variable X,

$$F(x) = \begin{cases} 0 & \text{if } x < -3 \\ 0.03 & \text{if } -3 \leq x < 1 \\ 0.20 & \text{if } 1 \leq x < 2.5 \\ 0.76 & \text{if } 2.5 \leq x < 7 \\ 1 & \text{if } 7 \leq x \end{cases}$$

a. Find the probability mass function. Show why it is a valid probability mass function.

b. What is the probability that X will be more than 2?

c. Given that X is at most 2, what is the probability that it is positive?

d. Given that X is positive, what is the probability that it will be at least 2?

e. Find $\mathbb{E}(X^3)$.

Exercise 13.8. True/false for masses. Which of the following statements are true about probability mass functions? Explain why each statement is true or false.

a. $P(X = x)$ can be 1 for a particular value of x.

b. The graph of $P(X = x)$ must be non-decreasing.

c. $P(X = x)$ can be negative for a particular value of x.

d. $P(X = x)$ can be 1 for more than one particular value of x.

e. $P(X = x)$ can be zero for a particular value of x.

f. $P(X = x)$ can be zero for infinitely many values of x.

g. $P(X = x)$ can be strictly positive for infinitely many values of x.

Exercise 13.9. True/false for CDFs. Which of the following statements are true about cumulative distribution functions? Explain why each statement is true or false.

a. $F_X(x)$ can be 1 for a particular value of x.

b. The graph of $F_X(x)$ must be non-decreasing.

c. $F_X(x)$ can be negative for a particular value of x.

d. $F_X(x)$ can be 1 for more than one particular value of x.

e. The graph of $F_X(x)$ must be continuous with no jumps.

f. $F_X(x)$ can be zero for a particular value of x.

g. $F_X(x)$ can be zero for infinitely many values of x.

h. $F_X(x)$ can be strictly positive for infinitely many values of x.

Exercise 13.10. **True/false for discrete random variables.** Which of the following statements are true for discrete random variables? Explain why each statement is true or false.

a. The mean can be negative.

b. The mean can be zero.

c. The standard deviation can be negative.

d. The standard deviation can be zero.

e. The standard deviation can be greater than the mean.

f. The variance can be negative.

g. The variance can be zero.

h. The expected value of X^2 is always positive.

i. The expected value of X^2 is always smaller than the square of the mean.

Exercise 13.11. **Trick coin game.** Suppose an acquaintance has a trick coin that comes up a head 75% of the time. He wants to play a game where he tosses the coin 3 times, and for each time the coin comes up a head, he will pay you $2. He wants to charge you a fee to play the game. If you are willing to play the game only if the game is fair, how much money would you be willing to pay (overall) to play the game?

Exercise 13.12. **Hidden bones.** Chet the dog likes to dig holes in the yard. He has buried bones in some of the holes, but others he dug just for fun. He covers them back up with a little mound of dirt when he is done digging. The yard has 20 holes that have been filled up, and 8 of them contain bones. Chet decides to randomly search 4 of the holes (without replacement) in hopes of finding some of his bones. Let X be the number of bones he finds.

a. Give the probability mass function for the number of bones Chet finds.

b. Give the cumulative distribution function for the number of bones he finds.

c. What is the probability he finds at most 3 bones?

d. What is the probability he finds at least 2 bones?

e. What is the probability he finds exactly 2 bones given that he finds at most 3 bones?

Exercise 13.13. Mystery CDF. Using the cumulative distribution function for the discrete random variable X given below:

$$F_X(x) = \begin{cases} 0 & \text{if } x < 30, \\ 0.2 & \text{if } 30 \leq x < 45, \\ 0.7 & \text{if } 45 \leq x < 60, \\ 1 & \text{if } 60 \leq x. \end{cases}$$

a. What is the probability that $X = 45$?

b. What is the probability mass function for X?

c. Is your answer to part b a valid probability mass function? Justify your answer.

d. What is the probability that X is between 25 and 50?

e. What is the probability that X is 70?

f. What is the probability that X is less than 70?

g. What is the probability that X is greater than 70?

h. What is the probability that X is between 60 and 70 (inclusive)?

i. What is the expected value of X?

j. What is the standard deviation of X?

Exercise 13.14. Realtor sales, part 1. Daily sales records for a realtor's office show that the realtor will sell 0, 1, or 2 homes in a week with the probabilities listed below:

X	0	1	2
$P(X = x)$	0.75	0.20	0.05

a. Is this a valid probability mass function? Why or why not?

b. What is the expected number of homes the realtor will sell in a week?

c. What is the standard deviation in the number of homes the realtor will sell in a week?

d. What is the expected number of homes the realtor will sell in a 12-week quarter?

e. What is the standard deviation in the number of homes the realtor will sell in a 12-week quarter? (Assume the weeks are independent.)

f. Do you think that the assumption about the weeks being independent is a good assumption? Why or why not? Does independence affect your answer to part d? To part e?

Exercise 13.15. Realtor sales, part 2. Using the realtor sales information from Exercise 13.14, the realtor makes a typical commission of $4000 on each home sold, but he needs to pay an assistant $900 each week.

a. What is the realtor's expected net profit each week?

b. What is the standard deviation in the realtor's weekly net profit?

c. During a 12-week quarter, what is the realtor's expected net profit?

d. During a 12-week quarter, what is the standard deviation in the realtor's net profit?

Exercise 13.16. Basketball score. Erika, a basketball player, scores an average of 10 points per game with a standard deviation of 2.6 points. Let X denote the number of points that she scores in a game.

a. What is the expected value of X^2?

b. What is the variance of $24X - 3$?

e. What is the expected value and standard deviation of her total points over 30 games, in which her performances are independent?

Exercise 13.17. Phone calls. A person makes an average of 3.12 phone calls a day with a standard deviation of 1.77 phone calls. Let X denote the number of calls made per day.

a. What is the expected value of X^2?

b. What is the expected value of $-12X + 10$?

c. Find the variance of $-12X + 10$.

Exercise 13.18. Rental car, part 1. The number of days a driver needs a rental car after having an accident, X, is a discrete random variable with probability mass function:

$$P(X = x) = \begin{cases} \frac{8-x}{28} & \text{for } x = 1, 2, \ldots, 7, \\ 0 & \text{otherwise} \end{cases}$$

a. Make a table which shows the values for X and $P(X = x)$.

b. Graph the probability mass function for X.

c. What is the cumulative distribution function for X?

d. Graph the cumulative distribution function for X.

e. What is the expected number of days that a driver will need a rental car?

f. What is the standard deviation of days that a driver will need a rental car?

Exercise 13.19. Rental car, part 2. An insurance company pays \$50 a day for rental car costs for up to 5 days after an accident and \$20 a day after that until the owner's car has been repaired. Using this information and the probability mass function from Exercise 13.18,

a. What is the probability mass function for the insurance payment?

b. What is the expected insurance payment?

c. What is the standard deviation for the insurance payment?

Exercise 13.20. Loss on claims, part 1. An insurance company models the amount N of loss (in thousands) if a homeowner has at least one small claim. There is a constant k such that the mass of N is

$$p(N = n) = \frac{k}{2n + 1} \qquad \text{for } n = 1, 2, 3, 4$$

with no other possible choices for N. Find the value of k to make this a probability mass function.

Exercise 13.21. Loss on claims, part 2. The probability mass function you found in Exercise 13.20 was only for homeowners who had at least one small claim. The insurance company believes the probability that a homeowner will have at least one small claim is 0.08. What is the probability mass function for amount of loss for all of their policy holders?

Exercise 13.22. Loss on claims, part 3. The insurance company in Exercise 13.20 and 13.21 has a \$2000 deductible. What is the expected net premium for this policy? (A customer is responsible for any claims up to the \$2000 deductible, but a premium is the amount the insurance company pays to the policy-holder for any costs above that \$2000 limit.)

Exercise 13.23. Bed tax. A community institutes a new 5% "bed tax" on all their hotel rooms to raise money for the local park system. Previously, the average nightly price of a hotel room was \$75 with a standard deviation of \$20. What is the new average price and standard deviation?

Exercise 13.24. Tour group. A tour group is planning to pick a town to stay in when they get tired of driving. Since they don't know exactly where they will be staying, they don't know exactly how much the hotel rooms will cost. They will need 12 rooms. They know that individual hotel rooms along their route have an average nightly price of \$75 with a standard deviation of \$20. What is their expected cost and standard deviation for the 12 rooms?

Exercise 13.25. Overbooking. An airline intentionally overbooks its flight because it knows that that probability an individual passenger will not show up is 0.05 (assume passengers are independent). The plane can hold 80 people, and the airline sold 81 tickets for the flight.

a. What is the probability that there will not be enough seats for the passengers who show up to fly?

b. Do you think it is reasonable to assume the passengers are independent? Why or why not?

c. The airline wants to predict their profit from this flight. Each passenger paid $200 for his or her ticket, and if someone does not show up, he or she is refunded only $150 of the ticket purchase price. What are the expected value and the standard deviation of the airline's income from this flight?

d. Using the story from part c, what is a customer's expected cost and the standard deviation of the cost?

Exercise 13.26. **Class sizes.** A probability mass function for the class sizes at a small college is given in the table below (assume that these are the only allowed class sizes):

Class size	Probability
10	0.05
20	0.15
30	0.45
40	0.10
50	0.10
60	0.10
70	0.05

a. What is the expected value of the class size for a class chosen at random?

b. What is the standard deviation of the class size for a class chosen at random?

c. What percentage of classes are within one standard deviation of the mean class size?

d. What is the cumulative distribution function for class size?

e. What is the probability that a class has between 30 and 60 students?

f. Given that a class has at least 30 students, what is the probability that it has strictly less than 60 students?

g. Given that a class has at least 30 students, what is the probability that it has at least 60 students?

h. Given that a class has at least 30 students, what is the probability that it has strictly more than 60 students?

i. Given that a class has at least 30 students, what is the probability that it has at most 60 students?

Part III

Named Discrete Random Variables

In the chapters we have covered so far on discrete random variables, there are some common themes, and perhaps you have even begun to recognize different "types" of random variables. In the chapters that follow, we will be even more precise about the kinds of discrete random variables and what they have in common. We will describe scenarios that usually lend themselves to a particular type of random variable, and we will systematically describe formulas to help make your life easier. In Chapters 14 through 20, we'll introduce you to these different "named" distributions. This will culminate with Chapter 21, where we review some strategies for comparing and contrasting all of these distributions. This summary gives precise techniques to help you pick the correct distribution for your situation.

There are many other discrete distributions than the seven we cover here, but these are the most common ones.

By the end of this part of the book, you should be able to:

1. Distinguish between Bernoulli, Binomial, Geometric, Negative Binomial, Poisson, Hypergeometric, and Discrete Uniform distributions when reading a story.

2. Identify the variable and the parameters in a story, and state in plain English what the variable and parameters mean.

3. Use the formulas for the mass, expected value, and variance to answer questions and find probabilities.

Math skills you will need: Binomial coefficients ("choose" parentheses), factorials, summation notation ($\sum$), e^x for Chapter 18, and $\ln x$ for one particular type of problem in Chapter 14.3.

Additional resources: Computer programs (such as Excel, Maple, Mathematica, Matlab, Minitab, R, SPSS, etc.) and calculators may be used to assist in the calculations.

Note about the problems in several of the sections: We realize the stories for some of the problems in these sections are fairly similar. We want you to notice important differences in wording that set these distributions apart. The context of the situation yields clues as to which random variable is relevant to the scenario. Throughout our discussion, we repeatedly emphasize how to build experience in recognizing which random variable is the right one for you to use.

Chapter 14

Bernoulli Random Variables

A pinch of probability is worth a pound of perhaps.

—"Such a Phrase as Drifts Through Dreams," reprinted in *Lanterns & Lances* by James Thurber (Harper, 1961)

Sometimes the most simple things are some of the most useful as well. Is a randomly chosen song a rock song or not? What is the probability of winning one game? Does a cereal box have the preferred prize inside?

14.1 Introduction

In our discussion of discrete random variables, we have already seen many types of Bernoulli random variables. Some outcomes are considered a "success" and some outcomes are considered a "failure." A **Bernoulli random variable** is always 1 or 0 to indicate these respective possibilities of success or failure. A Bernoulli random variable is often called an **indicator random variable** (or simply an **indicator**), because the 1 indicates success, while the 0 indicates failure. Using the terminology of Chapter 3, a Bernoulli random variable indicates the success or failure of a trial.

The Bernoulli distribution is the simplest named distribution. It is a building block for the Binomial, Geometric, and Negative Binomial distributions. We'll use the word "success" to mean "the thing that we're looking for," regardless of whether it is good or bad. E.g., when biologists study DNA to find which markers are associated with a rare disease, a "success" could occur when they identify such a marker, even though it is bad news for the relevant patient.

"Thumbs up" for success  "Thumbs down" for failure

Success probability is p, so the failure probability is $q = 1 - p$.

The notation for a Bernoulli distribution looks like:

$$X \sim \text{Bernoulli}(p),$$

where p denotes the probability of success.

Bernoulli random variables

The common thread: One success or one failure? For instance, you ask one person a yes/no question, where "yes" is a "success" and "no" is a "failure." Other examples include: flipping a coin (heads vs tails); seeing if one part of something is defective (yes or no); classifying a pet's size (large or small); etc.

Things to look for: One trial, which is either a success or a failure.

The variable:

$$X = \begin{cases} 0 & \text{if the outcome is a failure, or} \\ 1 & \text{if the outcome is a success.} \end{cases}$$

The parameter:

$$p = \text{the probability that the outcome is a success}$$
$$q = 1 - p = \text{the probability that the outcome is a failure}$$

Mass:

$$P(X = 1) = p$$
$$P(X = 0) = q$$

Expected value formula:

$$\mathbb{E}(X) = p$$

Variance formula:

$$\text{Var}(X) = pq$$

14.2 Examples

Example 14.1. Suppose 95% of people put peanut butter on first when making a peanut butter and jelly sandwich. You select a person at random to ask whether he or she puts the peanut butter on first.

a. Why is this a Bernoulli distribution situation? What is the parameter?

Single trial with success/failure outcome, $p = 0.95$

b. What do we consider a "success" for this trial?

Finding a person who puts peanut butter on first.

c. What is the probability the person you select puts jelly on first?

$$q = 1 - p = 0.05$$

d. What is the expected number of people who put the peanut butter on first if you ask one person?

$$\mathbb{E}(X) = p = 0.95$$

e. What is the standard deviation for the number of people who put the peanut butter on first if you ask one person?

$$\text{Var}(X) = pq = (0.95)(0.05) = 0.0475$$

$$\text{standard deviation} = \sigma_X = \sqrt{pq} = \sqrt{(0.95)(0.05)} = \sqrt{0.0475} = 0.2179$$

f. What does the mass look like for this story?

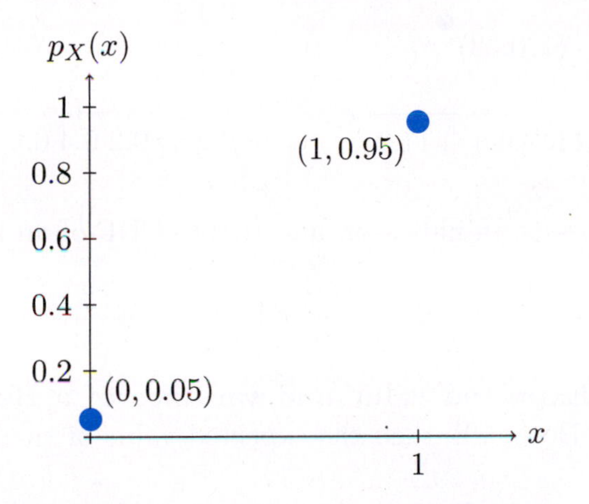

g. What does the CDF look like for this story?

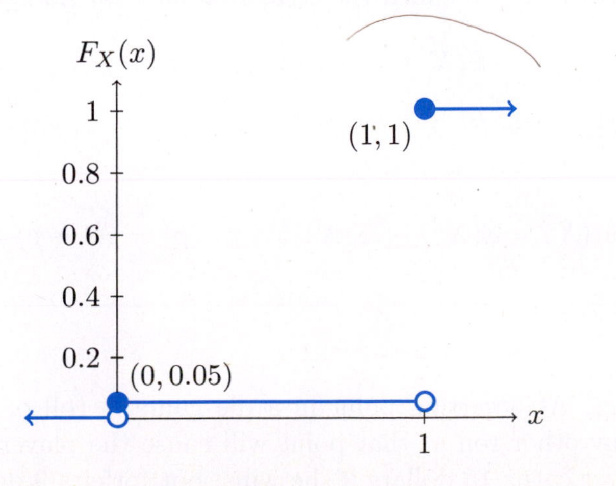

Example 14.2. Thirty-eight percent of the songs on a student's music player are rock songs. A student chooses a song at random, with all songs equally likely to be chosen. Let X indicate whether the selected song is a rock song.

a. Find the expected number and variance of X.

$$\mathbb{E}(X) = p = 0.38$$
$$\text{Var}(X) = pq = (0.38)(0.62) = 0.2356$$

b. How do the mass and CDF of X look?

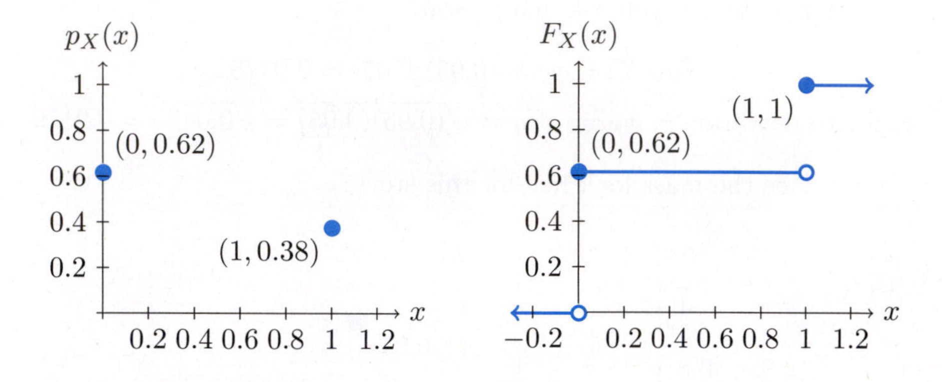

Left: Mass of an indicator, and Right: CDF of an indicator.

Remark 14.3. Expected value and variance of a Bernoulli random variable If X is Bernoulli, then the expected value of the Bernoulli random variable is

$$\mathbb{E}(X) = (0)(q) + (1)(p) = p$$

and $\mathbb{E}(X^2)$, which is also called the *second moment of* X, is

$$\mathbb{E}(X^2) = (0^2)(q) + (1^2)(p) = p.$$

Recall that $\text{Var}(X) = \mathbb{E}((X - \mathbb{E}(X))^2) = \mathbb{E}(X^2) - (\mathbb{E}(X))^2$, so the variance of X is

$$\text{Var}(X) = \mathbb{E}(X^2) - (\mathbb{E}(X))^2 = p - p^2 = p(1 - p) = pq.$$

Example 14.4. At a certain point in a die game, a roll of 5 or 6 is needed to win, and any other roll at that point will cause the player to lose. At that point, the player earns 15 dollars if she wins, but forfeits 9 dollars if she loses. What is the player's expected gain or loss at this point?

Let X be a Bernoulli random variable that indicates whether the player wins or loses. The probability of winning is 2/6, so

$$\mathbb{E}(X) = 2/6 = 1/3.$$

If $X = 1$, then the winnings are 15; if $X = 0$, then the winnings are -9. So the expected winnings are

$$\mathbb{E}(\text{winnings}) = (-9)(q) + (15)(p) = (-9)(2/3) + (15)(1/3) = -1$$

The formal way of writing this is to let $f(X)$ be the gain or loss, depending on X, so

$$\mathbb{E}(f(X)) = f(0)P(X = 0) + f(1)P(X = 1) = (-9)(2/3) + (15)(1/3) = -1$$

So the player expects to lose 1 dollar at this point in the game.

Example 14.5. A cereal company puts a Star Wars toy watch in each of its boxes as a sales promotion. Twenty percent of the cereal boxes contain a watch with Obi Wan Kenobi on it. You are a huge Obi Wan fan, so you decide to buy 1 box of the cereal in hopes that you will find an Obi Wan watch. You convince your brother and sister to each buy a box too.

a. How many Obi Wan watches does your family expect to obtain?

Let $X = 1$ if your box contains an Obi Wan watch, or $X = 0$ otherwise. Similarly, let Y be an indicator random variable that indicates whether your brother gets one; let Z be an analogous indicator for your sister.

Note: we do not even need to know whether our selections are independent to solve this problem about expected values! We will not rely on independence in our solution, because independence is not needed when taking the expected value of the sum of random variables.

The X, Y, Z are indicator variables describing (respectively) whether you, your brother, and your sister get Obi Wan watches. So the total number of watches obtained is $X + Y + Z$. So the expected number of watches obtained is

$$\mathbb{E}(X + Y + Z) = \mathbb{E}(X) + \mathbb{E}(Y) + \mathbb{E}(Z) = 0.2 + 0.2 + 0.2 = 0.6.$$

b. What is the variance of the number of Obi Wan watches that your family obtains?

In this situation, we do need to know about how X, Y, Z depend on each other. If the number of cereal boxes is sufficiently large that the values of X, Y, Z do not affect each other, we might be able to assume that X, Y, Z are independent. (We will consider such situations—with independence—much more in the next chapter. To handle situations where the X, Y, Z are dependent, we will have to wait until Chapter 19.) If X, Y, Z are independent, then we can add their variances:

$$\begin{aligned}\text{Var}(X+Y+Z) &= \text{Var}(X)+\text{Var}(Y)+\text{Var}(Z)\\ &= (0.2)(0.8)+(0.2)(0.8)+(0.2)(0.8)\\ &= (3)(0.16)\\ &= 0.48.\end{aligned}$$

14.3 Exercises

14.3.1 Practice

Exercise 14.1. Call from home. One out of every eight calls to your house is from a family member. You will record whether the next call is from a family member.

a. What do you consider a "success" in this story? What is its probability?

b. What do you consider a "failure" in this story? What is its probability?

c. Why is this a Bernoulli situation? What is the parameter?

d. Define in words what X is in terms of this story. What values can X take?

e. What is the probability that the next time the phone rings, it will be from a family member?

f. If the phone calls are independent, what is the probability the 3rd call today will be from a family member?

g. What is the mean of X?

h. What is the standard deviation of X?

Exercise 14.2. Dice. You roll a fair, six-sided die as part of a game. If you roll a 5, you will win the game.

a. What do you consider a "success" in this story? What is its probability?

b. What do you consider a "failure" in this story? What is its probability?

c. Why is this a Bernoulli situation? What is the parameter?

d. Define in words what X is in terms of this story. What values can X take?

e. Your friend will pay you $4 if you win the game. You owe your friend $1 if you lose the game. What are your expected winnings?

f. What is the variance of your expected winnings?

Exercise 14.3. Homework. Hui has a class of 300 students, and only 6 have done their homework assignment due today. He calls on a student at random to put a problem on the board to check whether he or she has done the assignment.

a. What does Hui consider a "success" in this story? What is the probability?

b. What does Hui consider a "failure" in this story? What is the probability?

c. Why is this a Bernoulli situation? What is the parameter?

d. Define X in terms of this story. What values can X take?

e. What is the probability that the student Hui selects is one who has done the assignment?

f. If the students are independent, what is the probability that the 3rd student Hui checks will have done the assignment?

Exercise 14.4. Blu-rays. Suppose that 1% of Blu-ray discs produced by a company are defective. You buy one of these discs and check to see if it is defective.

a. What do you consider a "success" in this story? What is the probability?

b. What do you consider a "failure" in this story? What is the probability?

c. Why is this a Bernoulli situation? What is the parameter?

d. Define X in terms of this story. What values can X take?

e. Draw the labeled graph of the mass for this story.

f. Draw the labeled graph of the CDF for this story.

Exercise 14.5. Shoes. Anne and Jane have shoes spread throughout the dorm room. Anne has 15 pairs of shoes; twenty percent of her shoe collection consists of sandals. Jane has 40 pairs of shoes; ten percent of her shoe collection consists of sandals.

a. A shoe is picked at random from the dorm room belonging to Anne and Jane; what is the probability that it is a sandal?

b. If a randomly chosen shoe is chosen from the room (with all shoes equally likely to be chosen), what is the probability that it belongs to Anne?

c. If a randomly chosen shoe is chosen from the room (with all shoes equally likely to be chosen), and upon examination this shoe is seen to be a sandal, what is the probability that it belongs to Anne?

Exercise 14.6. Movie date. Chris and Juanita always go to the movies on Friday night. Before meeting for their date, they each make a decision (independently, without consulting) of what genre of movie they prefer to see. Chris prefers an adventure movie with probability 70% and a romance with probability 30%; Juanita chooses adventure with probability 34% and a romance with probability 66%.

a. What is the probability that their choices agree?

b. What is the probability that they both choose an adventure movie?

Exercise 14.7. Studying. Let X be the number of nights that you spend studying in a 30-day month. Assume that you study, on a given night, with probability 0.65, independent of the other nights. Write X as the sum of thirty indicators (i.e., as the sum of 30 Bernoulli random variables).

a. Find $\mathbb{E}(X)$.

b. Find $\text{Var}(X)$.

14.3.2 Extensions

Exercise 14.8. Trucks and cars. On a certain highway, 7% of the vehicles have 18 wheels, and the other 93% of the vehicles have 4 wheels. (We ignore motorcycles, etc., for simplicity.) A child looks out the window and counts the wheels on the next vehicle to pass.

a. What is the expected number of wheels?

b. What is the variance of the number of wheels?

c. Show a labeled graph of the mass for this story.

d. Show a labeled graph of the CDF of this story.

Exercise 14.9. Japanese pan noodles. As in Exercises 3.2, 10.2, 11.2, and 12.3, at a certain local restaurant, students are known to prefer Japanese pan noodles 40% of the time (it is a very popular and tasty dish!).

a. Let X be an indicator for whether a randomly selected student orders Japanese pan noodles. Find $\mathbb{E}(X)$ and $\text{Var}(X)$.

b. The student is joined by his girlfriend; let Y indicate whether she buys Japanese pan noodles. Find $\mathbb{E}(Y)$.

c. Japanese pan noodles cost \$7, and all of the other noodle dishes at the restaurant cost \$6. What is the expected value of the bill for the couple?

Exercise 14.10. Winning and losing. Suppose that a person wins a game of chance with probability 0.40, and loses otherwise. If he wins, he earns 5 dollars, and if he loses, then he loses 4 dollars.

a. What is his expected gain or loss?

b. What is the variance of his gain or loss?

c. Find constants a, b such that if $X = 0$ when he loses and $X = 1$ when he wins, then $Y = aX + b$ is his earnings. (Hint: Solve $5 = a(1) + b$ and $-4 = a(0) + b$. Also: X is a Bernoulli.) Verify your results above by finding $\mathbb{E}(Y)$ and $\text{Var}(Y)$ with this method.

14.3.3 Advanced

Exercise 14.11. Reciprocal of a random variable. If X is a Bernoulli random variable, is $1/X$ a well-defined random variable? If not, why? If so, what is the mass?

Chapter 15

Binomial Random Variables

Not everything that can be counted counts, and not everything that counts can be counted.

—*Informal Sociology: A Casual Introduction to Sociological Thinking* by William Bruce Cameron (Random House, 1963)

You are playing a series of one-on-one basketball games with your friend. You plan on playing 5 games total, and your friend (who is a better player and wins about 68% of the time you play her), says that she will buy you one lunch for each game that you win. You will have to buy her lunch for each game she wins. What is the probability you will have to buy your friend lunch 5 days next week? More than half the school days next week? Only once?

15.1 Introduction

Binomial random variables are more general than Bernoullis. A Binomial is the number of successes in n independent trials. Equivalently, a Binomial is the sum of n independent Bernoulli random variables with the same probability of success p:

A Binomial is the sum of n independent Bernoullis.

$$\text{Binomial random variable}(n, p) = \text{sum of } n \text{ independent}$$
$$\text{Bernoulli}(p) \text{ random variables}$$
$$X = X_1 + \cdots + X_n$$

(In particular, if $n = 1$, a Binomial(n, p) is just a Bernoulli(p).)

For Binomial random variables, with parameters n and p, we must know in advance that there are n independent trials; the number of trials we conduct is not affected by how many are successes. We specify, in advance, how many trials will take place, regardless of how many succeed or fail.

For instance, if $n = 8$, we can conduct 8 independent trials and let $X_1, \ldots, X_8$ be the 8 Bernoulli random variables that show which of the 8 trials succeed or fail. If $X = X_1 + \cdots + X_8$, then X is a Binomial random variable that gives the total number of successes on these 8 trials. As an example:

"Thumbs up" for success; trials 1, 2, 5, 7 succeed, so $X_1 = 1$, $X_2 = 1$, $X_5 = 1$, and $X_7 = 1$; the other X_j's are 0, and

$$X = X_1 + \cdots + X_8 = 1 + 1 + 0 + 0 + 1 + 0 + 1 + 0 = 4$$

is a Binomial that gives the total number of successes.

The notation for the Binomial looks like:

$$X \sim \text{Binomial}(n, p),$$

where n is the total number of trials and p is the probability of success on each trial. We also need the Binomial notation:

Remark 15.1. Binomial coefficients
If there are n trials, the number of ways to have j successes (for $0 \leq j \leq n$) is:

$$\binom{n}{j} = \frac{(n)(n-1)(n-2)\cdots(n-j+1)}{j!} = \frac{n!}{j!(n-j)!}$$

where $n! = (n)(n-1)\cdots(1)$; $j! = (j)(j-1)\cdots(1)$; and $(n-j)! = (n-j)\cdots(1)$.

Note that $0!$ is defined to be equal to 1.

Also note that, for $j < 0$ or $j > n$, we define $\binom{n}{j} = 0$.

E.g., when $n = 8$, the number of ways to arrange j successes within the trials are:

$$\binom{8}{0} = \frac{8!}{0!8!} = 1 \qquad \binom{8}{1} = \frac{8!}{1!7!} = 8 \qquad \binom{8}{2} = \frac{8!}{2!6!} = 28$$

$$\binom{8}{3} = \frac{8!}{3!5!} = 56 \qquad \binom{8}{4} = \frac{8!}{4!4!} = 70 \qquad \binom{8}{5} = \frac{8!}{5!3!} = 56$$

$$\binom{8}{6} = \frac{8!}{6!2!} = 28 \qquad \binom{8}{7} = \frac{8!}{7!1!} = 8 \qquad \binom{8}{8} = \frac{8!}{8!0!} = 1$$

Binomial random variables

The common thread: Number of successes in a fixed number of independent trials with the same probability of success on each trial. For instance, you ask 10 randomly selected people one yes/no question each. Other examples include: number of 5's during 100 die rolls; number of defective items in a randomly sampled collection of 12 coming off an assembly line; number of seniors in a survey of 8500 students; etc.

Things to look for: Fixed number of independent trials; same success probabilities.

The variable:

$$X = \# \text{ of successes in the } n \text{ trials, so } 0 \leq X \leq n$$

The parameters:

$$\begin{aligned} n &= \text{the total number of trials (if } n = 1\text{, then the Binomial} \\ &\qquad \text{is just a Bernoulli)} \\ p &= \text{the probability that a given trial is a success} \\ q = 1 - p &= \text{the probability that a given trial is a failure} \end{aligned}$$

These values p and q must be the same for every trial.

Mass:

$$P(X = x) = \binom{n}{x} p^x q^{n-x}, \qquad x = 0, 1, \ldots, n$$

Expected value formula:

$$\mathbb{E}(X) = np$$

Variance formula:

$$\text{Var}(X) = npq$$

Where does the mass formula come from?

$$\begin{aligned} \binom{n}{x} &= \text{number of ways to arrange } x \text{ successes in } n \text{ trials} \\ p^x &= (\text{probability of a success on one trial})^{\# \text{ of successes}} \\ q^{n-x} &= (\text{probability of a failure on one trial})^{\# \text{ of failures}} \end{aligned}$$

We have seen examples of Binomial random variables several times already. See, for instance, Example 14.5b, in which X, Y, Z are *independent Bernoullis*, each with probability of success 0.2, so $X + Y + Z$ is Binomial with parameters $n = 3$ and $p = 0.2$. Binomial random variables are especially useful in problems that have sampling with replacement (the replacement assures us that the relevant probability of success does not change from trial to trial).

15.2 Examples

Example 15.2. In a certain city, we plan to predict each year—during a 31 year period—whether there is snow on New Year's Day. There is historically a 30% chance of snow that day. We assume the behavior on the New Year's Days during this 31 year period are independent from each other. We are interested in counting up the number of snowy New Year's Days during the 31 year period.

a. How many trials are there?

We plan to have trials on 31 independent days, so $n = 31$.

b. What does X represent in terms of this story? What values can it take?

Here, X is the number of snowy New Year's Days over a period of 31 years, so X can be any of the values $0, 1, \ldots, 31$.

c. Why is this a Binomial distribution problem?

We have a specified number (31) of independent trials, with equal probability of success (0.3) on each trial, and we are counting up the number of successes (# of snowy days).

d. What is the probability of exactly 5 snowy New Year's Days?

$$P(X = 5) = \binom{31}{5}(0.3)^5(0.7)^{26} = 0.03876$$

e. What is the probability of at least 1 snowy New Year's Day?

$$\begin{aligned}
P(X \geq 1) &= 1 - P(X = 0) \\
&= 1 - \binom{31}{0}(0.3)^0(0.7)^{31} \\
&= 1 - 0.000016 \\
&= 0.999984
\end{aligned}$$

Remark 15.3. In general, when computing the probability of "at least one" of something, it is easier to compute the probability of the *complement*, i.e., the probability that there are *zero occurrences* of the relevant phenomenon.

For example, in Example 15.2e, it is much easier to use the complement of the event than to add the probabilities corresponding to $1, 2, \ldots, 31$ snowy days.

f. How many snowy New Year's Days do we expect altogether?

$$\mathbb{E}(X) = 31(0.3) = 9.3 \text{ days}$$

g. What is the standard deviation in the number of snowy days? (Recall that the standard deviation measures the spread of the distribution of a random variable. A small standard deviation tells us that a distribution is well-concentrated around the expected value. A larger standard deviation indicates that a distribution is more spread out from the expected value.)

$$\text{Var}(X) = 31(0.3)(0.7) = 6.51$$
$$\sigma_X = \sqrt{6.51} = 2.5515 \text{ days}$$

h. What does the mass look like for this story?

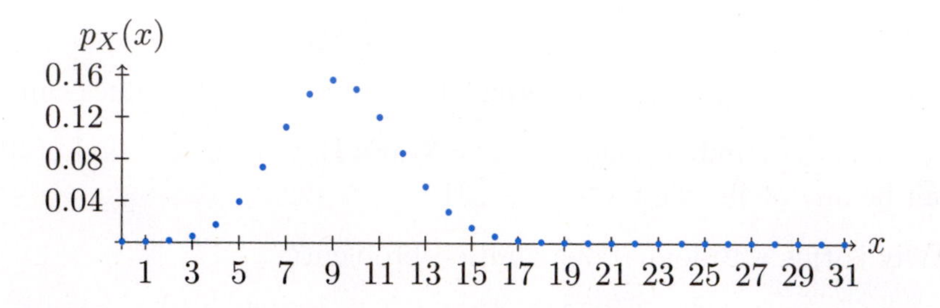

i. What does the CDF look like for this story?

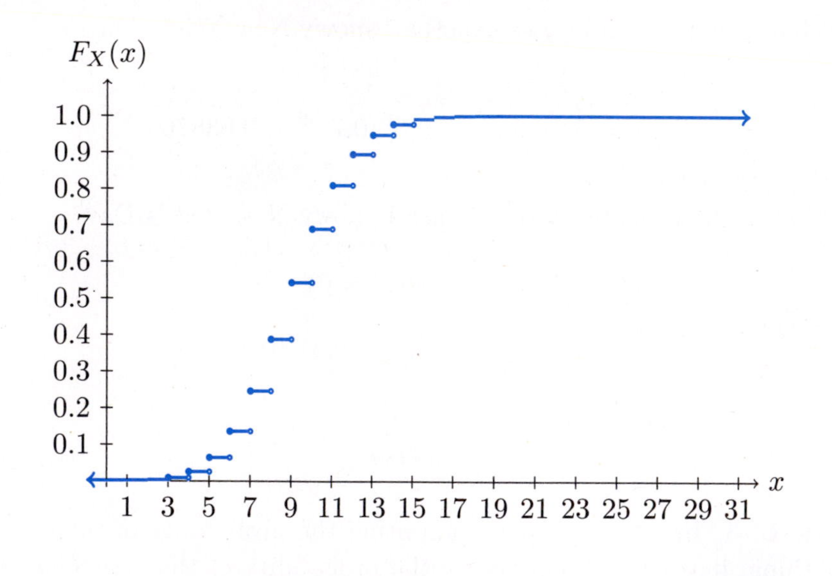

Example 15.4. Suppose a couple decides to have 4 children. Assume the probability is the same for having a boy or a girl, and they will not have any multiple births (so sexes are independent). Let X denote the number of girls that the couple has.

a. How do the mass and CDF of X look?

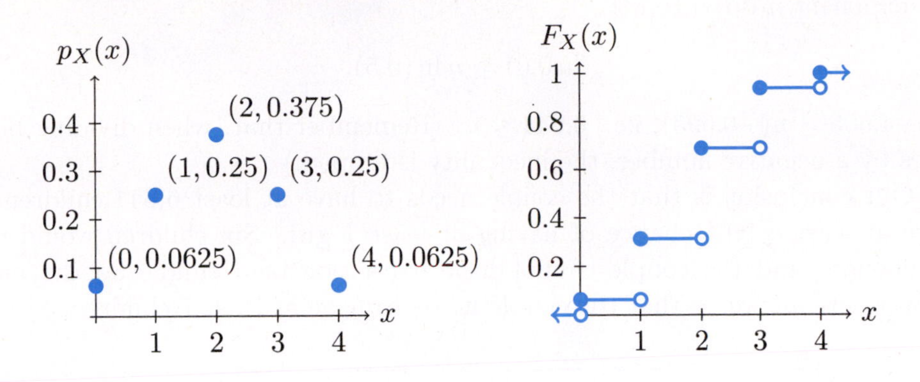

Left: Mass of the number of girls. Right: CDF of the number of girls.

b. What is the probability the couple has at least one girl?

$$\begin{aligned} P(X \geq 1) &= 1 - P(X = 0) \\ &= 1 - \binom{4}{0}(0.5)^0(0.5)^4 \\ &= 1 - 0.0625 \\ &= 0.9375 \end{aligned}$$

We used the complement, because computing $1 - P(X = 0)$ is easier than computing $P(X = 1) + P(X = 2) + P(X = 3) + P(X = 4)$. This concept, of computing the complement, is helpful in many similar situations.

c. Planning in advance of having any children, how many children would the couple have to have for there to be at least a 99% chance that at least one of their children is a girl? (Hint: This is a "backward" Binomial problem, in the sense that we know the desired probability, and the value of p is known to be 0.5, but we do not know the value of n.)

We need

$$P(X \geq 1) \geq 0.99.$$

Rewriting the left-hand side using the complement,

$$1 - P(X = 0) \geq 0.99,$$

or equivalently,

$$0.01 \geq P(X = 0). \tag{15.1}$$

We have

$$P(X = 0) = \binom{n}{0}(0.5)^0(0.5)^n = (0.5)^n,$$

since for any positive integer k, $\binom{k}{0} = 1$ and $k^0 = 1$. Substituting back into (15.1) gives

$$0.01 \geq (0.5)^n.$$

Taking a natural log of each side gives $\ln 0.01 \geq \ln{(0.5)^n}$, and then we can pull the exponent n down to get

$$\ln 0.01 \geq n \ln{(0.5)}.$$

So $-4.605 \geq n(-0.693)$, i.e., $6.644 \leq n$. (Remember that, when dividing both sides by a negative number, the inequality is flipped.)

Our conclusion is that the couple needs to have at least 6.644 children to have at least a 99% chance of having at least 1 girl. Six children would not be enough, and the couple cannot have a fraction of a child. So the most appropriate answer is that the couple needs to have at least 7 children.

Remark 15.5. Expected value and variance of a Binomial random variable If $X_1, \ldots, X_n$ are independent Bernoulli random variables, each with probability of success p, then

$$X = X_1 + \cdots + X_n$$

is a Binomial(n, p) random variable. So the expected value of X can be computed by using the linearity of the expected value:

$$\begin{aligned} \mathbb{E}(X) &= \mathbb{E}(X_1 + \cdots + X_n) \\ &= \mathbb{E}(X_1) + \cdots + \mathbb{E}(X_n) \\ &= p + \cdots + p \\ &= np. \end{aligned}$$

Since the X_j's are independent, then we can also add their variances to obtain the variance of X:

$$\begin{aligned} \mathrm{Var}(X) &= \mathrm{Var}(X_1 + \cdots + X_n) \\ &= \mathrm{Var}(X_1) + \cdots + \mathrm{Var}(X_n) \\ &= pq + \cdots + pq \\ &= npq. \end{aligned}$$

Remark 15.6. We stress that for the Binomial distribution (and also for the Geometric and Negative Binomial distributions, which we will see later) it is important to have independence, or at least relative independence. So what happens if you have a small population and you are sampling without replacement, i.e., the probabilities change between trials? We'll cover that in Chapter 19, with the Hypergeometric distribution.

Example 15.7. Roll a die n times. Let X denote the total number of 4's and 5's that appear.

Here X is a Binomial random variable, with each outcome of "4" or "5" treated as a "success." So $p = 2/6 = 1/3$, because 2 out of the 6 equally likely outcomes are successes. The mass of X is

$$\begin{aligned} p_X(x) &= \binom{n}{j}\left(\frac{1}{3}\right)^j\left(1-\frac{1}{3}\right)^{n-j} \\ &= \binom{n}{j}\left(\frac{1}{3}\right)^j\left(\frac{2}{3}\right)^{n-j} \\ &= \binom{n}{j}\frac{2^{n-j}}{3^n} \end{aligned}$$

for $j = 0, 1, 2, \ldots, n$, and otherwise $p_X(x) = 0$.

The expected value of X is

$$\mathbb{E}(X) = np = n/3,$$

and the variance of X is

$$\mathrm{Var}(X) = npq = (n)(1/3)(2/3) = 2n/9.$$

Example 15.8. (Refer to Example 14.1) Suppose 95% of people put peanut butter on first when making a peanut butter and jelly sandwich. Five people are independently asked about their sandwich-making habits. What is the probability that the majority of them put the peanut butter on first?

Let X be the number of people who put the peanut butter on first, so X is a Binomial random variable with mass

$$\begin{aligned} P(X = j) &= \binom{5}{j}(0.95)^j(1-0.95)^{5-j} \\ &= \binom{5}{j}(19/20)^j(1/20)^{5-j} \\ &= \binom{5}{j}\frac{19^j}{20^5}, \end{aligned}$$

for $j = 0, 1, 2, 3, 4, 5$, and $P(X = j) = 0$ otherwise. The majority of people put peanut butter on first if $X = 3$ or $X = 4$ or $X = 5$. So the desired probability is

$$\begin{aligned} P(X = 3 \text{ or } X = 4 \text{ or } X = 5) &= P(X = 3) + P(X = 4) + P(X = 5) \\ &= \binom{5}{3}\frac{19^3}{20^5} + \binom{5}{4}\frac{19^4}{20^5} + \binom{5}{5}\frac{19^5}{20^5} \\ &= 0.0214 + 0.2036 + 0.7738 \\ &= 0.9988 \end{aligned}$$

Example 15.9. Dominique, Raymond, and Samantha are independently using an app that randomly creates playlists for them. Forty percent of the songs this app has access to are rap songs. Dominique's playlist will have 30 songs, Raymond's 70, and Samantha's 90. What is the distribution for the total number of rap songs if the three people combine their playlists later? (Assume that there are no overlaps among their songs.)

The number of rap songs on their playlists are Binomial random variables X_1, X_2, X_3, each with $p = 0.40$, and with respective number of trials $n_1 = 30$, $n_2 = 70$, and $n_3 = 90$. So the total number of rap songs on their playlists is $Y = X_1 + X_2 + X_3$, which is also Binomial, with probability of success $p = 0.40$ on each trial, and $N = 30 + 70 + 90 = 190$ trials altogether.

Remark 15.10. Sums of Binomial random variables Consider independent Binomial random variables $X_1, X_2, \ldots, X_m$, for which the probabilities of success "p" are all the same, and for which the number of trials are (respectively) $n_1, n_2, \ldots, n_m$. Then $X_1 + X_2 + \cdots + X_m$ is the number of successes that occur in $n_1 + n_2 + \cdots + n_m$ independent trials. So the sum $Y = X_1 + X_2 + \cdots + X_m$ is also Binomial, with parameters $N = n_1 + n_2 + \cdots + n_m$ and p. So the sum Y has mass

$$p_Y(j) = \binom{N}{j} p^j q^{N-j}, \qquad \text{for integers } 0 \le j \le N$$

and $p_Y(j) = 0$ otherwise.

15.3 Exercises

15.3.1 Practice

Exercise 15.1. Skittles. Skittles candies come in the colors red, orange, yellow, green, and purple, with each color having equal probability. You are a quality control inspector, and your job is to count up the number of purple candies in a random sample of 25 candies from a large population of candies coming down the factory line.

a. What is a "success" in this situation? What is the probability of a success on a single trial?

b. What is a "failure" in this situation? What is the probability of a failure on a single trial?

c. Explain in words what X is in terms of the story. What values can it take?

d. Why is this a Binomial distribution situation? What are the parameters?

e. What is the probability that you will find exactly 5 purple candies in your sample?

f. What is the probability that you will find at least 2 purple candies in your sample?

g. What is the expected number of purple candies in the sample?

h. What is the standard deviation in the number of purple candies in the sample?

i. If purple candies cost the company 1.5 cents each, with an additional general production fee of 25 cents (not related to the number of purple candies), what is the expected cost of purple candies for this sample?

j. What is the standard deviation of the cost of purple candies in this sample?

Exercise 15.2. Colorblind. Approximately 8.33% of men are colorblind. You survey 8 men from the population of a large city and count the number who are colorblind.

a. What is a "success" in this situation? What is the probability of a success on a single trial?

b. What is a "failure" in this situation? What is the probability of a failure on a single trial?

c. Explain in words what X is in this situation and what values it can take.

d. Why is this a Binomial distribution situation? What are the parameters?

e. What is the probability exactly 2 will be colorblind?

f. What is the probability that 6 or fewer will be colorblind?

g. How many men would you have to survey in order to be at least 95% sure you would find at least 1 who is colorblind?

h. What is the expected value of X?

i. What is the variance of X?

j. Show the labeled graph of the mass for this story.

k. Show the labeled graph of the CDF for this story.

Exercise 15.3. Surgical delivery. Twenty percent of babies in a particular city are born by a surgical procedure called a Cesarean section (C-section). You randomly survey 9 parents of babies from the population of the large city and count the number of babies who are not born by C-section.

a. Explain in words what X is in this situation and what values it can take.

b. Why is this a Binomial distribution situation? What are the parameters?

c. What is the probability exactly 6 were not born by C-section?

d. What is the probability that between 4 and 6 babies were not born by C-section?

e. How many babies would you have to survey in order to be at least 90% sure you would find at least 1 who is not born by C-section?

f. What is the expected value of X?

g. What is the variance of X?

h. Show the labeled graph of the mass for this story.

i. Show the labeled graph of the CDF for this story.

Exercise 15.4. **Lost dog.** Gracie is looking for her lost dog. She will randomly ask people in her neighborhood if they have seen her dog. Assume that each neighbor is equally likely to have seen her dog and that the neighbors are independent. Is this a Binomial situation? Why or why not?

Exercise 15.5. **Cereal boxes.** A cereal company puts a Star Wars toy watch in each of its boxes as a sales promotion. Twenty percent of the cereal boxes contain a watch with Obi Wan Kenobi on it. You are a huge Obi Wan fan, so you really want one of these watches.

a. You decide to buy 100 boxes of cereal from the warehouse store to be on the safe side. What is the probability you don't find any Obi Wan watches?

b. What is the expected number of Obi Wan watches you will find in the 100 boxes?

c. If each box of cereal costs $3.50, and you believe the value of an Obi Wan watch will be approximately $50 in a few years, are you doing a smart thing by purchasing 100 boxes? (Think about the expected value for the cost and the profit.)

d. How many boxes of cereal would you need to buy to be at least 95% sure of finding at least 1 Obi Wan watch?

Exercise 15.6. **Mobile phone survey.** A recent survey states that 48% of mobile devices are iPhones. In order to learn more about how the iPhone works, a student starts asking random people on campus if they use an iPhone. Assume that the individuals on campus are independent.

a. If he surveys 20 people, what is the probability he finds exactly 2 people who use an iPhone?

b. If he surveys 20 people, what is the probability he finds at least 2 people who use an iPhone?

c. If he surveys 20 people, what is the probability he finds fewer than 2 people who use an iPhone?

d. How many students would he have to ask to be at least 90% certain of finding at least one person who uses an iPhone?

Exercise 15.7. **Spinner.** A spinner in a certain game lands on "0" 70% of the time and lands on "1" 30% of the time. How many times must a player spin so that the probability of having at least one result of "1" exceeds 95%?

Exercise 15.8. **Reshelving in library.** There are 7 books needing reshelving in the undergraduate library. Sixty-five percent of the library's collection consists of reference books. Let X be the number of reference books a student page reshelves out of the 7 on her cart.

a. What is the probability that all 7 of them are reference books?

b. What is the probability that the majority of these 7 books are reference books?

c. What is the expected number of reference books in this collection of 7 books?

d. What is the standard deviation of the number of reference books in this collection?

e. Show the labeled graph of the mass for this story.

f. Show the labeled graph of the CDF for this story.

Exercise 15.9. **Dice.** You have a 4-sided die with colors red, blue, yellow, and green for each face. You will roll the die 10 times. Let X be the number of rolls on which blue is the color selected.

a. What is the probability blue is selected exactly 3 times?

b. What is the probability blue is selected at most 3 times?

c. Given that blue is selected at most 3 times, what is the probability blue is selected at least 1 time?

d. Given that blue is selected at least 1 time, what is the probability blue is selected at most 3 times?

Exercise 15.10. **Blood type.** Suppose that 3% of people have AB+ blood type. How many people need to be sampled so that there is more than a 50% chance that at least one person in the group has AB+ blood type?

Exercise 15.11. **Field goals in football.** Jeff typically makes 80% of his field goals. Steve typically makes 60% of his field goals. Suppose they both have the opportunity to kick 3 field goals.

a. What is the probability Jeff will succeed in making at least 1 field goal?

b. What is the probability Steve will succeed in making at least 1 field goal?

c. What is the probability that both Steve and Jeff will succeed in making at least 1 field goal?

d. What is the probability that either Steve or Jeff will succeed in making at least 1 field goal?

e. Given that only one field goal total was scored, what is the probability that Jeff was the one who kicked it?

Exercise 15.12. Tennis. Suppose a tennis player hits an ace once out of every five serves. Suppose in a match the player performs 80 serves.

a. What is the expected number of aces?

b. What are the variance and standard deviation of the number of aces?

Exercise 15.13. Exam. On a multiple choice exam, a student decides to test his luck. His exam has 20 questions, each of which has 5 answer choices. The student decides to roll a die on each question and use the result on the die as his answer; any time that he rolls a 6, he just discards that roll and tries again. Let X be the number of questions he gets right on the exam altogether. What is the probability he gets a grade of A (overall score at least 90%) using this method?

Exercise 15.14. Dice. Two students decide to make bets about their plans for lunch during the next work week (Monday through Friday). They roll a six-sided die five times (once for each day). The agreement is that, for each day, when a 6 shows up, the first student has to pay for both lunches, and if a 1 shows up, the second student has to pay for both lunches. If neither of these events occurs, they will bring their lunch from home on that day.

a. How many days does the first student expect to buy lunch for the second student?

b. How many days do they expect to bring their lunches?

c. What is the mass of the number of days on which the first student buys lunch for the second student?

d. What is the mass of the number of days on which they bring their lunches?

Exercise 15.15. Winning and losing. Suppose that a person wins a game of chance with probability 0.40, and loses otherwise. If he wins, he earns 5 dollars, and if he loses, then he loses 4 dollars. Assume that he plays ten games independently. Let X denote the number of games that he wins. (Hint: His gain or loss is $5X + (-4)(10 - X) = 9X - 40$, since he loses $10 - X$ games.)

a. What is his expected gain or loss (altogether) during the ten games?

b. What is the variance of his gain or loss (altogether) during the ten games?

c. What is the probability that he wins \$32 or more during the ten games?

Exercise 15.16. Telemarketers. Assume that when your phone rings, the caller is a telemarketer with probability $1/8$, and that the probability of a telemarketer is independent from call to call. Let X denote the number of telemarketers during the next three calls.

a. What is the mass of X?

b. Draw a graph of the mass of X.

c. Draw a graph of the CDF of X.

Exercise 15.17. Hiking. You randomly text 20 of your friends to see who wants to go on a hiking trip. You think that they all respond to your requests independently of each other, and you estimate that each one is 7% likely to be available, interested in going hiking, and will actually text you back to accept the invitation.

a. Find the probability that at least 3 people would accept the invitation.

b. Find the expected number of people who would accept the invitation.

c. Find the variance of the number of people who would accept the invitation.

Exercise 15.18. Dining hall. Let X, Y, Z be (respectively) the number of nights that Alice, Bob, and Charlotte eat in the dining hall during a 7-day week. Assume that X, Y, Z are independent Binomial random variables that each have $n = 7$ and $p = 0.65$.

a. What is the distribution of $X + Y + Z$, i.e., the total number of meals eaten by these three people (altogether) during a week?

b. Find $\text{Var}(X + Y + Z)$.

Exercise 15.19. Hearts. You draw seven cards, *without replacement*, from a shuffled, standard deck of 52 playing cards. Let X be the number of hearts that are selected.

a. What is the expected number of hearts? Why?

b. Is X a Binomial random variable? Why or why not?

Exercise 15.20. Reserving a room. There is a big exam tonight, and all of the 400 students are invited to attend the help session. From past experience, the instructor finds that 60% of the students are likely to attend the help session. She wants to reserve an appropriately sized room.

a. What is the expected value and variance for the number of students who will attend?

b. How many seats must be in the reserved room, so that the professor can be at least 90% sure that everyone who wants to attend will have a seat? (A computer is necessary to solve this problem numerically.)

Exercise 15.21. Applying to schools. This year 557 students applied to school A, and only 40% of the students are typically accepted. At the same time, 903 students applied to school B, and 27% of these students are typically accepted. Let X_A and X_B be the number of students accepted to schools A and B, respectively. Which school do we expect to have more acceptances?

15.3.2 Extensions

Exercise 15.22. Saving energy. Despite lecturing your roommates on energy conservation, there is a 60% chance that the lights in a dorm room will be left on when nobody is home. Each day is independent. Suppose that, every day the light is left on in a dorm room, there are 1000 Watts of power used. Every day when the light is turned off, there are 200 Watts of power used. You keep track of X, the number of days the lights are left on over the next 30 days.

a. What is the expected amount of power used during the 30 days?

b. Find an expression for the probability that 16,400 Watts of power are used during the 30 days. You do not need to simplify or evaluate the expression.

Exercise 15.23. Selling encyclopedias. Samuel is an encyclopedia salesman. In order to make his life more interesting, when he encounters an intersection of two streets, he heads east 30% of the time and north the other 70%. He will walk 50 blocks (intersection-to-intersection) each day. West-East blocks contain 4 houses each and North-South blocks contain 6.

a. On a certain day, how many houses does he expect to visit?

b. Find an expression for the probability that he visits 290 or more houses. You do not need to simplify or evaluate the expression.

15.3.3 Advanced

Exercise 15.24. Ants. In the ant world, 98% of the ants are female, and 2% are male. In the queen's first batch of 100,000 offspring, let X be the total number of male births. Write an expression for the probability that 2100 or more of the ants are males. You do not need to simplify or evaluate your expression. Would it be difficult to calculate?

Exercise 15.25. Cereal for breakfast. Sixty percent of students usually eat cereal for breakfast. If n students eat in the dining halls for breakfast each day, let X_n be the number who have cereal. Find the limiting probability that at least one student has cereal, as n grows large, i.e., find $\lim_{n\to\infty} P(X_n > 0)$. Does your answer make sense intuitively? Why?

Chapter 16

Geometric Random Variables

If at first you don't succeed, try, try, try again.
—traced to Thomas H. Painter, Frederick Marryat, and perhaps others; popularized by William Edward Hickson

If at first you don't succeed, skydiving is not for you.
—Wendy Northcutt

You are playing a series of one-on-one basketball games with your friend. Instead of the 5-game series you played with your friend last week (described in the introduction to Chapter 15), this week you have decided that you will only play as many games as it takes for you to win your first game. You still have to buy your friend lunch for each game she wins, and then she will buy you lunch when you win your first game. What is the probability you will have to buy your friend a lunch exactly 4 days this week, before she buys your lunch? What is the probability you don't have to buy lunch for your friend at all? How many lunches do you expect to have to buy for your friend? How is this game different from last week's game?

16.1 Introduction

A Geometric random variable is the number of independent trials until (and including) the first success, when the probability of success on each trial is constant.

A Geometric random variable is the number of independent trials needed until the first success occurs. An equivalent interpretation is that a Geometric random variable is the number of independent Bernoulli random variables we need to check until the first one that indicates success. If $X_1, X_2, \ldots$ are independent Bernoulli(p) random variables, then

$$\text{Geometric random variable} = \text{smallest value of } M \text{ so that } X_M = 1$$
$$(\text{and } X_1 = \ldots = X_{M-1} = 0)$$

Geometric random variables

The common thread: Number of independent trials, each with the same probability of success, until the first success occurs. For instance, you ask people one yes/no question each, until you get the first "yes" response. You don't know how many trials you'll need in advance because you only stop asking when you find that 1st success. Other examples include: number of dice rolls until the first 5; number of parts coming off an assembly line until the first defective one; number of students surveyed until the first senior is discovered; etc.

Things to look for: Repeated independent trials until first success; same success probabilities.

The variable:

$$X = \# \text{ of trials until the first success, so } X \geq 1$$

The parameters:

$$p = \text{the probability that a given trial is a success}$$
$$q = 1 - p = \text{the probability that a given trial is a failure}$$

These values p and q must be the same for every trial.

Mass:

$$P(X = x) = q^{x-1}p, \qquad x = 1, 2, 3, \ldots$$

Expected value formula:

$$\mathbb{E}(X) = 1/p$$

Variance formula:

$$\text{Var}(X) = q/p^2$$

For instance, if a Geometric random variable X has the value 6, we can view this as five independent Bernoulli random variables that indicate failure, i.e., $X_1 = X_2 = X_3 = X_4 = X_5 = 0$, followed by a sixth independent Bernoulli that indicates success, i.e., $X_6 = 1$.

As an example:

"Thumbs up" for success; trials 1, 2, 3, 4, 5 fail, and trial 6 succeeds. So $X = 6$ is a Geometric that gives the total number of trials until the first success.

Geometric random variables only have one parameter, p, the probability of success of the independent trials. An easy way to distinguish Geometric random variables from Binomials is that a Geometric(p) random variable can be arbitrarily large (there is no upper limit n), whereas a Binomial(n, p) random variable is always between 0 and n. A Geometric random variable has to be at least 1, because at least one trial is needed to get the first success.

For a Geometric random variable with parameter p, the probability of success p must remain the same on every one of the independent trials.

We cannot know in advance how many independent trials will be needed to see the first success.

A key feature of a Geometric is the memoryless property; see Section 16.4.

The notation for the Geometric looks like:

$$X \sim \text{Geometric}(p),$$

where p is the probability of success on each trial.

Where does the mass formula come from?

$$\begin{aligned}
x - 1 &= \text{number of trials} - 1 \\
&= \text{number of failures} \\
q^{x-1} &= (\text{probability of a failure on one trial})^{\#\text{ of failures}} \\
p &= \text{probability of a success on one trial}
\end{aligned}$$

Example 16.1. Suppose 3% of pet owners give gifts to their pets on Valentine's Day. You plan to talk to randomly selected pet owners from a large population until you find one who gives a gift to his or her pet. Since you talk to them individually, assume their responses are independent.

a. What does X represent in this story? What values can X take?

The variable X is the number of pet owners we ask until we find the first one who gives a gift to his or her pet, so X can be $1, 2, \ldots$

b. Why is this an example of the Geometric distribution?

The trials are independent and have an equal probability of success (0.03). We are performing the trials—selecting pet owners—until we find our first success, i.e., a pet owner who gives his or her pet a Valentine's gift.

c. How many pet owners do we expect to have to ask until finding one who gives a Valentine's gift to his or her pet?

$$\mathbb{E}(X) = \frac{1}{0.03} = 33.33 \text{ pet owners}$$

d. What is the standard deviation of the number of people you have to talk to until you find one who gives a Valentine's gift to his or her pet?

$$\text{Var}(X) = \frac{1 - 0.03}{0.03^2} = 1077.78$$
$$\sigma_X = \sqrt{1077.78} = 32.8295 \text{ pet owners}$$

e. What does the mass look like for this story?

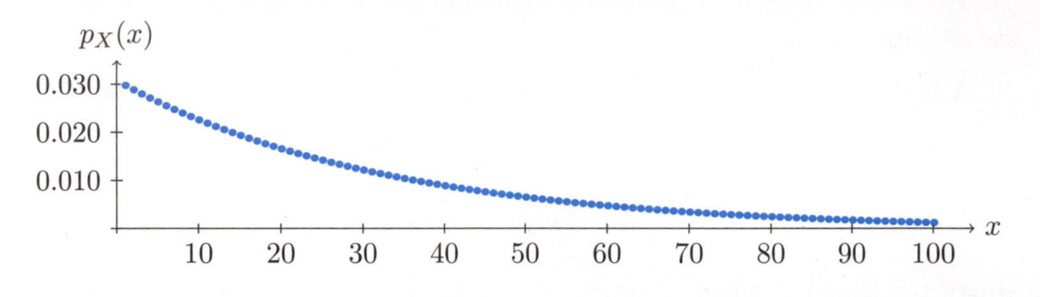

f. What does the CDF look like for this story?

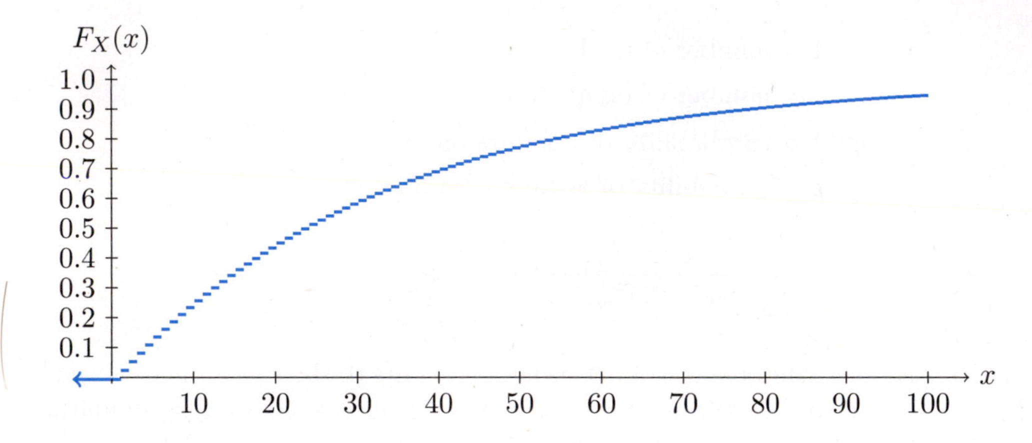

g. What is the probability you will have to talk to exactly 40 people total to find your first pet-gift-giver?

$$P(X = 40) = (0.97)^{39}(0.03) = 0.009146$$

h. If you decide in advance to talk to exactly 40 people, what is the probability you will have exactly one pet-gift-giver in the sample? How does this compare to the situation in part e?

$$P(X = 1) = \binom{40}{1}(0.97)^{39}(0.03)^1 = 40(0.009146) = 0.3658$$

There are 40 ways that the scenario in part h could happen (any of the 40 people could be the one pet-gift-giver), each with probability $(0.03)^1(0.97)^{39}$, but only 1 way that the scenario in part g could happen. Therefore the probability in part h is 40 times as large as the probability in part g.

The scenario in part h of the example above is a Binomial situation—not a Geometric situation—because we fixed the number of trials (40) in advance, and we are counting up the number of successes anywhere in those 40 trials. In scenario g we were asking for 39 failures in a row, with 1 success at the very end, i.e., the success *had to occur* as the last trial.

More generally, this illustrates why the Geometric formula does not have an "$\binom{n}{x}$" term, because we are not concerned with the number of ways to rearrange the successes and failures when working with a Geometric random variable. If you have 3 trials and 1 success, there is only one way to arrange these for the Geometric, because the success must come last:

but there are $\binom{3}{1} = 3$ ways to arrange the outcomes for the Binomial:

Remark 16.2. A frequent question about Geometric random variables is whether a Geometric random variable will always be finite. Could a Geometric random variable be infinite?

For example, if X is the number of fair coin flips until the first head, can X be infinite? The answer is that X will be finite with probability 1. So the event that an infinite number of coins are needed has probability 0. (In such a case, we might be tempted to say that X is infinite, although random variables are only supposed to take on real values.)

This is true much more generally: If the probability of success p on each trial is positive ($p > 0$), then a Geometric random variable will be finite with probability 1.

One way to see this is to calculate the probability that a Geometric random variable is finite. If X is a Geometric(p) random variable with $p > 0$, then

$$P(X \text{ is finite}) = \sum_{j \geq 1} P(X = j) = \sum_{j \geq 1} q^{j-1} p = p \sum_{j \geq 0} q^j = p/(1-q) = p/p = 1.$$

Another way to see this is, if X is infinite then for each n we know $n < X$, so

$$0 \leq P(X \text{ is infinite}) \leq P(n < X) = q^n,$$

but $\lim_{n \to \infty} q^n = 0$ since $q < 1$, so $P(X \text{ is infinite}) = 0$.

Remark 16.3. Expected value of a Geometric random variable If X is Geometric with probability of success p on each trial, then the expected value of the Geometric random variable is

$$\mathbb{E}(X) = \sum_{j\geq 1} jP(X = j) = \sum_{j\geq 1} jq^{j-1}p = p\sum_{j\geq 1} \frac{d}{dq}q^j.$$

It might seem unusual to differentiate with respect to "q," but the derivation lends itself to this change, and now we can switch the order of the summation and the differentiation, so

$$\mathbb{E}(X) = p\frac{d}{dq}\sum_{j\geq 1} q^j = p\frac{d}{dq}\frac{q}{1-q} = p\frac{(1-q)(1)-(q)(-1)}{(1-q)^2} = p\frac{1}{p^2} = 1/p.$$

Remark 16.4. Variance of a Geometric random variable We know that $\mathrm{Var}(X) = \mathbb{E}(X^2) - (\mathbb{E}(X))^2$, but $\mathbb{E}(X^2)$ will not be as nice for us to work with in this derivation as $\mathbb{E}((X)(X-1))$ will be. So we need to do a little algebraic manipulation: $X^2 = X^2 - X + X = (X)(X-1) + X$.

The second moment of X is computed in a similar manner if we observe that $(j)(j-1)q^{j-2}$ is a second derivative of q^j with respect to q. We get:

$$\begin{aligned}
\mathbb{E}((X)(X-1)) &= \sum_{j\geq 1} j(j-1)P(X = j) = \sum_{j\geq 1} j(j-1)q^{j-1}p \\
&= pq\sum_{j\geq 1} j(j-1)q^{j-2} = pq\sum_{j\geq 1} \frac{d^2}{dq^2}q^j \\
&= pq\frac{d^2}{dq^2}\sum_{j\geq 1} q^j = pq\frac{d^2}{dq^2}\frac{q}{1-q},
\end{aligned}$$

which we can calculate. This yields

$$\mathbb{E}((X)(X-1)) = pq\frac{2}{(1-q)^3} = \frac{2q}{p^2}.$$

This almost gives us the variance. Now we use the fact mentioned above, that $X^2 = (X)(X-1) + X$, and this yields

$$\mathbb{E}(X^2) = \mathbb{E}((X)(X-1)) + \mathbb{E}(X),$$

which we now substitute into the equation for the variance:

$$\begin{aligned}
\mathrm{Var}(X) &= \mathbb{E}(X^2) - (\mathbb{E}(X))^2 \\
&= \mathbb{E}((X)(X-1)) + \mathbb{E}(X) - (\mathbb{E}(X))^2 \\
&= \frac{2q}{p^2} + \frac{1}{p} - \frac{1}{p^2} \\
&= q/p^2.
\end{aligned}$$

Example 16.5. Draw a card from a well-shuffled deck until the ace of spades appears. If a draw is unsuccessful, then replace and reshuffle the deck before making the next selection. Let X be the number of draws needed until the ace of spades appears for the first time.

(Note that, in the case where the ace of spades never appears, we have not specified the value of X, but this omission does not affect the probabilities associated with the problem, because the probability of the ace of spades never appearing is zero.)

The mass of X is $p_X(j) = (51/52)^{j-1}(1/52)$ for positive integers j, and $p_X(x) = 0$ otherwise. So X is a Geometric random variable with $p = 1/52$. The expected number of cards until the first ace of spades appears is $\mathbb{E}(X) = \frac{1}{1/52} = 52$. The variance of X is $\text{Var}(X) = \frac{51/52}{(1/52)^2} = 2652$.

Example 16.6. Suppose that a basketball player makes 80% of her free throws successfully. She attempts to make a basket as many times as necessary, until scoring the first basket. Let X denote the number of necessary attempts.

(Again, the probability that she never scores is 0; we have not specified the value of X when she never scores.)

The mass of X is $p_X(j) = (0.20)^{j-1}(0.80)$ for positive integers j, and $p_X(x) = 0$ otherwise. Thus, X is Geometric with $p = 0.80$. So $\mathbb{E}(X) = 1/0.80 = 1.25$ and $\text{Var}(X) = 0.20/0.80^2 = 0.3125$.

Example 16.7. People are selected randomly and independently for a drug test. Each person passes the test 98% of the time. What is the expected number of people who take the test until the first person fails?

Let X be the number of people tested, until the first person fails the test. Then $P(X = j) = (0.98)^{j-1}(0.02)$ for each positive integer j. Thus, X is Geometric with $p = 0.02$. So we expect $\mathbb{E}(X) = 1/0.02 = 50$ people to be tested until the first person fails the test. The variance of X is $\text{Var}(X) = 0.98/0.02^2 = 2450$.

16.2 Special Features of the Geometric Distribution

We discuss several common methods for using inequality symbols with the Geometric distribution. Hint: It's much easier to work with the Geometric inequalities if you rephrase things in terms of a Geometric random variable being greater than a specific number.

What is the probability we need more than 5 trials to get the 1st success?

$$P(X > 5) = P(\text{1st 5 rolls all failures}) = q^5$$

What is the probability we need at least 5 trials to get the 1st success?

$$\begin{aligned} P(X \geq 5) &= P(\text{1st 4 rolls all failures, 5th roll could be a success or a failure}) \\ &= P(X > 4) \\ &= q^4 \end{aligned}$$

What is the probability we need at most 5 trials to get the 1st success?

$$P(X \leq 5) = 1 - P(X > 5) = 1 - q^5$$

What is the probability we need fewer than 5 trials to get the 1st success?

$$P(X < 5) = 1 - P(X \geq 5) = 1 - P(X > 4) = 1 - q^4$$

16.3 The Number of Failures

Geometric random variables are defined to be the number of trials required until the first success occurs, and the success itself is included as one of these trials. So, for instance, if X is Geometric, then $X = 10$ corresponds to the fact that the 10th trial is the first success. As an example, X might be the number of coin flips until the first head occurs, including the head itself.

Sometimes we are more interested in the number of failures that occur until the first success, i.e., we do not want to take the success itself into account. For instance, let Y be the number of failures that occur before the first success, not including the success itself. E.g., if Y is the number of tails that occur before the first head (not including the head itself), then Y is the number of failures. Whenever a Geometric random variable X is equal to j, there is an analogous random variable Y, the number of failures, that is equal to $j-1$ in the same case.

For the number of failures Y to be exactly j, there must be exactly j failures in a row, followed immediately by a success, so, the mass of such a random variable Y is

$$p_Y(j) = q^j p \qquad \text{for } j = 0, 1, 2, 3, \ldots,$$

and $p_Y(y) = 0$ otherwise.

With these definitions, Y and $X-1$ have the same distribution, or equivalently, $Y+1$ and X have the same distribution. For instance, $p_Y(3) = q^3 p = p_X(4)$.

Notice that the number of failures is always one less than the number of trials until the first success, so the expected values must differ by 1 too. E.g., $\mathbb{E}(Y) = \mathbb{E}(X-1) = \mathbb{E}(X) - 1 = \frac{1}{p} - 1 = \frac{1-p}{p} = q/p$.

Also notice that, since the number of failures and the number of trials until the first success always differ by 1, they have the same variance. (Remember that the addition or subtraction of a constant from a random variable does not affect the variance.) Thus $\mathrm{Var}(Y) = \mathrm{Var}(X-1) = \mathrm{Var}(X)$.

16.4 Geometric Memoryless Property

If you have conditional probabilities using the “$>$” symbol on both sides, you can use this shortcut. (Food for thought: this really works seamlessly only for “$>$,” not for “$<$”; why?)

Given that you need more than 2 trials to get your 1st success, what is the probability you will need more than 5 trials total?

$$P(X > 5 \mid X > 2) = P(X > 3) = q^3$$

Trial			?	?	?
Outcome	No	No	?	?	?
Past/future?	Past	Past	Future	Future	Future

Why does this work? Each trial is independent. Once some trials have passed, the next trials do not depend on what has already happened in the past. You already know the first 2 trials were failures. What you don't know is what will happen on the next 3 trials.

Here is the same explanation using conditional probability:

$$P(X > 5 \mid X > 2) = \frac{P(X > 5 \text{ and } X > 2)}{P(X > 2)},$$

but we only have both $X > 5$ and $X > 2$ if we have $X > 5$. So

$$P(X > 5 \mid X > 2) = \frac{P(X > 5)}{P(X > 2)} = q^5/q^2 = q^3 = P(X > 3).$$

Example 16.8. Given that more than j trials are needed to get the first success, what is the probability that more than k trials are needed?

Of course we have $P(X > k \mid X > j) = 0$ if $k < j$. So we now consider $k \geq j$. In such a case, the first j trials are already given to be failures, so only trials number $j+1, j+2, \ldots, k$ are needed to be failures for the first k trials to all be failures. So

$$P(X > k \mid X > j) = P(X > k - j) = q^{k-j}.$$

Here is a more rigorous explanation using conditional probabilities:

$$\begin{aligned} P(X > k \mid X > j) &= \frac{P(X > k \text{ and } X > j)}{P(X > j)} \\ &= \frac{P(X > k)}{P(X > j)} \\ &= q^k/q^j \\ &= q^{k-j} \\ &= P(X > k - j) \end{aligned}$$

Example 16.9. Memoryless Geometric property. As in Exercises 8.5, 10.18, and 11.4, the Super Breakfast Challenge (SBC) is known to be very difficult to consume. Only 10% of people are able to eat all of the SBC.

Let X be the number of people required until a customer is able to eat the whole SBC.

a. How many people are needed, on average, until the first successful customer?

Since X is Geometric with $p = 0.10$, then $\mathbb{E}(X) = 1/0.10 = 10$.

b. What is the variance of the number of people needed?

The variance of the number of customers needed is $\text{Var}(X) = 0.90/0.10^2 = 90$.

c. What is the probability that more than 4 customers are needed, until the first success?

$$P(X > 4) = P(\text{1st 4 were unsuccessful}) = q^4 = (0.90)^4 = 0.6561$$

d. Given that the first 4 are unsuccessful, what is the probability at least 8 are needed?

$$\begin{aligned} P(X \geq 8 \mid X > 4) &= P(X > 7 \mid X > 4) \\ &= P(X > 3) \\ &= P(\text{1st 3 are unsuccessful}) \\ &= (0.90)^3 \\ &= 0.729 \end{aligned}$$

Remark 16.10. A note of caution:
The memoryless property does not work with any of the other discrete random variables that we have learned. The memoryless property will also work with one of the continuous random variables (the Exponential random variable, which will be covered in the second half of the book). Do not try to use the memoryless property with any other discrete distributions.

16.5 Random Variables That Are Not Geometric

One of the things to notice in particular about Geometric random variables is that the probability of success p must not change during the series of independent trials. So, as an example, consider a shuffled deck of cards, in which we draw cards repeatedly—without replacement—until the ace of spades appears. Let X denote the number of draws necessary. Then X is not a Geometric random variable. There are a few ways to see this. One way to see this immediately is that $X \leq 52$ always, so $P(X = 53) = 0$, and thus X cannot be Geometric. Another way to see this is $P(X = j) = 1/52$, because $X = j$ if and only if the jth card in the deck was the ace of spaces. So X does not have the right kind of mass to be a Geometric random variable.

16.6 Exercises

16.6.1 Practice

Exercise 16.1. **Skittles.** Skittles candies come in the colors red, orange, yellow, green, and purple, with each color having equal probability. Due to a dye mix-up there are a few rainbow-striped candies coming down the line. There is a 5% chance that a candy is rainbow striped. You are a quality control inspector, and your job is to find the first rainbow-striped candy coming down the line. (The population is so big that we can assume relative independence.)

a. Explain in words what X is in terms of the story. What values can it take?

b. Why is this a Geometric distribution situation? What is the parameter?

c. What is the probability that you will need to check exactly 10 candies to find the first rainbow one?

d. What is the probability that you will have to check fewer than 12 candies to find the first rainbow one?

e. What is the probability you will have to check more than 8 candies to find the first rainbow one?

f. Given that you have to check at least 3 candies, what is the probability that you will need to check a total of more than 9 for the first rainbow one?

g. What is the expected number of candies you will have to check until you find the first rainbow one?

h. What is the standard deviation in the number of candies you will have to check until you find your first rainbow one?

Exercise 16.2. **Colorblind.** Approximately 8.33% of men are colorblind. You survey men from a large population until you find one who is colorblind.

a. Explain in words what X is in this situation and what values it can take.

b. Why is this a Geometric distribution situation? What is the parameter?

c. What is the probability you will have to survey at most 16 men until you find the first one who is colorblind?

d. What is the probability you will have to ask exactly 12 men until you find the first who is colorblind?

e. What is the expected value of X?

f. What is the variance of X?

g. Show the labeled graph of the mass for this story.

h. Show the labeled graph of the CDF for this story.

Exercise 16.3. **Surgical delivery.** Twenty percent of babies in a particular city are born by a surgical procedure called a Cesarean section (C-section). You are interested in gathering information about hospital experiences of parents who did not have their baby by C-section. You survey randomly parents of babies from the population of the large city until you find the first one who had a baby **not** born by C-section.

a. Explain in words what X is in this situation and what values it can take.

b. Why is this a Geometric distribution situation? What is the parameter?

c. What is the probability that you have to survey parents of at least 5 babies until you find the first one not born by C-section?

d. Given that you have to survey parents of more than 1 baby to find the first one not born by C-section, what is the probability that you have to survey at least 5 parents of babies?

e. What is the probability that you must survey between 4 and 6 (inclusive) parents of babies to find one that was not born by C-section?

f. What is the expected value of X?

g. What is the variance of X?

h. Show the labeled graph of the mass for this story.

i. Show the labeled graph of the CDF for this story.

Exercise 16.4. Mobile phone survey. (Refer to Exercise 15.6) A recent survey states that 48% of mobile devices are iPhones. In order to learn more about how the iPhone works, a student starts asking his friends if they use an iPhone. Let X be the number of friends he will need to ask until he finds one who uses an iPhone.

a. What is the expected value of X?

b. What is the standard deviation of X?

c. Show the labeled graph of the mass for this story.

d. Show the labeled graph of the CDF for this story.

Exercise 16.5. Radio airplay. A certain radio station plays songs from the 1970's, 80's, and 90's. We know that 20% of the songs on the station are from the 70's; 37% are from the 80's; and 43% of the songs are from the 90's. Let X be the number of songs you listen to until you hear the first one from the 1990's.

a. What is the expected value of X?

b. What is the standard deviation of X?

c. Show the labeled graph of the mass for this story.

d. Show the labeled graph of the CDF for this story.

Exercise 16.6. Homecoming. A young woman realizes that Homecoming is quickly approaching, and she needs to find a date. She estimates that 72% of the male students (in a large population) would be willing to accept her invitation. She plans to start randomly asking men for a date until someone accepts.

a. What is the expected number of men she will need to invite until she has a date for Homecoming?

b. Show the labeled graph of the mass for this story.

c. Show the labeled graph of the CDF for this story.

Exercise 16.7. Chores. Sarah and Thomas play a card game to determine who will have to take out the trash (the loser gets this unpopular chore) on Monday. They use a standard 52-card deck by taking turns randomly drawing cards from a shuffled deck, with replacement, until somebody draws an ace. Whoever gets the ace does not have to do the chore.

a. How long do they expect to have to play the game?

b. What is the probability that it takes at least four cards for them to find an ace?

c. If they play this game for six weeks, what is the probability it takes at least four cards for them to find an ace on each of those 6 Mondays? What distribution is being used here? What are the parameters?

Exercise 16.8. Random guessing. On a certain online math assignment, a student is allowed to submit an unlimited number of answers before moving on to the next problem. The problem is a free response question, and the student believes that his guessed answer is correct about 7% of the time. (There are so many possible guesses that he thinks this ratio stays the same every time.) Under these assumptions, let X be the number of times he needs to submit a different random guessed answer until he gets the question correct.

a. How many tries should he expect if he wants to get this question right without actually learning the material?

b. What is the probability it takes him at least 20 tries to get the question right?

c. If there are 10 homework questions of a similar nature on this assignment, and he uses the same random technique with all of the questions, what is the probability it takes him at least 20 tries to get each of these questions right? What distribution is being used here? What are the parameters?

d. If each homework question takes him 2 minutes to read (once) and 30 seconds for each random entry attempt, what is the expected time it will take him to get one question correct? What about the 10 question assignment? (Do you think he would have been better off just learning the material and doing the homework properly?)

Exercise 16.9. Shopping before Thanksgiving. Studies have shown that 28% of people do all of their Christmas shopping before Thanksgiving each year. Let X be the number of people you have to ask until you find somebody who has finished all of their Christmas shopping before Thanksgiving, assuming each person is independent of the others.

a. What is the expected number of people you will have to ask until you find somebody who has finished her Christmas shopping before Thanksgiving?

b. If each interview takes approximately 5 minutes, how much time should you expect to have to spend total?

c. What is the probability that you will have to ask more than 10 people?

Exercise 16.10. Lucky Charms. Michael reaches into a very large box and pulls out Lucky Charms. If percentages of the pieces are: 50% regular cereal, 6.25% each for hearts, stars, horseshoes, clovers, blue moons, pots of gold, rainbows, and red balloons, and he only wants blue moons. Let X be the number of individual pieces he has to pull out until he gets a blue moon.

a. What is the probability that X is more than 8?

b. What is the probability that X is at least 8?

c. What is the probability that X is less than 8?

d. What is the probability that X is at most 8?

Exercise 16.11. Vampire difficulties. Edward and his friend cannot go to school when it's sunny outside because they are vampires. Forks, Washington is experiencing a sunny period these days, and there is an 80% chance each day that it will be sunny. Assume that the weather is independent from day to day. Let X be the number of days until the first day Edward can go outside.

a. What is the probability that Edward will have to wait between 10 to 12 days (inclusive) to go back to school?

b. Given that he has already waited 5 sunny days, what is the probability that his waiting time altogether is between 10 to 12 days (inclusive) to go back to school?

c. What is the expected number of days that he will have to wait to go back to school?

Exercise 16.12. Blindfolded basketball. A basketball player shoots free throws until he makes one. However, because his coach wants him to practice his technique and the feel of the motions, his coach has the player blindfold himself, which makes each throw independent from the others. When he is blindfolded, he has only a 2% chance of making a free throw on any particular try.

a. What is the probability that he will have to throw over 100 balls until he makes his first basket?

b. What are the expected value and standard deviation for the number of throws he will need to make until he gets his first basket?

c. What would you have to change about this story to turn it into a Binomial question?

Exercise 16.13. Winning and losing. Suppose that a person wins a game of chance with probability 0.40, and loses otherwise. He plays the game until he wins for the first time, and then he stops. Assume that the games are independent of each other. Let X denote the number of games that he must play until (and including) his first win.

a. How many games does he expect to play until (and including) his first win?

b. What is the variance of the number of games he plays until (and including) his first win?

c. What is the probability that he plays 4 or more games altogether?

Exercise 16.14. Winning and losing (continued) Continue to use the scenario from the previous problem. As before, let X denote the number of games that he must play until (and including) his first win. Assume that, if he wins, he earns 5 dollars, and if he loses, then he loses 4 dollars. (Also assume that he is allowed to borrow money, i.e., having a negative amount of money is not a problem here.)

a. Find a formula for his gain or loss, in terms of X, i.e., if Y denotes his gain or loss in dollars, write Y in terms of X.

b. What is his expected gain or loss (altogether) during the X games, i.e., what is $\mathbb{E}(Y)$?

c. What is the variance of his gain or loss (altogether) during the X games, i.e., what is $\text{Var}(Y)$?

Exercise 16.15. Dating. You randomly call friends who could be potential partners for a dance. You think that they all respond to your requests independently of each other, and you estimate that each one is 7% likely to accept your request. Let X denote the number of phone calls that you make to successfully get a date.

a. Find the expected number of people you need to call, i.e., $\mathbb{E}(X)$.

b. Find the variance of the number of people you need to call, i.e., $\text{Var}(X)$.

c. Given that the first 3 people do not accept your invitation, let Y denote the additional number of people you need to call (Y does not include those first 3 people); i.e., suppose $X > 3$ is given, then let $Y = X - 3$. Under these conditions, what is the mass of Y?

Exercise 16.16. Hearts. You draw cards, one at a time, *with replacement* (i.e., placing them randomly back into the deck after they are drawn), from a shuffled, standard deck of 52 playing cards. Let X be the number of cards that are drawn to get the first heart that appears.

a. How many cards do you expect to draw, to see the first heart?

b. Now suppose that you draw five cards (again, with replacement), and none of them are hearts. How many additional cards (not including the first five) do you expect to draw to see the first heart?

16.6.2 Extensions

Exercise 16.17. Telemarketers. Assume that when your phone rings, the caller is a telemarketer with probability 1/8, and that the probability of a telemarketer is independent from call to call. Let X denote the number of telemarketers during the next three calls. If n is a nonnegative integer, what is $P(X > n)$?

Exercise 16.18. Looking for a wife. Prince Charming has to go around town asking if the glass slipper fits until he finds a woman whose foot fits properly in the slipper so that he will know who to marry. (We do not endorse this technique for finding a wife.) The probability of a glass slipper fitting a randomly selected woman is 0.12, and the probabilities are independent for the various women. Let X be the number of women he has to visit until he finds a woman who fits the slipper. Assume that there are an unlimited supply of eligible women in the town.

a. Given that he has already checked with 4 women without success, what is the probability he will still need to check with at least 5 more? (He's getting impatient.)

b. Given that he has already checked with 4 women without success, what is the probability he will succeed with the very next woman?

c. What is the probability he will succeed on his first try?

d. What is the expected number of women he will have to try?

e. If he takes an entourage with him everywhere he goes, and he has to pay the entourage \$100 for the day plus \$10 for every visit he makes, how much does he expect to pay if he does all his visits on one day?

Exercise 16.19. Geometric versus Binomial. Why does the memoryless property work for the Geometric distribution but not for the Binomial distribution?

16.6.3 Advanced

Exercise 16.20. Consider two independent Geometric random variables, X and Y, with parameters p_1 and p_2, respectively. Give a general formula (in terms of p_1 and p_2) for the probability that X and Y are equal, i.e., for $P(X = Y)$.

Exercise 16.21. Use the probability mass function to justify the fact that, if X is a Geometric random variable, then $P(X > x) = q^x$. (In Section 16.2, we justified this intuitively, but we did not use the PMF to prove it.)

Chapter 17

Negative Binomial Random Variables

The toughest thing about success is that you've got to keep on being a success.
—Irving Berlin

A coach wants to put together an intramural basketball team, from people living in a large dorm. She estimates that 12% of people in the dorm like to play basketball. She goes door to door to ask people if they would be interested in playing on the team. How many dorm residents does she expect to interview before finding 5 people to create the team? How is this expected value different than if she was only trying to find one person to join a team? What is the probability that she needs to talk to 20 people, in order to find 5 people who will join the team?

17.1 Introduction

A Negative Binomial random variable is the number of independent trials until r successes have occurred.

A Negative Binomial random variable is the number of independent trials required until a certain number of successes have occurred. For instance, a Negative Binomial random variable could be the number of independent trials until the 3rd success occurs. The successes do not have to be consecutive (they usually are not). A Negative Binomial random variable can be interpreted as the sum of several independent Geometric random variables. For example, if X, Y, Z are independent Geometric random variables with the same parameter, then $X + Y + Z$ is a Negative Binomial random variable for the number of trials until the third success.

Negative Binomial random variables

The common thread: Number of independent trials, each with probability p of success, until the rth success occurs. For instance, you randomly ask people one yes/no question each, until you get the rth "yes" response. The successes do not need to be consecutive. Other examples include: number of dice rolls until the third 5; number of cars observed until the tenth black car; etc.

Things to look for: Repeated independent trials until rth success; same success probabilities.

The variable:

$$X = \# \text{ of trials until the } r\text{th success, so } X \geq r$$

X can't be less than r because you need at least r trials to get r successes.

The parameters:

$$\begin{aligned} r &= \text{the desired number of successes} \\ p &= \text{the probability that a given trial is a success} \\ q = 1 - p &= \text{the probability that a given trial is a failure} \end{aligned}$$

These values p and q must be the same for every trial.

Mass:

$$P(X = x) = \binom{x-1}{r-1} q^{x-r} p^r, \qquad x = r, r+1, r+2, \ldots$$

Expected value formula:

$$\mathbb{E}(X) = r/p$$

Variance formula:

$$\text{Var}(X) = qr/p^2$$

Negative Binomial random variables have two parameters: p, the probability of success of the independent trials, and r, the desired number of successes. The probability of success p must stay the same on each trial. Negative Binomial random variables are always r or larger, because it takes at least r trials to have r successes.

$$\text{Negative Binomial random variable} = j \text{ if exactly } r-1 \text{ of } X_1, \ldots, X_{j-1} \text{ are 1 (i.e., successes) and also } X_j = 1$$

After each individual success, the search for the next success starts again, independent of what came before. For this reason, a Negative Binomial with parameters r and p is exactly the sum of r independent Geometric random variables, each with parameter p. Therefore, the expected value and variance of Negative Binomial random variables are easy to determine from the expected value and variance of Geometric random variables.

For instance, if a Negative Binomial random variable X with $r = 3$ has the value 13, this means that the 13th trial is a success, and exactly 2 of the earlier 12 trials are a success too. As an example:

The notation for the Negative Binomial looks like:

$$X \sim \text{NegBinomial}(r, p)$$

where p is the probability of success on each trial, and we want r successes to occur.

Where does the mass formula come from?

$$\begin{aligned} P(X = x) &= \binom{x-1}{r-1} p^r q^{x-r} \\ x - r &= \text{number of failures} \\ q^{x-r} &= (\text{probability of a failure on one trial})^{\#\text{ of failures}} \\ p^r &= (\text{probability of a success on one trial})^{\#\text{ of successes}} \\ \binom{x-1}{r-1} &= \text{number of possible arrangements of the successes} \\ &= \#\text{ of ways to arrange } r-1 \text{ successes in } x-1 \text{ trials} \end{aligned}$$

We use $\binom{x-1}{r-1}$ instead of $\binom{x}{r}$ because the last success is always the last trial, so we only need to consider the ways to arrange $r - 1$ successes within the first $x - 1$ trials.

17.2 Examples

Example 17.1. A coach needs to build a team of 5 basketball players quickly for a campus league. She decides to ask randomly selected people whether they played basketball in high school. Assume that 12% of the people have played basketball in high school, that all people with basketball experience would be willing to play on the team, and that the school is very large, so there is near independence among the responses she gets.

a. What does X represent in terms of this story? What values can it take?

The variable X is the number of people she has to ask until finding the 5th member of the team, so X can be $5, 6, \ldots$ (She needs to ask at least 5 people since she is looking for 5 successes.)

b. Why is this a Negative Binomial situation? What are the parameters?

She is counting up number of trials until getting the 5th success, which is a person who played basketball in high school ($r = 5$). Each trial (person) is independent and each person has the same probability of success ($p = 0.12$).

c. What is the probability that she finds the 5th member of the team on the 20th person she asks?

$$P(X = 20) = \binom{19}{4}(0.12)^5(0.88)^{15} = 0.01418$$

d. What is the expected number of people she needs to ask in order to form her team?

$$\mathbb{E}(X) = \frac{5}{0.12} = 41.6667 \text{ people}$$

e. What is the standard deviation in the number of people she will need to ask in order to form the team?

$$\begin{aligned} \text{Var}(X) &= \frac{5(1 - 0.12)}{0.12^2} = 305.5556 \\ \sigma_X &= \sqrt{305.5556} = 17.4801 \text{ people} \end{aligned}$$

Example 17.2. Comparing Binomial and Negative Binomial. In the basketball team example above, we compare two different situations:

question	What is the probability there will be 5 basketball players in the first 20 people she asks?
distribution	Binomial ($n = 20, p = 0.12$)
meaning of X	$X = \#$ of basketball players $= 5$
probability	$P(X = 5) = \binom{20}{5}(0.12)^5(0.88)^{15} = 0.0574$

question	What is the probability the 20th person she asks will be the 5th player?
distribution	Negative Binomial $(r = 5, p = 0.12)$
meaning of X	$X = \#$ of people you ask $= 20$
probability	$P(X = 20) = \binom{19}{4}(0.12)^5(0.88)^{15} = 0.0143$

Remark 17.3. Expected value and variance of a Negative Binomial random variable If $X_1, X_2, \ldots, X_r$ are independent Geometric random variables, each with parameter p, then

$$X = X_1 + X_2 + \cdots + X_r$$

is a Negative Binomial random variable with parameters r and p. So

$$\mathbb{E}(X) = \mathbb{E}(X_1 + \cdots + X_r) = \mathbb{E}(X_1) + \cdots + \mathbb{E}(X_r) = \frac{1}{p} + \cdots + \frac{1}{p} = \frac{r}{p},$$

and

$$\mathrm{Var}(X) = \mathrm{Var}(X_1 + \cdots + X_r) = \mathrm{Var}(X_1) + \cdots + \mathrm{Var}(X_r) = \frac{q}{p^2} + \cdots + \frac{q}{p^2} = \frac{qr}{p^2}.$$

This makes sense when you think that a Negative Binomial random variable is just the sum of r independent Geometric random variables. For the basketball example, you have the sum of 5 Geometric random variables since you need 5 basketball players.

	X_1			X_2	X_3						X_4								X_5	
person	1	2	3	4	5	6	7	8	9	10	11	12	13	14	15	16	17	18	19	20
answer	N	N	Y	Y	N	N	N	N	N	Y	N	N	N	N	N	N	N	Y	N	Y

Example 17.4. Roll a die until the third "5" appears.

	X_1			X_2	X_3			
die roll	1	2	3	4	5	6	7	8
result	2	3	5	5	4	3	3	5

If we let X denote the number of rolls needed until the third "5" appears, then X is a Negative Binomial random variable with parameters 3 and 1/6. We could write $X = X_1 + X_2 + X_3$, where X_1 is the number of rolls until the first

"5" appears, and X_2 is the number of rolls after that point until the second "5" appears, and X_3 is the number of rolls after that point until the third "5" appears, etc. Then each X_j is Geometric with parameter 1/6, and the X_j's are independent.

Example 17.5. Dominique, Raymond, and Samantha are independently using an app that randomly creates playlists for them. Forty percent of the songs this app has access to are rap songs. Dominique randomly adds songs to her playlist until she gets 5 rap songs. Raymond randomly adds songs until he gets 12 rap songs. Samantha randomly adds songs until she gets 10 rap songs. What is the distribution for the total number of songs if the three people combine their playlists later?

The number of songs on each of their playlists are Negative Binomial random variables X_1, X_2, X_3, each with probability of success (i.e., probability of a rap song) $p = 0.40$, and with respective numbers of successes $r_1 = 5$, $r_2 = 12$, and $r_3 = 10$. So the total number of songs on their playlists is $Y = X_1 + X_2 + X_3$, which is also Negative Binomial, with probability of success $p = 0.40$ on each trial, and $R = 5 + 12 + 10 = 27$ successes altogether.

Remark 17.6. Sums of Negative Binomial random variables
Suppose $X_1, X_2, \ldots, X_m$ are independent Negative Binomial random variables, each with a common probability of success "p" on each trial, and for which the desired numbers of successes are (respectively) $r_1, r_2, \ldots, r_m$. Then $X_1 + X_2 + \cdots + X_m$ is the number of trials needed to reach $r_1 + r_2 + \cdots + r_m$ successes. Thus, the sum $Y = X_1 + X_2 + \cdots + X_m$ is also Negative Binomial, with parameters $R = r_1 + r_2 + \cdots + r_m$ and p. So the sum Y has mass

$$p_Y(y) = \binom{y-1}{R-1} q^{y-R} p^R, \qquad \text{for integers } y \geq R$$

and $p_Y(y) = 0$ otherwise.

17.3 Exercises

17.3.1 Practice

Exercise 17.1. Skittles. Skittles candies come in the colors red, orange, yellow, green, and purple, with each of these colors having equal probability. Due to a dye mix-up there are a few rainbow-striped candies coming down the line. There is a 5% chance that a candy is rainbow striped. You are a quality control inspector, and your job is to find the 3rd rainbow-striped candy. (The population is so big, and we randomly sample the Skittles, so we can assume relative independence.)

a. Explain in words what X is in terms of the story. What values can it take?

b. Why is this a Negative Binomial distribution situation? What are the parameters?

c. What is the probability that you will need to check exactly 20 candies to find the 3rd rainbow one?

d. What is the probability that you will have to check fewer than 6 candies to find the 3rd rainbow one?

e. What is the probability you will have to check more than 6 candies to find the 3rd rainbow one?

f. What is the expected number of candies you will have to check to find the 3rd rainbow one?

g. What is the standard deviation of the number of candies you will have to check to find your 3rd rainbow one?

Exercise 17.2. Colorblind. Approximately 8.33% of men are colorblind. You randomly survey men from a large population until you find 2 who are colorblind.

a. Explain in words what X is in this situation and what values it can take.

b. Why is this a Negative Binomial distribution situation? What are the parameters?

c. What is the probability you will have to ask exactly 12 men to find the 2nd who is colorblind?

d. What is the probability that you will have to ask between 12 and 14 men to find the 2nd who is colorblind?

e. What is the expected value of X?

f. What is the variance of X?

g. Show the labeled graph of the mass for this story. Draw an arrow to indicate where the mean is.

h. Show the labeled graph of the CDF for this story.

Exercise 17.3. Surgical delivery. Twenty percent of babies in a particular city are born by a surgical procedure called a Cesarean section (C-section). You are interested in gathering information about hospital experiences of parents who did not have their baby by C-section. You randomly survey parents of babies from the population of the large city until you find 7 babies not born by C-section.

a. Explain in words what X is in this situation and what values it can take.

b. Why is this a Negative Binomial distribution situation? What are the parameters?

c. What is the probability that you will have to survey exactly 10 parents to find the seventh one not born by C-section?

d. What is the probability that you have to survey parents of at least 10 babies to find the seventh one not born by C-section? (Hint: Think about the complement and what X is allowed to be.)

e. What is the expected value of X?

f. What is the variance of X?

g. Show the labeled graph of the mass for this story.

h. Show the labeled graph of the CDF for this story.

Exercise 17.4. Consumer panel. A public relations intern realizes that she forgot to assemble the consumer panel her boss asked her to do. She panics and decides to randomly ask (independent) people if they will work on the panel for an hour. Since she is willing to pay them for their work, she believes she will have a 75% chance of people agreeing to work with her.

a. What is the expected number of people she will need to ask until she finds 5 who will agree to participate?

b. What is the standard deviation in the number of people she will need to ask until she finds 5 who will agree to participate?

c. Show the labeled graph of the mass for this story.

d. Show the labeled graph of the CDF for this story.

Exercise 17.5. Surfing. Harry is a surfer who can successfully ride about 70% of waves. Assume that his wave-riding ability is independent, from wave to wave. Let X be the number of waves that pass until he successfully catches his 8th wave. What is the probability that X is between 10 and 12 (inclusive)?

Exercise 17.6. Dart board. You throw darts at a board with 20 equally likely spaces. You throw until you hit "1," then you throw until you hit "2," then you throw until you hit "3," etc., and finally you throw until you hit "20." Let X be the number of throws you make to achieve this goal.

a. What are the expected value and standard deviation of X?

b. Let Y be the number of throws needed until the 20th time hitting 1. Do X and Y have the same distribution? Why or why not?

Exercise 17.7. Interviewing students. In the very large lecture class CHEM 115, 85% of students passed the first exam. I want to start a study group for students who failed the first test, with 10 students in the group. Because of FERPA privacy laws, I'm not allowed to get this information from the instructor. In surveying CHEM 115 students, let X be the number I have to interview to find 10 who failed the first exam. What are the expected value and standard deviation of X?

Exercise 17.8. Still looking for a wife. Twelve percent of single women in the kingdom have feet which will fit into a glass slipper. Prince Charming thinks he must continue finding women who fit such a slipper, so that he has a collection to choose from. He would like 10 women who fit the slipper to compete on a "Bachelor"-type show for his hand in marriage.

a. How many women should he expect to have to check until he finds 10 women who fit the glass slipper?

b. What is the probability he has to check 75 women?

c. Given that he has to check between 75 and 80 women, what is the probability he has to check exactly 75 women?

Exercise 17.9. Replaying a video game. You vow to replay a tough level in Super Mario World until you win 3 times. Assume that you have a 25% chance of winning each time you play, and each round is independent (your skill does not improve from game to game, because there is a lot of luck involved). Let X denote the number of times you have to play. Each attempt takes 5 minutes.

a. What is the probability it will take you more than an hour to win 3 times?

b. How many minutes do you expect to need to win 3 times?

17.3.2 Extensions

Exercise 17.10. Missing an early class. My alarm clock wakes me up only 64% of the time. My probability class meets at 8:30 am, and if I don't hear my alarm, I'll miss class. My teacher takes off one percent of our grade for every class that we miss. Let X be the number of class days that pass until I have lost 10% of my grade.

a. What is the expected value of X?

b. What is the probability that X is 30?

c. If there are 48 class days over the semester, what is the probability that I will have lost only 9% of my grade? What distribution is this? How do you know?

d. Let Y be the number of days until I lose my first percentage off my grade. What distribution is this? How do you know?

Exercise 17.11. Zombies. During a zombie apocalypse, one human finds that about 1 out of every 3 shots he makes actually kills a zombie. Let X be the number of shots he has to take until he kills his fifth zombie.

a. Given that it takes between 14 and 16 shots (inclusive) to kill his fifth zombie, what is the probability that it takes at least 15 shots?

b. Given that it takes him 12 tries to get his 4th success, what is the probability that he will need exactly 3 more tries to get his 5th success? What distribution is this? How do you know?

Exercise 17.12. Monopoly. Philip and Callum are playing Monopoly. Callum has an 80% chance of winning whenever he plays Philip. If they play Monopoly until somebody wins 3 games (assuming no ties and that the games are independent):

a. What is the probability that Callum wins the series in exactly 3 games? In 4 games? In 5 games?

b. What is the probability Philip wins the series in exactly 3 games? In 4 games? In 5 games?

c. Given that the series took 5 games, what is the probability that Callum won?

d. Given that Callum wins the series, what is the expected number of games that the boys play?

Chapter 18

Poisson Random Variables

Math is like war, people! If you fall behind in your unit, you will die!
—attributed to Mr. Williams, math teacher from Poland Seminary High School in Poland, Ohio, as remembered by Catharine Patrone, Director of Student Services for the Honors College at Purdue University

---

You are an epidemiologist trying to find people who have a rare disease you would like to study. The disease is so rare that only 1 out of every 20,000 people have it. In a city of 100,000 people, what is the probability that exactly 4 people have the disease?

18.1 Introduction

How do you pronounce Poisson? Pwah-sow(n), with the accent on the second syllable (the n is implicit). Poisson is French for "fish," but it is actually named after Siméon Poisson.

Instead of counting up the number of trials or the number of successes, Poisson random variables have a different motivation. When you know the *average rate* of occurrences of some event, the Poisson distribution is often correct for describing the number of events that actually occur. For instance, there might be a Poisson number of cars passing a building during a 1-hour period, or a Poisson number of raindrops landing on a sidewalk square in five minutes, or a Poisson number of shoppers in a store during a given 3.5-hour afternoon, etc.

We use λ as the average number of occurrences of an event during a fixed time period. We can think of λ as an *average rate*, because it depends on the time period. For instance, suppose that there are an average of 10 cars passing by a building in an hour. We use $\lambda = 10$ if we want the average rate of cars passing by the building in an hour, but we use $\lambda = 5$ if we want to know the average number of cars passing in a 30-minute period, or $\lambda = 40$ for the average number of cars passing in a 4-hour period. To use Poisson random variables in these simple settings, we must assume that the *average rate* is proportional to the length of time we observe. This is a very different use of random variables than we have encountered so far.

Poisson random variables can occur in other related ways too. For instance, while reading a novel, the number of errors per page can be treated as a Poisson random variable. In this case, the flow of words corresponds to the flow of time, and an error on a page corresponds to an event.

The notation for a Poisson distribution looks like:

The λ is the average rate of events.

$$X \sim \text{Poisson}(\lambda),$$

where λ denotes the average rate of events.

Poisson random variables

The common thread: How many events in a given period? You know the average rate of events, i.e., number of events that occur in a period (this average rate must be proportional to the length of the period). You count the actual number of events that occur in such a time period. For instance, you measure the number of cars that pass in a 40-minute period, the number of raindrops that fall in one minute on a particular sidewalk square, the number of times that cell phones in a room are ringing, etc.

Things to look for: A number of events, for which the average rate is known.

The variable:

$$X = \# \text{ of events that occur during the specified period}$$

The parameter:

$$\lambda = \text{the average rate of events that occur during the specified period}$$

Mass:

$$P(X = x) = \frac{e^{-\lambda}\lambda^x}{x!}, \qquad x = 0, 1, 2, 3, \ldots$$

Expected value formula:

$$\mathbb{E}(X) = \lambda$$

Variance formula:

$$\text{Var}(X) = \lambda$$

Poisson random variables are not poison.

Example 18.1. Let X denote the number of errors on a randomly selected page of a book. Suppose that X has mass

$$p_X(j) = P(X = j) = \frac{e^{-0.2}(0.2)^j}{j!}.$$

Then X is a Poisson random variable.

In this case, for instance, the expected number of errors on a randomly selected page is $\mathbb{E}(X) = 0.2$, and the variance of the number of errors on a randomly selected page is also $\text{Var}(X) = 0.2$.

For example, the probability of exactly 3 errors on a randomly selected page of the book is

$$p_X(3) = \frac{e^{-0.2}(0.2)^3}{3!} = 0.00109.$$

Example 18.2. Customers arrive at a checkout counter at an average rate of 8 per hour. The cashier will count how many come in a particular time frame.

a. Why is this story a Poisson situation? What is the parameter?

We are not looking for successes in individual trials. We are given a rate (customers per hour), and we are asked to count up the number of customers who come in a particular time frame. The average rate is $\lambda = 8$ customers/hour.

b. What is the probability that exactly 2 customers arrive in the next hour?

$$P(X = 2) = \frac{8^2 e^{-8}}{2!} = 0.01073$$

c. What is the probability that no more than 3 customers arrive in the next hour?

$$\begin{aligned}
P(X \le 3) &= P(X = 0) + P(X = 1) + P(X = 2) + P(X = 3) \\
&= \frac{8^0 e^{-8}}{0!} + \frac{8^1 e^{-8}}{1!} + \frac{8^2 e^{-8}}{2!} + \frac{8^3 e^{-8}}{3!} \\
&= \left(\frac{8^0}{0!} + \frac{8^1}{1!} + \frac{8^2}{2!} + \frac{8^3}{3!}\right) e^{-8} \\
&= 126.333 e^{-8} \\
&= 0.04238
\end{aligned}$$

d. Given that at least 1 customer arrives in the next hour, what is the probability that more than 3 arrive? (Notice also that the Poisson is not memoryless.)

$$\begin{aligned}
P(X > 3 \mid X \ge 1) &= \frac{P(X > 3 \cap X \ge 1)}{P(X \ge 1)} \\
&= \frac{P(X > 3)}{P(X \ge 1)} \qquad (X > 3 \ \& \ X \ge 1 \text{ only if } X > 3) \\
&= \frac{1 - P(X \le 3)}{1 - P(X = 0)} \qquad \text{computing the complement} \\
&= \frac{1 - 0.04238}{1 - 0.0003355} \\
&= 0.9579
\end{aligned}$$

In general, for Poisson random variables, if we need to compute $P(X > a)$, and we only have a calculator, then we need to compute the complement:

$$P(X > a) = 1 - P(X \leq a) = 1 - P(X = 0) - P(X = 1) - \cdots - P(X = a).$$

A direct calculation is infeasible because $P(X > a) = \sum_{j=a+1}^{\infty} P(X = j)$ is an infinite sum that we cannot simplify (except by using the complement, as suggested above).

e. What does the mass look like for this story?

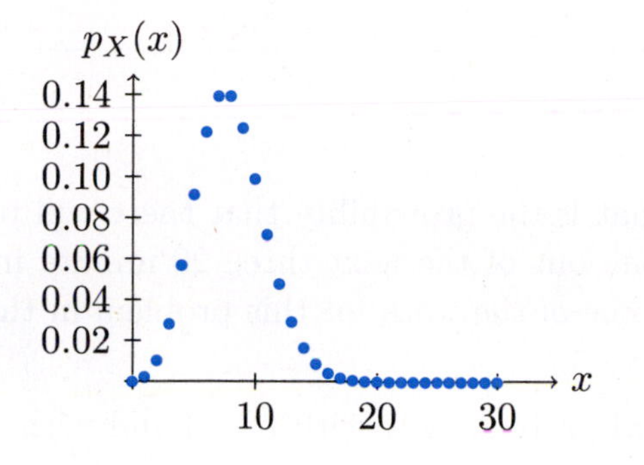

f. What does the CDF look like for this story?

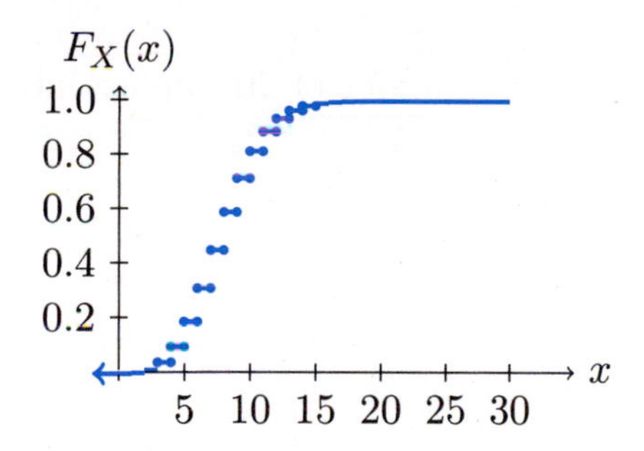

Example 18.2 (continued) Now we continue to study customer arrivals that have a Poisson distribution, but we suppose that the length of time changes. It is important to make sure that the expected number of arrivals in the time interval is correct.

g. What is the probability that there will be exactly 14 customer arrivals in the next 2 hours?

Be careful: Make sure your units for λ match the question you are being asked.

First, we have changed our time interval, so we need to change our λ to 16 customers/2-hour interval.

$$P(X = 14) = \frac{16^{14} e^{-16}}{14!} = 0.09302$$

h. How many customers do you expect in the next 2 hours?

$$\mathbb{E}(X) = \lambda = 16 \text{ customers}$$

i. What is the probability that there will be exactly 6 customer arrivals in the next 20 minutes?

Since 20 minutes is 1/3 of an hour, we need to change the scale: we change λ to 2.6667 customers/20-minute interval.

$$P(X=6) = \frac{2.667^6 e^{-2.6667}}{6!} = 0.03472$$

Example 18.3. What is the probability that there will be exactly 6 customer arrivals in exactly one out of the next three 20-minute intervals? (Hint: You have already done some of the work for this problem in the previous question.)

This is a Binomial problem now with $n = 3$ and with

$$p = \text{the answer to Example 18.2i} = 0.03472.$$

The random variable X is the number of 20-minute intervals in which exactly 6 customers arrive.

$$P(X=1) = \binom{3}{1}(0.03472)^1(0.96528)^2 = 0.09705$$

Remark 18.4. Expected value of a Poisson random variable If X is Poisson with average rate λ of events, then the expected value of the Poisson random variable is

$$\mathbb{E}(X) = \sum_{j\geq 0} jP(X=j) = \sum_{j\geq 0} j\frac{e^{-\lambda}\lambda^j}{j!}.$$

The $j = 0$ term is 0, so drop it, and this leaves

$$\mathbb{E}(X) = \sum_{j\geq 1} j\frac{e^{-\lambda}\lambda^j}{j!} = \lambda\sum_{j\geq 1}\frac{e^{-\lambda}\lambda^{j-1}}{(j-1)!}.$$

Then we shift the indices by 1, to obtain

$$\mathbb{E}(X) = \lambda\sum_{j\geq 0}\frac{e^{-\lambda}\lambda^j}{j!} = \lambda e^{-\lambda}e^{\lambda} = \lambda.$$

Remark 18.5. Variance of a Poisson random variable As with Geometric random variables, we use $X^2 = (X)(X-1) + X$ to make the second moment easier. We compute

$$\begin{aligned}\mathrm{Var}(X) &= \mathbb{E}(X^2) - (\mathbb{E}(X))^2 \\ &= \mathbb{E}((X)(X-1)) + \mathbb{E}(X) - (\mathbb{E}(X))^2 \\ &= \mathbb{E}((X)(X-1)) + \lambda - \lambda^2.\end{aligned}$$

Now we think about how the simplification of $j/j! = 1/(j-1)!$ was performed for the mean. We will use $(j)(j-1)/j! = 1/(j-2)!$ below. We write

$$\mathbb{E}((X)(X-1)) = \sum_{j\geq 0} (j)(j-1)P(X=j) = \sum_{j\geq 0} (j)(j-1)\frac{e^{-\lambda}\lambda^j}{j!}.$$

The $j = 0$ and $j = 1$ terms are 0, so we drop them, and get

$$\mathbb{E}((X)(X-1)) = \sum_{j\geq 2} (j)(j-1)\frac{e^{-\lambda}\lambda^j}{j!} = \lambda^2 \sum_{j\geq 2} \frac{e^{-\lambda}\lambda^{j-2}}{(j-2)!}.$$

Finally, we shift the indices by 2, to conclude

$$\mathbb{E}((X)(X-1)) = \lambda^2 \sum_{j\geq 0} \frac{e^{-\lambda}\lambda^j}{j!} = \lambda^2 e^{-\lambda} e^{\lambda} = \lambda^2.$$

Thus

$$\mathrm{Var}(X) = \lambda^2 + \lambda - \lambda^2 = \lambda.$$

18.2 Sums of Independent Poisson Random Variables

Remark 18.6. The sum of independent Poisson random variables is Poisson too. If X and Y are both Poisson random variables with parameters λ_1 and λ_2, respectively, then we can show $X + Y$ is also Poisson with parameter $\lambda_1 + \lambda_2$. We define $Z = X + Y$ and $\lambda = \lambda_1 + \lambda_2$, and it suffices to prove that Z has mass

$$p_Z(j) = \frac{e^{-\lambda}\lambda^j}{j!} \qquad \text{for } j = 0, 1, 2, \ldots$$

and $p_Z(z) = 0$ otherwise.

To prove this, observe Z can only take on nonnegative integer values, since X and Y are each known to take on only nonnegative integer values. Also, in order to have $Z = j$, we need to have $X = i$ for some i with $0 \leq i \leq j$, and then we also need $Y = j - i$. So we just compute

$$p_Z(j) = \sum_{i=0}^{j} P(X = i \text{ and } Y = j - i) = \sum_{i=0}^{j} P(X = i)P(Y = j - i),$$

where the last equality comes from the independence of X and Y. Moreover, we know the masses of X and Y and can use them, as follows

$$p_Z(j) = \sum_{i=0}^{j} \frac{e^{-\lambda_1}\lambda_1^i}{i!}\frac{e^{-\lambda_2}\lambda_2^{j-i}}{(j-i)!}.$$

Next we factor out $e^{-\lambda_1-\lambda_2}$, and multiply and divide by $j!$. Using $\lambda = \lambda_1 + \lambda_2$, this gives

$$p_Z(j) = \frac{e^{-\lambda_1-\lambda_2}}{j!}\sum_{i=0}^{j}\frac{j!}{i!(j-i)!}\lambda_1^i\lambda_2^{j-i} = \frac{e^{-\lambda}}{j!}\sum_{i=0}^{j}\binom{j}{i}\lambda_1^i\lambda_2^{j-i}.$$

By the Binomial theorem, we conclude

$$p_Z(j) = \frac{e^{-\lambda}}{j!}(\lambda_1+\lambda_2)^j = \frac{e^{-\lambda}\lambda^j}{j!}.$$

Thus, we have established the following:

Theorem 18.7. Sum of independent Poisson random variables
If X and Y are independent Poisson random variables with parameters λ_1 and λ_2, then $X+Y$ is also Poisson with parameter $\lambda_1+\lambda_2$.

Applying this reasoning repeatedly, we get the following:

Corollary 18.8. Sum of independent Poisson random variables
If $X_1, X_2, \ldots, X_n$ are independent Poisson random variables with parameters $\lambda_1, \lambda_2, \ldots, \lambda_n$, then $X_1 + X_2 + \cdots + X_n$ is also Poisson with parameter $\lambda_1 + \lambda_2 + \cdots + \lambda_n$.

Example 18.9. With the knowledge that the sum of two independent Poisson random variables is also a Poisson, we can return to Example 18.2g, and let Y be the number of customers during hour 1, and Z be the number of customers during hour 2. Then Y, Z are independent Poisson random variables, each with mean 8. So $X = Y + Z$ is the total number of customers during the two hours, and X is also Poisson, with mean $8 + 8 = 16$.

18.3 Using the Poisson as an Approximation to the Binomial

The Poisson distribution can also be used as an approximation to the Binomial distribution in certain cases. Here's how it works.

When does this work? If n is really big (say, for instance, $n \geq 1000$) and p is really close to 0 or really close to 1. How "big" and how "close" are needed? If n is large and npq is near 1, or perhaps (as a rule of thumb) within a factor of 10 away from 1, i.e.,

$$1/10 \leq npq \leq 10,$$

then the Poisson approximation to the Binomial is usually very appropriate. For instance, if $n = 10{,}000$ and $p = \frac{1}{10{,}000}$, and $q = \frac{9999}{10{,}000}$, then $nqp = 0.9999$, so the Poisson approximation to the Binomial should be good.

What parameters are involved? The expected value of a Poisson is λ, which is an average rate. The expected value of a Binomial is n (number of trials) times p (probability of success on a single trial). So we must set the expected values equal.

How do you make the switch?

$$\lambda = np$$

We're setting

$$\mathbb{E}(X) \text{ for a Poisson} = \mathbb{E}(X) \text{ for a Binomial.}$$

Remark 18.10. Poisson approximation to Binomial random variables If X is a Binomial random variable with parameters n and p, and if n is a really large positive integer, and if npq is somewhat close to 1, then many of the attributes of X are approximately the same as the attributes of the Poisson random variable with parameter $\lambda = np$.
(The reasoning behind Poisson approximation is explained in Remark 18.13.)

Example 18.11. Poisson approximation to the Binomial. In a city with a population of 100,000, a person has a 0.00005 chance of developing a rare disease. What is the probability that exactly 4 people have this disease?

a. If X is Binomial with $n = 100{,}000$ and $p = 0.00005$, then

$$P(X = 4) = \binom{100{,}000}{4}(0.00005)^4(0.99995)^{99{,}996} = 0.175470002.$$

Our calculators don't want the calculation above. Does yours? On the other hand, we can do the approximation in the next part on any hand-held calculator, and it agrees with the actual answer to several decimal places of accuracy.

b. If Y is Poisson with $\lambda = np = (100{,}000)(0.00005) = 5$, then

$$P(Y = 4) = \frac{5^4 e^{-5}}{4!} = 0.175467370.$$

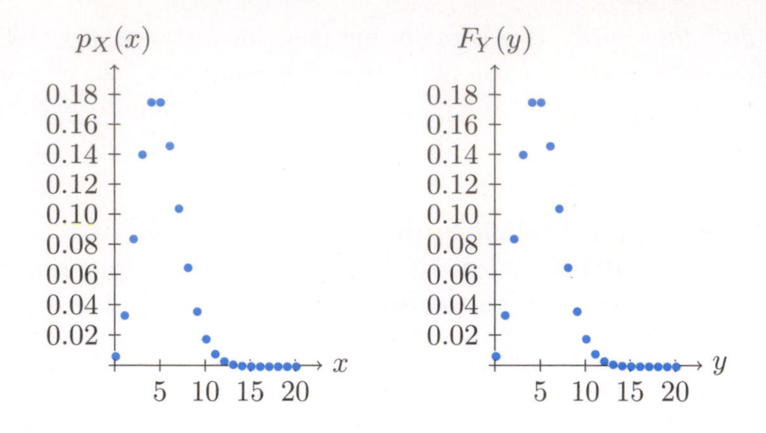

FIGURE 18.1: Left: The mass of X, a Binomial with $n = 100{,}000$ and $p = 0.00005$. Right: The mass of Y, a Poisson with $\lambda = 5$.

c. What is the expected number of people in this city who have this disease?

Using Binomial or Poisson, $\mathbb{E}(X) = \mathbb{E}(Y) = 5$.

d. What do the masses look like for this story? Show for both the Binomial and for the Poisson.

See Figure 18.1. As you can see from these graphs, the masses look practically identical for the Poisson and the Binomial with these parameters, and the differences are extremely small.

Example 18.12. If there are $n = 10000$ words in a short story, and if each word has a probability of $p = 1/3000$ of getting misspelled, then the number of misspelled words in the entire short story is approximately Poisson, with expected value $np = 10/3 = 3.3333$.

In this example, if we let X be the number of misspelled words, then X is a Binomial random variable with $n = 10000$ and $p = 1/3000$. So the exact probability that there is at most 1 misspelled word on the page is $p_X(0) + p_X(1) = \binom{n}{0}p^0(1-p)^{10000} + \binom{n}{1}p^1(1-p)^{9999} = 0.15454$. The number of errors, however, is approximately a Poisson random variable Y with parameter $\lambda = np = 10/3$. So the probability that there is at most 1 misspelled word on the page in the story is approximately

$$p_Y(0) + p_Y(1) = \frac{e^{-10/3}(10/3)^0}{0!} + \frac{e^{-10/3}(10/3)^1}{1!} = e^{-10/3}(13/3) = 0.15459.$$

Remark 18.13. We briefly explain why a Poisson random variable Y with parameter $\lambda = np$ has many of the same properties as a Binomial random variable X with parameters n and p, when n is large and npq is moderate.

The exact mass of X at j is

$$p_X(j) = \binom{n}{j} p^j q^{n-j} = \frac{(n)(n-1)(n-2)\cdots(n-j+1)}{j!} p^j (1-p)^{n-j}.$$

When n is much larger than j, then n and $n-1$ and $n-2$ and ... and $n-j+1$ are all approximately near n, and thus $(n)(n-1)(n-2)\cdots(n-j+1) \approx n^j$. Also

$$(1-p)^{n-j} = \left(1 - \frac{np}{n}\right)^{n-j}.$$

Now we substitute $\lambda = np$. From calculus, we learned that $\left(1 - \frac{c}{n}\right)^n \to e^{-c}$ as $n \to \infty$. Again, if j is much smaller than n, this means that

$$(1-p)^{n-j} = \left(1 - \frac{np}{n}\right)^{n-j} = \left(1 - \frac{\lambda}{n}\right)^{n-j} \approx e^{-\lambda}$$

Putting all of this together, and again using $np = \lambda$, we have

$$\begin{aligned} p_X(j) &= \binom{n}{j} p^j q^{n-j} \\ &= \frac{(n)(n-1)(n-2)\cdots(n-j+1)}{j!} p^j (1-p)^{n-j} \\ &\approx \frac{n^j}{j!} p^j e^{-\lambda} = \frac{e^{-\lambda}\lambda^j}{j!} = p_Y(j) \end{aligned}$$

So the mass of X at j is approximately equal to the mass of Y at j.

18.4 Exercises

18.4.1 Practice

Exercise 18.1. Marriage licenses. According to the *Guinness Book of World Records* (2005), there are 280 marriage licenses issued in Las Vegas per day. Let X be the number of marriage licenses issued in the next 5 minutes.

a. What values can X take? Why is this a Poisson situation? What is the parameter?

b. What is the exact probability that exactly 300 marriage licenses will be issued tomorrow (a decimal approximation is not needed)?

c. What is the expected number of marriage licenses issued during the next *hour*?

d. What is the standard deviation of the number of marriage licenses issued in the next hour?

e. What is the probability that between 10 and 12 (inclusive) marriage licenses will be issued in the next hour?

f. What is the probability that between 10 and 12 marriage licenses will be issued in 12 out of the next 24 hours? What distribution are you using now and what are the parameters?

Exercise 18.2. Left-handed. Approximately 6.85 left-handed people are killed each day by using an object or machinery designed for right-handed people. Let X be the number of left-handed people killed this way in one day.

a. What values can X take? Why is this a Poisson situation? What is the parameter?

b. What is the probability that exactly 7 left-handed people will be killed using a right-handed object tomorrow?

c. What is the expected number of left-handed people killed using a right-handed object over the next *week*?

d. What is the standard deviation of the number of left-handed people killed using a right-handed object over the next *week*?

e. What is the probability that at least 2 left-handed people will be killed using a right-handed object tomorrow?

f. What is the probability that you will have to check more than 3 days until you find the first day which has at least 2 left-handed people killed using a right-handed object? What distribution are you using now and what is the parameter?

Exercise 18.3. Laughing. Children laugh an average of 16.67 times an hour and adults laugh an average of 0.6667 times an hour. Let X be the number of times a child laughs in an hour, and let Y be the number of times an adult laughs in an hour.

a. What is the probability that a child will laugh exactly 20 times in the next hour? What is the corresponding probability that an adult will laugh exactly 20 times in the next hour?

b. Why are these situations Poisson?

c. Given that an adult laughs more than 1 time in an hour, what is the probability that the adult laughs at least 3 times in the next hour?

d. Given that an adult laughs less than 3 times in the next hour, what is the probability that the adult laughs once or less?

e. Show the labeled graph of the mass for adults in this story.

f. Show the labeled graph of the mass for children in this story.

g. Show the labeled graph of the CDF for adults in this story.

h. Show the labeled graph of the CDF for children in this story.

(The graphs for e and g should look relatively clear after just a few points, but perhaps two dozen points will be needed to get a good understanding of the graphs for f and h. A graphing calculator or computer might make the graphs in f and h quicker to computer.)

Exercise 18.4. Pumpkin carving. According to the *Guinness Book of World Records* (2005), the fastest pumpkin-carver on record, Steven Clarke, carved 42 pumpkins an hour. Assume this is his average rate. Let X be the number of pumpkins Steven carves in an hour. (We suppose that the carver can steadily maintain work at his record rate.)

a. Assume that this a Poisson situation. What is the parameter?

b. What is the probability Steven will carve exactly 40 pumpkins in the next hour?

c. Given that he has carved at least 3 pumpkins in a 5-minute interval, what is the probability that he will carve at least 4 pumpkins during that 5-minute interval?

d. Given that he carves fewer than 4 pumpkins in a 5-minute interval, what is the probability he carves fewer than 2 pumpkins during that 5-minute interval?

e. What is the expected number of pumpkins he can carve in 3 minutes at this pace?

f. Show the labeled graph of the mass for pumpkins carved in a 3-minute interval.

g. Show the labeled graph of the CDF for pumpkins carved in a 3-minute interval.

Exercise 18.5. Quadruplets. The probability of a mother giving birth to quadruplets is 1 in 729,000. An obstetrician checks the records of 1,000,000 mothers, to see how many mothers of quadruplets are in her database.

a. Which distribution is technically appropriate in this situation? What are the parameters?

b. Write a formula (using the distribution in part a) for the probability that there are exactly 3 mothers of quadruplets, but do not solve.

c. What is the expected number of mothers of quadruplets in this sample?

d. Which distribution would be a good approximation to the distribution in part a? What is the parameter?

e. Approximate the probability in part b by using the approximate distribution in part d, and solve.

Exercise 18.6. Roseate spoonbills. Roseate spoonbills are common along the Gulf Coast of the U.S. See `nationalzoo.si.edu/Animals/Birds/Facts/fact-rosespoonbill.cfm` for more information. Suppose that the number of roseate spoonbills that fly overhead in 1 hour (at a nature preserve) has a Poisson distribution with mean 2. Also suppose that the number of roseate spoonbills is independent from hour to hour. For example, the number of birds between noon and 1 PM does not affect the number of birds between 1 PM and 2 PM.

a. A bird watcher sits and looks for the birds for 3 hours. How many of these birds does she expect to see?

b. What is the probability that she sees exactly 5 of these birds during a 3-hour time period?

c. What is the probability that she sees exactly 1 bird in each of 3 separate 1-hour time periods? Which distribution is this, and what are the parameters?

Exercise 18.7. Lottery. In a person's lifetime, he enters a local daily raffle 10,000 times, but his chances of winning each time are only 1 in 5000.

Approximate the probability that he wins at least one time during his lifetime.

Exercise 18.8. Rare disease. There is a certain disease in America which is very rare. Each person has a probability of 1 in 100,000,000 of having the disease. Assume that there are 310,000,000 people in America.

a. Approximate the probability that nobody in America actually has the disease.

b. Approximate the probability that at most 3 people (i.e., 3 or less) in America have the disease.

Exercise 18.9. Cars. Cars pass an intersection at an average rate of 4 cars per 5-minute interval. Let X be the number of cars that pass in the next 5 minutes.

a. Show the labeled graph of the mass for X.

b. Show the labeled graph of the CDF for X.

Exercise 18.10. Customers. Pedestrians walk by your store front frequently, but they only come into your store about once every 10 minutes on average. Let X be the number of pedestrians who enter the store in the next hour.

a. Show the labeled graph of the mass for X.

b. Show the labeled graph of the CDF for X.

Exercise 18.11. Beetles. Beetles walk across a lunch counter at an average rate of 5 beetles per hour.

a. What is the probability that more than 2 beetles will cross the counter in the next half hour?

b. What is the probability that more than 2 beetles will cross the counter in each of the next two half hours?

c. What is the probability that more than 4 beetles will cross the counter in the next hour?

Exercise 18.12. New car contest. People are competing to win a new car. The task, which only 1 in 5000 people can achieve, is to hit a golf ball into a cup 250 feet away. One hundred thousand people sign up to compete.

a. How many people will win, on average? What is the standard deviation of the number of people who will win?

b. What is the probability that exactly 18 people win?

c. What is the probability that between 18 and 20 people win?

Exercise 18.13. Bakery. Customers arrive at a bakery at an average rate of 6 per half-hour. For simplicity, suppose each customer spends $2.50 on their purchase.

a. What is the probability that exactly 3 customers will arrive in the next 10 minutes?

b. What is the probability that at least 3 customers will arrive in the next 10 minutes?

c. Given that at least 3 customers will arrive in the next 10 minutes, what is the probability that exactly 3 will arrive?

d. What is the expected amount of money the bakery will make in the next 10 minutes?

e. What is the standard deviation in the amount of money the bakery will make in the next 10 minutes?

Exercise 18.14. Toy defects. Workers at a factory produce a toy with a defect about once every 4 hours on average. Each toy costs the factory approximately $7 in labor and supplies.

a. What is the expected number of toys with defects at the end of a 40-hour work week?

b. What is the standard deviation in the number of toys with defects at the end of a 40-hour work week?

c. What is the expected cost to the factory for toys with defects at the end of a 40-hour work week?

d. What is the standard deviation in cost to the factory for toys with defects at the end of the 40-hour work week?

Exercise 18.15. Toy defects (continued) Workers at a factory produce a toy with a defect about once every 4 hours on average.

a. What is the probability there will be exactly 5 defects in 24 hours?

b. What is the probability that there will be exactly 5 defects in each of the next seven 24-hour periods?

c. What is the probability there will be a total of $5 \times 7 = 35$ defects in the next week (168 hours)? Why is this answer different from part b?

Exercise 18.16. Albino fish. There are approximately 11,000 fish in a lake. Each fish has a 1 in 5500 chance of being albino. Let X be the number of albino fish.

a. What is the expected number of albino fish in the lake?

b. What is the approximate probability that there will be exactly 4 albino fish in the lake?

c. Given that there are at least 2 albino fish in the lake, what is the probability that there are exactly 4?

Exercise 18.17. Telemarketing. A telemarketing firm uses a computer connected to an online phone book for Metropolis. A large number of messages are sent, but on average only 10 people per hour listen to such messages. Let X be the number of people who listen to these messages during a given hour.

a. What is the probability that there will be exactly 8 successful calls in the next hour?

b. How many hours are expected to pass until the first hour with exactly 8 people listening to the messages? What distribution is this?

Exercise 18.18. Hungry customers. At a certain hot dog stand, during the working day, the number of people who arrive to eat is Poisson, with an average of 1 person every 2 minutes.

a. What is the probability that exactly 3 people arrive during the next 10 minutes?

b. What is the probability that nobody arrives during the next 10 minutes?

c. What is the probability that at least 3 people arrive during the next 10 minutes?

Exercise 18.19. Errors in a book. An author has carefully edited his book, but as all careful readers know, all books have some errors. Suppose that the number of errors *per page* is Poisson, with an average of 0.04 errors per page.

a. How many errors are expected in the next 100 pages?

b. What is the probability of exactly 5 errors in the next 100 pages?

Exercise 18.20. Telemarketers. Suppose that, on average, 3 telemarketers call your house during a 7-day period.

a. What is the mass of the number of telemarketers calling your house during one day?

b. What is the probability that no telemarketers will call your house tomorrow?

c. What is the probability that exactly 2 telemarketers will call your house tomorrow?

Exercise 18.21. Baseball fans. The number of Yankees fans shopping at a sports store in a given hour is Poisson with mean 8. The number of Red Sox fans shopping at the same store is Poisson with mean 6 per hour. Assume that the numbers of fans of the two types are independent. In particular, there is no person who is simultaneously a fan of both teams.

a. In a 3-hour period, how many Yankees and Red Sox fans do we expect altogether?

b. Find the probability that exactly one person enters the store during the next 20 minutes who likes the Yankees or Red Sox.

Exercise 18.22. Website visitors. Suppose that the number of men who visit a website is Poisson, with mean 12 per minute, and the number of women who visit the same site is also Poisson, with mean 15 per minute. Assume that the number of men and women are independent.

a. During the next 10 seconds, what is the probability that 1 man and 2 women visit the site?

b. What is the expected number of people who visit the site in the next 5 minutes?

b. What is the variance of the total number of people who visit the site in the next 5 minutes?

Exercise 18.23. Superfans. At the local stadium, there are 60,000 fans attending a football game. It is well known that only a few people at the game will be impartial (i.e., will not care about the outcome of the game). Suppose that each person at the game has probability $\frac{1}{10{,}000}$ of being impartial (and that the impartiality of a fan has no bearing on the other fans).

a. Give an exact formula for the probability that 8 of the people at the game are impartial. You do *not* have to evaluate the formula on your calculator.

b. Use a Poisson estimation for the probability above.

c. Use your calculator to evaluate the Poisson expression that you gave in part b.

Exercise 18.24. Shoppers. During the holiday rush, there are 100,000 shoppers in a certain region. Each of these shoppers is extremely likely to make a purchase. Suppose that a person makes a purchase with probability 49,999/50,000 and declines to make a purchase with probability 1/50,000. Let X be the number of people who decline to make a purchase.

a. Give an exact formula for the probability that $P(X \leq 3)$. You do *not* have to evaluate the formula on your calculator.

b. Use a Poisson estimation for the probability above.

c. Use your calculator to evaluate the Poisson expression that you gave in part b.

18.4.2 Extensions

Exercise 18.25. Bakery (continued) If it costs the bakery owner \$20 to keep the bakery open each hour (staff, electricity, etc.), each customer spends \$2.50, and customers arrive at the bakery at an average rate of 6 per half-hour, will the bakery be able to stay in business (i.e., will the bakery make more money than it spends in an hour)?

Exercise 18.26. Verifying a mass. Verify that the mass of a Poisson random variable is really a mass, i.e., verify that the terms of the mass sum to 1.

18.4.3 Advanced

Exercise 18.27. If X is a Poisson random variable with $\lambda = 3$, find $\mathbb{E}(5^X)$.

Chapter 19

Hypergeometric Random Variables

And will you succeed? Yes indeed, yes indeed! Ninety-eight and three-quarters percent guaranteed.

—*Oh, the Places You'll Go!* by Dr. Seuss (Random House, 1990)

In a Lotto game, there are 40 balls labeled 1 to 40. You pick 5 different numbers for your lottery ticket. That night on TV, the state lottery office will randomly select 5 different balls (each with a unique number) to be the winning ticket. You could win \$1 million if you picked all 5 winning numbers, but you can win smaller prizes for picking 3 or 4 correct numbers. What is the probability that you pick all 5 winning numbers? What is the probability that you win a prize of some sort? How many numbers do you expect to get right?

19.1 Introduction

A Hypergeometric random variable is the number of desirable items we pick when selecting some items without replacement from a mixed collection of desirable and undesirable items.

Hypergeometric random variables have three parameters, N, M, n:

$$\begin{aligned} N &\text{ total items are available} \\ M &\text{ of the items are desirable} \\ N - M &\text{ of the items are undesirable} \\ n &\text{ items are selected} \end{aligned}$$

Hypergeometric random variables

The common thread: We choose some items from a collection, some of which are desirable and the others are undesirable. We see how many desirable items we get. For instance, we see how many peanut butter cookies we get when we select 5 cookies from a jar containing peanut butter and lemon cookies. Other examples include: We randomly choose some shoes and see how many are blue; we see how many parts on an assembly line are working correctly; etc.

Things to look for: We choose a collection of items *without replacement* and see how many of the chosen items are desirable. The selections are dependent.

The variable:

$$X = \# \text{ of desirable items selected}$$

The parameters:

$$\begin{aligned} N &\text{ total items are available} \\ M &\text{ of the items are desirable} \\ N - M &\text{ of the items are undesirable} \\ n &\text{ items are selected} \end{aligned}$$

Mass:

$$P(X = x) = \frac{\binom{M}{x}\binom{N-M}{n-x}}{\binom{N}{n}}.$$

Expected value formula:

$$\mathbb{E}(X) = n\frac{M}{N}$$

Variance formula:

$$\text{Var}(X) = n\frac{M}{N}\left(1 - \frac{M}{N}\right)\frac{N-n}{N-1}$$

Where does the mass formula come from?

$$\binom{M}{x} = \text{number of ways to select } x \text{ out of the } M \text{ desirable items}$$

$$\binom{N-M}{n-x} = \text{number of ways to select } n-x \text{ out of the } N-M \text{ undesirables}$$

$$\binom{N}{n} = \text{number of ways to select } n \text{ out of the } N \text{ items altogether}$$

$$\begin{aligned} \text{Hypergeometric random variable} &= \text{number of desirable items we received} \\ &= X_1 + X_2 + \cdots + X_n, \end{aligned}$$

where $X_j = 1$ if the jth item we choose is desirable. We emphasize that **the X_j's are dependent**. This is the first named random variable that we have seen that has dependent trials. Binomial, Geometric, and Negative Binomial random variables all use independent trials with the same probability of success on each trial.

The notation for the Hypergeometric looks like:

$$X \sim \text{Hypergeometric}(M, N, n),$$

where X is the number of desirable items we select.

19.2 Examples

Example 19.1. In a box of 64 crayons, 60 are in good condition, and 4 are broken. If a child randomly selects two crayons from the box, what is the probability that both are in good condition? What is the probability that exactly one of the two selected crayons is broken? What is the probability that both are broken?

Let X be the number of crayons selected which are in good condition. Then X is a Hypergeometric random variable with $N = 64$ crayons altogether, $M = 60$ crayons in good condition, and $N - M = 4$ crayons that are broken. The child selects $n = 2$ crayons. So the probability that both crayons selected are in good condition is

$$p_X(2) = P(X = 2) = \frac{\binom{60}{2}\binom{4}{0}}{\binom{64}{2}} = \frac{(60)(59)/2}{(64)(63)/2} = \frac{(60)(59)}{(64)(63)} = \frac{1770}{2016} = 0.877976.$$

The probability that exactly one of the two selected crayons is broken is

$$p_X(1) = P(X = 1) = \frac{\binom{60}{1}\binom{4}{1}}{\binom{64}{2}} = \frac{(60)(4)}{(64)(63)/2} = \frac{240}{2016} = 0.119048.$$

The probability that both crayons selected are broken is

$$p_X(0) = P(X = 0) = \frac{\binom{60}{0}\binom{4}{2}}{\binom{64}{2}} = \frac{(4)(3)/2}{(64)(63)/2} = \frac{6}{2016} = 0.002976.$$

Notice that these 3 probabilities add up to 1 because they cover every possible way of selecting 2 crayons.

Example 19.2. A college student is running late for his class and does not have time to pack his backpack carefully. He has 12 folders on his desk, 4 of which include homework assignments due today. Without taking time to look, he accidentally grabs just 3 folders from his stack. When he gets to class, he counts how many of them contain his homework assignments. Assume that all of the outcomes are equally likely.

a. Explain in words what X is in this story. What values can it take?

The random variable X is the number of homework assignment folders he grabs. Thus, X can be 0, 1, 2, or 3 since only three folders are sampled. (Even if more folders were to be sampled (i.e., if n was larger) the X could not—even under these new assumptions—be larger than 4, since there are only 4 desirable folders.)

b. Why is this a Hypergeometric distribution? What are the parameters?

Sampling without replacement; $M = 4, N = 12, n = 3$.

c. What is the probability 2 of them contain his homework assignments?

$$P(X = 2) = \frac{\binom{4}{2}\binom{8}{1}}{\binom{12}{3}} = \frac{(6)(8)}{220} = 0.2182$$

d. What is the probability at least 2 of the 3 folders contain his homework assignments?

$$\begin{aligned} P(X \geq 2) &= P(X = 2) + P(X = 3) \\ &= \frac{\binom{4}{2}\binom{8}{1}}{\binom{12}{3}} + \frac{\binom{4}{3}\binom{8}{0}}{\binom{12}{3}} \\ &= 0.21818 + 0.01818 \\ &= 0.2364 \end{aligned}$$

e. What is the probability that at least 1 of the folders out of the 3 he grabbed contain his homework?

$$P(X \geq 1) = 1 - \frac{\binom{4}{0}\binom{8}{3}}{\binom{12}{3}} = 1 - \frac{56}{220} = 1 - 0.2545 = 0.7455$$

f. On a different day, in the same situation (this student should really invest in a reliable alarm clock), the student grabbed 8 folders at random. What is the probability he got all 4 homework assignment folders?

Now $n = 8$, but there are only 4 possible "successes" available.

$$P(X = 4) = \frac{\binom{4}{4}\binom{8}{4}}{\binom{12}{8}} = \frac{(1)(70)}{495} = 0.1414$$

g. When grabbing the 8 folders, what is the probability he grabbed at least 4 homework assignment folders?

This will be the same as the answer for part f since there are only 4 successes possible.

h. When grabbing 8 folders, what is the average number of homework assignments he will get?

$$\mathbb{E}(X) = 8\left(\frac{4}{12}\right) = 2.6667 \text{ homework assignments}$$

i. What is the standard deviation in the number of homework assignments he will get when he grabs 8 folders?

$$\text{Var}(X) = 8\left(\frac{4}{12}\right)\left(1 - \frac{8}{12}\right)\left(\frac{12-8}{12-1}\right) = 0.3232$$
$$\sigma_X = \sqrt{0.3232} = 0.5685 \text{ homework assignments}$$

Example 19.3. Suppose that there are 100 roseate spoonbills altogether in Indiana. Also suppose that 40 of them have been observed in 2009 by an ornithologist (a computerized tagging system allows the observer to know when the observances are unique). In 2010 he observes 32 birds. What is the probability that j out of these 32 birds were already seen previously, in 2009?

In this case, there are $N = 100$ roseate spoonbills altogether. The "desirable" birds are the ones that were previously seen in 2009, i.e., $M = 40$; the undesirable birds are the ones that were not seen in 2009, i.e., $N - M = 60$. We use X to denote the number of the birds out of the $n = 32$ seen in 2010 which had been previously seen in 2009. So the probability that exactly j of the 32 birds this year were seen in 2009 is

$$p_X(j) = P(X = j) = \frac{\binom{40}{j}\binom{60}{32-j}}{\binom{100}{32}}.$$

Remark 19.4. Expected value and variance of a Hypergeometric random variable If X is Hypergeometric, then X can be viewed as the sum of Bernoulli random variables **that are dependent**, where $X_j = 1$ if the jth item selected is desirable, and $X_j = 0$ otherwise. Thus

$$X = X_1 + X_2 + \cdots + X_n.$$

Any of the N items is equally likely to be chosen on the jth draw, and exactly M of them are desirable, so $\mathbb{E}(X_j) = M/N$ for each j. (This is worrisome to some students at first, but just keep in mind that we are momentarily only focused on the jth draw.) Even though the X_j's are dependent, the expected value is still linear, i.e., the expected value of the sum is equal to the sum of the expected values:

$$\mathbb{E}(X) = \mathbb{E}(X_1 + \cdots + X_n) = \mathbb{E}(X_1) + \cdots + \mathbb{E}(X_n) = \frac{M}{N} + \cdots + \frac{M}{N} = n\frac{M}{N}.$$

For the second moment:

$$\begin{aligned}\mathbb{E}(X^2) &= \mathbb{E}((X_1 + \cdots + X_n)(X_1 + \cdots + X_n)) \\ &= \sum_{j=1}^{n} \mathbb{E}(X_j X_j) + 2 \sum_{1 \le i < j \le n} \mathbb{E}(X_i X_j)\end{aligned}$$

Each X_j is either 0 or 1, so $X_j X_j = X_j$ always. Thus $\mathbb{E}(X_j X_j) = \mathbb{E}(X_j) = M/N$.

Also, for $i \neq j$, we see $X_i X_j = 1$ if and only if $X_i = 1$ and $X_j = 1$. So

$$\begin{aligned}\mathbb{E}(X_i X_j) &= P(X_i X_j = 1) \\ &= P(X_i = 1 \text{ and } X_j = 1) \\ &= P(X_i = 1) P(X_j = 1 \mid X_i = 1).\end{aligned}$$

As before, $P(X_i = 1) = M/N$. Given that $X_i = 1$, there are $M - 1$ desirable items out of $N - 1$ other items that could be chosen on the jth draw, so

$$P(X_j = 1 \mid X_i = 1) = (M - 1)/(N - 1).$$

There are n terms of the form $\mathbb{E}(X_j X_j)$. The other $n^2 - n$ terms are of the form $\mathbb{E}(X_i X_j)$. So

$$\mathbb{E}(X^2) = n\frac{M}{N} + (n^2 - n)\frac{M}{N}\frac{M-1}{N-1}.$$

Thus

$$\mathrm{Var}(X) = \mathbb{E}(X^2) - (\mathbb{E}(X))^2 = n\frac{M}{N} + (n^2 - n)\frac{M}{N}\frac{M-1}{N-1} - \left(n\frac{M}{N}\right)^2,$$

which simplifies to

$$\mathrm{Var}(X) = n\frac{M}{N}\left(1 - \frac{M}{N}\right)\frac{N-n}{N-1}.$$

19.3 Binomial Approximation to the Hypergeometric

The Binomial distribution can be used as an approximation to the Hypergeometric distribution in certain cases.

When does this work? If N is really big (e.g., in the hundreds or thousands, such an approximation works great) and n is relatively small, compared to N. Remember that N is the population size and n is the sample size.

Why does this work? If your population size is really big, then even though you are sampling without replacement for a Hypergeometric, your population doesn't really notice that you've taken a sample. The probability of getting a success on your second selection is approximately the same as your probability of getting a success on your first selection. For a big population, sampling with replacement and sampling without replacement give similar results because it is very unlikely that the same person or item would be selected twice.

What parameters are involved?

Binomial needs an n (number of trials) and a p (probability of success on a single trial).

Hypergeometric has the number of trials n already, and

$$\frac{M}{N} = \frac{\#\text{ successes in population}}{\text{total }\#\text{ in population}} \approx p$$

for a large population.

Example 19.5. Binomial approximation to the Hypergeometric. In a state lottery, one million tickets are issued, out of which only 50,000 are winning tickets. If a man buys 10 tickets, what is the probability exactly one of them is a winning ticket?

a. As a Hypergeometric, $n = 10$, $N = 1{,}000{,}000$, $M = 50{,}000$,

$$P(X = 1) = \frac{\binom{50{,}000}{1}\binom{950{,}000}{9}}{\binom{1{,}000{,}000}{10}}.$$

Our calculators do not want to do this! Does yours?

b. As a Binomial, $n = 10$, $p = \frac{M}{N} = \frac{50{,}000}{1{,}000{,}000} = 0.05$,

$$P(X = 1) = \binom{10}{1}(0.05)^1(0.95)^9 = 0.3151.$$

c. As a Hypergeometric, what is the expected number of winning tickets this man will have? What is the variance?

$$\mathbb{E}(X) = 10\left(\frac{50{,}000}{1{,}000{,}000}\right) = 0.5$$

$$\begin{aligned}\text{Var}(X) &= 10\left(\frac{50{,}000}{1{,}000{,}000}\right)\left(1 - \frac{50{,}000}{1{,}000{,}000}\right)\left(\frac{1{,}000{,}000 - 10}{1{,}000{,}000 - 1}\right)\\ &= 10(0.05)(0.95)(0.999991)\\ &= 0.4749957.\end{aligned}$$

d. As a Binomial, what is the expected number of winning tickets this man will have? What is the variance?

$$\begin{aligned}\mathbb{E}(X) &= 10(0.05) = 0.5\\ \text{Var}(X) &= 10(0.05)(0.95) = 0.475\end{aligned}$$

The answers for Binomial and Hypergeometric are almost exactly the same!

The masses for the Hypergeometric and for the Binomial approximation in this story agree to five decimal places of accuracy.

19.4 Exercises

19.4.1 Practice

Exercise 19.1. Vending machine. You are hungry and decide to go to patronize your office's vending machine. There are 6 bags of potato chips, 7 bags of pretzels, and 5 packs of chocolate chip cookies. Unfortunately there is something wrong with the buttons on the vending machine, and it will not let you type in your selection. However, it will drop 3 snacks out at random. You are hoping for 2 bags of cookies, and you don't care what the other snack will be.

a. What is a success in this story? What is a failure?

b. Explain in words what X is in terms of this story. What values can it take?

c. Why is this a Hypergeometric situation? What are the parameters?

d. What is the probability you get the 2 bags of cookies?

e. What is the probability that you get at most 1 bag of cookies?

f. What is the expected number of bags of cookies you will get?

g. What is the standard deviation in the number of bags of cookies you will get?

h. What is the expected number of bags of pretzels you will get?

i. What is the expected number of bags of potato chips you will get?

Exercise 19.2. SUV. You are going on a trip with some friends, and you need to rent 4 large cars to get all of you to the beach. There are 15 SUVs and 10 minivans on the lot. The rental car dealer will pick 4 cars at random. You are hoping to get SUVs because you think they look more stylish.

a. What is a success in this story? What is a failure?

b. Explain in words what X is in terms of this story. What values can it take?

c. Why is this a Hypergeometric situation? What are the parameters?

d. What is the probability you get 3 SUVs?

e. Given that you get at least 1 SUV, what is the probability you get at least 3 SUVs?

f. What is the expected number of SUVs you will get?

g. What is the standard deviation in the number of SUVs you will get?

h. What is the expected number of minivans you will get?

i. What is the standard deviation in the number of minivans you will get?

Exercise 19.3. Skittles. Skittles candies come in the colors red, orange, yellow, green, and purple. Green is your favorite color. Assume that the colors are evenly distributed in a bag of 45 candies. You will eat 10 candies, counting the number of green ones as you eat.

a. What is a success in this story? What is a failure?

b. Explain in words what X is in terms of this story. What values can it take?

c. Why is this a Hypergeometric situation? What are the parameters?

d. What is the probability you eat more than 2 green candies?

e. Given that you eat more than 2 green candies, what is the probability you eat exactly 9 green candies?

f. What is the expected number of green Skittles you will get?

Exercise 19.4. Bouquet. You are picking a bouquet of 20 flowers for your mother at random from a garden with 25 dahlias, 35 daisies, and 42 bachelor's buttons. Your mother likes daisies the best, and you want to end up with 15 of those in your bouquet.

a. What is a success here? What is a failure?

b. Explain in words what X is in terms of this story. What values can it take?

c. Why is this a Hypergeometric situation? What are the parameters?

d. What is the probability you pick exactly 15 daisies in your bouquet?

e. What is the probability that you end up with all daisies in your bouquet?

f. What is the expected number of daisies you will pick?

Exercise 19.5. Mislabeled DVDs. Vera wants to buy 20 copies of the DVD of "Bridget Jones' Diary" as a graduation gift for the women in her sorority. She buys them from a large internet company with 50,000 of the DVDs in stock. However 10,000 of the DVDs labeled as "Bridget Jones" were really "Terminator" and just mislabeled, leaving only 40,000 correctly labeled DVDs.

a. Explain in words what X is in terms of this story. What values can it take?

b. Why is this a Hypergeometric situation? What are the parameters?

c. What are the expected number of movies she receives which are truly "Bridget Jones"?

d. What is the probability that 18 of the DVDs she purchases are actually "Bridget Jones" and not "Terminator"? Just set up the equation, but don't try to solve it.

e. What would be a good approximation to use in this situation? What are the parameters?

f. What is the approximate probability that 18 of the DVDs she purchases are actually "Bridget Jones" and not "Terminator"? Set up the equation and solve.

Exercise 19.6. Cat fur vaccine. Dr. I.M.A. Genius has developed an experimental vaccine for an allergy to cat fur. He needs to test it out on some people to make sure it works. He has a list of 4000 people from his clinic, and 1250 of those have an allergy to cat fur. His receptionist randomly calls in 50 different people from his list, and he checks how many of them have cat fur allergies.

a. Explain in words what X is in terms of this story. What values can it take?

b. Why is this a Hypergeometric situation? What are the parameters?

c. What are the expected number of people in his sample who have cat fur allergies?

d. What is the probability that exactly 20 of the people in his sample have cat fur allergies? Just set up the equation, but don't try to solve it.

e. What would be a good approximation to use in this situation? What are the parameters?

f. What is the approximate probability that 20 of the people in his sample have cat fur allergies? Set up the equation and solve.

Exercise 19.7. Student volunteers. You have 5 freshmen, 6 sophomores, 10 juniors, and 2 seniors in your probability class. The teacher randomly selects 3 students to go to the board to do problems.

a. What is the probability all 3 students will be juniors?

b. What is the probability all 3 students will be from the same year?

c. What is the probability that at least one student will be a junior?

d. Given that at least one student will be a freshman, what is the probability that all 3 students will be freshmen?

e. What is the expected number of juniors selected each day?

f. What is the standard deviation of the number of juniors selected each day?

g. Assume that, every day, the teacher uses the same random method of selecting 3 students from the whole class. It is possible for the same student to be called on 2 days in a row but not twice in one class. What is the probability that the teacher will select only juniors for 2 days in a row? What distribution is this?

Exercise 19.8. Popsicles. In your freezer, there is a box of popsicles with 5 grape, 5 cherry, and 5 lime. Your freezer light isn't working, so you reach in and randomly pull out 4 popsicles for you and your friends on a hot afternoon. You are hoping to get all 4 the same flavor so that there won't be fighting. What is the probability of this happening?

Exercise 19.9. Ramen noodles. There are 20 packs of ramen noodle packages in a variety pack box, with 10 chicken and 10 beef flavored. What is the probability that you grab one of each when you reach in to pull out two packages of noodles without looking?

Exercise 19.10. Playlist. Suppose you have a playlist called "I Love the 90s" which contains 35 songs, including 10 tracks from the Spin Doctors' "Pocket Full of Kryptonite." If I shuffle the songs, what is the probability that 3 songs from the album (no repeats) shuffle to the top 5?

Exercise 19.11. Random beverages. At a barbecue there is a cooler with 50 beverages inside: 30 bottles of water and 20 cans of soda pop. I grab 4 beverages out of the cooler at random to share with some friends.

a. What is the probability of 3 of those 4 beverages being water?

b. What is the expected number of water bottles that I grab out of 4 total bottles?

Exercise 19.12. Granola bars. I have 6 chocolate chunk granola bars, 10 raspberry granola bars, and 8 chocolate chip granola bars. I grab 3 without looking.

a. What is the probability that I grab exactly 2 that are chocolate (either chunk or chip)?

b. What is the probability that I grab fewer than 2 that are chocolate?

c. What is the expected number of chocolate granola bars out of the 3 that I grab?

Exercise 19.13. Lucky Charms. Michael reaches into a bowl of 100 pieces of Lucky Charms cereal. He desires blue moons, which make up 5% of the charms. If his cereal bowl contains 40 charms, what is the probability he won't get any blue moons? What number of blue moons should he expect to get?

Exercise 19.14. Random pants. Henry has 10 pants: 4 dress slacks and 6 jeans. In a hurry while getting ready for a trip, he asks his kids to throw 3 pants in a suitcase for him without specifying which kind of pants he needs.

a. What is the probability the kids correctly throw in 2 pairs of dress slacks and 1 pair of jeans?

b. What is the probability the kids throw in 2 pairs of jeans and 1 pair of dress slacks?

c. Are the probabilities in parts a and b the same? Do they add up to 1? Should they add up to 1? Why or why not?

Exercise 19.15. Corn and beans. As a prank, your roommate removed all the labels from all the cans in your pantry and shuffled them around. Now you have no idea what is in a can when you get ready to cook dinner! You know that you have 8 cans of corn and 5 cans of beans. You don't want to waste any of the cans, so the best thing to do is to randomly open 2 cans every night and just eat whatever is in those cans.

a. What is the probability that on the first night you get one of each type of can?

b. If you got one of each type of can on the first night, what is the probability you got one of each type of can on the second night?

c. Are these probabilities in parts a and b the same? Should they be? Why or why not?

Exercise 19.16. Hungry customers. At a certain restaurant, during the working day, there are 12 customers. Seven of them have pizza, and the other five have burgers. Suppose that a person is conducting a survey of three customers at the restaurant, and he conducts the survey without replacement, i.e., he does not talk to the same customer more than once. All possible selections of three people for the survey are equally likely.

Let X be the number of people in the survey who are eating pizza.

a. What is the mass of X?

b. Evaluate the mass at the four values where the mass is strictly positive. (Make sure that your four numbers sum to 1.)

c. What is the average number of people in the survey who are eating pizza?

Exercise 19.17. **Harmonicas.** Carlos "Coyote" Jones owns quite a few harmonicas. In particular, he has 7 professional harmonicas and 12 cheaper harmonicas. They have relatively similar shapes, so when he reaches into his harmonica container without looking, he does not notice a difference between them. Suppose that he grabs 8 harmonicas, without replacement, and all selections are equally likely.

a. How many professional harmonicas does he expect to select?

b. What is the variance of the number of professional harmonicas that he selects?

c. What is the probability that exactly 5 out of the 8 harmonicas are professional?

d. What is the probability that he selects *exactly* 5 professional harmonicas, given that he selects *at least* 5 professional harmonicas?

19.4.2 Extensions

Exercise 19.18. **Capture-recapture sampling.** A wildlife biologist is trying to estimate population size of a pack of hyenas. She tags 20 hyenas and then releases them back into the wild. She revisits the same area 1 year later, and examines 50 hyenas. She notes that 10 of the hyenas have the tags from 1 year earlier. Estimate the population size, and explain your reasoning. Why would this be considered a form of Hypergeometric distribution?

Exercise 19.19. **Spades.** Given a standard deck of cards, you are dealt a hand of 13.

a. What is the probability that more than half the spades in the deck wind up in your hand?

b. How many spades are you expected to get in this hand?

Chapter 20

Discrete Uniform Random Variables

Uncertainty and expectation are the joys of life.

—*Love for Love, Act IV, Scene XX* by William Congreve (play that premiered on April 30, 1695)

When did you first start thinking about probability when you were young? Was it while playing a board game that involved rolling a die? A die roll is a fairly simple use of probability because each side is equally likely to come up. The values 1 to 6 are equally likely to appear.

20.1 Introduction

Think about rolling a die 200 times. We did this, and we show the results below. Our results should show a fairly similar number of 1s, 2s, 3s, 4s, 5s, and 6s.

die value	1	2	3	4	5	6
number of occurrences	36	36	28	31	38	31
percent of occurrences	0.18	0.18	0.14	0.155	0.19	0.155

Using the computer to simulate rolling a million dice, the results might be the following:

die value	1	2	3	4	5	6
occurrences	166863	166786	166513	166890	166753	166195
percent	0.1669	0.1668	0.1665	0.1669	0.1668	0.1662

(The more we roll, the more evenly distributed the results would be.) The result of a single die toss follows the Discrete Uniform distribution because each of the 6 possible outcomes is equally likely to occur.

For our purposes, we'll assume that any Discrete Uniform distribution, if not numbered consecutively with integers already, could be relabeled this way.

The Discrete Uniform is one of the simplest distributions.

Notation for the Discrete Uniform looks like:

$$X \sim \text{D.Uniform}(N)$$

Discrete Uniform random variables

The common thread: We choose one item out of a collection, to see which one we get. All possibilities are equally likely. For instance, the number of tickets (out of a fixed set of 157 shuffled tickets) that we need to check until we find the unique one with the special bonus prize. Other examples include: We choose exactly one student in a classroom; one book from a bookshelf; one employee in a company; etc.

Things to look for: We know the number of possible outcomes, all equally likely. We record which outcome is selected.

The variable:

$$X = \text{the choice of the outcome}$$

You usually order your outcomes and number them $1, 2, \ldots, N$.

It is also possible to have a Discrete Uniform distribution with X having values between a and b (an example would be raffle tickets numbered between 120 and 185). You could work with the numbers as they are or you could re-label these tickets as 1 to 66.

The parameters:

$$N = \text{the number of possible outcomes.}$$

Mass:

$$P(X = x) = \frac{1}{N} \qquad x = 1, 2, \ldots, N$$

Expected value formula:

$$\mathbb{E}(X) = \frac{N+1}{2}$$

Variance formula:

$$\text{Var}(X) = \frac{N^2 - 1}{12}$$

Where does the mass formula come from? All of the $p_X(x)$ values are the same, and we know that they also sum to 1:

$$p_X(1) + p_X(2) + \cdots + p_X(N) = 1.$$

So we must have $p_X(x) = 1/N$ for each $1 \le x \le N$.

Note: For some of these Discrete Uniform distribution situations, the expected value and the variance won't make much sense to the story. The Gilbreth family example below is one of those, but the die tossing example does make sense.

20.2 Examples

Example 20.1. The Gilbreth family (from Cheaper by the Dozen fame) has 12 children (in order from oldest to youngest): Anne, Ernestine, Mary, Martha, Frank, Bill, Lillian, Fred, Dan, Jack, Bob, and Jane. Mr. Gilbreth needs some help with a project, so he whistles for a child to come help him. Each of the children is equally likely to appear. We are numbering the children so that 1 is for Anne (the oldest) and 12 is for Jane (the youngest).

a. What is the probability Anne comes to help?

$P(X = 1) = 1/12 = 0.08333$

b. What is the probability Bob comes to help?

$P(X = 11) = 1/12 = 0.08333$

c. What is the probability Martha does not come to help?

$P(X \neq 4) = 1 - P(X = 4) = 11/12 = 0.91667$

d. Why is this a Discrete Uniform distribution, and what is the parameter?

All 12 outcomes are equally likely, and we are only selecting one at random. $N = 12$, the total number of possible outcomes.

e. Explain in words what X is in terms of the story. What values can it take on?

The random variable X is the age order of the person who comes to help. So X can be $1, 2, \ldots, 12$.

Example 20.2. For a die toss on a fair, 6-sided die, the value of the roll is a Discrete Uniform, between 1 and 6. Let X be the value of a die roll.

a. $\mathbb{E}(X) = (N + 1)/2 = (6 + 1)/2 = 3.5$

b. $\text{Var}(X) = (N^2 - 1)/12 = (6^2 - 1)/12 = 35/12 = 2.9167$

c. The mass and CDF of X were already shown back in Figures 8.1 and 8.2.

d. The experimental results above give very similar results to the theoretical distribution. As the number of experiments (here, number of rolls) grows, the closer the experimental results will get to the theoretical results (this will be investigated further in Chapter 37, when we study the Central Limit Theorem).

Example 20.3. As in Example 10.8, a student shuffles a deck of cards thoroughly (one time) and then selects cards from the deck *without replacement* until the ace of spades appears. If we let X denote the number of cards that the student expects to draw, until the ace of spades appears, then X is a Discrete Uniform with $n = 52$, and $P(X = j) = 1/52$ for $1 \leq j \leq 52$.

20.3 Exercises

20.3.1 Practice

Exercise 20.1. Skittles. Skittles candies come in 5 different colors: red, orange, yellow, green, and purple. You have a bowl of the candies, so you reach in and grab one. Each of the five candies is equally likely to appear.

a. Why is this a Discrete Uniform distribution situation? What is the parameter?

b. What is the probability the candy you grab is a purple one?

c. Explain in words what X is in terms of the story. What values can it take if you number the candies in the order listed above?

d. What is the probability the candy you grab is not a purple one?

e. Show the labeled graph of the mass for the colors of candies.

f. Show the labeled graph of the CDF for the colors of candies.

Exercise 20.2. New ID. A computer is randomly selecting a 4-digit ID number for you, and there are 10,000 possibilities, each equally likely. (0000 would be the smallest possible number, and 9999 would be the biggest.)

a. What is the probability your new ID number happens to match your birthdate (e.g., 0822 for August 22nd)?

b. Why is this a Discrete Uniform distribution situation? What is the parameter?

c. Explain in words what X is in terms of the story. What values can it take?

d. What is the average ID number?

e. What is the standard deviation of the ID numbers?

f. Describe in words what the graph of the mass for the ID numbers would look like.

g. Describe in words what the graph of the CDF for the ID numbers would look like.

20.3.2 Extensions

Exercise 20.3. Let X be a Bernoulli random variable with $p = 1/2$. Let Y be Discrete Uniform on $\{1, 2\}$. Explain (intuitively) why they have different expected values but the same variance. No calculation should be needed.

20.3.3 Advanced

Exercise 20.4. Prove that the expected value of a Discrete Uniform random variable on the set $\{1, 2, \ldots, N\}$ is $\mathbb{E}(X) = (N + 1)/2$. Also prove that the variance is $\text{Var}(X) = (N^2 - 1)/12$.

Chapter 21

Review of Named Discrete Random Variables

21.1 Summing-up: How To Tell Random Variables Apart

These distributions are all starting to sound confusingly similar, aren't they? To help you sort them all out, we present **Frequently Confused Distributions**:

Bernoulli vs. Geometric vs. Binomial

- **Bernoulli** is a single yes/no trial. We use $X = 1$ if you get a success, or $X = 0$ if you get a failure.
- **Geometric** is continuing to do more independent Bernoulli trials until you get your 1st success. The X is the number of trials you have to do.
- **Binomial** is doing a pre-set number of independent Bernoulli trials and then counting up the number of successes. The X is the number of successes.
- Below is a comparison of results for 1 success in 5 trials for the Geometric and Binomial distributions.

Binomial	Geometric
Y N N N N	
N Y N N N	
N N Y N N	
N N N Y N	
N N N N Y	N N N N Y
$\binom{5}{1} = 5$ arrangements	only 1 arrangement since the success must come on the last trial

Summary of Named Discrete Random Variables

Name	Mass	Expected value	Variance	Parameters	What X is	When used
Bernoulli	$p_X(1) = p$ $p_X(0) = q$	p	pq	$p =$ prob. succ./trial	0 or 1 (no or yes)	1 success or failure
Binomial	$\binom{n}{x} p^x q^{n-x}$	np	npq	$n = \#$ trials; $p =$ prob. succ./trial	$0, 1, 2, \ldots, n$ (successes)	successes in n trials
Geometric	$q^{x-1} p$	$1/p$	q/p^2	$p =$ prob. succ./trial	$1, 2, 3, \ldots$ (trials)	# trials to 1st succ.
Negative Binomial	$\binom{x-1}{r-1} q^{x-r} p^r$	r/p	qr/p^2	$p =$ prob. succ./trial; $r = \#$ of succ. needed	$r, r+1, \ldots$ (trials)	# trials to rth succ.
Poisson	$e^{-\lambda} \lambda^x / x!$	λ	λ	λ rate	$0, 1, 2, 3, \ldots$ (events)	# events in period
Hyper-geometric	$\frac{\binom{M}{x}\binom{N-M}{n-x}}{\binom{N}{n}}$	$n\frac{M}{N}\left(1 - \frac{M}{N}\right)\frac{N-n}{N-1}$	nM/N	M good, $N - M$ bad; n selected	$1, 2, \ldots, M$	# of good selected
Discrete Uniform	$1/N$	$(N+1)/2$	$(N^2 - 1)/12$	N outcomes	$1, 2, \ldots, N$	equally likely

Geometric vs. Negative Binomial

- **Geometric** is doing trials until you get your 1st success.
- **Negative Binomial** is doing trials until you get your 2nd or 4th or 15th or rth success. A Negative Binomial is the sum of several Geometrics.

Binomial vs. Negative Binomial

- **Binomial** has the number of trials set in advance, and you wait until all the trials are over to count your successes. Successes can come in any order.
- **Negative Binomial** has the number of successes set in advance, and you do as many trials as you necessary in order to get that last success. The final success has to be on the last trial.
- Below is a table comparing possible arrangements of outcomes for 6 trials with 2 successes for both the Binomial and the Negative Binomial.

Binomial	Negative Binomial
Y Y N N N N	
Y N Y N N N	
Y N N Y N N	
Y N N N Y N	
Y N N N N Y	Y N N N N Y
N Y Y N N N	
N Y N Y N N	
N Y N N Y N	
N Y N N N Y	N Y N N N Y
N N Y Y N N	
N N Y N Y N	
N N Y N N Y	N N Y N N Y
N N N Y Y N	
N N N Y N Y	N N N Y N Y
N N N N Y Y	N N N N Y Y
$\binom{6}{2} = 15$ arrangements since any 2 of 6 can be success	$\binom{5}{1} = 5$ arrangements since 2nd success must come on last trial

Binomial vs. Hypergeometric

- The X is the number of successes in n trials for both.
- **Binomial** is sampling with replacement. You will know the sample size and the probability of success on a single trial. The probability of success will be the same for each trial, and the trials are independent.

- **Hypergeometric** is sampling without replacement. You will know the population size, the sample size, and the population number of successes. The probability of success will not be the same for each trial because for each trial, the population size will be decreased by 1.

- If N (population size) is really big for a Hypergeometric and n (sample size) is much smaller, you may want to use the **Binomial approximation to the Hypergeometric**. In a large population, even if you do sampling without replacement, the results will be very similar to sampling with replacement. Use the same n, but set $p = M/N$ to make the switch. The X stays the same.

Binomial vs. Poisson

- The X is the number of successes (or arrivals or counts) for both.

- **Binomial** is used when you know the number of trials (n) and the probability of success on each trial (p). The probability of success will be the same for each trial, and the trials are independent from each other.

- **Poisson** is used when you know the average rate of arrival for the counts (λ). You have a set interval instead of a set number of trials.

- If your sample size (n) is really big and your probability of success on a single trial (p) is really small, you may want to use the **Poisson approximation to the Binomial**. To make this switch, find the average for the Binomial $\mathbb{E}(X) = np$, and set that equal to λ. The X stays the same.

Bernoulli vs. Discrete Uniform

- **Bernoulli** is a single yes/no trial with a defined probability of success. The probability of success and the probability of failure do not have to be the same, but they do need to add up to 1.

- **Discrete Uniform** has one or more possible outcomes, and all of the outcomes are equally likely.

- A Bernoulli trial with a probability of success $p = 0.5$, like a coin toss, is similar to a Discrete Uniform distribution with $N = 2$. The only thing different is that the outcomes are shifted to $\{0, 1\}$ for the Bernoulli, instead of the $\{1, 2\}$ in a Discrete Uniform. So the expected value is different (because of the shift), but the variance does not change.

Poisson vs. Discrete Uniform

- **Poisson** counts up the number of successes in an interval when you know the average rate of arrival.

- **Discrete Uniform** has only one success, and it must come from one of the N possible outcomes.

21.2 Exercises

For each of the following situations, state which distribution (and approximate distribution, if applicable) would be most appropriate, and why you think so.

Exercise 21.1. Let X be the number of ice cream cones in your sample which are broken if you sample 50 of them from a large, independent population, and 12% of the cones in the entire population are broken.

Exercise 21.2. Let X be the number of ice cream cones you need to sample (again, from a large population) in order to find your 4th broken one, if they come from a large, independent population, and 12% of the cones in the entire population are broken.

Exercise 21.3. Let X be the number of broken cones you would find in the next hour if broken cones come down the assembly line at a rate of 2 broken cones per minute.

Exercise 21.4. Let X indicate whether the next ice cream cone is broken if 12% of the cones in a large, independent population are broken.

Exercise 21.5. Let X be the number of broken ice cream cones in your sample if you check 30 from a box (without replacement). Twelve are broken out of the 100 in the box total.

Exercise 21.6. Let X be the number of broken ice cream cones you need to sample in order to find your first broken one if they come from a large, independent population, and 12% of the cones in the entire population are broken.

Exercise 21.7. Let X be the number on the box you randomly select, if you are choosing 1 box from 7 numbered boxes of ice cream cones.

Exercise 21.8. Let X be the number of broken ice cream cones in your sample if you check 30 from a shipment (without replacement). The lot has 1200 broken cones out of a total of 10,000.

Exercise 21.9. Let X be the number of undercooked ice cream cones in your shipment of 10,000 if you sample from a large population. Undercooked ice cream cones have a 0.005% chance of occurring in general.

21.3 Review Problems

As you answer each question, also state which distribution you are using and what the parameters are.

Exercise 21.10. Chinese checkers. Philip and Callum play a game of Chinese checkers. Each time they play a game, Philip has a 0.7 chance of winning. Assume the games are independent.

a. In a tournament, Philip and Callum will play 8 games. What is the probability Philip wins exactly 6 games?

b. Philip and Callum decide to keep playing until Callum wins 3 times. What is the probability they play a total of 12 games?

c. How many 8-game tournaments will they expect to have to play until Philip wins exactly 6 games?

Exercise 21.11. Car dealer. According to past experience, 20% of customers coming into Frank's car dealership will buy one of his cars. Assume customers are independent and sampled from a very large population. Assume his dealership is open 7 days a week.

a. What is the probability the next person who walks in the door will buy a car?

b. If 10 customers come in to the dealership today, what is the probability at least 2 of them will buy cars?

c. What is the probability that the 4th customer coming in today is the first one who will buy a car?

d. What is the expected number of customers he needs to come into the dealership, to sell his 3rd car?

e. If 10 customers come to his dealership each day, what is the probability that he will sell at least 2 cars in each of 3 days out of the next week?

f. If customers are equally likely to want to buy cars painted red, brown, blue, black, or white, what is the probability that the next customer who buys a car picks a black car?

Exercise 21.12. Babies. On average, 9 babies are born per 24-hour day at the local hospital.

a. What is the probability that at least one baby will be born today on the 8 am to 4 pm shift?

b. What is the probability that at least one baby will be born in the next hour?

c. What is the probability that exactly 4 babies will be born on the 8 am to 4 pm shift?

d. What is the probability that exactly 4 babies will be born on each of the next three 8-hour shifts?

e. What is the probability that you would have to wait for four 8-hour shifts until you got the first one with exactly 4 babies born?

f. What is the probability that exactly 12 babies will be born total in the next 24 hours?

Exercise 21.13. Scholarship. There are 5 juniors and 10 seniors (one of which is Amelia), trying to win a scholarship to a summer music program. Only 3 students can win, and the winners will be selected randomly by pulling names out of a hat.

a. If no one can win more than once, what is the probability that all 3 winners will be seniors?

b. If students can win more than once and the students are independent from each other, what is the probability that all 3 winners will be seniors?

c. If students can win more than once and the students are independent from each other, what is the probability that the name-puller will call the first senior on the third name?

d. What is the probability that the first person to win a scholarship is a senior?

e. What is the probability that the third person to win a scholarship is a senior?

Exercise 21.14. Cookies. Chocolate chips are randomly distributed onto on a bakery's cookies with a rate, on average, of 11 chips per cookie. Assume that the cookies are independent.

a. What is the probability that a randomly selected cookie contains exactly 12 chips?

b. On average, how many cookies would you have to check until found a cookie with exactly 12 chips?

c. On average, how many cookies would you have to check until you found the 5th cookie with exactly 12 chips?

d. If you sample 30 cookies (with replacement) what is the probability there will be at least one cookie with exactly 12 chips?

e. You have a batch of 100 cookies, and 10 of these cookies have exactly 12 chips. If you randomly eat 5 cookies from this 100-cookie batch, what is the probability that at least 1 of the cookies you eat will have exactly 12 chips?

Exercise 21.15. Donuts. The probability that a student has a donut for breakfast is 0.08.

a. If I ask students (at random) from a large university campus whether they had a donut for breakfast, what is the probability the twentieth person I ask will be the first one who had a donut for breakfast?

b. How many students do I expect to have to ask until I find the fourth one who had a donut for breakfast?

c. If I ask 150 students whether they had a donut for breakfast, what is the expected number of students who did?

d. What is the probability the next student I ask did not have a donut for breakfast?

Exercise 21.16. Cards. Stacey is playing a game of cards using a standard 52-card deck (13 each of hearts, clubs, diamonds, and spades).

a. What is the probability the first card she is dealt will be the ace of spades?

b. She is dealt 7 cards at the beginning of the game to hold in her hand. What is the probability that all 7 will be hearts?

c. What is the expected number of hearts she will have in her 7 cards? What is the standard deviation?

Exercise 21.17. Rain. Suppose rain is falling at an average rate of 30 drops per square inch per minute.

a. What is the probability that a particular square inch is hit by exactly 4 drops in the next 10-second period?

b. How many 10-second intervals do you expect to observe this square inch until you find a 10-second interval with exactly 4 raindrops?

c. What is the probability you would have to observe ten 10-second intervals to find three of them with exactly 4 raindrops?

Exercise 21.18. Guessing on an exam. An exam consists of 20 multiple-choice questions with 5 possible answer choices per question. You haven't studied, so you decide to make random guesses for each answer choice. Assume each question's answer is independent from the others.

a. What is the probability that a person who randomly guesses on each question gets exactly 5 questions correct given that they got at least 1 question correct?

b. What is the expected number of correct answers on the exam? What is the standard deviation?

c. What is the probability I will have to grade more than 12 exams to find one with exactly 5 correct answers, assuming all of my students were randomly guessing on all the questions?

d. What is the expected number of correct answers if a person takes 6 exams? What is the standard deviation in the number of correct answers if a person takes 6 exams?

e. If you have to pay $2 to take the exam, but your parents pay you 50 cents for every correct answer, what is your expected net profit? What is the standard deviation in your net profit?

Exercise 21.19. Socks. You are doing laundry, and you are trying to find any small new red socks which may have fallen into the pile of white clothes in the laundry basket by mistake. It's really dark in your basement laundry room, so you have to randomly sample items of clothing.

a. In a single laundry basket, there are 35 white socks and 10 red socks. If you sample 5 socks at random (without replacement), what is the probability that at least one of them will be a red sock?

b. Now suppose instead of being in your basement laundry room, you are dealing with this problem at a huge cotton clothing manufacturer's pre-wash room, where there are 35,000 white socks and 1000 red socks. You sample 5 items of clothing at one time. What is the approximate probability at least one of them will be a red sock?

c. Using your answer to part b, how many 5-item samples of clothing would you expect to have to take to find the first 5-item sample with at least one red sock?

Exercise 21.20. Coffee beans. You are in charge of coffee bean quality control. In a very large batch of coffee beans, you estimate that the chance a bean is roasted incorrectly is 0.008. You have your employees sample 1000 beans from the seemingly endless supply, to see how many incorrectly roasted beans they find. (The roast of the beans is assumed to be independent, since the quantity of beans is large, and the beans are well-mixed before they get inspected.)

a. What is the expected number of incorrectly roasted beans in the 1000-item sample? What is the standard deviation?

b. Set up the exact probability (but do not solve) that at least 2 beans from the sample are incorrectly roasted.

c. What is the approximate probability that at least 2 beans from the sample are incorrectly roasted.

Exercise 21.21. Dogs. Frédérique has 7 dogs: Molly, Ted, Fido, Rocket, Max, Sandy, and Nellie. Assume each dog is independent from the other.

a. One of the dogs has stolen her sandwich off her plate at lunch, and she wants to figure out which one. They are each equally likely to have done it, so she is going to randomly check a dog to see if he or she has bread crumbs on his or her fur. What is the probability Rocket was the one who did it?

b. What is the expected number of sandwiches that would be stolen until Rocket was the one who did it, assuming each dog is equally likely to steal her lunch, and that the sandwich thieves are working independently, from day to day (only 1 dog can steal the sandwich of the day).

c. Each time she leaves the table, there is a 0.25 chance that a dog will steal her lunch. Out of 20 lunchtimes in which she is interrupted, what is the probability that her lunch will be stolen exactly 4 times?

d. If Frédérique stays at the table for the entire lunchtime, there is only a 0.0001 chance a dog will steal her lunch. Out of 10 years worth of lunches (3650) what is the approximate probability a dog will steal her lunch exactly 4 times?

Exercise 21.22. Slug. The number of vehicles crossing a line on an interstate is Poisson with an average of 5 per hour. A slug arrives at the interstate and will wait until no vehicles have crossed the line for 5 minutes until it attempts to get to the other side of the interstate. (This particular slug can count and tell time.) Assume the cars are independent.

a. For a given 5-minute period, what is the expected number of cars to cross the line?

b. What is the probability that no vehicles will pass in a given 5-minute period?

c. How many 5-minute intervals will the slug have to wait until the first one with no vehicles?

d. It will take the slug 10 minutes to cross the road. What is the probability the slug will be run over by at least one car? (Start the clock and the car-counting when the slug starts crossing the road.)

Exercise 21.23. Mice experiment. There are 5 white mice, 6 brown mice, and 7 gray mice in a cage. For your psychology experiment, you reach in and randomly select 3 to run in a maze.

a. What is the expected and standard deviation in the number of white mice which are selected?

b. What is the probability that at least 2 of the mice selected are white?

c. What if you had 5000 white mice, 6000 brown mice, and 7000 gray mice to choose from? You still need 3 mice to run in the maze. What is the approximate probability that at least 2 of the mice selected are white?

Exercise 21.24. Left-handed desk. Fred is a left-handed person who walks into a large lecture room for a 4-hour exam. He's a little absent-minded at the moment because he's worried about the exam. He needs to find a left-handed desk to sit in for the exam or else his back will get cramped up before the exam is over, but he won't know if the desk is left-handed until he actually sits in one, because the desk part folds underneath the chair. Ten percent of the desks in the exam room are for left-handed people, and the choice of left-or-right handedness is independent from desk to desk, throughout the room. Assume Fred is the first student to walk in the door.

a. What is the probability that Fred finds his first left-handed desk on his 5th try?

b. If Fred keeps sampling desks to find the left-handed desk which "feels lucky," what is the probability he finds his 3rd left-handed desk on his 8th try?

c. What is the average number of desks Fred has to try until finding his 2nd left-handed desk?

d. If the first 9 desks tested are not left-handed, what is the probability that he has to keep looking for more than 12 desks total to find his first left-handed desk?

Exercise 21.25. Ming and Shaheed each roll a dice (repeatedly, in rounds) until their results match, and then they stop. Let X be the number of rounds in which the sum of the two dice was 3, and let Y be the number of rounds in which the sum of the two dice was 9. Find the conditional PMF of X, given $X + Y = 10$.

Exercise 21.26. Bethany has a fair six-sided die (with sides $1, \ldots, 6$). Angelica has a fair coin, with "1" on one side and "4" on the other side. Each day, Bethany rolls the die one time, and Angelica flips the coin one time (the results are independent). Let N denote the first day on which Bethany's die has a *strictly larger* result than Angelica's coin. Let X denote the value of Bethany's die on day N. Find the expected value of X.

Exercise 21.27. Let X be a Poisson random variable with mean λ. Let Y be a Geometric random variable with parameter p, i.e., Y has mass $P(Y = j) = p(1-p)^{j-1}$ for integers $j \geq 1$. (Here, $\lambda > 0$ and $0 < p < 1$.) Also X and Y are independent. Find $P(Y > X)$.

Part IV

Counting

We have talked about basic probability ideas in the previous chapters, and we have discussed the difference between sampling with or without replacement and whether order does/doesn't matter. However, in this part, we want to focus purely on counting problems, including rearrangements. You will see familiar ideas from the Binomial and Hypergeometric discrete distributions, but we will expand on those ideas in more complex situations. These problems can be both challenging and fun. We suggest you draw pictures or even create little models for yourself if you are having trouble seeing the story clearly in your mind. Much of counting is common sense once you can visualize the rules for a particular story.

By the end of this part of the book, you should be able to:

1. Characterize the differences between sampling with/without replacement.
2. Characterize the differences between sampling in which order matters, or in which order does not matter.
3. Perform probability calculations for sampling with/without replacement.
4. Perform probability calculations for sampling, either depending on the order of selection, or when the order of selection doesn't matter.
5. Calculate the number of possible rearrangements of dice, playing cards, seating, passwords, etc.

Math skills you will need: Binomial coefficients ("choose" parentheses), factorials.

Additional resources: Calculators may be used to assist in the calculations.

Chapter 22

Introduction to Counting

The hardest arithmetic to master is that which enables us to count our blessings.
—*Reflections On The Human Condition* by Eric Hoffer (Harper, 1973)

How many unique results can appear on 3 differently colored dice? How many unique results are possible if the dice are indistinguishable?

22.1 Introduction

The concept of "counting," as it relates to probability, is usually performed in the context of a sample space S with finitely many outcomes. Moreover, the outcomes are usually all equally likely. We have seen a few examples of such situations, during our study of outcomes, events, and discrete random variables. For instance, if we shuffle a deck of 52 cards, and we remove one card, then there are 52 possible outcomes, and they are each equally likely to occur. Each event that contains just one outcome has probability $1/52$. If the event contains 13 of the outcomes (for instance, the 13 outcomes in which a heart is selected), then the event has probability $13/52$. More generally, if the event has j outcomes, then the event has probability $j/52$. Recall that the general situation was established in Corollary 2.10:

Corollary 2.10 *If sample space S has n* equally likely *outcomes, and A is an event with j outcomes, then event A has probability j/n of occurring, i.e., $P(A) = j/n$.*

We explore some ways this corollary can be applied. Counting (combinatorics) is an area of study, with courses, books, and deep theorems of its own. We admittedly only scratch the surface. If the study of counting interests you, we encourage you to pursue further study in combinatorics. Combinatorics has a long history of development, often in tandem with probability theory.

Example 22.1. Roll two dice. The sample space of all 36 equally likely outcomes is given in Figure 22.1. Some students have difficulty distinguishing how often each outcome happens. One remedy for this is to imagine that one of the dice is painted red and one is painted green, or to think of one die as getting rolled first and the other to get rolled second. The probability that the sum of the dice is 8 or larger is $15/36 = 5/12$.

	1	2	3	4	5	6
1	2	3	4	5	6	7
2	3	4	5	6	7	8
3	4	5	6	7	8	9
4	5	6	7	8	9	10
5	6	7	8	9	10	11
6	7	8	9	10	11	12

FIGURE 22.1: Bubbles appear around outcomes corresponding to sums of 8 or larger on two dice. E.g., all outcomes with a sum of 10 are in a bubble.

Example 22.2. What is the probability that the results on the two dice differ by 2 or less, i.e., if the two results are X and Y, what is $P(|X - Y| \leq 2)$?

Since 24 of the 36 outcomes differ by 2 or less, the probability is $24/36 = 2/3$. See Figure 22.2.

	1	2	3	4	5	6
1	0	1	2	3	4	5
2	1	0	1	2	3	4
3	2	1	0	1	2	3
4	3	2	1	0	1	2
5	4	3	2	1	0	1
6	5	4	3	2	1	0

FIGURE 22.2: A loop around 24 outcomes on 2 dice that differ by 2 or less.

Example 22.3. Let X and Y denote the results on two dice. Define $Z = \max(X, Y)$. Find the mass of Z.

The value of Z is between 1 and 6. We can just add the number of possible outcomes that correspond to each value of Z. For instance, $Z = 5$ if the outcome is contained in the event

$$\{(1,5),(2,5),(3,5),(4,5),(5,1),(5,2),(5,3),(5,4),(5,5)\}.$$

Thus $P(Z = 5) = 9/36$. Figure 22.3 shows the values of Z for each outcome. In general, the mass of Z is

z	1	2	3	4	5	6
$p_Z(z)$	1/36	3/36	5/36	7/36	9/36	11/36

	1	2	3	4	5	6
1	1	2	3	4	5	6
2	2	2	3	4	5	6
3	3	3	3	4	5	6
4	4	4	4	4	5	6
5	5	5	5	5	5	6
6	6	6	6	6	6	6

FIGURE 22.3: The maximum of two dice. A loop is drawn around each set of outcomes for which the maximum is the same.

One of the most helpful rules of thumb in counting problems is the method of multiplying the number of outcomes when two or more things are happening simultaneously.

Example 22.4. If there are 4 colors of pants that you could pick from, and 3 colors of shirts, then there are 12 ways that you can put on both your pants and shirt.

More generally, if there are two things happening simultaneously, and the first thing has n possibilities, and (for each of the first things) the second thing always has m possibilities, then there are nm possibilities altogether.

Theorem 22.5. Multiplying possibilities to get the total number of outcomes (two simultaneous processes)
Suppose two processes are happening, and the first one has n possibilities, and for each such possibility, there are m possibilities for the second process. Then there are nm possibilities altogether for the pair of results from the two processes.

Continuing to generalize, suppose now that we not only must put on our pants and shirt, but also our shoes.

Example 22.6. If there are 4 pants, and for each such choice, 3 shirts, and for each pair of pants/shirt, there are 5 styles of shoe, then there are $(4)(3)(5) = 60$ possible outfits altogether.

In the most general view, with j processes happening all at once:

Theorem 22.7. Multiplying possibilities to get the total number of outcomes (j simultaneous processes)
Suppose j processes are happening, in which there are n_1 possibilities for the first process, and for each such possibility, there are n_2 possibilities for the second process, and etc., etc., there are always n_j possibilities for the jth process. Then there are $n_1 \times n_2 \times \cdots \times n_j$ possibilities altogether.

Example 22.8. A playlist has 10 rock songs, 3 blues songs, and 7 hip-hop songs. The mp3 player is working in a "shuffle" mode in which each of the 20 songs are equally likely to appear each time, and none of the song choices affect any of the other song choices. In particular, repetitions are certainly allowed. What is the probability that the listener hears two consecutive hip-hop songs followed by a rock song?

There are $20 \times 20 \times 20 = 8000$ equally likely outcomes for the sequence of three songs. Exactly $7 \times 7 \times 10 = 490$ have the form (hip-hop, hip-hop, rock). So the probability of hip-hop, hip-hop, rock is $490/8000 = 0.06125$.

Example 22.9. While building a loft in your dorm room, you find that there are 8 holes remaining but only 6 screws available. The 8 holes are arranged in a line, from top to bottom. You randomly pick 6 holes to fill, with all possibilities equally likely. What is the probability that you do not fill the bottom hole?

There are $\binom{8}{6} = \frac{8!}{6!2!} = 28$ ways to pick which 6 holes will be filled. In exactly $\binom{7}{1} = 7$ of these ways, the last hole will remain empty. So the probability that the last hole remains empty is $7/28 = 1/4$.

Example 22.10. Chris has an 18-pack of Gatorade sports drink: 6 orange, 6 lemon-lime, and 6 fruit punch. What is the probability that he pulls out six bottles that match (i.e., have the same type), followed by another six that match (i.e., have the same type), followed by another six that match?

There are 18! equally likely ways that he can choose the flavors. To be successful (i.e., to pull them out in the way that is described), there are $3! = 6$ ways that the types can be ordered. There are 6! ways that the bottles of the first type chosen can be arranged, and then there are 6! ways that the second type chosen can be arranged, and then there are 6! ways that the third type chosen can be arranged. So the desired probability is

$$\frac{3!6!6!6!}{18!} = \frac{1}{2858856} = 3.4979 \times 10^{-7}.$$

22.2 Order and Replacement in Sampling

In this section, we describe four related general scenarios for handling various selection methods. For each method, we give the number of "unique" ways that the selection can be performed. Note that the uniqueness of the ways depends on whether we are regarding the order of selection.

The distinction of "with" or "without" replacement is analogous to Binomial versus Hypergeometric random variables. For a Binomial random variable, the probability p does not change between selections; this is analogous to sampling with replacement of each selection after it is done, so that the ratios for the selections stay the same from trial to trail. For a Hypergeometric random variable, on the other hand, there are a fixed number of possible good and bad outcomes; once a trial is done, the selected item is not restored; this is sampling without replacement. Throughout this section, for consistency, we use

$$n = \# \text{ of objects available to choose from,}$$
$$r = \# \text{ of objects we actually choose.}$$

	Sampling with replacement	Sampling without replacement
Order matters	n^r If I roll one red die, one green die, and one blue die, how many possible unique results can I get? ($r = 3$ dice, $n = 6$ possibilities) General: you have an unlimited supply and you care about "when" objects are selected (and how many times).	$\frac{n!}{(n-r)!}$ **(permutation)** If I have 10 students, how many unique ways can 4 different students go to the board for 4 different homework problems? ($r = 4$ chosen, $n = 10$ possibilities) General: you care about "when" not just "if" something is selected. Nothing can be selected more than once.
Order does not matter	$\binom{n+r-1}{r}$ If I roll 3 identical dice one time each, how many possible unique results can I get? ($r = 3$ dice, $n = 6$ possibilities) General: you have an unlimited supply, and you only care "if" not "when" something is selected.	$\binom{n}{r} = \frac{n!}{r!(n-r)!}$ **(combination)** If I have 10 students, how many ways can 4 different students go to the board for a problem? ($r = 4$ chosen, $n = 10$ possibilities) General: you only care "if" not "when" objects are selected. Nothing can be chosen more than once.

The upper left box has n^r ways because we have r choices, each of which has n possibilities, so there are $(n)(n)\cdots(n) = n^r$ ways. The upper right box has n first choices, $n-1$ second choices, ..., $n-r+1$ for the rth choice, so $(n)(n-1)\cdots(n-r+1) = n!/(n-r)!$ ways altogether. In the lower right box, the count is the same, but the $r!$ different arrangements of each are viewed as the same, so the count is $n!/((n-r)!r!)$. Finally, in the lower left box, we have n types of items to choose from r types, so we can just put $r-1$ dividing lines between n unlabeled balls. The number of balls before the first line are of type 1; the number of balls between the first and second lines are of type 2; etc., the number of balls between the $(r-2)$nd and $(r-1)$st lines are of type $r-1$, and the number of balls after the $(r-1)$st line are of type r. It is helpful to try this on your own, with some small n's and r's.

Sampling with replacement, order matters: n^r

Example 22.11. If I roll one red die, one green die, and one blue die, how many possible unique results can I get?

Each die has 6 possible values ($n = 6$), and the 3 dice have distinguishable colors ($r = 3$), which means a "Red 1, Blue 2, Green 3" would be considered a different result than "Red 3, Blue 2, Green 1," for example. Also, dice use sampling with replacement because each time you roll a die, you have the exact same options with the same probabilities that you had the first time. The red die result has no effect on the blue or green dice results. Therefore the number of possible unique results I can get will be $6^3 = 216$.

Example 22.12. If I choose 5 cards from a deck, with replacement in between the draws, keeping track of the order, then I am choosing 5 cards, each 1 through 52 (numbers are reusable). How many different combinations are possible, keeping track of the order?

The 52 numbers are reusable, so for example, I could choose $\{A\spadesuit, A\spadesuit, A\spadesuit, A\spadesuit, 2\heartsuit\}$, or $\{2\heartsuit, A\spadesuit, A\spadesuit, A\spadesuit, A\spadesuit\}$; the order distinguishes these two. Therefore I have $n = 52$ for my options for each card, and $r = 5$ because I need to select five cards. There are $52^5 = 380204032$ possible selections of the cards, with replacement, and keeping track of the order of selection.

Sampling with replacement, order does not matter: $\binom{n+r-1}{r}$

Example 22.13. If I roll 3 indistinguishable dice, each one time, how many possible unique results can I get?

This time we still have 6 options on each die ($n = 6$), and we still have 3 dice ($r = 3$), but since the dice are identical, a result of "1, 2, 3" would be the same as "2, 3, 1" or "3, 1, 2" because we cannot tell the dice apart. The number of unique results will be $\binom{6+3-1}{3} = \binom{8}{3} = 56$. Compare this answer to what we found in the example with the red, blue, and green distinguishable dice (216 unique outcomes). Losing the distinction of the ordering causes there to be fewer unique sets of values.

Example 22.14. If I choose 5 cards from a deck, with replacement in between the draws, but this time I do not keep track of the order, then I am choosing $r = 5$ cards from the $n = 52$ available (numbers are reusable). How many different combinations are possible, without keeping track of the order?

The 52 numbers are reusable, so for example, I could choose $\{A\spadesuit, A\spadesuit, A\spadesuit, A\spadesuit, 2\heartsuit\}$, or $\{2\heartsuit, A\spadesuit, A\spadesuit, A\spadesuit, A\spadesuit\}$, but these are considered as the same possibility; the order does not distinguish these two. Therefore I have $n = 52$ for my options for each card, and $r = 5$ because I need to select five cards. There are $\binom{52+5-1}{5} = 3819816$ possible selections of the cards, with replacement, but without keeping track of the order of selection.

Sampling without replacement, order matters: $\frac{n!}{(n-r)!}$

Example 22.15. If I have 10 students, how many unique ways can 4 different students go to the board to do 4 different homework problems?

There are $n = 10$ students in the room, and each student can only be selected once (sampling without replacement). I need $r = 4$ different students to do 4 different homework problems, so the order that the students are chosen are distinct. The number of unique ways I can select these 4 students is $\frac{10!}{(10-4)!} = \frac{10!}{6!} = (10)(9)(8)(7) = 5040$.

Example 22.16. If I choose 5 cards from a deck, without replacement in between the draws, keeping track of the order, then I am choosing 5 cards, each 1 through 52 (numbers are not reusable). How many different combinations are possible, keeping track of the order?

The 52 numbers are not reusable, e.g., I could choose $\{A\spadesuit, 7\spadesuit, 10\spadesuit, 3\heartsuit, 7\heartsuit\}$, or $\{10\spadesuit, 3\heartsuit, A\spadesuit, 7\heartsuit, 7\spadesuit\}$; the order distinguishes these two. Therefore I have $n = 52$ for my options for each card, and $r = 5$ because I need to select five cards. There are $\frac{52!}{(52-5)!} = \frac{52!}{47!} = (52)(51)(50)(49)(48) = 311875200$ possible selections of the cards, without replacement, and keeping track of the order of selection. This is a smaller number of possibilities, as compared to the situation "with replacement."

Sampling without replacement, order does not matter: $\binom{n}{r}$

Example 22.17. If I have 10 students, how many unique ways can 4 different students go to the board to work on the same problem?

There are $n = 10$ students in the room, and each student can only be selected once (sampling without replacement). I need $r = 4$ different students to do the same problem, so the order that the students are chosen makes no difference. All that matters is which 4 of the 10 students are selected. The number of unique ways I could choose 4 of these 10 students is $\binom{10}{4} = \frac{10!}{4!(10-4)!} = \frac{(10)(9)(8)(7)}{4!} = 210$. This is closely related to the previous situation, in which the order of the students mattered. Grouping together the selections from before according to the students, we see that $4! = 24$ of the previous selections corresponds to just 1 selection here. So there is a factor of 4! fewer distinct possibilities when we ignore the order in which the students are chosen.

Example 22.18. If I choose 5 cards from a deck, without replacement in between the draws and without keeping track of the order, then I am choosing $r = 5$ cards from the $n = 52$ available (numbers are not reusable). How many different combinations are possible, when we do not keep track of the order?

The 52 numbers are not reusable, e.g., I could choose $\{A\spadesuit, 7\spadesuit, 10\spadesuit, 3\heartsuit, 7\heartsuit\}$, or $\{10\spadesuit, 3\heartsuit, A\spadesuit, 7\heartsuit, 7\spadesuit\}$; but these are considered as the same choice, because the order does not distinguish these two. Therefore I just choose 5 out of the 52 cards, so there are $\frac{52!}{5!(52-5)!} = \frac{52!}{5!47!} = \frac{(52)(51)(50)(49)(48)}{5!} = 2598960$ possible selections of the cards, without replacement, and without keeping track of the order of selection. Again, this is a factor of 5! fewer possibilities than the previous situation, because we ignore the orderings of the cards.

22.3 Counting: Seating Arrangements

Example 22.19. If Alice and Alan (a couple) and Barbara and Bob (another couple) and Christine and Charlie (another couple) sit in a row of chairs, what is the probability that each of the 3 couples sit together?

There are $6! = 720$ ways that the 6 people can be seated altogether. (This is true in all of the questions with 6 people, so we won't repeat this fact.)

Grouping the couples together again, there are 3! ways that the couples can be seated (i.e., either A's/B's/C's, or A's/C's/B's, or B's/A's/C's, etc... 3! ways total). For each such way, there are 2 ways that the A's can be arranged among themselves, and 2 ways that the B's can be arranged among themselves, and 2 ways that the C's can be arranged among themselves. So the total probability that the couples are seated together is $3!(2)(2)(2)/720 = 1/15$.

Example 22.20. If n couples sit in a row of chairs, what is the probability that each of the n couples sits together?

There are $(2n)!$ ways that the $2n$ people can be seated altogether. (This is true in all of the questions with $2n$ people, so we won't repeat this fact.)

Grouping the couples together again, there are $n!$ ways that the couples can be seated. For each such way, there are 2 ways within *each couple* that the men and woman can be arranged in their two reserved seats. So the total probability that the couples are seated together is $n!2^n/(2n)!$.

Example 22.21. If Alice, Barbara, and Christine (three women) and Alan, Bob, and Charlie (three men) sit in a row of chairs, what is the probability that the women all sit together (the men may or may not be in a group)?

There are 3! ways that the men can be arranged. The women, as a group, can be collectively put into any of the 4 gaps between the men, including the spaces on the left- or right-hand ends. Once the women are placed as a group, there are 3! arrangements among just the women themselves. So the total probability that the women sit together as a group is $(3!)(4)(3!)/6! = 1/5$.

women sofa	man	man	man
man	women sofa	man	man
man	man	women sofa	man
man	man	man	women sofa

FIGURE 22.4: Ways to arrange 3 men and 3 women, if the women should sit together in a group.

This can be visualized by thinking about the women sitting together on a sofa, and the men on individual chairs. There are four places where the sofa for the women can go, as in Figure 22.4.

Example 22.22. If n women and n men sit in a row of chairs, what is the probability that the women all sit together?

There are $n!$ ways to arrange the men. The women, as a group, can be put into any of the $n+1$ gaps between the men, including the left- or right-hand ends. There are $n!$ arrangements among the women themselves. So the probability is $n!(n+1)n!/(2n)!$. (Compare with Example 22.21, when $n = 3$.)

Example 22.23. If Alice, Barbara, and Christine (three women) and Alan, Bob, and Charlie (three men) sit in a row of chairs, what is the probability that the women all sit together and all the men sit together?

Either the women sit collectively on the right or on the left (i.e., 2 ways). Once the women are placed as a group, there are 3! arrangements among just the women themselves, and there are 3! arrangements among just the men themselves. So the probability is $(2)(3!)(3!)/6! = 1/10$.

Example 22.24. If n women and n men sit in a row of chairs, what is the probability that the women all sit together and all the men sit together?

Either the women sit collectively on the right or on the left (i.e., 2 ways). Once the women are placed as a group, there are $n!$ arrangements among just the women themselves, and there are $n!$ arrangements among just the men themselves. So the probability is $(2)(n!)(n!)/(2n)!$.

Example 22.25. If Alice, Barbara, and Christine (three women) and Alan, Bob, and Charlie (three men) sit in a row of chairs, what is the probability that none of the women are adjacent and none of the men are adjacent?

Either the leftmost chair is for a woman or a man (i.e., 2 ways). Afterwards, the sexes of the people are determined. This leaves 3! arrangements among the women, and 3! arrangements among the men. So the probability is $(2)(3!)(3!)/6! = 1/10$.

Example 22.26. If n women and n men sit in a row of chairs, what is the probability none of the women are adjacent and none of the men are adjacent?

Either the leftmost chair is for a woman or a man (i.e., 2 ways). Afterwards, the sexes of the people are determined. Once they are determined, there are $n!$ arrangements among just the women themselves, and there are $n!$ arrangements among just the men themselves. So the probability is $(2)(n!)(n!)/(2n)!$.

Example 22.27. If n couples sit in a row of chairs, what is the expected number of couples that sit together?

Method #1. Let X denote the number of couples that sit together. Let X_j indicate whether the jth man sits next to his wife, so that

$$X_j = 1 \text{ if the } j\text{th man sits next to his wife, or } X_j = 0 \text{ otherwise.}$$

Then we always have $X = X_1 + X_2 + \cdots + X_n$, so

$$\mathbb{E}(X) = \mathbb{E}(X_1 + \cdots + X_n) = \mathbb{E}(X_1) + \cdots + \mathbb{E}(X_n).$$

We know $\mathbb{E}(X_j) = P(X_j = 1)$, i.e., $\mathbb{E}(X_j)$ is just equal to the probability that the jth man sits next to his wife. This equals the probability that he sits on the either end of the row and his wife sits in the unique seat next to him, plus the probability that he sits in the interior of the row and his wife sits in either of the seats next to him. The probability that the man sits at either end of the row is $2/(2n)$, and, given that he sits at either end of the row, the probability that his wife sits next to him is $1/(2n-1)$. On the other hand, the probability that the man sits in the interior of the row is $(2n-2)/(2n)$, and, given that he sits in the interior of the row, the probability that his wife sits next to him is $2/(2n-1)$. So the probability that he sits next to his wife is

$$\mathbb{E}(X_j) = \left(\frac{2}{2n}\right)\left(\frac{1}{2n-1}\right) + \left(\frac{2n-2}{2n}\right)\left(\frac{2}{2n-1}\right),$$

which simplifies to

$$\mathbb{E}(X_j) = \frac{2+4n-4}{(2n)(2n-1)} = \frac{4n-2}{(2n)(2n-1)} = \frac{1}{n}.$$

So the expected number of couples that sit together is

$$\mathbb{E}(X) = \frac{1}{n} + \frac{1}{n} + \cdots + \frac{1}{n} = n(1/n) = 1.$$

Method #2. Let X denote the number of couples that sit together. Let X_j indicate whether the person in the jth chair has her/his partner to the right, so that

$$X_j = \begin{cases} 1 & \text{if the person in the } j\text{th chair has her/his partner to the right,} \\ 0 & \text{otherwise,} \end{cases}$$

Then we always have $X = X_1 + X_2 + \cdots + X_{2n-1}$, so

$$\mathbb{E}(X) = \mathbb{E}(X_1 + \cdots + X_{2n-1}) = \mathbb{E}(X_1) + \cdots + \mathbb{E}(X_{2n-1}).$$

We know $\mathbb{E}(X_j) = P(X_j = 1)$, i.e., $\mathbb{E}(X_j)$ is just equal to the probability that the jth person has her or his partner to the right. No matter who is in the jth chair, the probability that the person's partner is found to the right is $1/(2n-1)$. So

$$\mathbb{E}(X_j) = 1/(2n-1).$$

Thus, the expected number of couples that sit together is

$$\mathbb{E}(X) = \frac{1}{2n-1} + \frac{1}{2n-1} + \cdots + \frac{1}{2n-1} = \frac{2n-1}{2n-1} = 1.$$

Example 22.28. If n couples sit in a circle of chairs (as opposed to a row, in the previous example), what is the expected number of couples that sit together?

Method #1. Let X denote the number of couples that sit together. Let X_j indicate whether the jth man sits next to his wife, so that

$$X_j = 1 \text{ if the } j\text{th man sits next to his wife, or } X_j = 0 \text{ otherwise.}$$

Then we always have $X = X_1 + X_2 + \cdots + X_n$, so

$$\mathbb{E}(X) = \mathbb{E}(X_1 + \cdots + X_n) = \mathbb{E}(X_1) + \cdots + \mathbb{E}(X_n).$$

We know $\mathbb{E}(X_j) = P(X_j = 1)$, i.e., $\mathbb{E}(X_j)$ is just equal to the probability that the jth man sits next to his wife. Regardless of where the jth man sits, his wife will sit next to him with probability $2/(2n-1)$. So

$$\mathbb{E}(X_j) = 2/(2n-1).$$

Thus, the expected number of couples that sit together is

$$\mathbb{E}(X) = \frac{2}{2n-1} + \cdots + \frac{2}{2n-1} = \frac{2n}{2n-1}.$$

Method #2. Let X denote the number of couples that sit together. Let X_j indicate whether the person in the jth chair has her/his partner to the right, so that

$$X_j = \begin{cases} 1 & \text{if the person in the } j\text{th chair has her/his partner to the right,} \\ 0 & \text{otherwise.} \end{cases}$$

Then we always have $X = X_1 + X_2 + \cdots + X_{2n}$ (we include the $(2n)$th chair, since this scenario includes people in a circle, so the partner could be in the 1st chair; this was not the case when the chairs were in a row), so

$$\mathbb{E}(X) = \mathbb{E}(X_1 + \cdots + X_{2n}) = \mathbb{E}(X_1) + \cdots + \mathbb{E}(X_{2n}).$$

We know $\mathbb{E}(X_j) = P(X_j = 1)$, i.e., $\mathbb{E}(X_j)$ is just equal to the probability that the jth person has her or his partner to the right. No matter who is in the jth chair, the probability that the person's partner is found to the right is $1/(2n-1)$. So

$$\mathbb{E}(X_j) = 1/(2n-1).$$

So the expected number of couples that sit together is

$$\mathbb{E}(X) = \frac{1}{2n-1} + \frac{1}{2n-1} + \cdots + \frac{1}{2n-1} = \frac{2n}{2n-1}.$$

Note that it is slightly more likely for a man to sit next to his wife when the row of chairs wraps around into a circle. If the man and the woman were on opposite ends of a row, they will not be next to each other, but if the row is wrapped into a circle, then they are next to each other.

22.4 Exercises

22.4.1 Practice

Exercise 22.1. Raffle tickets. There are 30 raffle tickets in a bowl. Three winning tickets will be selected. Each ticket can win at most one prize. How many ways can the prizes be distributed if the following additional information is known?

a. All 3 winners receive goldfish (the goldfish are indistinguishable).

b. The 1st winner receives a car, the 2nd a bicycle, and the 3rd a goldfish.

Exercise 22.2. Defective robots. Among 7 robots produced by a factory one day, 3 are defective. If 3 robots are purchased by the local toy store, find the probability that at least one will be nondefective.

Exercise 22.3. Rearrangements. Consider the ways that the letters in the word "Mississippi" can be rearranged (the 4 i's are indistinguishable, the 4 s's are indistinguishable, and the 2 p's are indistinguishable).

a. What is the probability that the S's are grouped together?

b. What is the probability that the S's are grouped together and the P's are grouped together?

c. What is the probability S is the 1st letter and I is the 5th letter?

d. What is the probability S or P is in the 1st spot?

Exercise 22.4. Cloudy days. In a certain city, the weather is cloudy on a given day with probability 0.55, and is sunny with probability 0.45. The weather is measured on several consecutive days (which are deemed to be sufficiently independent for this problem).

a. What is the probability of 3 consecutive cloudy days, followed by a sunny day?

b. What is the probability that exactly 1 out of 4 consecutive days will be sunny?

c. What is the probability that at least 1 out of 4 consecutive days will be sunny?

d. Given that the 1st 3 days were cloudy, what is the probability the 4th day is sunny?

Exercise 22.5. Mochas. Suppose that typically 3/10 of the customers order "grande" mocha (made with skim milk and extra whipped cream). Assume that customers are independent.

a. What is the probability that exactly 2 of the next 5 customers will order a "grande" mocha?

b. What is the probability that the 1st 3 customers do not order "grande" mochas and the 4th and 5th customers do order "grande" mochas?

Exercise 22.6. Jelly beans. A student buys a bag of jelly beans at the store. She eats most of them, but 20 remain at the end of the day. Exactly 13 of these jelly beans are fruity, and the other 7 are root beer favored. Her boyfriend grabs 5 of the 20 jelly beans.

a. What is the probability that exactly 3 of the 5 jelly beans that he grabs are fruity?

b. What is the expected number of fruity jelly beans that he grabs?

Exercise 22.7. Stamps. A professor buys postage stamps and puts them in an envelope. Exactly 30 of the stamps are for letters, and exactly 10 of the stamps are for postcards. The professor's wife grabs two stamps. All outcomes are equally likely.

a. What is the probability that both stamps she selects are for letters?

b. What is the probability that both stamps she selects are for postcards?

c. What is the probability that she gets one of each type?
(Hint: These three answers should sum to 1 altogether.)

Exercise 22.8. Rolling dice. Roll five dice. What is the probability that all five of the values that appear are distinct, i.e., there are no repetitions among the five dice?

Exercise 22.9. Letters. A player is given ten letters in a game:

"B,E,G,M,N,P,Q,R,S,U"

He randomly arranges all 10 of the letters on the table in front of him, in a row, without paying attention (e.g., with his eyes closed). All of the possible orders are equally likely. What is the probability the *first 5 letters* in the row are "S,U,P,E,R," in that exact order?

Exercise 22.10. **Hiring.** An employer has a very large pool of applicants for 8 jobs. The employer needs to report only the sex of the 8 people who are hired (not the names or any other distinguishing features), to a gender-equity review board. How many ways of hiring men and women for these positions are there if:

a. The jobs are all identical?

b. Each of the 8 jobs are unique?

Exercise 22.11. **Socks.** Running late for class, you grab 2 socks out of your drawer without looking at what color they are. In the drawer you have one pair each of black, red, green, brown, blue and white socks, but they are not folded as pairs—it's a big jumbled mess in your drawer! What is the probability that you grab 2 socks of the same color?

Exercise 22.12. **Gumballs.** You fill a mini gumball machine with 60 gumballs: red cherry, pink bubblegum, green lime, and orange orange, 15 of each. Over time you eat all of the gumballs, two at a time. What is the probability of being left with one pink gumball and one green gumball at the end?

Exercise 22.13. **Rock block.** On a certain radio station, 70% of the songs are rock songs, and 30% of the songs are pop songs. The songs are selected independently. Each "block" of songs (a "block" is a set of songs between commercials) contains 10 songs. The DJ says that the next "block" of songs has at least 8 rock songs. Given this information, what is the probability that *all 10 songs* in that "block" will be rock songs?

Exercise 22.14. **Ramen noodles.** There are 20 ramen noodle packages in a bag. There are 10 beef flavored, and the other 10 are chicken flavored. What is the probability of getting at least one chicken and at least one beef flavor when you grab 3 packages randomly?

Exercise 22.15. **Badminton.** A coach is choosing 4 students to go to a badminton competition (the order of selection does not matter). There are 20 boys and 15 girls to choose from, and the coach chooses randomly so everyone has an equal chance of being picked, but nobody can be picked more than once.

a. What is the probability that all 4 chosen will be boys?

b. What is the probability that all 4 chosen will be girls?

c. Should these two probabilities add up to 1? Why or why not?

Exercise 22.16. Family photos. Your computer has 437 family photos. You decide to take 30 random photos (without replacement) from the collection (each is equally likely), to use as a slideshow on the TV when guests come over. Of your family photos, 42 are from your favorite vacation. What is the expected number of photos from your favorite vacation in the slideshow?

Exercise 22.17. Action figures. In a box there are 31 Power Ranger action figures. Seven of them are red and the other 24 are blue. Timmy closes his eyes and randomly picks 8 action figures out of the box. What is the probability 6 of them are red?

Exercise 22.18. Cereal. On the cereal shelf of the house you share with several roommates, you have 4 types of cereal with chocolate (Reese's Puffs, Cocoa Puffs, Count Chocula, and Cookie Crisp) and 3 types of cereal without chocolate (Corn Chex, Cheerios, and Captain Crunch). You haven't had your coffee yet, so you will blindly choose 3 boxes of cereal to mix together for breakfast. What is the probability you will get at least one chocolate cereal mixed in?

Exercise 22.19. Deck of cards. In a game of chance, you are allowed to shuffle a standard deck of cards and then choose 3 cards randomly. If two or more of them are hearts, then you win.

a. What is the probability of winning the game?

b. What is the expected number of hearts drawn?

Exercise 22.20. Pizza parlor. You make pizzas for a local pizza parlor. On a busy Friday night, you've had 12 orders in the last hour. The orders are placed independently. Customers are known to prefer cheese pizza 37% of the time, meaty pizza 47% of the time, and veggie pizza the other 16% of the time. What is the probability that exactly 3 of the pizzas are veggie, 4 are cheese, and 5 are meaty?

22.4.2 Extensions

Exercise 22.21. Daughters and mothers. If you randomly assign daughters a, b, c, d, e to mothers A, B, C, D, E, and you let X be the number of daughters who are correctly assigned to their mother, find the mass of X.

Exercise 22.22. Balls in boxes. Consider 6 identical balls and 3 distinguishable boxes.

a. In how many ways can we put the 6 identical balls into the 3 distinguishable boxes?

b. If each box gets at least 1 ball?

c. If boxes can be empty?

Exercise 22.23. Bears. A little girl randomly arranges 30 bears—consisting of 10 red bears, 10 yellow bears, and 10 blue bears—into 10 bowls of 3 bears each. (All outcomes are equally likely.)

Let X denote the number of bowls that have at least one blue bear in the bowl. (The little girl loves blue bears.) Find the expected value of X.

Exercise 22.24. More bears. A total of 30 bears—consisting of 10 red bears, 10 yellow bears, and 10 blue bears—are randomly arranged in a bucket. A child begins grabbing the bears at random, with all selections equally likely. The bears are selected "without replacement," i.e., she never puts the bears back after she grabs them.

How many bears does the child expect to grab *before* the first red bear appears? (Please do not include the red bear itself; only include the bears that appear strictly before the first red bear appears.)

Exercise 22.25. Picking letters at random. Five friends, named Ali, Bob, Charlie, Daniel, and Ebony, each pick a letter from the alphabet on their own (i.e., independently), with all possible outcomes equally likely. For instance, they might pick (starting with Ali's choice) "M,D,P,Z,P."

a. What is the probability that they choose 5 distinct letters?

b. What is the probability that they choose 5 distinct letters *and that the letters are in increasing alphabetic order starting from Ali's choice and ending at Ebony's choice*? For instance, starting with Ali's choice, "D,H,P,T,X" would be such an alphabetical ordering.

Exercise 22.26. Bridge. If you are playing bridge with a standard 52-card deck, what is the probability that a player will get 5 clubs, 4 diamonds, 3 hearts, and 1 spade in a 13-card deal?

Exercise 22.27. Getting dressed. Running late on Wednesday for probability class, Shaheed realizes that snow is forecast, thus he decides to wear two T-shirts layered together. He has 10 T-shirts that are basically identical except for color. He has 5 colors available, with 2 shirts of each color. One of these colors is black. What is the probability that he grabs at least one black shirt?

Exercise 22.28. Fire dance. There are 48 candles for the ring of fire dance. Exactly half are orange, and the other half are black. Each dancer can choose two candles to hold during the dance. What is the probability that 5 dancers choose 2 black candles each? The candles are not replaced after a dancer chooses, so this would leave 24 untouched orange candles and only 14 black candles after the selections are made.

Exercise 22.29. Work clothes. Pierre works 5 days a week. He has 12 shirts, 8 pants, 8 ties, and 4 jackets that he can wear to work. Of these, 4 shirts, 3 pants, 2 ties, and 2 jackets are blue. Each day he randomly selects one of each item to wear. Assume the selections are independent, and assume his butler launders his clothes every night so he has a full closet each morning.

a. What is the probability Pierre's entire outfit next Monday will be blue?

b. What is the probability Pierre will wear at least 1 entirely blue outfit during his next 5-day work week?

c. What is the probability that Pierre will wear entirely blue outfits on Monday and Friday while wearing outfits which are not entirely blue on Tuesday through Thursday?

d. What is the probability Pierre will wear an entirely blue outfit on exactly 2 of the 5 days next week?

Exercise 22.30. Lunch choices. According to an ad in the newspaper for the local café, their Bagel Sandwiches can be "made your way" with "100s of choices for you!" They advertise 7 flavors of bagels, 5 kinds of meat, 3 kinds of cheese, 9 dressings, and 3 veggie selections.

a. If you select a sandwich made with one of each: bagel, meat, cheese, dressing, and veggie (and you don't care about the order), how many different combinations of sandwich do you really have to choose from?

b. Now assume you only eat whole wheat bagels. The same 20 toppings listed above can be used. How many combinations do you have?

c. If you can choose 3 items to go on your bagel, how many different types of sandwiches can you have if the different items are reusable (tomato and double ham would be ok)?

d. If you can choose 3 items but they are not reusable, how many are possible?

Exercise 22.31. Pizza toppings. A pizza place offers the choice of the following toppings: extra cheese, mushrooms, pepperoni, ham, sausage, onions, and green peppers. Assume that each pizza must have at least 1 topping and that order of toppings is irrelevant. Also assume that toppings cannot be reused (double sausage is impossible).

a. How many 3-topping pizzas are possible?

b. If I randomly order a 3-topping pizza, what is the probability that sausage is a topping?

c. If I randomly order a 3-topping pizza, what is the probability that sausage and pepperoni are toppings?

d. Now suppose that you allow toppings to be reused. For instance, double sausage is now possible. (We have the possibility to double, triple, or even quadruple a topping, if desired.) How many different 4-topping pizzas are there?

Exercise 22.32. Bookshelf. A student has 3 math books, 4 history books, 2 chemistry books, and 1 Latin book. He wants to arrange them on a bookshelf.

a. If all the books have different titles, in how many distinct ways can he arrange them?

b. Throughout parts b–e, assume that all the books from a particular topic have the same title (for example, 3 indistinguishable copies of "Calculus" by Carey). In how many distinct ways can he arrange his books?

c. If he groups the identical math books together, and he groups the identical history books together, and he groups the identical chemistry books together, in how many distinct ways can he arrange his books?

d. If he groups the identical history books together (but isn't picky about the other books), in how many distinct ways can he arrange his books?

e. Given that the identical history books are grouped together, what is the probability that he has grouped each of the identical books (the situation in part c)?

Exercise 22.33. Movie theater. Ten people are going to a private screening of a movie in a small theater. There are two rows in the theater. The first row has 4 seats, and the second row has 6 seats.

a. How many ways can you arrange the seating of the 10 people altogether?

b. How many seating arrangements are possible if 3 of the 10 people are in a family, and that family wants to sit together?

Exercise 22.34. Beach books. You have a row of 20 books on your desk. You're about to go on vacation and want to grab some good reading material, preferably a good novel or some other work of fiction. Of the 20 books, 12 are fiction and 8 are nonfiction. You grab 5 books at random.

a. What is the expected number of fiction books you will have?

b. What is the probability that 2 or more of your 5 books are fiction?

Exercise 22.35. Chair circle. A group of men and women sit in a circle. There are 20 chairs in the circle, and 10 pairs of married individuals. What is the probability that a man will sit directly across from his wife if everyone sits randomly?

Exercise 22.36. License plates. On a "Save The Wetlands" license plate, there are always two letters (e.g., "SW") followed by four digits, and thus 10,000 combinations of plates are available for each pair of letters. If the four digits are selected randomly, and all 10,000 possibilities are equally likely, what is the probability that the four digits are distinct and in ascending order?

Exercise 22.37. Socks. In my sock drawer there are 21 white socks, 8 black socks, and 4 brown socks.

a. If I randomly pull out 6 socks to take with me on a trip, what is the probability that I pull out one pair of each color?

b. What is the probability all the socks are the same color?

c. What is the probability that I pull out 2 socks of one color and 4 socks of a second color?

Exercise 22.38. **Coins.** My friend Alejandro has 35 coins, of which 26 are quarters and the other 9 are pennies. If he gives me 4 coins at random, what is the probability that I will have enough money to buy my favorite 89 cent candy bar? How much money do I expect to get from Alejandro?

22.4.3 Seating Arrangement Problems

In this chapter we covered many questions about couples who are sitting in a row of chairs. Now we consider some questions in which the couples sit in a circle of chairs.

Exercise 22.39. Rodrigo and Edna are divorced. They are sitting at a circular table with $n \geq 2$ chairs altogether, and all seating arrangements are equally likely. What is the probability that they are *not* sitting next to each other?

Exercise 22.40. If Alice and Alan (a couple) and Barbara and Bob (another couple) sit in a circle of chairs, what is the probability that each of the couples sit together?

Exercise 22.41. If Alice and Alan (a couple) and Barbara and Bob (another couple) and Christine and Charlie (another couple) sit in a circle of chairs, what is the probability that each of the 3 couples sit together?

Exercise 22.42. If n couples sit in a circle of chairs, what is the probability that each of the n couples sits together?

Exercise 22.43. If Alice and Barbara (two girls) and Alan and Bob (two boys) sit in a circle of chairs, what is the probability that the girls both sit together and both boys sit together?

Exercise 22.44. If Alice, Barbara, and Christine (three girls) and Alan, Bob, and Charlie (three boys) sit in a circle of chairs, what is the probability that the girls all sit together and the boys all sit together?

Exercise 22.45. If n girls and n boys sit in a circle of chairs, what is the probability that the girls all sit together and the boys all sit together?

Exercise 22.46. If Alice and Barbara (two girls) and Alan and Bob (two boys) sit in a circle of chairs, what is the probability that none of the girls are adjacent and none of the boys are adjacent?

Exercise 22.47. If Alice, Barbara, and Christine (three girls) and Alan, Bob, and Charlie (three boys) sit in a circle of chairs, what is the probability that none of the girls are adjacent and none of the boys are adjacent?

Exercise 22.48. If n girls and n boys sit in a circle of chairs, what is the probability that none of the girls are adjacent and none of the boys are adjacent?

Chapter 23

Two Case Studies in Counting

23.1 Poker Hands

In all of the "hands" below, we are describing the probability of an event associated with a selection of 5 cards from a well-shuffled deck, so that in every case, all $\binom{52}{5} = \frac{52!}{5!47!} = 2{,}598{,}960$ possible hands are equally likely. We do not take the order of the dealing of the cards into consideration. (If the reader wants to take the order in account, then multiply the numerator by 5! and the denominator by 5!, for each probability in this section, and these 5!'s will cancel out.)

23.1.1 Straight Flush

A straight flush consists of 5 cards from the same suit that can be placed in order. There are four possible suits for a straight flush. Within each suit, there are ten types of straights:

$$\begin{array}{ccccc} (A,2,3,4,5) & (2,3,4,5,6) & (3,4,5,6,7) & (4,5,6,7,8) & (5,6,7,8,9) \\ (6,7,8,9,10) & (7,8,9,10,J) & (8,9,10,J,Q) & (9,10,J,Q,K) & (10,J,Q,K,A) \end{array}$$

Since there are 4 suits, there are $(4)(10) = 40$ types of possible straight flushes. So the probability of getting a straight flush in one 5-card hand is $40/2598960$.

23.1.2 Four of a Kind

A four of a kind consists of four cards of the same rank (a quadruple, e.g., four 9's) and also one other card from another rank. There are 13 possible ranks for the quadruple:

$$A, 2, 3, 4, 5, 6, 7, 8, 9, 10, J, Q, K;$$

for each such choice, there remain 12 possible ranks for the single card. Once the rank of the single card is chosen, it can have any of four possible suits. So there are $(13)(12)(4) = 624$ types of possible four of a kinds. Thus, the probability of getting a four of a kind in one 5-card hand is $624/2598960$.

23.1.3 Full House

A full house consists of a triple of cards from one rank and a pair of cards from another rank. There are 13 possible ranks for the triple; for each such choice, there remain 12 possible ranks for the pair. Once the ranks have been selected, there are $\binom{4}{3} = 4$ ways to pick the triple, and $\binom{4}{2} = 6$ ways to pick the pair. So there are $(13)(12)(4)(6) = 3744$ types of possible full houses. Therefore, the probability of getting a full house in one 5-card hand is 3744/2598960.

23.1.4 Flush

A flush consists of five cards from the same suit. There are 4 possible suits; for each suit, there are $\binom{13}{5} = 1287$ possible cards from that suit. So there are $(4)(1287) = 5148$ types of possible flushes. Usually the straight flushes are removed from this classification (because they have a classification all on their own); therefore, there are $5148 - 40 = 5108$ types of flushes that are not straight flushes. so the probability of getting a flush (but not a straight flush) in one 5-card hand is 5108/2598960.

23.1.5 Straight

A straight consists of five cards that can be placed in order. There are 10 possible sets of values for the cards in the straight, as seen in the straight flush example above; for each type of straight, there are $\binom{4}{1} = 4$ ways to pick a card from each of the ranks, so there are $(4)(4)(4)(4)(4) = 4^5$ ways to pick the cards altogether. So there are $(10)(4^5) = 10{,}240$ types of possible straights. Usually the straight flushes are removed from this classification (because they have a classification all on their own), which leaves $10{,}240 - 40 = 10{,}200$ types of straights that are not straight flushes. Thus, the probability of getting a straight (but not a straight flush) in one 5-card hand is 10200/2598960.

23.1.6 Three of a Kind

A three of a kind consists of a triple of cards from one rank and two other cards from two different, other ranks. There are 13 possible ranks for the triple; for each such choice, there remain $\binom{12}{2} = 66$ possible ranks for the other two cards. Once the ranks have been selected, there are $\binom{4}{3} = 4$ ways to pick the triple, and $\binom{4}{1} = 4$ ways to pick the greater of the two non-matched cards, and $\binom{4}{1} = 4$ ways to pick the lesser of the two non-matched cards. Combining these, we have $(13)(66)(4)(4)(4) = 54{,}912$ types of possible three of a kinds. So the probability of getting three of a kind in one 5-card hand is 54912/2598960.

23.1.7 Two Pair

Two pair consists of a pair of cards from one rank, another pair of cards from a different rank, and one other cards from a third rank. There are $\binom{13}{2} = 78$ possible ranks for the two pairs; and then there always remains 11 ways to pick the rank for the non-matched card. for each such choice, there are $\binom{4}{2} = 6$ ways to pick the pair of cards in the greater-valued pair, and $\binom{4}{2} = 6$ ways to pick the pair of cards in the lesser-valued pair. There are $\binom{4}{1} = 4$ ways to pick the non-matched card. This gives us $(78)(11)(6)(6)(4) = 123{,}552$ types of possible two pairs, and therefore the probability of getting two pairs in one 5-card hand is $123552/2598960$.

23.1.8 One Pair

One pair consists of a pair of cards from one rank, and then three other unmatched cards, all from unique ranks. There are 13 possible ranks for the one pair; and then there always remains $\binom{12}{3} = 220$ ways to pick the ranks for the three non-matched cards. For each such choice, there are $\binom{4}{2} = 6$ ways to pick the pair of cards in the pair, and $\binom{4}{1} = 4$ ways to pick each of the non-matched cards. So there are $(13)(220)(6)(4)(4)(4) = 1{,}098{,}240$ types of possible one pairs. Finally, the probability of getting one pair in one 5-card hand is $1098240/2598960$.

23.2 Yahtzee

In Yahtzee, five dice (each is a standard die, with 6 sides) are rolled, so that there are $6^5 = 7776$ equally likely outcomes. It is usually helpful to think of the five dice separately, to keep the outcomes clear in one's mind. For instance, one might think of five differently colored dice, or five dice that are rolled by five numbered players 1, 2, 3, 4, 5, etc. Thus, the scenario is a bit different than in the poker hands case study above.

In Yahtzee, we are allowed to roll the dice up to three times, but for simplicity here, we will only roll each die one time.

23.2.1 Upper Section

Fix a desired value, for instance, "3," that a player wants to appear on a die. The number of dice X that show a specific value (in our case, "3") is Binomial with parameters $n = 5$ and $p = 1/6$. So

$$P(X = j) = \binom{5}{j}(1/6)^j(5/6)^{5-j} = \frac{5!5^{5-j}}{j!(5-j)!6^5}.$$

For another method of computing this probability, note that all 6^5 outcomes are equally likely. In order to have exactly j dice show a certain value (for instance, the value "3"), there are $\binom{5}{j}$ ways to pick which dice will show that value. The $5 - j$ dice that will not equal the desired value can appear in 5 ways each (e.g., the other $5 - j$ dice can each be anything except "3"). This means that there are $\binom{5}{j}5^{5-j}$ possible outcomes in which there are exactly j

occurrences of a specific desired value. The desired probability is

$$\frac{\binom{5}{j}5^{5-j}}{6^5} = \frac{5!5^{5-j}}{j!(5-j)!6^5}.$$

23.2.2 Three of a Kind

To obtain three of a kind, there must be exactly 3 occurrences of some value.

Fix a desired value, for instance, "1," that a player wants to get for the three of a kind. The player needs 3 of the 5 dice to equal value "1." Then the other two dice need to have two different, distinct values. Thus, the probability of getting a three of a kind that consists of exactly three occurrences of value "1" (without being a full house, i.e., without the other pair matching) is

$$\binom{5}{3}(1/6)^3(5/6)(4/6) = \frac{5!(5)(4)}{3!(5-3)!6^5} = 25/972 = 0.02572.$$

There are 6 possible values for the three of a kind, and these are all disjoint, so the total desired probability is

$$(6)\left(\frac{25}{972}\right) = \frac{25}{162} = 0.1543.$$

For another method of computing this probability, note that all 6^5 outcomes are equally likely. There are six ways to choose the value for the three of a kind, e.g., "1." To get exactly three values equal to this specific value, there are $\binom{5}{3}$ ways to pick the dice that must show this value; there are (5)(4) ways that the other two dice can show two other values that are distinct. Combining these, we see that there are $(6)\binom{5}{3}(5)(4)$ possible outcomes in which there are exactly 3 occurrences of one of the values, without the other pair matching. Therefore, the desired probability is

$$\frac{(6)\binom{5}{3}(5)(4)}{6^5} = \frac{25}{162} = 0.1543.$$

23.2.3 Four of a Kind

To obtain four of a kind, there must be exactly 4 occurrences of some value.

Fix a desired value, for instance, "3," that a player wants to get for the four of a kind. The number of dice X that show that specific value (in our case, "3") is Binomial with parameters $n = 5$ and $p = 1/6$. So we have

$$P(X = 4) = \binom{5}{4}(1/6)^4(5/6)^{5-4} = \frac{5!5}{4!(5-4)!6^5} = 25/7776 = 0.003215.$$

There are 6 possible values for the four of a kind, and these are all disjoint; thus, the total desired probability is

$$(6)\left(\frac{25}{7776}\right) = \frac{25}{1296} = 0.01929.$$

For another method of computing this probability, note that all 6^5 outcomes are equally likely. There are six ways to fix a desired value of j. In order to have exactly four values equal to this specific j, there are ways to pick which of the $\binom{5}{4}$ dice show this value; the remaining $5 - 4 = 1$ die can show any of the other 5 values. So there are $(6)\binom{5}{4}5$ possible outcomes in which there are exactly 4 occurrences of one of the values. So the chance of a four of a kind is

$$\frac{(6)\binom{5}{4}5}{6^5} = \frac{25}{1296} = 0.01929.$$

23.2.4 Full House

To obtain a full house, there must be three occurrences of one value, and two occurrences of a different value.

There are 6 ways to pick the value for the triple (e.g., "5"s), and 5 remaining ways to pick the value for the pair (e.g., "1"s). Once these values are chosen, there are $\binom{5}{3} = 10$ ways to decide which dice will get the triple (and the other two will get the pair). So the desired probability is

$$\frac{(6)(5)\binom{5}{3}}{6^5} = \frac{300}{7776} = \frac{25}{648} = 0.03858.$$

23.2.5 Small Straight

To obtain a small straight, which is four dice in consecutive order, there are 3 ways to choose which kind of small straight, i.e., either "1 through 4," or "2 through 5," or "3 through 6."

For the type "1 through 4," the other die cannot be a 5, or then we have a large straight. We either have 1–4, along with 6; or we have 1–4, with one value duplicated. There are 5! ways to have 1–4 along with 6. To have 1–4, with one value duplicated, there are 4 ways to choose which value is duplicated, and then there are $\binom{5}{2} = 10$ ways to choose which dice get that repeated value, and then there are 3! ways to choose how the other 3 values are arranged. So there are

$$5! + (4)\binom{5}{2}(3!) = 360 \text{ ways.}$$

For the type "2 through 5," the other die cannot be 1 or 6, or then we have a large straight. We have 2–5, with one value duplicated. Thus, there are 4 ways to choose which value is duplicated, and then there are $\binom{5}{2} = 10$ ways to choose which dice get that repeated value, and then there are 3! ways to choose how the other 3 values are arranged. So, altogether, there are

$$(4)\binom{5}{2}(3!) = 240 \text{ ways.}$$

For the type "3 through 6," the other die cannot be a 2, or then we have a large straight. We either have 3–6, along with 1; or we have 3–6, with one value duplicated. There are 5! ways to have 3–6 along with 1. To have 3–6, with one value duplicated, there are 4 ways to choose which value is duplicated, and then there are $\binom{5}{2} = 10$ ways to choose which dice get that repeated value, and then there are 3! ways to choose how the other 3 values are arranged. This yields

$$5! + (4)\binom{5}{2}(3!) = 360 \text{ ways.}$$

Altogether, there are $360 + 240 + 360 = 960$ ways that the results can be arranged on the dice. So the desired probability is

$$\frac{960}{6^5} = \frac{10}{81} = 0.1235.$$

23.2.6 Large Straight

To obtain a large straight, which is five dice in consecutive order, there are 2 ways to choose which kind of large straight, i.e., either 1 through 5, or 2 through 6. Once the kind has been chosen, there are 5! ways that the results can be arranged among the 5 distinct dice. Thus, the desired probability is

$$\frac{(2)(5!)}{6^5} = \frac{5}{162} = 0.03086.$$

23.2.7 Yahtzee

To obtain a "yahtzee," which is just five of a kind, there must be exactly 5 occurrences of some value.

Fix a desired value (e.g., "3") that a player wants to get for the yahtzee, i.e., for the Yahtzee. The number of dice X that show that specific value (in our case, "3") is Binomial with parameters $n = 5$ and $p = 1/6$. So

$$P(X = 5) = \binom{5}{5}(1/6)^5(5/6)^{5-5} = \frac{5!}{5!(5-5)!6^5} = 1/6^5 = 1/7776 = 0.0001286.$$

There are 6 possible values for the yahtzee, and these are all disjoint. So we conclude that the total desired probability is

$$(6)\left(\frac{1}{7776}\right) = \frac{1}{1296} = 0.0007716.$$

For another method of computing this probability, note that all 6^5 outcomes are equally likely. There are six ways to fix a desired value of j. Once the desired value of the yahtzee is chosen, then all of the dice must equal the value. So the desired probability is

$$\frac{6}{6^5} = \frac{1}{1296} = 0.0007716.$$

Part V

Continuous Random Variables

You have learned how to calculate probabilities for discrete random variables. Earlier in the text we discussed the differences between discrete and continuous random variables. In the chapters that follow, you will learn how to calculate probabilities for continuous random variables. Afterwards, we will apply these rules and formulas in the next part of the book with five different types of named continuous random variables.

By the end of this part of the book, you should be able to:

1. Distinguish between discrete and continuous random variables when reading a story.
2. Calculate probabilities for continuous random variables.
3. Calculate and graph a density (i.e., probability density function, PDF).
4. Calculate and graph a CDF (i.e., a cumulative distribution function)
5. Calculate the mean, variance, standard deviation, and median of the continuous random variable.

Math skills you will need: integrals of one and two variables (in particular, integration by parts), derivatives, e^x, and $\ln x$.

Additional resources: Computer programs (such as Excel, Maple, Mathematica, Matlab, Minitab, R, SPSS, etc.) and calculators may be used to assist in the calculations.

Chapter 24

Continuous Random Variables and PDFs

The true logic of this world is in the calculus of probabilities.
—James Clerk Maxwell

The number of passengers on a randomly chosen bus is an integer, but the speed of the bus is a real-valued number. The height of a tree is real-valued, but the number of leaves is discrete.

24.1 Introduction

In Chapter 7, you already learned the difference between a discrete random variable and a continuous random variable. Here's a review:

Discrete means that you could list list the specific possible outcomes that the variable can take (there are either a finite number of them, or a countably infinite number). An example is $X \in \{0, 1, 2, 3, 4, 5, \ldots\}$, for the number of 4-leaf clovers you will find in a 1-acre pasture this afternoon.

Continuous means that the variable takes on a range of values. You could usually state the beginning and end points, but you would have infinitely many possibilities of answers within that range. An example is $62.8 \leq X \leq 67.0$, for how many ounces of soda will actually be in the next 64-ounce bottle you open.

We give a table with an overview of the differences between discrete and continuous random variables. As you can see, many of the formulae are very similar in structure. A key difference is that for discrete random variables you use summations, and for continuous random variables you use integrals.

	Discrete	Continuous
probability function	mass (probability mass function; PMF)	density (probability density function; PDF)
	$0 \le p_X(x) \le 1$	$0 \le f_X(x)$ (not necessarily ≤ 1)
	$\sum_x p_X(x) = 1$	$\int_{-\infty}^{\infty} f_X(x)\,dx = 1$
	$P(0 \le X \le 2)$ $= P(X=0) + P(X=1) +$ $P(X=2)$ if X is integer valued	$P(0 \le X \le 2)$ $= \int_0^2 f_X(x)\,dx$
	$P(X \le 3) \ne P(X < 3)$ when $P(X=3) \ne 0$	$P(X \le 3) = P(X < 3)$ since $P(X=3) = 0$ always
cumulative distribution function (CDF) $F_X(x)$	$F_X(a) = P(X \le a)$ $= \sum_{x \le a} P(X = a)$ graph of CDF is a step function with jumps of the same size as the mass, from 0 to 1	$F_X(a) = P(X \le a)$ $= \int_{-\infty}^{a} f_X(x)\,dx$ graph of CDF is nonnegative and continuous, rising up from 0 to 1
examples	counting: defects, hits, die values, coin heads/tails, people, card arrangements, trials until success, etc.	lifetimes, waiting times, height, weight, length, proportions, areas, volumes, physical quantities, etc.
named distributions	Bernoulli, Binomial, Geometric, Negative Binomial, Poisson, Hypergeometric, Discrete Uniform	Continuous Uniform, Exponential, Gamma, Beta, Normal
expected value	$\mathbb{E}(X) = \sum_x x p_X(x)$ $\mathbb{E}(g(X)) = \sum_x g(x) p_X(x)$	$\mathbb{E}(X) = \int_{-\infty}^{\infty} x f_X(x)\,dx$ $\mathbb{E}(g(X)) = \int_{-\infty}^{\infty} g(x) f_X(x)\,dx$
$\mathbb{E}(X^2)$	$\mathbb{E}(X^2) = \sum_x x^2 p_X(x)$	$\mathbb{E}(X^2) = \int_{-\infty}^{\infty} x^2 f_X(x)\,dx$
variance	$\mathrm{Var}(X) =$ $\mathbb{E}(X^2) - (\mathbb{E}(X))^2$	$\mathrm{Var}(X) =$ $\mathbb{E}(X^2) - (\mathbb{E}(X))^2$
std. dev.	$\sigma_X = \sqrt{\mathrm{Var}(X)}$	$\sigma_X = \sqrt{\mathrm{Var}(X)}$

A *continuous random variable* takes on values within an interval of finite or infinite length. For instance, X could be a waiting time, such as the time until the phone rings, or the time until the next email arrives, or the time until you win a lottery game. In all of these cases, X could take on any nonnegative real number. As another example, X could be the height of a randomly chosen person, in which case X could be any real number from (say) 0 to 9 (height given in feet). Another example is that X is the horizontal distance to the left or right of the location where a dart lands, in comparison to the center of a dart board.

The probabilities associated with a continuous random variable are obtained from the **probability density function**, often abbreviated as "PDF" or just called the **density**. Densities have some things in common with masses, but some things are different too. A density is only used with a continuous random variable. A mass is only used with a discrete random variable. Every density has the following properties:

Definition 24.1. Density of a Continuous Random Variable
If X is a continuous random variable, and $f_X(x)$ is the density of X, then:
1. The density is always nonnegative, i.e., $f_X(x) \geq 0$ for all real numbers x.
2. The density, integrated from a to b, gives the probability that X is between a and b, i.e.,

$$P(a \leq X \leq b) = \int_a^b f_X(x)\,dx.$$

To calculate probabilities from a density, we just integrate the density over the range of interest. E.g., we find the probability that X is between 3.2 and 5.82 by integrating the density over $[3.2, 5.82]$,

$$P(3.2 \leq X \leq 5.82) = \int_{3.2}^{5.82} f_X(x)\,dx.$$

Remark 24.2. Integral of a Density over the entire real line is 1
If X is a continuous random variable, and if $f_X(x)$ is the density of X, then

$$P(-\infty < X < \infty) = 1,$$

so the integral over the whole real line must be 1:

$$\int_{-\infty}^{\infty} f_X(x)\,dx = 1.$$

How are the mass and density alike?

- Both are building blocks for probabilities, the CDF, the expected value, and the variance.

- Both are always nonnegative:

  - Mass (discrete): $0 \leq p_X(x) \leq 1$
  - Density (continuous): $0 \leq f_X(x)$

- The sum of the mass over all possible x values is 1. The integral of the mass over all possible x values is 1.

  - Mass (discrete): $\sum_x p_X(x) = 1$
  - Density (continuous): $\int_{-\infty}^{\infty} f_X(x)\,dx = 1$

How are the mass and density different?

- Continuous versus isolated points:
  - The mass consists of positive probabilities for individual points (corresponding to the possible outcomes)
  - The densities we consider are piecewise continuous on the real line.[1]
- Adding versus integrating:
  - Summing values of the mass gives probabilities.
  - Integrating the density over an interval gives probabilities.
- Maximum values:
  - The mass has to be between 0 and 1 for every x value.
  - The density has no upper limit. E.g., if $f_X(x) = 10$ for $0 \leq x \leq 1/10$, and $f_X(x) = 0$ otherwise, then $f_X(x)$ is a density.
- At a single point:
  - The mass $P(X = x) = p_X(x)$ is strictly positive if x corresponds to a possible outcome of the random variable ($p_X(x) = 0$ otherwise).
  - For continuous X, the probability $P(X = x)$ is 0 for any specific value x. This corresponds to integrating a random variable over just one point, i.e.,

$$P(X = x) = P(x \leq X \leq x) = \int_x^x f_X(x)\,dx = 0.$$

Remark 24.3. Probability that a Continuous Random Variable is equal to a specific value
If X is a continuous random variable, and if a is any particular value, then X has probability 0 of being equal to that particular value. To see this, we just write

$$P(X = a) = P(a \leq X \leq a) = \int_a^a f_X(x)\,dx = 0.$$

Does it matter whether we include the endpoints of an interval when computing probabilities with continuous random variables? No! Since $P(X = a) = 0$ and $P(X = b) = 0$ for any values a, b, then all four of the

[1] In more advanced courses, densities can be defined using the notion of *measure*, but we do not consider such generalities at this level of presentation. See books such as Billingsley [1] or Durrett [2] for a more advanced treatment.

following values are the same:

$$P(a \le X \le b) = \int_a^b f_X(x)\,dx$$

$$P(a \le X < b) = \int_a^b f_X(x)\,dx$$

$$P(a < X \le b) = \int_a^b f_X(x)\,dx$$

$$P(a < X < b) = \int_a^b f_X(x)\,dx$$

Any nonnegative function for which "the integral of the function over the whole real line is 1" is the density of a continuous random variable.

Remark 24.4. Creating a Continuous Random Variable
If $g(x)$ is any integrable, nonnegative function, and $\int_{-\infty}^{\infty} g(x)\,dx = 1$, then $g(x)$ is the density for a continuous random variable. We can write $g(x) = f_X(x)$ and then the associated random variable X has the property that

$$P(a \le X \le b) = \int_a^b f_X(x)\,dx.$$

The density and CDF are related just like derivatives and integrals are related.

Remark 24.5. Densities and CDFs
The CDF is the integral of the density, i.e.,

$$F_X(a) = \int_{-\infty}^a f_X(x)\,dx,$$

so the density is the derivative of the CDF:

$$f_X(x) = \frac{d}{dx} F_X(x).$$

The second equation only holds at x for which the CDF is differentiable.

24.2 Examples

Example 24.6. If X is a continuous random variable with density

$$f_X(x) = \begin{cases} \frac{1}{26}(4x+1) & \text{if } 2 \le x \le 4, \\ 0 & \text{otherwise,} \end{cases}$$

what is $P(1 \le X \le 3)$?

The density of X is shown on the left of Figure 24.1.

Remember that $P(1 \le X \le 3)$ is the region under the curve of the density. So we need to integrate the density over the range $1 \le x \le 3$. Since the density is 0 in the left-hand-portion of this range, i.e., for $1 \le x \le 2$, then it suffices to integrate the density only over the x's for which the density is nonzero, i.e., for $2 \le x \le 3$. Thus

$$\begin{aligned} P(1 \le X \le 3) &= P(2 \le X \le 3) \\ &= \int_2^3 \frac{1}{26}(4x+1)\,dx \\ &= \frac{1}{26}(2x^2+x)\Big|_{x=2}^{3} \\ &= \frac{1}{26}(((2)(3^2)+3)-((2)(2^2)+2)) \\ &= 11/26 \\ &= 0.4231 \end{aligned}$$

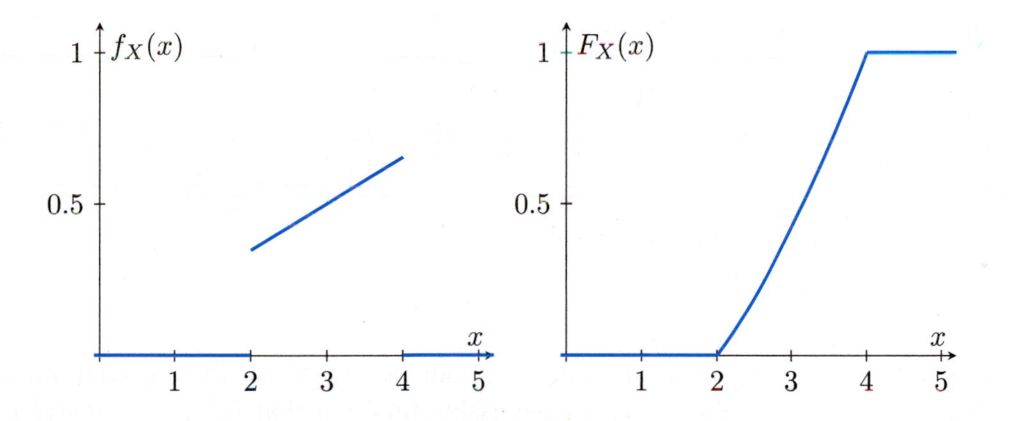

Figure 24.1: Left: The density $f_X(x) = \frac{1}{26}(4x+1)$. Right: The CDF $F_X(x) = \frac{1}{26}(2x^2+x-10)$, of a random variable defined on $[2,4]$.

Example 24.6 (continued) If X is a continuous random variable with density

$$f_X(x) = \begin{cases} \frac{1}{26}(4x+1) & \text{if } 2 \le x \le 4, \\ 0 & \text{otherwise,} \end{cases}$$

what is the CDF?

Remember that the CDF of a random variable X is

$$F_X(x) = P(X \le x).$$

Since X is always between 2 and 4, then the CDF is

$$F_X(x) = P(X \leq x) = 0 \qquad \text{for } x < 2,$$

and

$$F_X(x) = P(X \leq x) = 1 \qquad \text{for } 4 < x.$$

The most interesting values of the CDF happen for $2 \leq x \leq 4$. We compute the density from $-\infty$ to a, where a is in the interval $[2, 4]$, but the density is 0 for $x < 2$. Thus, it suffices to integrate over the interval $[2, a]$.

$$\begin{aligned} P(X \leq a) &= \int_2^a \frac{1}{26}(4x+1)\,dx \\ &= \frac{1}{26}(2x^2 + x)\Big|_{x=2}^{a} \\ &= \frac{1}{26}\big(((2)(a^2)+a) - ((2)(2^2)+2)\big) \\ &= \frac{1}{26}(2a^2 + a - 10) \end{aligned}$$

Thus the CDF of X is:

$$F_X(x) = \begin{cases} 0 & \text{if } x < 2, \\ \frac{1}{26}(2x^2 + x - 10) & \text{if } 2 \leq x \leq 4, \\ 1 & \text{if } 4 < x. \end{cases}$$

The CDF of X is shown on the right of Figure 24.1.

Notice that the CDF for a continuous random variable—just like the CDF for a discrete random variable—is a nondecreasing function that is between 0 and 1. On the other hand, the CDF for a discrete random variable has a jump at each point where the random variable takes its mass, but the CDF for a continuous random variable is continuous, with a limit of 0 as $x \to -\infty$ and a limit of 1 as $x \to \infty$.

Example 24.6 (continued) If X is a continuous random variable with density

$$f_X(x) = \begin{cases} \frac{1}{26}(4x+1) & \text{if } 2 \leq x \leq 4, \\ 0 & \text{otherwise,} \end{cases}$$

we calculate some probabilities using the CDF.

The probability that X is 3 or smaller is

$$P(X \leq 3) = F_X(3) = \frac{1}{26}((2)(3^2) + 3 - 10) = 11/26 = 0.4231.$$

The probability that X is between 2.8 and 4.3 is

$$\begin{aligned} P(2.8 \leq X \leq 4.3) &= P(X \leq 4.3) - P(X \leq 2.8) \\ &= F_X(4.3) - F_X(2.8) \\ &= 1 - \frac{1}{26}((2)(2.8^2) + 2.8 - 10) \\ &= 0.6738 \end{aligned}$$

The probability that X is at least 3 is

$$P(X \geq 3) = 1 - P(X \leq 3) = 1 - F_X(3) = 1 - \frac{1}{26}((2)(3^2) + 3 - 10) = 0.5769.$$

Given that X is less than 3, what is the probability X is greater than 2.8?

$$\begin{aligned} P(X > 2.8 \mid X < 3) &= \frac{P(X > 2.8 \text{ and } X < 3)}{P(X < 3)} \\ &= \frac{P(X \leq 3) - P(X \leq 2.8)}{P(X \leq 3)} \\ &= \frac{F_X(3) - F_X(2.8)}{F_X(3)} \end{aligned}$$

We already know $F_X(3) = 0.4231$. Also, $F_X(2.8) = \frac{1}{26}((2)(2.8^2) + 2.8 - 10) = 0.3262$. Thus

$$P(X > 2.8 \mid X < 3) = \frac{0.4231 - 0.3262}{0.4231} = 0.229.$$

Example 24.7. If X is a continuous random variable with CDF

$$F_X(x) = \begin{cases} 1 - e^{-x/4} & \text{if } x > 0, \\ 0 & \text{otherwise,} \end{cases}$$

what is the density $f_X(x)$?

The CDF of X is shown on the right of Figure 24.2.

As before, a key thing to remember is that, since the CDF is the integral of the density, then the density is the derivative of the CDF. Thus, for $x > 0$,

$$f_X(x) = \frac{d}{dx}F_X(x) = \frac{d}{dx}(1 - e^{-x/4}) = \frac{1}{4}e^{-x/4},$$

and $f_X(x) = 0$ otherwise.

Now calculate some probabilities using the CDF.

The probability that X is less than 4 is

$$P(X < 4) = F_X(4) = 1 - e^{-4/4} = 1 - e^{-1} = 0.6321.$$

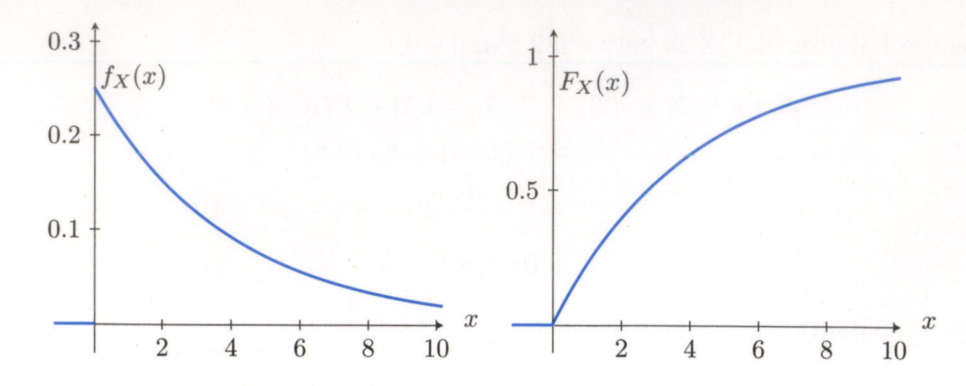

FIGURE 24.2: Left: The density $f_X(x) = \frac{1}{4}e^{-x/4}$. Right: The CDF $F_X(x) = 1 - e^{-x/4}$, of a random variable with nonnegative density for $x > 0$.

The probability that X is at least 1 is

$$P(X > 1) = 1 - P(X \le 1) = 1 - F_X(1) = 1 - (1 - e^{-1/4}) = e^{-1/4} = 0.7788.$$

Given that X is less than 4, what is the probability that X is also less than 1?

$$P(X < 1 \mid X < 4) = \frac{P(X < 1 \text{ and } X < 4)}{P(X < 4)} = \frac{P(X < 1)}{P(X < 4)} = \frac{F_X(1)}{F_X(4)}.$$

Using the closed form for the CDF, we conclude:

$$P(X < 1 \mid X < 4) = \frac{1 - e^{-1/4}}{1 - e^{-4/4}} = 0.3499.$$

Example 24.8. Suppose a random variable X has a density given by:

$$f_X(x) = \begin{cases} k(x^3 + 1)(3 - x) & \text{for } 1 \le x \le 3, \\ 0 & \text{otherwise.} \end{cases}$$

Find the constant k so that this function is a valid density.

The integral of $f_X(x)$ over all x needs to evaluate to 1, for $f_X(x)$ to be a density. In this case, the density is zero for x outside the range $[1, 3]$. So we just integrate over $[1, 3]$ and solve for k, as follows:

$$\begin{aligned}
1 &= \int_1^3 k(x^3 + 1)(3 - x)\,dx \\
&= \int_1^3 k(-x^4 + 3x^3 - x + 3)\,dx \\
&= k\left(-\frac{x^5}{5} + \frac{3x^4}{4} - \frac{x^2}{2} + 3x\right)\Bigg|_{x=1}^{3} \\
&= k\left(\left(-\frac{3^5}{5} + \frac{3(3^4)}{4} - \frac{3^2}{2} + (3)(3)\right) - \left(-\frac{1^5}{5} + \frac{3(1^4)}{4} - \frac{1^2}{2} + (3)(1)\right)\right) \\
&= (k)(68/5)
\end{aligned}$$

Thus $k = 5/68 = 0.0735$.

Example 24.9. Sometimes a density can have more than two pieces. As an example:

$$f_X(x) = \begin{cases} 3/4 & \text{if } 0 \le x \le 1, \\ 1/4 & \text{if } 3 \le x \le 4, \\ 0 & \text{otherwise.} \end{cases}$$

The density is shown on the left of Figure 24.3. Find the CDF.

We integrate to find the CDF in five disjoint regions:

For $a < 0$, we have

$$F_X(a) = \int_{-\infty}^{a} 0\,dx = 0.$$

For $0 \le a \le 1$, we have

$$F_X(a) = \int_{-\infty}^{0} 0\,dx + \int_{0}^{a} 3/4\,dx = \frac{3}{4}a.$$

For $1 \le a \le 3$, we have

$$F_X(a) = \int_{-\infty}^{0} 0\,dx + \int_{0}^{1} 3/4\,dx + \int_{1}^{a} 0\,dx = \frac{3}{4}.$$

For $3 \le a \le 4$, we have

$$F_X(a) = \int_{-\infty}^{0} 0\,dx + \int_{0}^{1} 3/4\,dx + \int_{1}^{3} 0\,dx + \int_{3}^{a} 1/4\,dx = \frac{3}{4} + \frac{1}{4}(a-3).$$

For $4 \le a$, we have

$$F_X(a) = \int_{-\infty}^{0} 0\,dx + \int_{0}^{1} 3/4\,dx + \int_{1}^{3} 0\,dx + \int_{3}^{4} 1/4\,dx + \int_{4}^{a} 0\,dx = \frac{3}{4} + \frac{1}{4} = 1.$$

Thus, the CDF of X is

$$F_X(x) = \begin{cases} 0 & \text{if } x < 0, \\ \frac{3}{4}x & \text{if } 0 \le x \le 1, \\ \frac{3}{4} & \text{if } 1 < x < 3, \\ \frac{3}{4} + \frac{1}{4}(x-3) & \text{if } 3 \le x \le 4, \\ 1 & \text{if } 4 < x. \end{cases}$$

The CDF is shown on the right of Figure 24.3. By the way, the median can be read directly from figure for this CDF. We have $F_X(x) = 1/2$ when $x = 2/3$, since (looking at the region where $0 \le x \le 1$ and $F_X(x) = \frac{3}{4}x$), we can solve $F_X(x) = \frac{3}{4}x = 1/2$ to get $x = 2/3$.

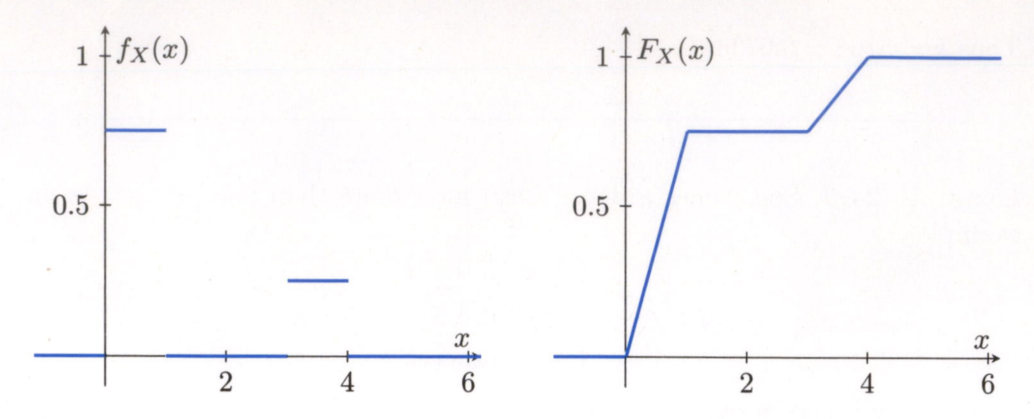

FIGURE 24.3: Left: The density $f_X(x)$. Right: The CDF $F_X(x)$, of the random variable described in Example 24.9, in which the density is nonnegative on two disjoint regions.

Example 24.10. Check to make sure that the following function is a valid density:

$$f_X(x) = \frac{3}{4}(x)(2 - x) \qquad \text{for } 0 \le x \le 2,$$

and $f_X(x) = 0$ otherwise. See Figure 24.4. Also, find $P(X \le 1)$.

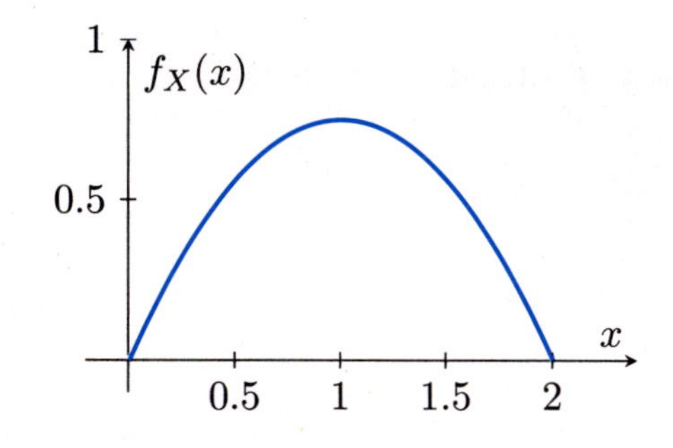

FIGURE 24.4: The density function $f_X(x) = \frac{3}{4}(x)(2 - x)$ for $0 \le x \le 2$.

The only time when $f_X(x)$ is not equal to zero is for $0 < x < 2$. In this region, $x > 0$ and $2 - x > 0$ so $f_X(x) > 0$ too. Thus, $f_X(x)$ is always nonnegative. Since X is always between 0 and 2, we calculate

$$\int_{-\infty}^{\infty} f_X(x)\,dx = \int_0^2 \frac{3}{4}(x)(2 - x)\,dx = 1,$$

so $f_X(x)$ is a density (nonnegative, and integrates to 1).

Now we find the probability that $X \le 1$. Since $f_X(x)$ is 0 when $-\infty < x < 0$, we compute

$$P(X \le 1) = \int_{-\infty}^{1} f_X(x)\,dx = \int_0^1 \frac{3}{4}(x)(2-x)\,dx = 1/2.$$

Initially we might believe that the probability is 1/2 since we included half of the interval from 0 to 2 in the previous computation, but the situation is not always so simple. For instance, the following example does not have a symmetric density:

Example 24.11. Suppose that a random variable X has density

$$f_X(x) = \frac{3}{8}(x)(2-x)(3-x) \qquad \text{for } 0 \le x \le 2,$$

and $f_X(x) = 0$ otherwise. Find $P(X \le 1)$. See Figure 24.5.

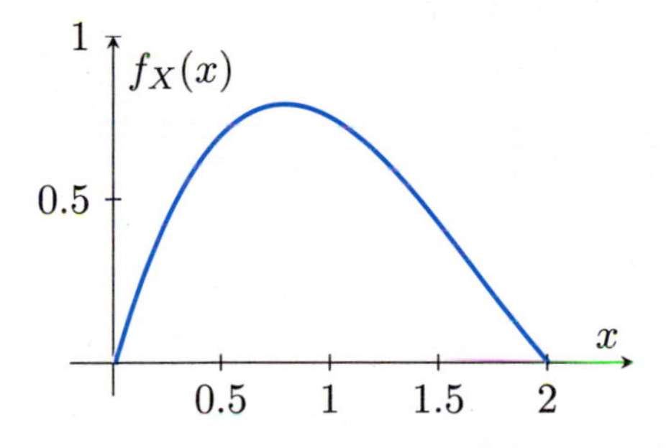

FIGURE 24.5: The density function $f_X(x) = \frac{3}{8}(x)(2-x)(3-x)$ for $0 \le x \le 2$.

We compute

$$P(X \le 1) = \int_{-\infty}^{1} f_X(x)\,dx = \int_0^1 \frac{3}{8}(x)(2-x)(3-x) = 19/32.$$

Example 24.12. One of the most frequently used types of continuous random variables are constant on an interval, and zero otherwise. For instance, consider the density

$$f_X(x) = 1/6 \qquad \text{for } 4 < x < 10,$$

and $f_X(x) = 0$ otherwise.

We notice that $f_X(x) \geq 0$ for all x, and also

$$\int_{-\infty}^{\infty} f_X(x)\,dx = \int_4^{10} 1/6\,dx = (1/6)(6) = 1.$$

So this function really is a density. In fact, for any constants a and b inside the interval, i.e., with $4 \leq a \leq b \leq 10$, we have

$$P(a \leq X \leq b) = \int_a^b 1/6\,dx = (b-a)/6.$$

For example,

$$P(2 \leq X \leq 5) = \int_2^5 1/6\,dx = (5-2)/6 = 1/2.$$

Such nice facts work because, *when we integrate a constant value over an interval, the result is just the constant times the length of the interval.* For instance, when we integrate 1/6 on the interval [2, 5], the result is just 1/6 times the length of the interval [2, 5] (i.e., 3). So the result is (1/6)(3) = 1/2. This is an example of the Uniform density, to be considered more in Chapter 31.

Example 24.13. Let

$$f_X(x) = 3e^{-3x} \qquad \text{for } x > 0,$$

and $f_X(x) = 0$ otherwise. See Figure 24.6.

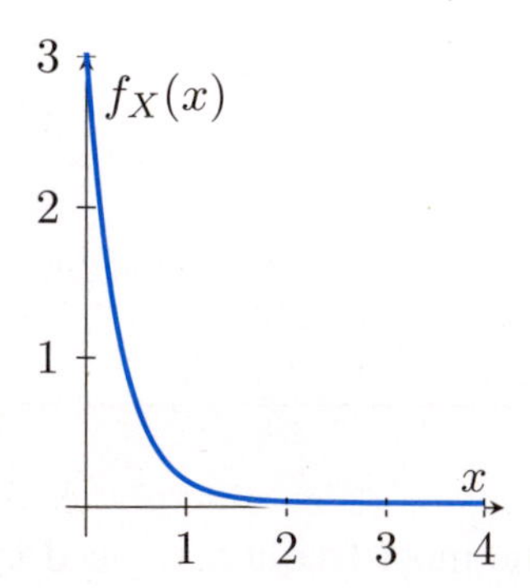

FIGURE 24.6: The density $f_X(x) = 3e^{-3x}$ for $x > 0$.

In such a case, X assumes only nonnegative values. The density $f_X(x)$ of

X is always nonnegative, i.e., $f_X(x) \geq 0$ for all x. Also, we compute

$$\begin{aligned}
\int_{-\infty}^{\infty} f_X(x)\,dx &= \int_0^{\infty} 3e^{-3x}\,dx \\
&= 3\left.\frac{e^{-3x}}{-3}\right|_{x=0}^{\infty} dx \\
&= -e^{-3x}\Big|_{x=0}^{\infty} \\
&= \lim_{x\to\infty}(-e^{-3x}) - (-e^{-(3)(0)}) \\
&= 0+1 \\
&= 1
\end{aligned}$$

So $f_X(x)$ is a density.

The probability that $X \leq 2$ is

$$\begin{aligned}
P(X \leq 2) &= \int_{-\infty}^{2} f_X(x)\,dx \\
&= \int_0^{2} 3e^{-3x}\,dx \\
&= 3\left.\frac{e^{-3x}}{-3}\right|_{x=0}^{2} \\
&= -e^{-3x}\Big|_{x=0}^{2} \\
&= (-e^{-(3)(2)}) - (-e^{-(3)(0)}) \\
&= 1 - e^{-6} \\
&= 0.997521
\end{aligned}$$

The density considered here is Exponential, to be revisited in Chapter 32.

24.3 Exercises

24.3.1 Practice

Exercise 24.1. Identify from the information below whether X is a discrete or continuous random variable.

a. Let X be the height of a randomly selected 8-year-old child.

b. Let X be a random variable that takes values on the nonnegative integers, i.e., $X \in \{0, 1, 2, 3, \ldots\}$.

c. Let X be a random variable that takes values on all nonnegative real numbers, i.e., $X \geq 0$.

d. Let X be the number of 8-year-old children who will attend the 1:30 showing of the movie "Shrek 3."

Exercise 24.2. Identify from the information below whether X is a discrete or continuous random variable.

a. Let X be a random variable that takes values on the set $\{3, 4, 5, 6, 7\}$

b. Let X be the number of defective light bulbs in a shipment.

c. Let X be a random variable that takes values on the real interval $[3, 7]$

Exercise 24.3. If you know $P(X > 2) = 0.3$, fill in the chart for the other values you know. Write "???" if there is not enough information to figure out a value.

	X is discrete	X is continuous
$P(X \geq 2)$		
$P(X < 2)$		
$P(X \leq 2)$		
$P(X = 2)$		

Exercise 24.4. If you know $P(X \leq 5) = 0.9$, fill in the chart for the other values you know. Write "???" if there is not enough information to figure out a value.

	X is discrete	X is continuous
$P(X < 5)$		
$P(X > 5)$		
$P(X \geq 5)$		
$P(X = 5)$		

Exercise 24.5. Let X have density $f_X(x) = kx^2(1-x)^2$ for $0 \leq x \leq 1$, and $f_X(x) = 0$ otherwise, where k is constant.

a. Find the value of k.

b. Find $P(X \geq 3/4)$.

Exercise 24.6. Let X have density $f_X(x) = \frac{\sqrt{3(x+2)}}{6}$ for $-2 \leq x \leq 1$, and $f_X(x) = 0$ otherwise. Find the probability that X is positive.

Exercise 24.7. Let X have density

$$f_X(x) = \frac{1}{5}, \qquad \text{for } 2 \leq x \leq 7,$$

and $f_X(x) = 0$ otherwise.

a. Find $P(X > 3)$.

b. Find the CDF $F_X(x)$.

c. Find $P(X < 5)$.

d. Find $P(X > 8)$.

e. Find $P(5 \le X \le 6.5)$.

f. Find $P(X < 1)$.

g. Graph the density $f_X(x)$.

h. Graph the CDF $F_X(x)$.

Exercise 24.8. Let X be the waiting time (in minutes) until a student's friend arrives. Suppose that X has density

$$f_X(x) = \frac{1}{3}e^{-x/3}, \qquad \text{for } 0 < x,$$

and $f_X(x) = 0$ otherwise.

a. Find $P(3 \le X \le 6)$.

b. Find the CDF $F_X(x)$.

c. Find $P(X \ge 24)$.

d. Find $P(X \le -3)$.

e. Find $P(3 \le X \le 12)$.

f. Find $P(X < 300)$.

g. Graph the density $f_X(x)$.

h. Graph the CDF $F_X(x)$.

Exercise 24.9. Consider the function from Example 24.12, i.e., $f_X(x) = 1/6$ for $4 < x < 10$, and $f_X(x) = 0$ otherwise.

a. What is the cumulative distribution function $F_X(x)$?

b. Draw a picture of the cumulative distribution function $F_X(x)$.

Exercise 24.10. Consider the function from Example 24.13, i.e., $f_X(x) = 3e^{-3x}$ for $x > 0$, and $f_X(x) = 0$ otherwise.

a. What is the cumulative distribution function $F_X(x)$?

b. Draw a picture of the cumulative distribution function $F_X(x)$.

Exercise 24.11. Let

$$f_X(x) = x/4, \qquad \text{for } 0 \le x \le \sqrt{8},$$

and $f_X(x) = 0$ otherwise. What is the probability that X exceeds 1?

Exercise 24.12. Let

$$f_X(x) = \frac{3}{64}x^2, \qquad \text{for } 0 \le x \le 4,$$

and $f_X(x) = 0$ otherwise. Find $P(X > 2)$.

Exercise 24.13. Let

$$f_Y(y) = 12y^2(1-y), \qquad \text{for } 0 \le y \le 1,$$

and $f_Y(y) = 0$ otherwise. Find the CDF of Y.

Exercise 24.14. Suppose that X has CDF

$$F_X(x) = \begin{cases} 0 & \text{if } x < 3, \\ \frac{1}{171}(x^3 - 6x - 9) & \text{if } 3 \le x \le 6, \\ 1 & \text{if } 6 < x. \end{cases}$$

Find the density $f_X(x)$.

24.3.2 Extensions

Exercise 24.15. Suppose that X has density

$$f_X(x) = \frac{1}{2}\sin x, \qquad \text{for } 0 \le x \le \pi,$$

and $f_X(x) = 0$ otherwise.

a. Find $P(X \le \pi/6)$.

b. Find $P(\pi/3 \le X \le 2\pi/3)$.

c. Find $P(\pi/3 \le X)$.

Exercise 24.16. Let

$$f_X(x) = \frac{\ln x}{e^2}, \qquad \text{for } e \le x \le e^2,$$

and $f_X(x) = 0$ otherwise. Find the CDF of X.

Exercise 24.17. Let $c > 0$ be a fixed constant. Let

$$f_X(x) = ce^{-cx}, \qquad \text{for } 0 < x,$$

and $f_X(x) = 0$ otherwise. Find a general expression for the CDF $F_X(x)$, for any $x > 0$.

Exercise 24.18. Let

$$F_Y(y) = \sqrt{y}, \qquad \text{for } 0 \le y \le 1;$$

and $F_Y(y) = 0$ for $y < 0$, and $F_Y(y) = 1$ for $y > 1$. Find the density of Y.

Exercise 24.19. Suppose a density $f_X(x)$ increases linearly from $(16, 0)$ to $(24, 1/4)$.

a. Find the CDF of X.

b. What is the value of a so that $P(X > a) = 0.75$?

Exercise 24.20. What is the constant k that makes the following function a valid density?

$$f_X(x) = \begin{cases} kx^9(1-x)^2 & \text{if } 0 \le x \le 1, \\ 0 & \text{otherwise,} \end{cases}$$

Exercise 24.21. What is the constant k that makes the following function a valid density?

$$f_X(x) = \begin{cases} k\left(\frac{1}{x} - 3x^2\right) & \text{if } 1 \le x \le 20, \\ 0 & \text{otherwise,} \end{cases}$$

Exercise 24.22. Suppose that X has CDF

$$F_X(x) = \begin{cases} 0 & \text{if } x < 0, \\ \frac{7}{8}x^2 & \text{if } 0 \le x \le 1, \\ \frac{7}{8} & \text{if } 1 < x < 7, \\ \frac{7}{8} + \frac{1}{8}(x-7) & \text{if } 7 \le x \le 8, \\ 1 & \text{if } 8 < x. \end{cases}$$

a. Find the density $f_X(x)$.

b. Find the median of X.

Exercise 24.23. Again consider the function from Example 24.13, i.e., let $f_X(x) = 3e^{-3x}$ for $x > 0$, and $f_X(x) = 0$ otherwise. Evaluate $\int_{-\infty}^{\infty} x f_X(x)\,dx$.

Exercise 24.24. Let X be the lifetime (in years) of a carbon-14 atom before it decays. Then X has density

$$f_X(x) = \frac{\ln 2}{5730} \exp\left(-\frac{\ln 2}{5730}x\right), \qquad \text{for } 0 < x,$$

and $f_X(x) = 0$ otherwise. Find the length of time a (in years) so that

$$P(X \le a) = 1/2 \qquad \text{and} \qquad P(X > a) = 1/2.$$

24.3.3 Advanced

Exercise 24.25. Let X have density

$$f_X(x) = \frac{1}{2}x^2 e^{-x}, \qquad \text{for } x > 0,$$

and $f_X(x) = 0$ otherwise. Find $P(X < 2)$.

Exercise 24.26. Let k be a fixed, positive integer. Let X have density

$$f_X(x) = x^k e^{-x}, \qquad \text{for } x > 0,$$

and $f_X(x) = 0$ otherwise. Find the value of k so that $f_X(x)$ is a valid density function.

Exercise 24.27. Comparing two CDFs. Is it possible to define two random variables X and Y in such a way that the CDF for X is always strictly larger than the CDF for Y? In other words, is it possible to have $F_X(x) > F_Y(y)$ always? If yes, then give an example. If not, then explain why not.

Exercise 24.28. Moments of a Continuous Uniform random variable. Let X be a Continuous Uniform random variable on the interval $[a, b]$. Compute $\mathbb{E}(X^n)$.

Chapter 25

Joint Densities

To expect the unexpected shows a thoroughly modern intellect.

—*An Ideal Husband, Act 3* by Oscar Wilde (play that premiered on January 3, 1895)

You use a social networking application that sends out notifications for personal messages and notifications for your friends posting new pictures. You know the distribution for wait time for a personal message and the distribution of wait time for a new-picture notification. How long do you expect to have to wait for either a personal message or a picture notification?

25.1 Introduction

Just as with discrete random variables, we can handle more than one random variable at a time. We can have a joint probability density function, also called a joint density, associated with two continuous random variables:

Definition 25.1. Joint probability density function
The joint probability density function—also called a joint density—of a pair of continuous random variables X and Y is $f_{X,Y}(x, y)$, and it has the following properties:

1. The joint density is always nonnegative, i.e.,

$$f_{X,Y}(x, y) \geq 0 \text{ for all } x, y.$$

2. The joint density can be integrated to get probabilities, i.e., if A and B are sets of real numbers, then

$$P(X \in A \text{ and } Y \in B) = \int_A \int_B f_{X,Y}(x, y)\, dy\, dx.$$

We also have a notion of a joint cumulative distribution function, also called a joint CDF:

Definition 25.2. Joint cumulative distribution function
If X and Y are random variables, then the joint cumulative distribution function (also called joint CDF) of X and Y is the probability that X does not exceed x and Y does not exceed y, i.e.,

$$F_{X,Y}(x,y) = P(X \leq x \text{ and } Y \leq y).$$

If we are given the joint density $f_{X,Y}(x,y)$, we know how to compute the joint CDF.

Remark 25.3. Obtaining the joint CDF from the joint density of a pair of continuous random variables
If X and Y are a pair of continuous random variables, with joint density $f_{X,Y}(x,y)$, then X and Y have joint CDF

$$F_{X,Y}(a,b) = \int_{-\infty}^{a} \int_{-\infty}^{b} f_{X,Y}(x,y)\, dy\, dx.$$

The observation in the box above can also be viewed as follows: Since

$$F_{X,Y}(a,b) = P(X \leq a \text{ and } Y \leq b),$$

then to get $-\infty < x \leq a$ and $-\infty < y \leq b$, we can use $A = (-\infty, a]$ and $B = (-\infty, b]$, and we can integrate the joint density in each of the variables:

$$F_{X,Y}(a,b) = \int_{-\infty}^{a} \int_{-\infty}^{b} f_{X,Y}(x,y)\, dy\, dx.$$

We can create pairs of random variables by using joint densities that satisfy properties like those in the previous chapter. Any nonnegative function of x and y, whose integral over all x and all y is 1, will describe a pair of continuous random variables.

Remark 25.4. Creating a Pair of Continuous Random Variables
If $f(x,y)$ is any nonnegative function, and $\int_{-\infty}^{\infty} \int_{-\infty}^{\infty} f(x,y)\, dy\, dx = 1$, then $f(x,y)$ is the joint density for a pair of continuous random variables. We can write $f(x,y) = f_{X,Y}(x,y)$ and then the associated random variables X and Y have the property that

$$P(a \leq X \leq b \text{ and } c \leq Y \leq d) = \int_{a}^{b} \int_{c}^{d} f_{X,Y}(x,y)\, dy\, dx.$$

Note: The reason $f(x,y)$ has to be a nonnegative function in the box above is that densities are always nonnegative. This doesn't mean that the associated random variables have to be nonnegative. For instance, if $f(x,y) = 1/4$ for $-1 \leq x \leq 1$ and $-1 \leq y \leq 1$, and $f(x,y) = 0$ otherwise, then $f(x,y)$ is nonnegative, but the random variables X and Y associated with this density could be positive or negative, i.e., $-1 \leq X \leq 1$ and $-1 \leq Y \leq 1$.

25.2 Examples

First we present a series of four examples that all focus on the same pair of random variables. In Example 25.5, we verify that $f_{X,Y}(x,y)$ is a joint density. In Example 25.6, we consider the joint CDF and we calculate a probability. In Example 25.7, we study another random variable, Z, that is defined to be the minimum of X and Y. Finally, in Example 25.8, we compute the probability that Y is bigger than X, and we do this in two different ways.

Example 25.5. Let X be the time (in minutes) that Maxine waits for a traffic light to turn green, and let Y be the time (in minutes, at a different intersection) that Daniella waits for a traffic light to turn green. Suppose that X and Y have joint density

$$f_{X,Y}(x,y) = 15e^{-3x-5y}, \qquad \text{for } x > 0 \text{ and } y > 0,$$

and $f_{X,Y}(x,y) = 0$ otherwise. See Figure 25.1.

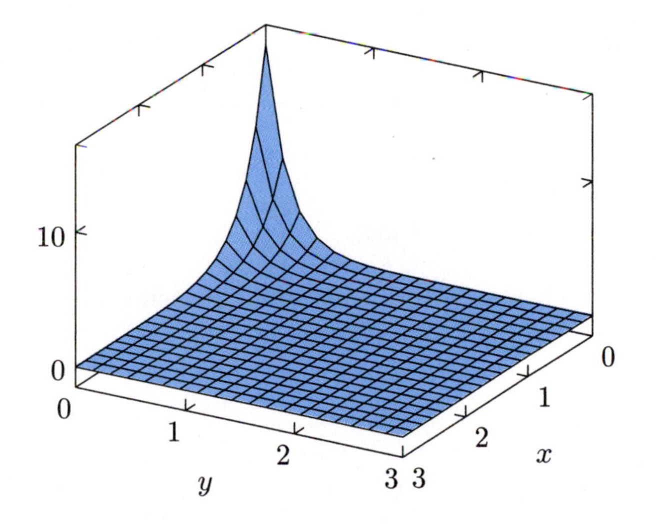

FIGURE 25.1: The joint density $f_{X,Y}(x,y) = 15e^{-3x-5y}$, for $x > 0$ and $y > 0$.

We first check that $f_{X,Y}(x,y)$ is a joint density. We see that $f_{X,Y}(x,y)$ is nonnegative. Next, we check that the integral over all x's and y's is 1:

$$\int_{-\infty}^{\infty}\int_{-\infty}^{\infty} f_{X,Y}(x,y)\,dy\,dx = \int_{0}^{\infty}\int_{0}^{\infty} 15e^{-3x-5y}\,dy\,dx$$

We can factor the 15 into 3 times 5, anticipating the division by 3 and 5 that will occur in the respective divisions. We get

$$\begin{aligned}\int_{-\infty}^{\infty}\int_{-\infty}^{\infty} f_{X,Y}(x,y)\,dy\,dx &= \int_0^{\infty} 3e^{-3x}\,dx \int_0^{\infty} 5e^{-5y}\,dy \\ &= \left(-e^{-3x}\Big|_{x=0}^{\infty}\right)\left(-e^{-5y}\Big|_{y=0}^{\infty}\right) \\ &= 1\end{aligned}$$

So we have verified that $f_{X,Y}(x,y)$ is a joint density.

Example 25.6. Now we compute $F_{X,Y}(1/2, 1/4) = P(X \le 1/2 \text{ and } Y \le 1/4)$, i.e., the probability that Maxine waits 1/2 of a minute or less, and Daniella waits 1/4 of a minute or less. In the computation, we take advantage of the fact that the integration in x does not affect the integration in y, so we split the integral into two parts.

$$\begin{aligned}F_{X,Y}(1/2,1/4) &= \int_{-\infty}^{1/2}\int_{-\infty}^{1/4} f_{X,Y}(x,y)\,dy\,dx \\ &= \int_0^{1/2}\int_0^{1/4} 15e^{-3x-5y}\,dy\,dx \\ &= \left(\int_0^{1/2} 3e^{-3x}\,dx\right)\left(\int_0^{1/4} 5e^{-5y}\,dy\right) \\ &= \left(-e^{-3x}\Big|_{x=0}^{1/2}\right)\left(-5e^{-5y}\Big|_{y=0}^{1/4}\right) \\ &= (1-e^{-3/2})(1-e^{-5/4}) \\ &= 0.5543\end{aligned}$$

We can compute much more than just joint CDFs using joint densities. For instance, we could compute $P(X > 1/10 \text{ and } Y > 2/7)$, i.e., the probability that Maxine waits more than 1/10 of a minute and Daniella waits more than 2/7 of a minute, as follows:

$$\begin{aligned}P(X > 1/10 \text{ and } Y > 2/7) &= \int_{1/10}^{\infty}\int_{2/7}^{\infty} f_{X,Y}(x,y)\,dy\,dx \\ &= \int_{1/10}^{\infty}\int_{2/7}^{\infty} 15e^{-3x-5y}\,dy\,dx \\ &= \left(\int_{1/10}^{\infty} 3e^{-3x}\,dx\right)\left(\int_{2/7}^{\infty} 5e^{-5y}\,dy\right) \\ &= \left(-e^{-3x}\Big|_{x=1/10}^{\infty}\right)\left(-5e^{-5y}\Big|_{y=2/7}^{\infty}\right) \\ &= \left(e^{-3/10}\right)\left(e^{-10/7}\right) \\ &= e^{-121/70} \\ &= 0.1775\end{aligned}$$

Example 25.7. Using the same X and Y as above—the waiting times for Maxine and Daniella at their respective traffic lights—we let $Z = \min(X, Y)$. Notice that Z is a waiting time too, because Z is the waiting time until the first one of the pair (Maxine or Daniella) has her own light turn green. So Z is the waiting time until Maxine or Daniella gets to depart. We find the cumulative distribution function of Z.

First of all, since $X > 0$ always and $Y > 0$ always, then $Z > 0$ always too. So, for $a \le 0$, we have $F_Z(a) = P(Z \le a) = 0$.

Now we turn our attention to $a > 0$. Since $Z = \min(X, Y)$, then $F_Z(a)$ is equal to $P(\min(X, Y) \le a)$. This is a difficult probability to calculate, because we could either have $X \le a$ or $Y \le a$ or both. It is much easier to calculate the complementary probability,

$$P(Z > a) = P(\min(X, Y) > a) = P(X > a \text{ and } Y > a).$$

We compute, for $a > 0$,

$$\begin{aligned}
F_Z(a) &= P(Z \le a) \\
&= 1 - P(Z > a) \\
&= 1 - P(\min(X, Y) > a) \\
&= 1 - P(X > a \text{ and } Y > a) \\
&= 1 - \int_a^\infty \int_a^\infty 15e^{-3x-5y}\,dy\,dx \\
&= 1 - \int_a^\infty -3e^{-3x-5y}\Big|_{y=a}^{\infty}\,dx \\
&= 1 - \int_a^\infty 3e^{-3x-5a}\,dx \\
&= 1 + e^{-3x-5a}\Big|_{x=a}^{\infty} \\
&= 1 - e^{-8a}
\end{aligned}$$

So, in summary, the cumulative distribution function of Z is

$$F_Z(z) = 1 - e^{-8z} \qquad \text{for } z > 0,$$

and $F_Z(z) = 0$ otherwise.

Example 25.8. With joint densities, we can calculate probabilities about all kinds of relationships between X and Y. For instance, we could calculate the probability that Daniella waits longer at her light than Maxine, e.g., $P(Y > X)$. We have two ways to set up the integral:

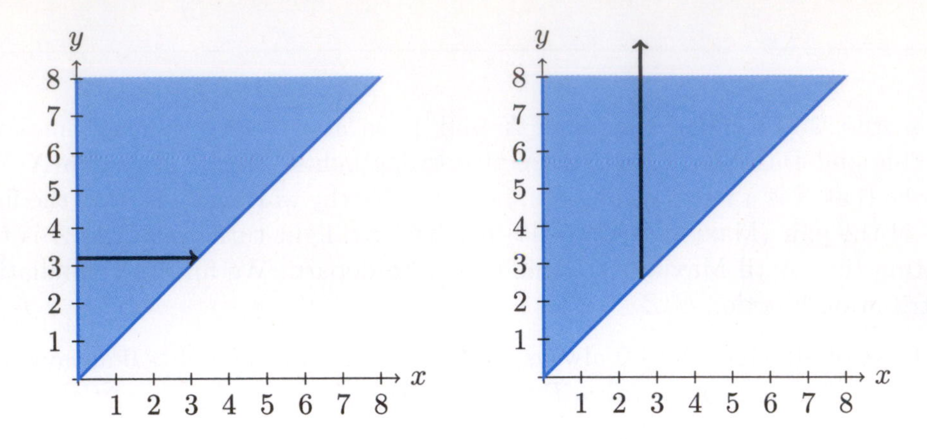

FIGURE 25.2: Left: Setting up the integral for $P(Y > X)$, with y as the outer integral and x as the inner integral. The y is fixed (e.g., $y = 3.2$), and x ranges from 0 to y. Right: Setting up the integral for $P(Y > X)$, with x as the outer integral and y as the inner integral. The x is fixed (e.g., $x = 2.6$), and y ranges from x to ∞.

Method #1: We can integrate for the outer integral over all y's; for the inner integral, y is fixed, and we integrate over all of the x's that are smaller than y, i.e., $0 \leq x \leq y$, as shown on the left of Figure 25.2. We get

$$P(Y > X) = \int_0^\infty \int_0^y 15e^{-3x-5y}\,dx\,dy = \int_0^\infty -5e^{-3x-5y}\Big|_{x=0}^{y}\,dy$$

which yields

$$P(Y > X) = \int_0^\infty (5e^{-5y} - 5e^{-8y})\,dy = (-e^{-5y} + (5/8)e^{-8y})\Big|_{y=0}^{\infty} = 3/8.$$

Method #2: We can integrate for the outer integral over all x's; for the inner integral, x is fixed, and we integrate over all of the y's that are larger than x, i.e., $x \leq y$, as shown on the right of Figure 25.2. We get

$$P(Y > X) = \int_0^\infty \int_x^\infty 15e^{-3x-5y}\,dy\,dx = \int_0^\infty -3e^{-3x-5y}\Big|_{y=x}^{\infty}\,dx$$

which yields

$$P(Y > X) = \int_0^\infty 3e^{-3x-5x}\,dx = \int_0^\infty 3e^{-8x}\,dx = -(3/8)e^{-8x}\Big|_{x=0}^{\infty} = 3/8.$$

This agrees with the result computed using *Method #1.*

Example 25.9. Suppose that a person throws a dart at a square dart board. Let X and Y denote, respectively, the x- and y-coordinates (in feet) of the location where the dart lands; the middle of the dart board is at $(0, 0)$. Suppose that the dart board is two feet wide and two feet tall, so that the dart only lands on the dart board if $-1 \le X \le 1$ and $-1 \le Y \le 1$. Also suppose that the person always hits the dart board, and moreover, X and Y have joint density:

$$f_{X,Y}(x, y) = \frac{9}{16}(1 - x^2)(1 - y^2) \qquad \text{for } -1 \le x \le 1,\ -1 \le y \le 1,$$

and $f_{X,Y}(x, y) = 0$ otherwise. See Figure 25.3.

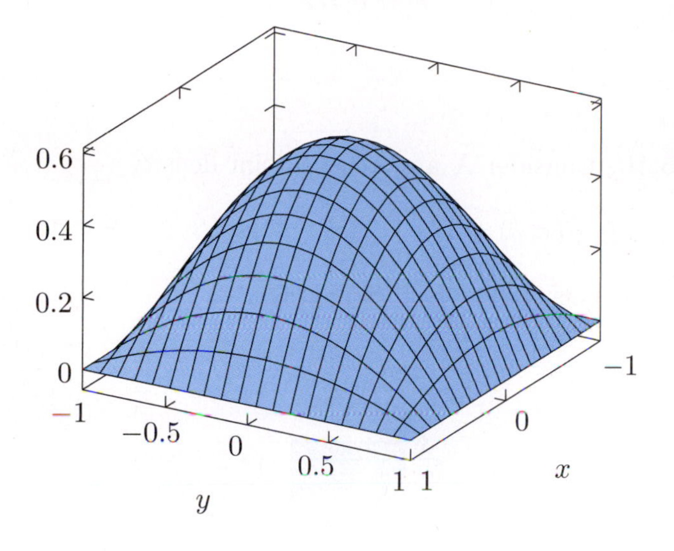

FIGURE 25.3: Joint PDF $f_{X,Y}(x, y) = \frac{9}{16}(1 - x^2)(1 - y^2)$, for $-1 \le x \le 1$, $-1 \le y \le 1$.

To check that $f_{X,Y}(x, y)$ is a joint density, we see that $f_{X,Y}(x, y) \ge 0$ always; the function is 0 except possibly when $-1 \le x \le 1$, $-1 \le y \le 1$, in which case $1 - x^2$ and $1 - y^2$ are both nonnegative. Also

$$\begin{aligned}
\int_{-\infty}^{\infty}\int_{-\infty}^{\infty} f_{X,Y}(x, y)\, dy\, dx &= \int_{-1}^{1}\int_{-1}^{1} \frac{9}{16}(1 - x^2)(1 - y^2)\, dy\, dx \\
&= \frac{9}{16}\int_{-1}^{1}(1 - x^2)\, dx \int_{-1}^{1}(1 - y^2)\, dy \\
&= \frac{9}{16}\,(x - x^3/3)\big|_{x=-1}^{1}\,(y - y^3/3)\big|_{y=-1}^{1} \\
&= (9/16)(4/3)(4/3) \\
&= 1
\end{aligned}$$

So $f_{X,Y}(x, y)$ is a joint density. Now we find the probability that the x- and y-coordinates of the location where the dart lands are each within a 1/2 foot from the center of the board. We compute

$$
\begin{aligned}
P\left(-\tfrac{1}{2} \le X \le \tfrac{1}{2}, -\tfrac{1}{2} \le Y \le \tfrac{1}{2}\right) &= \int_{-1/2}^{1/2}\int_{-1/2}^{1/2} \frac{9}{16}(1-x^2)(1-y^2)\,dy\,dx \\
&= \frac{9}{16}\int_{-1/2}^{1/2}(1-x^2)\,dx \int_{-1/2}^{1/2}(1-y^2)\,dy \\
&= \frac{9}{16}\,(x - x^3/3)\Big|_{x=-1/2}^{1/2}\,(y-y^3/3)\Big|_{y=-1/2}^{1/2} \\
&= 121/256 \\
&= 0.4727
\end{aligned}
$$

Example 25.10. Consider X and Y with joint density

$$f_{X,Y}(x,y) = 1/9 \qquad \text{for } 0 \le x \le 3,\ 0 \le y \le 3,$$

and $f_{X,Y}(x,y) = 0$ otherwise. See Figure 25.4.

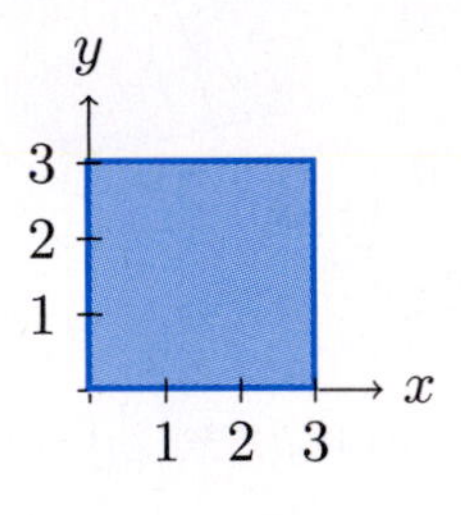

FIGURE 25.4: The joint density $f_{X,Y}(x,y) = 1/9$ for $0 \le x \le 3$, $0 \le y \le 3$; and $f_{X,Y}(x,y) = 0$ otherwise.

To see that $f_{X,Y}(x,y)$ is a joint density, we note that $f_{X,Y}(x,y) \ge 0$ for all x, y, and we could compute that

$$\int_0^3\int_0^3 1/9\,dy\,dx = 1.$$

For two-variable densities, a probability is calculated as the area under the curve, i.e., the density is the volume of a solid. *If the density is constant, such and integration is easy to perform.* In such a case, the integral is equal to the integrand times the area over which we are integrating. In this example, the integrand is just the constant 1/9, and the area over which we integrate is 9, so the integral is $(1/9)(9) = 1$. So $f_{X,Y}(x,y)$ is a joint density.

Now we find the probability that the minimum of X and Y is less than 1. In other words, we find $P(\min(X,Y) \le 1)$. The area for which $\min(X,Y) \le 1$ is 5, and seen in Figure 25.5. So the desired probability is just $(1/9)(5) = 5/9$. There is no need to perform the integration using complicated methods. Anytime that the integrand is constant, we just multiply the integrand (in this case, 1/9) by the area of the integration region (in this case, 5).

If a reader insists on performing the integration using the tools learned in calculus, we caution that the integral will need to be broken into at least two parts. See Figure 25.5. We could integrate over the region in the middle of Figure 25.5, and then integrate over the region on the right side of Figure 25.5, and add the results together, since these two regions are disjoint, and their union is the entire region over which we need to integrate:

$$\begin{aligned} P(\min(X,Y) \leq 1) &= \int_0^1 \int_0^3 1/9\, dy\, dx + \int_1^3 \int_0^1 1/9\, dy\, dx \\ &= (1)(3)(1/3) + (2)(1)(1/9) \\ &= 5/9 \end{aligned}$$

The final answer, as expected, is 5/9.

We emphasize the utility of the first method, which can always be used easily when the integrand is constant and the area of the region of integration can be easily determined.

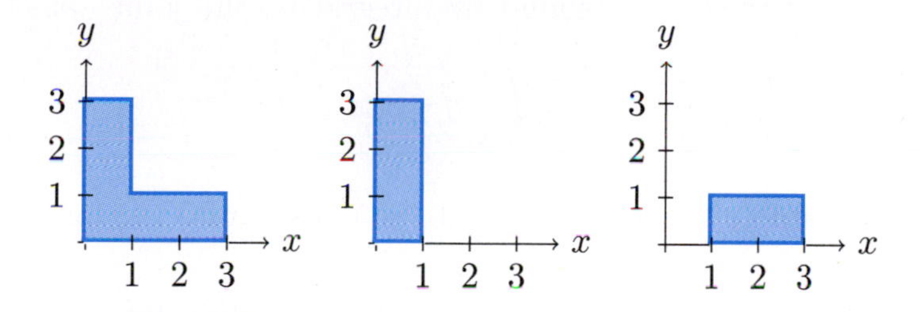

FIGURE 25.5: Left: The region where $\min(X,Y) \leq 1$. The region can be split into two disjoint parts. Middle: The region where $\min(X,Y) \leq 1$ and $0 \leq X \leq 1$. Right: The region where $\min(X,Y) \leq 1$ and $1 \leq X \leq 3$.

Example 25.11. Consider the joint density

$$f_{X,Y}(x,y) = \begin{cases} 9e^{-3x} & \text{for } 0 < y < x; \\ 0 & \text{otherwise.} \end{cases}$$

Find the density of X and the density of Y.

The CDF of X is

$$F_X(a) = P(X \leq a) = \int_{-\infty}^{a} \int_{-\infty}^{\infty} f_{X,Y}(x,y)\, dy\, dx,$$

Differentiating on both sides with respect to a yields

$$f_X(a) = \int_{-\infty}^{\infty} f_{X,Y}(a, y)\, dy,$$

or equivalently,

$$f_X(x) = \int_{-\infty}^{\infty} f_{X,Y}(x, y)\, dy.$$

This idea works in general—nothing here is specific to the X and Y in the example:

Theorem 25.12. Finding the density of one random variable, using the joint density
If two continuous random variables X and Y have joint density $f_{X,Y}(x, y)$, the density of X can be retrieved by integrating over all y's:

$$f_X(x) = \int_{-\infty}^{\infty} f_{X,Y}(x, y)\, dy.$$

Similarly, the density of Y is found by integrating the joint density over all x's:

$$f_Y(y) = \int_{-\infty}^{\infty} f_{X,Y}(x, y)\, dx.$$

Returning to Example 25.11, we see that the density of X is, for $x > 0$,

$$f_X(x) = \int_{-\infty}^{\infty} f_{X,Y}(x, y)\, dy = \int_0^x 9e^{-3x}\, dy = 9xe^{-3x}.$$

Similarly, the density of Y is, for $y > 0$,

$$f_Y(y) = \int_{-\infty}^{\infty} f_{X,Y}(x, y)\, dx = \int_y^{\infty} 9e^{-3x}\, dx = 3e^{-3y}.$$

25.3 Exercises

25.3.1 Practice

Exercise 25.1. Consider a pair of random variables X, Y with constant joint density on the triangle with vertices at $(0, 0)$, $(3, 0)$, and $(0, 3)$. Find $P(X+Y > 2)$.

Exercise 25.2. Consider a pair of random variables X, Y with constant joint density on the quadrilateral with vertices $(0, 0)$, $(2, 0)$, $(2, 6)$, $(0, 12)$. Find $P(Y \geq 3X)$.

Exercise 25.3. Let X, Y have joint density

$$f_{X,Y}(x,y) = \begin{cases} 14e^{-2x-7y} & \text{if } x > 0 \text{ and } y > 0, \\ 0 & \text{otherwise.} \end{cases}$$

Find $P(X > Y)$.

Exercise 25.4. Suppose X, Y have joint density

$$f_{X,Y}(x,y) = \begin{cases} 1/16 & \text{if } -2 \le x \le 2 \text{ and } -2 \le y \le 2, \\ 0 & \text{otherwise.} \end{cases}$$

Find $P(|X - Y| \le 1)$.

Exercise 25.5. Suppose X, Y have joint density

$$f_{X,Y}(x,y) = \begin{cases} \frac{1}{9}(3-x)(2-y) & \text{if } 0 \le x \le 3 \text{ and } 0 \le y \le 2, \\ 0 & \text{otherwise.} \end{cases}$$

Find $P(Y > X)$.

Exercise 25.6. Consider random variables X and Y with joint density

$$f_{X,Y}(x,y) = \frac{1}{8}(1-x^2)(3-y) \qquad \text{for } -1 \le x \le 1 \text{ and } -1 \le y \le 1,$$

and $f_{X,Y}(x,y) = 0$ otherwise. Find the probability that X and Y are both negative.

Exercise 25.7. Let X, Y have joint density

$$f_{X,Y}(x,y) = \frac{3}{38750}(x^3 + y^2), \qquad \text{for } 0 \le x \le 10 \text{ and } 0 \le y \le 5,$$

and $f_{X,Y}(x,y) = 0$ otherwise. Find $P(X \le 5, Y \le 2)$.

Exercise 25.8. Suppose X, Y are jointly distributed with joint density

$$f_{X,Y}(x,y) = \frac{1}{304}(x+1)(y^2+1), \qquad \text{for } 0 \le x, y \le 4,$$

and $f_{X,Y}(x,y) = 0$ otherwise.

a. Find $P(1 \le X \le 2,\ 3 \le Y \le 4)$.

b. Find the density of X.

c. Find the density of Y.

Exercise 25.9. Consider X, Y with joint density

$$f_{X,Y}(x,y) = \frac{9}{64}x^2y^2, \qquad \text{for } 0 \le x \le 2, 0 \le y \le 2,$$

and $f_{X,Y}(x,y) = 0$ otherwise.

a. Find $P(X < 1, Y < 1)$.

b. Find $P(X > 1, Y < 1)$.

c. Convince yourself that, by symmetry, $P(X < 1, Y > 1)$ is the same as the value from part b.

d. Find $P(X > 1, Y > 1)$.

e. Check that the answers from parts a–d sum to 1.

Exercise 25.10. Consider X, Y with joint density

$$f_{X,Y}(x, y) = \frac{1}{2500}(10 - x)(10 - y), \qquad \text{for } 0 \le x \le 10,\ 0 \le y \le 10,$$

and $f_{X,Y}(x, y) = 0$ otherwise.

a. Find $P(X < 9, Y < 9)$.

b. Find $P(X < 9, Y > 9)$.

c. Convince yourself that, by symmetry, $P(X > 9, Y < 9)$ is the same as the value from part b.

d. Find $P(X > 9, Y > 9)$.

e. Check that the answers from parts a–d sum to 1.

Exercise 25.11. Let X, Y have joint density

$$f_{X,Y}(x, y) = x + y, \qquad \text{for } 0 \le x, y \le 1,$$

and $f_{X,Y}(x, y) = 0$ otherwise. Find the joint CDF, i.e., $F_{X,Y}(x, y) = P(X \le x, Y \le y)$.

Exercise 25.12. Let X, Y have joint density

$$f_{X,Y}(x, y) = \frac{6}{7}(x + y)^2, \qquad \text{for } 0 \le x, y \le 1,$$

and $f_{X,Y}(x, y) = 0$ otherwise. Find $P(X \le 1/2, Y \le 1/2)$.

Exercise 25.13. Consider a pair of random variables X, Y with joint density

$$f_{X,Y}(x, y) = 8e^{-2x-4y}, \qquad \text{for } 0 < x,\ 0 < y,$$

and $f_{X,Y}(x, y) = 0$ otherwise. Find $P(X \le 4, Y \le 8)$.

Exercise 25.14. Let X, Y have constant density on the square where $0 \le X \le 4$, $0 \le Y \le 4$. Find $P(X + Y < 4)$.

Exercise 25.15. If X and Y have joint density

$$f_{X,Y}(x, y) = \frac{9}{32}x^2y^2, \qquad \text{for } |x| \le 1 \text{ and } |y| \le 2,$$

and $f_{X,Y}(x, y) = 0$ otherwise, find the probability that X and Y are both positive.

Exercise 25.16. Consider X, Y with joint density

$$f_{X,Y}(x,y) = 10e^{-2x-5y}, \qquad \text{for } 0 < x,\ 0 < y,$$

and $f_{X,Y}(x,y) = 0$ otherwise. Find $P(X < 2, Y < 1)$.

25.3.2 Extensions

Exercise 25.17. Consider the random variables X and Y defined in Example 25.5, i.e., X is Maxine's waiting time, and Y is Daniella's waiting time. Let $W = \max(X, Y)$, i.e., W is either Maxine's or Daniella's waiting time, whichever is larger! Find $F_W(w) = P(W \leq w)$, the cumulative distribution function of W. This is equal to $P(\max(X, Y) \leq w)$, i.e., the probability that Maxine waits less than w minutes and Daniella waits less than w minutes.

Exercise 25.18. Freddy and Jane have entered a game in which they each win between 0 and 2 dollars. If X is the amount Freddy wins, and Y is the amount that Jane wins, they believe that the joint density of their winnings will be

$$f_{X,Y}(x,y) = \frac{1}{4}xy \qquad \text{for } 0 \leq x \leq 2 \text{ and } 0 \leq y \leq 2,$$

and $f_{X,Y}(x,y) = 0$ otherwise. Find the probability that their combined winnings exceed 2, i.e., find $P(X + Y > 2)$.

Exercise 25.19. If X and Y have joint distribution

$$f_{X,Y}(x,y) = \sin x \cos y, \qquad \text{for } 0 \leq x, y \leq \pi/2,$$

and $f_{X,Y}(x,y) = 0$ otherwise, then find the probability that X and Y are both smaller than $\pi/6$.

Exercise 25.20. If X and Y have joint distribution

$$f_{X,Y}(x,y) = \frac{1}{4}\cos x \sin y + \frac{1}{4}, \qquad \text{for } -1 \leq x, y \leq 1,$$

and $f_{X,Y}(x,y) = 0$ otherwise, then find the probability that $0 \leq X \leq 1/2$ and $Y \leq 2X$.

Exercise 25.21. Suppose that X, Y are jointly distributed with $f_{X,Y}(x,y) = 1/10$, for (x,y) in the triangle with vertices at the origin, $(2,0)$, and $(0,10)$, and $f_{X,Y}(x,y) = 0$ otherwise. Find the probability that $Y \leq 2$.

25.3.3 Advanced

Exercise 25.22. Consider X, Y with joint density

$$f_{X,Y}(x,y) = \frac{\operatorname{sech}^2 x}{(y+1)^2}, \qquad \text{for } x \geq 0 \text{ and } y \geq 0,$$

and $f_{X,Y}(x, y) = 0$ otherwise. Find $P(X \leq 2, Y \leq 2)$. (The function "sech" is the hyperbolic secant, which should be familiar to many readers who recently completed a course in calculus. Other readers may choose to skip this exercise, because sech will make very few appearances in the densities that we consider in this text.)

Exercise 25.23. Consider X, Y with joint density

$$f_{X,Y}(x, y) = \frac{1}{4\pi}, \qquad \text{for } x^2 + y^2 \leq 4,$$

and $f_{X,Y}(x, y) = 0$ otherwise. Find the probability that (X, Y) is at most 1 unit from the origin and is located in the first quadrant, i.e., that $X \geq 0$ and $Y \geq 0$ and $X^2 + Y^2 \leq 1$.

Exercise 25.24. Suppose that the joint PDF of X and Y is $f(x, y) = 2e^{-x-y}$ for $0 < y < x < \infty$, and $f(x, y) = 0$ otherwise.

a. Find $f_X(x)$, the density of X.

b. Find $f_Y(Y)$, the density of Y.

Chapter 26

Independent Continuous Random Variables

I'm trying to handle all this unpredictability
In all probability
—"Long Shot," sung by Kelly Clarkson, from the album *All I Ever Wanted*; song written by Katy Perry, Glen Ballard, Matt Thiessen

---

Are a person's height and weight independent or dependent random variables? How about a person's height and GPA?

26.1 Introduction

Two continuous random variables are independent if they satisfy the continuous versions of Definition 9.13, using densities instead of masses.

Definition 26.1. Independent continuous random variables: Several equivalent formulations

1. Joint density can be factored into a function of x times a function of y. These functions of x and y can also be normalized so that they are the densities of X and Y, respectively: $f_{X,Y}(x, y) = f_X(x)f_Y(y)$ for all x and y.

2. Joint CDF can be factored into a function of x times a function of y. Again, we can normalize these, so that they are the CDFs of X and Y, respectively: $F_{X,Y}(x, y) = F_X(x)F_Y(y)$ for all x and y.

3. We will define conditional densities in the next chapter, but we will go ahead and state the way they will be used for independence: (3a) Density of X equals conditional density $f_{X|Y}(x \mid y)$ of X given $Y = y$, or (3b) Density of Y equals conditional density $f_{Y|X}(y \mid x)$ of Y given $X = x$.

26.2 Examples

One of the trickiest things about the above statement of independence is remembering to check that the factoring works for all x's and y's. For instance, we briefly reconsider Exercise 25.6:

Example 26.2. Consider random variables X and Y with joint density

$$f_{X,Y}(x,y) = \frac{1}{8}(1-x^2)(3-y) \qquad \text{for } -1 \le x \le 1 \text{ and } -1 \le y \le 1,$$

and $f_{X,Y}(x,y) = 0$ otherwise.

The region where the density is nonzero is a square, and thus the factorization of $f_{X,Y}(x,y)$ into the "x stuff" and the "y stuff" works for all x's and y's. So, in this case, X and Y are independent.

Since the X and Y in this example are independent, then we can find the density of X by itself; all we need to do is determine the constant at the front of the density of X. We compute

$$\int_{-1}^{1} (1-x^2)\,dx = (x - x^3/3)\big|_{x=-1}^{1} = 4/3.$$

So we need a constant of $3/4$ at the front of the density of X, so that the integral will turn out to be 1. Thus, the density of X is

$$f_X(x) = (3/4)(1-x^2) \qquad \text{for } -1 \le x \le 1,$$

and $f_X(x) = 0$ otherwise. Similarly, we compute

$$\int_{-1}^{1} (3-y)\,dy = (3y - y^2/2)\big|_{y=-1}^{1} = 6.$$

So, as before, we need a constant at the front of the density of Y; in this case, we need a $1/6$ in the density of Y, so that the integral will turn out to be 1. Thus, the density of Y is

$$f_Y(y) = (1/6)(3-y) \qquad \text{for } -1 \le y \le 1,$$

and $f_Y(y) = 0$ otherwise. Finally, we check that

$$\begin{aligned} f_{X,Y}(x,y) &= \frac{1}{8}(1-x^2)(3-y) \\ &= (3/4)(1-x^2)(1/6)(3-y) \\ &= f_X(x)f_Y(y) \qquad \text{for all } -1 \le x \le 1 \text{ and } -1 \le y \le 1, \end{aligned}$$

and otherwise $f_{X,Y}(x,y) = 0$, $f_X(x) = 0$, and $f_Y(y) = 0$.

Example 26.3. In Exercise 25.18, we have two continuous random variables X and Y with joint density

$$f_{X,Y}(x,y) = \frac{1}{4}xy \qquad \text{for } 0 \le x \le 2 \text{ and } 0 \le y \le 2,$$

and $f_{X,Y}(x,y) = 0$ otherwise.

In this example, the region where the density is nonzero is again square, so the factorization of $f_{X,Y}(x,y)$ into the "x stuff" and the "y stuff" again works for all x's and y's. Therefore, again in this case, X and Y are independent.

We easily compute that

$$\int_0^2 x\,dx = x^2/2|_{x=0}^2 = 2.$$

So we need a correction factor of 1/2 at the front to make this into a density. So the density of X must be

$$f_X(x) = (1/2)(x) \qquad \text{for } 0 \le x \le 2,$$

and $f_X(x) = 0$ otherwise. Since the entire problem is symmetric in terms of X and Y, the density of Y must also be

$$f_Y(y) = (1/2)(y) \qquad \text{for } 0 \le y \le 2,$$

and $f_Y(y) = 0$ otherwise.

Notice that $f_{X,Y}(x,y) = xy/4 = (1/2)(x)(1/2)(y) = f_X(x)f_Y(y)$.

Example 26.4. Just because the joint density appears to factor, we must use caution. Consider, for instance, the joint density

$$f_{X,Y}(x,y) = \frac{3}{2}xy \qquad \text{for } 0 \le x \text{ and } 0 \le y \text{ and } x+y \le 2,$$

and $f_{X,Y}(x,y) = 0$ otherwise.

Important: *We might initially be inclined to say that $f_{X,Y}(x,y)$ can be factored in a way that works for all x and y, but this is not the case.* For instance, if $x = 1/2$, then y can be between 0 and 3/2; but, if $x = 1$, then y can be between 0 and 1. So Y is not independent of X in this example. Nonetheless, we can still use the technique of finding the density of one random variable, using the joint density (given in the box at the end of the previous chapter), i.e.,

$$f_X(x) = \int_{-\infty}^{\infty} f_{X,Y}(x,y)\,dy.$$

In this example, we know $f_X(x) = 0$ for $x < 0$ or $x > 2$, since X is always between 0 and 2. Now we consider $0 \le x \le 2$, and we get

$$f_X(x) = \int_0^{2-x} \frac{3}{2}xy\,dy = \frac{3}{2}xy^2/2|_{y=0}^{2-x} = \frac{3}{4}x(2-x)^2.$$

So we conclude that the density of X is

$$f_X(x) = \frac{3}{4}x(2-x)^2 \qquad \text{for } 0 \le x \le 2,$$

and $f_X(x) = 0$ otherwise. In particular, note that X and Y are dependent, since $f_X(x)$ is not a factor of $f_{X,Y}(x,y)$.

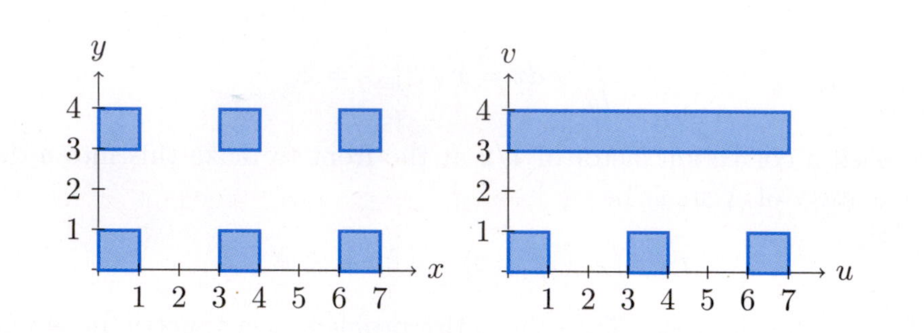

FIGURE 26.1: Left: A region S containing six rectangles. Right: A region T containing four rectangles.

For two continuous random variables to be independent, the domains where the random variables are defined must be rectangles (this is a necessary—but not sufficient—condition). *If X and Y are defined in a range that is a circle, or a triangle, or some other non-rectangular shape, then X and Y must be dependent.* Also, X and Y can be defined on several rectangles and be independent. For instance, consider a region S with six rectangles, as on the left side of Figure 26.1.

Let $f_{X,Y}(x,y)$ be constant on the region S, so

$$f_{X,Y}(x,y) = 1/6 \qquad \text{for } (x,y) \in S,$$

and $f_{X,Y}(x,y) = 0$ otherwise. Then X and Y are independent. On the other hand, consider another pair of random variables, say U and V, which are defined on a region T with four rectangles, as on the right side of Figure 26.1. Let $f_{U,V}(u,v)$ be constant on the region T, so

$$f_{U,V}(u,v) = 1/10 \qquad \text{for } (u,v) \in T,$$

and $f_{U,V}(u,v) = 0$ otherwise. Then U and V are *dependent.* For instance, if $v = 0.5$ then either u is in the range $[0,1]$ or $[3,4]$ or $[6,7]$, but on the other hand, if $v = 3.5$, then u can be anywhere in the range $[0, 7]$. So U and V are dependent. So, the random variables must be defined in rectangles, and if more than one rectangle is present, the rectangles must be in a grid shape.

One more note: We stated earlier that, "For two continuous random variables to be independent, the domains where the random variables are defined must be rectangles (this is a necessary—but not sufficient—condition)." We are just emphasizing that having a rectangle-shaped domain of definition is not enough for independence either. The density must (of course) also factor, for the random variables to be independent. Consider, for instance, Exercise 25.11, in which X and Y are defined in the unit square $0 \le x, y \le 1$, with density $f_{X,Y}(x, y) = x + y$. Despite the fact that the domain of definition is a rectangle (moreover, a square), the joint density does not factor, so X and Y are dependent.

Remark 26.5. How to check for independence, using a joint density:
1. The joint density $f_{X,Y}(x, y)$ must be defined on a rectangle, or on several rectangles arranged in a grid shape.
2. The joint density $f_{X,Y}(x, y)$ must factor into a product, with all of the x's in one part and all of the y's in the other part. Properly normalized, the x part is the density $f_X(x)$ of X in such a case, and the y part is the density $f_Y(y)$ of Y.

Example 26.6. Each time a pitcher delivers a fastball, the speed is distributed between 90 and 100 miles per hour, with density 1/10. Assume that the speeds of pitches are independent.

Consider two fastballs thrown by the pitcher. Find the probability that at least one of them is 93 miles per hour or faster.

Let X and Y denote the speeds of the two pitches. Then

$$f_X(x) = 1/10 \qquad \text{for } 90 \le x \le 100,$$

and $f_X(x) = 0$ otherwise. Similarly,

$$f_Y(y) = 1/10 \qquad \text{for } 90 \le y \le 100,$$

and $f_Y(y) = 0$ otherwise. Since X and Y are independent, it follows that the joint density of X and Y is

$$f_{X,Y}(x, y) = 1/100 \qquad \text{for } 90 \le x \le 100,\ 90 \le y \le 100,$$

and $f_{X,Y}(x, y) = 0$ otherwise.

So the desired probability is

$$\begin{aligned}
P(X \ge 93 \text{ or } Y \ge 93) &= 1 - P(X \le 93 \text{ and } Y \le 93) \\
&= 1 - \int_{90}^{93} \int_{90}^{93} 1/100 \, dy \, dx \\
&= 1 - 9/100 \\
&= 91/100
\end{aligned}$$

Another method for computing the desired probability is the following: Let Z denote the number of the two fastballs that exceed 93 miles per hour. Then Z is either 0 or 1 or 2, and moreover, Z is Binomial with $n = 2$ and with probability of success $p = \int_{93}^{100} 1/10\,dx = 7/10$ on each throw. So the probability of at least one fastball over 93 miles per hour is

$$\begin{aligned} P(Z \geq 1) &= 1 - P(Z = 0) \\ &= 1 - \binom{2}{0} p^0 (1-p)^2 \\ &= 1 - (1 - 7/10)^2 \\ &= 1 - (3/10)^2 \\ &= 1 - 9/100 \\ &= 91/100. \end{aligned}$$

Example 26.7. The lifetime (in years) of a music player—before it permanently fails—is a random variable X with density

$$f_X(x) = \frac{1}{3}e^{-x/3} \qquad \text{for } x > 0,$$

and $f_X(x) = 0$ otherwise.

a. Consider 2 music players that are assumed to have independent lifetimes, X and Y, respectively. Then the joint density of X and Y is

$$f_{X,Y}(x, y) = \frac{1}{3}e^{-x/3}\frac{1}{3}e^{-y/3} = \frac{1}{9}e^{-(x+y)/3} \qquad \text{for } x > 0 \text{ and } y > 0,$$

and $f_{X,Y}(x, y) = 0$ otherwise.

b. The probability that a music player permanently fails within the first ten years is

$$\int_0^{10} \frac{1}{3}e^{-x/3}\,dx = -e^{-x/3}\Big|_{x=0}^{10} = 1 - e^{-10/3} = 0.9643260.$$

c. Now suppose that each of the 23 students in a class has a music player. What is the probability that all 23 of the music players permanently fail within the first ten years? The probability is

$$\left(\int_0^{10} \frac{1}{3}e^{-x/3}\,dx\right)^{23} = \left(e^{-x/3}\Big|_{x=0}^{10}\right)^{23} = \left(1 - e^{-10/3}\right)^{23} = 0.4336599.$$

Another way to see this is the following: The number of music players that permanently fail within the first ten years is a Binomial random variable with

$n = 23$ and $p = 1 - e^{-10/3} = 0.9643260$. So the probability that all 23 fail within the first ten years is

$$\binom{23}{23} p^{23}(1-p)^0 = \left(1 - e^{-10/3}\right)^{23} = 0.4336599.$$

d. Now suppose that we check music players until we find one that dies within only one year! The number of music players that we need to check is Geometric with probability of success $\int_0^1 \frac{1}{3}e^{-x/3}\,dx = e^{-x/3}\big|_{x=0}^{1} = 1 - e^{-1/3} = 0.2834687$. So the expected number of music players that we would need to check, until we find one that dies within only one year, is

$$\frac{1}{1 - e^{-1/3}} = 3.527726.$$

26.3 Exercises

26.3.1 Practice

Exercise 26.1. Suppose X, Y have joint density

$$f_{X,Y}(x,y) = \begin{cases} \frac{1}{9}(3-x)(2-y) & \text{if } 0 \le x \le 3 \text{ and } 0 \le y \le 2, \\ 0 & \text{otherwise.} \end{cases}$$

a. Are X and Y independent? Why or why not?

b. Find the density $f_X(x)$ of X.

c. Find the density $f_Y(y)$ of Y.

Exercise 26.2. If two random variables X and Y have joint density

$$f_{X,Y}(x,y) = \frac{12}{7}(xy + x^2), \qquad \text{for } 0 \le x \le 1,\ 0 \le y \le 1,$$

and $f_{X,Y}(x,y) = 0$ otherwise, are X and Y independent? Why or not?

Exercise 26.3. When X, Y have joint density

$$f_{X,Y}(x,y) = \frac{x^2}{y \ln 2}, \qquad \text{for } 0 \le x \le 1,\ 1/2 \le y \le 4,$$

and $f_{X,Y}(x,y) = 0$ otherwise, find:

a. The density $f_X(x)$ of X.

b. The density $f_Y(y)$ of Y.

Exercise 26.4. Let X, Y have joint density

$$f_{X,Y}(x,y) = 2x/45, \qquad \text{for } 0 \le x \le 3,\ 0 \le y \le 5$$

and $f_{X,Y}(x,y) = 0$ otherwise.

a. Are X and Y independent?

b. Find the density $f_X(x)$ of X.

c. Find the density $f_Y(y)$ of Y.

Exercise 26.5. Let X, Y have joint density

$$f_{X,Y}(x, y) = 3xy/1250, \qquad \text{for } 0 \le x,\ 0 \le y,\ x + y \le 10$$

and $f_{X,Y}(x, y) = 0$ otherwise.

a. Are X and Y independent?

b. Find the density $f_X(x)$ of X.

c. Find the density $f_Y(y)$ of Y.

26.3.2 Extensions

Exercise 26.6. When an emergency occurs, the response time (in hours) of the first police car is a random variable X with density

$$f_X(x) = 12e^{-12x} \qquad \text{for } x > 0,$$

and $f_X(x) = 0$ otherwise. The response time (in hours) of the first fire engine is a random variable Y with density

$$f_Y(y) = 10e^{-10y} \qquad \text{for } y > 0,$$

and $f_Y(y) = 0$ otherwise. Assume X and Y are independent.

Find the probability that the first police car arrives before the first fire engine.

Exercise 26.7. Police cars are randomly stationed throughout the town. When an emergency occurs, the distance a police car must travel north or south is a random variable X with density

$$f_X(x) = 1/6 \qquad \text{for } 0 \le x \le 6,$$

and $f_X(x) = 0$ otherwise. The distance a police car must travel east or west is a random variable Y with density

$$f_Y(y) = 1/6 \qquad \text{for } 0 \le y \le 6,$$

and $f_Y(y) = 0$ otherwise. Assume that X and Y are independent. So $X + Y$ is the total distance traveled. Find the probability that $X + Y \le 4$.

(Hint: First find the joint density $f_{X,Y}(x, y)$ of X and Y, and draw a picture for where the joint density is defined. If you write an integral for the probability, the integrand is constant. So you can compute the desired area in your picture, divided by the total area.)

Exercise 26.8. Let

$$f_{X,Y}(x,y) = \left(\frac{9}{64}\sqrt{2}\right)(x^2)(\sqrt{y}), \qquad \text{for } 0 \le x \le 2,\ 0 \le y \le 2,$$

and $f_{X,Y}(x,y) = 0$ otherwise.

a. Are X and Y independent? Why?

b. Find $P(X \le 1)$.

c. Find $P(Y \le 1)$.

d. Find $P(X + Y \le 1)$.

Exercise 26.9. Suppose X and Y have joint density

$$f_{X,Y}(x,y) = e^{-x-y}, \qquad \text{for } 0 < x \text{ and } 0 < y,$$

and $f_{X,Y}(x,y) = 0$ otherwise.

a. Are X and Y independent?

b. Find $P(X < 2, Y < 2)$.

c. Compare with the situation in which X and Y have joint density

$$f_{X,Y}(x,y) = 2e^{-x-y}, \qquad \text{for } 0 < x < y,$$

and $f_{X,Y}(x,y) = 0$ otherwise. Find $P(X < 2, Y < 2)$ in this case. Are X and Y independent in this new case? Why or why not? Can you see whether X and Y are independent by only looking at the joint density of X and Y?

Exercise 26.10. Suppose that X, Y have constant joint density on the triangle with corners at $(4, 0)$, $(0, 4)$, and the origin.

a. Find $P(X < 3, Y < 3)$.

b. Are X and Y independent?

Exercise 26.11. Suppose X and Y have joint density

$$f_{X,Y}(x,y) = 18x^2y^5, \qquad \text{for } 0 < x < 1,\ 0 < y < 1$$

and $f_{X,Y}(x,y) = 0$ otherwise.

a. Are X and Y independent?

b. Find $P(X < 1/2)$.

c. Find $P(Y < 1/2)$.

d. Find $P(X + Y < 1)$.

Exercise 26.12. Suppose X and Y have joint density

$$f_{X,Y}(x, y) = 1512x^2y^5, \qquad \text{for } 0 \leq x,\ 0 \leq y,\ x + y < 1$$

and $f_{X,Y}(x, y) = 0$ otherwise.

a. Are X and Y independent?

b. Find $P(X < 1/2)$.

c. Find $P(Y < 1/2)$.

d. Find $P(X + Y < 1/2)$.

Exercise 26.13. Let X and Y have constant joint density on the parallelogram with corners at the origin, $(1, 1)$, $(1, 2)$, and $(0, 1)$.

a. Find the joint density $f_{X,Y}(x, y)$.

b. Find $P(Y < 3/2)$.

Exercise 26.14. Suppose X and Y have joint density

$$f_{X,Y}(x, y) = \frac{8 - 2x^2}{3x^2y^3}, \qquad \text{for } 1 \leq x \leq 2,\ 1/2 \leq y \leq 1$$

and $f_{X,Y}(x, y) = 0$ otherwise.

a. Are X and Y independent?

b. Find $P(X < 3/2, Y < 3/4)$.

Exercise 26.15. Let X, Y, Z have joint density

$$f_{X,Y,Z}(x, y, z) = 6, \qquad \text{for } 0 < x < y < z < 1$$

and $f_{X,Y,Z}(x, y, z) = 0$ otherwise.

a. Are X and Y and Z independent?

b. Find the density $f_X(x)$ of X.

c. Find the density $f_Y(y)$ of Y.

d. Find the density $f_Z(z)$ of Z.

Exercise 26.16. Consider two random variables X and Y with joint density

$$f_{X,Y}(x, y) = x/8 + y/8, \qquad \text{for } 0 < x < 2,\ 0 < y < 2$$

and $f_{X,Y}(x, y) = 0$ otherwise.

a. Are X and Y independent?

b. Find $P(X > 1)$ and $P(Y > 1)$.

c. Find $P(X + Y < 2)$.

Exercise 26.17. If X and Y have joint density

$$f_{X,Y}(x,y) = 75xe^{-5x-3y}, \qquad \text{for } 0 < x,\ 0 < y$$

and $f_{X,Y}(x,y) = 0$ otherwise:

a. Are X and Y independent?

b. Find $P(X < Y)$.

Exercise 26.18. Suppose X and Y have constant density on the region in Figure 26.2.

a. Are X and Y independent?

b. Find $P(X + Y < 2)$.

c. Find $P(X + Y < 2.5)$.

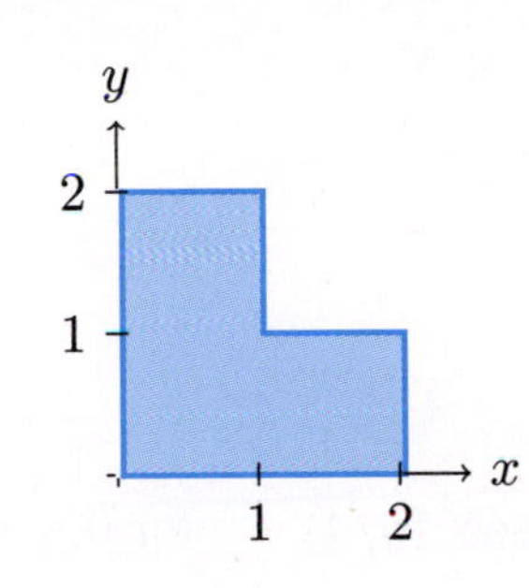

FIGURE 26.2: Region where X and Y have positive density.

Exercise 26.19. If the pair of random variables X and Y have

$$f_{X,Y}(x,y) = 180x^2y^2, \qquad \text{for } 0 < x,\ 0 < y, \text{ and } x + y < 1$$

and $f_{X,Y}(x,y) = 0$ otherwise, find $P(X + Y > 1/2)$.

Exercise 26.20. Let X, Y have joint density

$$f_{X,Y}(x,y) = \frac{9}{32}\sqrt{xy}, \qquad \text{for } 0 \le x \le 1,\ 0 \le y \le 4$$

and $f_{X,Y}(x,y) = 0$ otherwise.

a. Are X and Y independent?

b. Find $P(X < Y)$.

c. Find $P(X < 2Y)$.

d. Find $P(X < 4Y)$.

Exercise 26.21. Let X, Y have joint density

$$f_{X,Y}(x, y) = \frac{3}{32}(x - y), \qquad \text{for } 0 \le y \le x \le 4$$

and $f_{X,Y}(x, y) = 0$ otherwise.

a. Are X and Y independent?

b. Find $P(X \le 2)$.

c. Find $P(Y \le 2)$.

Exercise 26.22. Let X, Y have joint density

$$f_{X,Y}(x, y) = \frac{(8 - 2x)(2 - y)}{256}, \qquad \text{for } -2 \le x \le 2,\ -2 \le y \le 2$$

and $f_{X,Y}(x, y) = 0$ otherwise.

a. Are X and Y independent?

b. Find the density $f_X(x)$ of X.

c. Find the density $f_Y(y)$ of Y.

26.3.3 Advanced

Exercise 26.23. Let

$$f_{X,Y}(x, y) = \frac{1}{8}\sin x \sec^2(y/4) \qquad \text{for } 0 \le x \le \pi,\ 0 \le y \le \pi,$$

and $f_{X,Y}(x, y) = 0$ otherwise. Find:

a. The density $f_X(x)$ of X.

b. The density $f_Y(y)$ of Y.

c. $P(X < \pi/2, Y < \pi/2)$.

Exercise 26.24. Suppose X and Y have joint density

$$f_{X,Y}(x, y) = \frac{2\text{sech}^2(1/x)}{y^3}, \qquad \text{for } 0 < x < y,$$

and $f_{X,Y}(x, y) = 0$ otherwise.

a. Show that $f_{X,Y}(x, y)$ is a joint density, i.e., the integral over all x's and y's is 1.

b. Are X and Y independent?

Exercise 26.25. When X and Y have joint density

$$f_{X,Y}(x, y) = \frac{2x^2y^2}{e^{x+2y}}, \qquad \text{for } 0 < x,\ 0 < y$$

and $f_{X,Y}(x, y) = 0$ otherwise, find $P(X < Y)$.

Chapter 27

Conditional Distributions

The probability of meeting someone you know increases when you are with someone you don't want to be seen with.

—Anonymous

A dancer is twirling randomly on a semi-circular stage with a radius of 3 feet. When the music stops, she is 2 feet from the edge of the stage closest to the audience. The spotlight is set up to shine on the middle of the stage, but if the dancer is too far stage left or stage right (more than 2.5 feet from the middle), she will be poorly lit. What is the probability that the dancer will end up in the dark?

27.1 Introduction

Conditional densities are defined similarly to conditional masses. The conditional density of X given $Y = y$ tells us how X behaves when $Y = y$. We can use the conditional density to get conditional probabilities. Later, once we delve into expected values, we will use conditional densities to get expected values for one variable that are conditional on knowing the value of the other variable.

The conditional density of X given Y, written $f_{X|Y}(x \mid y)$, is only defined when $f_Y(y) > 0$. The conditional density is defined as follows:

Definition 27.1. Conditional density
The conditional density of a continuous random variable X, given $Y = y$, is defined as the joint density of X and Y divided by the density of Y:

$$f_{X|Y}(x \mid y) = \frac{f_{X,Y}(x,y)}{f_Y(y)}.$$

For the conditional density of X given Y to make sense, we must use Y values such that $f_Y(y) > 0$.

Equivalently, when $f_Y(y) > 0$, then $f_{X|Y}(x \mid y)$ is the unique function that gives

$$f_{X,Y}(x,y) = f_Y(y) f_{X|Y}(x \mid y).$$

To explain how the conditional density can actually be used to obtain conditional probabilities, consider the following example:

The probability

$$P(X \in A \mid Y = 3) = \int_A f_{X|Y}(x \mid 3)\, dx$$

only depends on A. For instance, if A is the interval from 0 to 10, we have

$$P(0 \le X \le 10 \mid Y = 3) = \int_0^{10} f_{X|Y}(x \mid 3)\, dx.$$

More generally,

$$P(X \in A \mid Y = y) = \int_A f_{X|Y}(x \mid y)\, dx.$$

The value "y" should be thought of as "fixed," and then we are interested in whether X is in some interval A. For instance, if we think of "y" as fixed, we might ask whether X is found in the interval from 0 to 10:

$$P(0 \le X \le 10 \mid Y = y) = \int_0^{10} f_{X|Y}(x \mid y)\, dx.$$

Some specific examples should help to clarify this idea.

27.2 Examples

Example 27.2. A bird lands in a grassy region described as follows: $x \ge 0$, and $y \ge 0$, and $x + y \le 10$. This region is shown in Figure 27.1.

Let X and Y be the coordinates of the bird's landing. Assume that X and Y have the joint density

$$f_{X,Y}(x,y) = 1/50 \qquad \text{for } 0 \le x \text{ and } 0 \le y \text{ and } x + y \le 10,$$

and $f_{X,Y}(x,y) = 0$ otherwise.

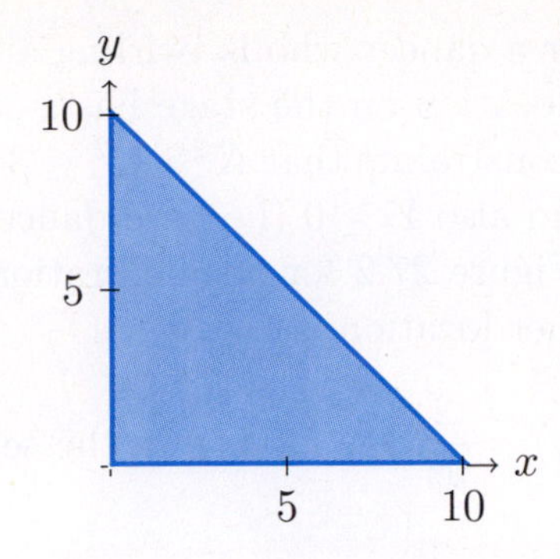

FIGURE 27.1: The grassy region where a bird lands.

If we know that the bird lands somewhere on the line where $Y = 2$, we can find the conditional density of X given that $Y = 2$. First we need the density of Y. For $0 \le y \le 10$, we integrate over all relevant x's—here, over x from 0 to $10 - y$. After the integral with respect to x is performed, and we substitute in $10 - y$ and 0 for x, there will be no x's remaining. In other words,

We can think of this as integrating x out of the picture.

$$\begin{aligned} f_Y(y) &= \int_{-\infty}^{\infty} f_{X,Y}(x, y)\,dx \\ &= \int_0^{10-y} 1/50\,dx \\ &= (10 - y)/50 \end{aligned}$$

We know that $0 \le X \le 10 - 2 = 8$ when $Y = 2$. So, for $0 \le x \le 8$, we compute

$$\begin{aligned} f_{X|Y}(x \mid 2) &= \frac{f_{X,Y}(x, 2)}{f_Y(2)} \\ &= \frac{1/50}{(10 - 2)/50} \\ &= 1/8 \end{aligned}$$

Given that the bird's landing y-coordinate is 2, what is the probability that the x-coordinate is between 0 and 5?

The probability is

$$P(0 \le X \le 5 \mid Y = 2) = \int_0^5 f_{X|Y}(x \mid 2)\,dx = \int_0^5 1/8\,dx = 5/8.$$

This makes sense, because in this problem, all of the integrands are constant. We are asking where the x-coordinate will be, and we know that the x-coordinate is somewhere from 0 to 8, and all of the parts of the line are treated equally, in a sense.

Example 27.3. Consider a dancer who is twirling around a semicircle shaped stage. Suppose that her location on the stage has x- and y-coordinates X and Y, respectively, with the constraints that $X^2+Y^2 \leq 3^2$ (i.e., X and Y are found in a circle of radius 3), and also $Y \geq 0$ (i.e., the dancer is in a semicircle at the front of the stage). See Figure 27.2 for an illustration of this region. Suppose that the joint density of her location is

$$f_{X,Y}(x,y) = \frac{2}{9\pi} \qquad \text{for } x, y \text{ in the semicircle,}$$

and $f_{X,Y}(x,y) = 0$ otherwise.

Given that the dancer's y-coordinate is exactly 2, what is the density of the x-coordinate of her location?

We first compute that, for $0 \leq y \leq 3$, we have $-\sqrt{9-y^2} \leq x \leq \sqrt{9-y^2}$. The density of Y for $0 \leq y \leq 3$ is thus:

$$f_Y(y) = \int_{-\infty}^{\infty} f_{X,Y}(x,y)\,dx = \int_{-\sqrt{9-y^2}}^{\sqrt{9-y^2}} \frac{2}{9\pi}\,dx = \frac{4\sqrt{9-y^2}}{9\pi}$$

and $f_Y(y) = 0$ otherwise.

Given that the dancer's y-coordinate is 2, her x-coordinate must be between $-\sqrt{9-2^2} = -\sqrt{5}$ and $\sqrt{9-2^2} = \sqrt{5}$. So, for $-\sqrt{5} \leq x \leq \sqrt{5}$, we have

$$f_{X|Y}(x \mid 2) = \frac{f_{X,Y}(x,2)}{f_Y(2)} = \frac{2/(9\pi)}{4\sqrt{9-2^2}\ /\ (9\pi)} = \frac{1}{2\sqrt{5}}$$

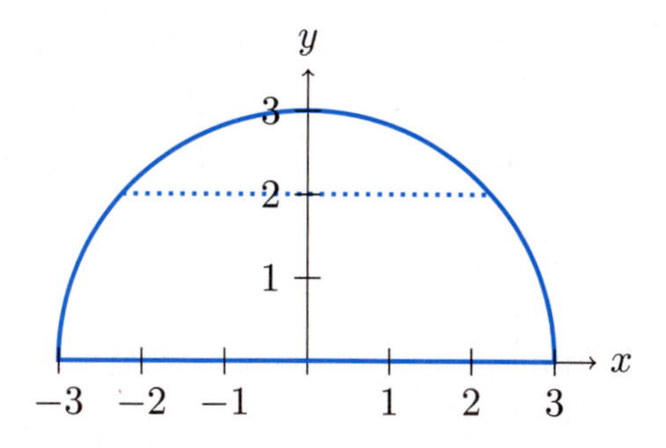

Figure 27.2: A stage where a dancer is twirling around.

We give more examples in Chapter 31 about constant joint densities and conditional distributions (constant densities correspond to Continuous Uniform random variables).

Example 27.4. Consider the joint density from Example 26.4:

$$f_{X,Y}(x,y) = \frac{3}{2}xy \qquad \text{for } 0 \le x \text{ and } 0 \le y \text{ and } x+y \le 2,$$

and $f_{X,Y}(x,y) = 0$ otherwise.

Suppose we want to compute the density of X given Y. Then we need to know the density of Y. For $0 \le y \le 2$, we integrate x out of the picture, as follows:

$$\begin{aligned} f_Y(y) &= \int_{-\infty}^{\infty} f_{X,Y}(x,y)\,dx \\ &= \int_0^{2-y} \frac{3}{2}xy\,dx \\ &= \frac{3}{2}\frac{x^2}{2}y\Big|_{x=0}^{2-y} \\ &= \frac{3}{2}\frac{(2-y)^2}{2}y \end{aligned}$$

It follows that the conditional density of X given Y, for $0 \le y \le 2$ is

$$f_{X|Y}(x \mid y) = \frac{\frac{3}{2}xy}{\frac{3}{2}\frac{(2-y)^2}{2}y} = \frac{2x}{(2-y)^2} \qquad \text{for } 0 \le x \le 2-y,$$

and $f_{X|Y}(x \mid y) = 0$ otherwise.

As a specific example, suppose that we know $Y = 1/2$. In such a case, we also know that $0 \le X \le 3/2$. (See Figure 27.3.) What is the conditional

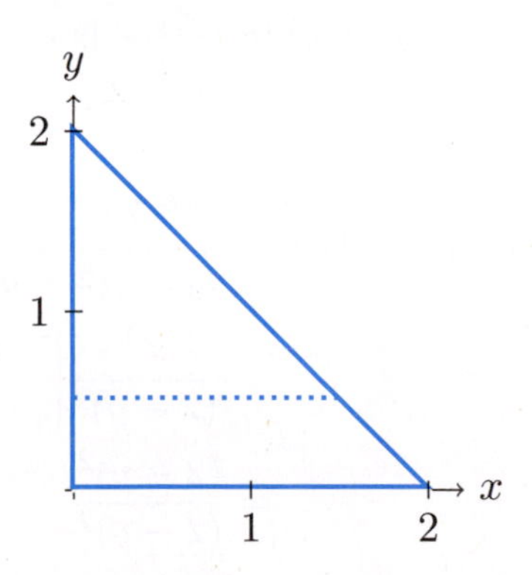

FIGURE 27.3: Region in which $Y = 1/2$ is given; so X must be between 0 and $2 - Y = 3/2$.

probability that $X \leq 1$, given $Y = 1/2$? We can compute

$$\begin{aligned}
P(X \leq 1 \mid Y = 1/2) &= \int_{-\infty}^{1} f_{X|Y}(x \mid 1/2)\,dx \\
&= \int_{0}^{1} \frac{2x}{(2-1/2)^2}\,dx \\
&= \left.\frac{x^2}{(3/2)^2}\right|_0^1 \\
&= \frac{1^2}{9/4} \\
&= 4/9
\end{aligned}$$

As another specific example, suppose again that we know $Y = 1/2$. In such a case, we also know that $0 \leq X \leq 3/2$. What is the conditional probability that $X \geq 4/5$, given $Y = 1/2$? We can compute

$$\begin{aligned}
P(X \geq 4/5) &= \int_{4/5}^{\infty} f_{X|Y}(x \mid 1/2)\,dx \\
&= \int_{4/5}^{3/2} \frac{2x}{(2-1/2)^2}\,dx \\
&= \left.\frac{x^2}{(3/2)^2}\right|_{4/5}^{3/2} \\
&= \frac{(3/2)^2 - (4/5)^2}{9/4} \\
&= 161/225 \\
&= 0.7156
\end{aligned}$$

Finally, we can verify that $f_{X|Y}(x \mid y) = \frac{2x}{(2-y)^2}$ is a density for $0 \leq y \leq 2$ and for $0 \leq x \leq 2-y$. To see this, we note that $\frac{2x}{(2-y)^2} \geq 0$ and also

$$\begin{aligned}
\int_{-\infty}^{\infty} f_{X|Y}(x \mid y)\,dx &= \int_{0}^{2-y} \frac{2x}{(2-y)^2}\,dx \\
&= \left.\frac{x^2}{(2-y)^2}\right|_0^{2-y} \\
&= \frac{(2-y)^2}{(2-y)^2} \\
&= 1.
\end{aligned}$$

So $f_{X|Y}(x \mid y)$ is a density.

Remark 27.5. Conditional density for independent random variables As discussed in the previous chapter, X and Y are independent random variables if and only if the conditional density of X given Y is exactly the same as the density of X.

To see this: The conditional density of a continuous random variable X, given $Y = y$, is defined as:

$$f_{X|Y}(x \mid y) = \frac{f_{X,Y}(x,y)}{f_Y(y)},$$

but X and Y are independent if and only if

$$f_{X,Y}(x,y) = f_X(x) f_Y(y),$$

and in this case, the equation for the conditional density simplifies to

$$f_{X|Y}(x \mid y) = \frac{f_{X,Y}(x,y)}{f_Y(y)} = \frac{f_X(x) f_Y(y)}{f_Y(y)} = f_X(x).$$

Thus, X and Y are independent if and only if the conditional density $f_{X|Y}(x \mid y)$ of X given Y is equal to the density $f_X(x)$ of X.

27.3 Exercises

27.3.1 Practice

For the joint densities in Exercises 27.1 to 27.10, find (a) the density $f_Y(y)$ of Y, and (b) the conditional density $f_{X|Y}(x \mid y)$ of X given Y.

Exercise 27.1. Let

$$f_{X,Y}(x,y) = \frac{15}{256} x^2 y, \qquad \text{for } 0 \le x,\ 0 \le y,\ x + y \le 4,$$

and $f_{X,Y}(x,y) = 0$ otherwise.

Exercise 27.2. Let $f_{X,Y}(x,y)$ be constant on the region shown in Figure 27.4.

Exercise 27.3. Let

$$f_{X,Y}(x,y) = \frac{3}{80}(x^2 + y), \qquad \text{for } 0 \le x \le 2,\ 0 \le y \le 4,$$

and $f_{X,Y}(x,y) = 0$ otherwise.

Exercise 27.4. Let

$$f_{X,Y}(x,y) = \frac{3}{32}(4 - x) y, \qquad \text{for } 0 < y < x < 4,$$

and $f_{X,Y}(x,y) = 0$ otherwise.

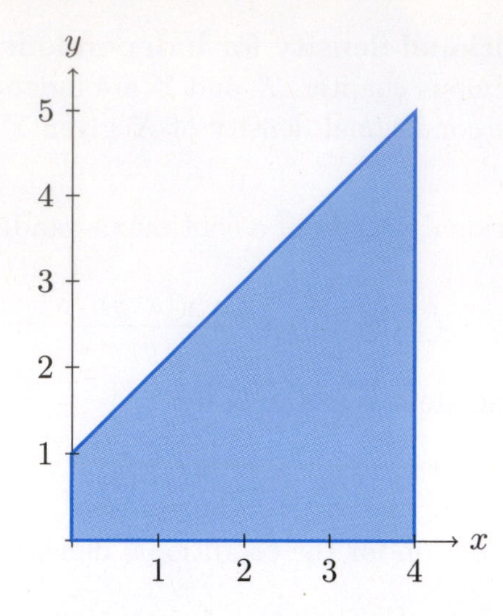

Figure 27.4: Region where X and Y have positive density.

Exercise 27.5. Let

$$f_{X,Y}(x,y) = 5e^{-x-3y}, \qquad \text{for } 0 < y < x/2,$$

and $f_{X,Y}(x,y) = 0$ otherwise.

Exercise 27.6. Let

$$f_{X,Y}(x,y) = \frac{9}{100}e^{-x/4-y/5}, \qquad \text{for } 0 < x < y,$$

and $f_{X,Y}(x,y) = 0$ otherwise.

Exercise 27.7. Let

$$f_{X,Y}(x,y) = \frac{6}{7}(x+y)^2, \qquad \text{for } 0 \le x \le 1,\ 0 \le y \le 1,$$

and $f_{X,Y}(x,y) = 0$ otherwise.

Exercise 27.8. Let

$$f_{X,Y}(x,y) = \frac{12}{5}(1-x^2), \qquad \text{for } 0 \le x,\ 0 \le y,\ x+y \le 1,$$

and $f_{X,Y}(x,y) = 0$ otherwise.

Exercise 27.9. Let $f_{X,Y}(x,y)$ be constant on the region where x and y are nonnegative and $x + y \le 30$.

Exercise 27.10. Let

$$f_{X,Y}(x,y) = \frac{3}{2}xy, \qquad \text{for } 0 \le x,\ 0 \le y,\ x+y \le 2,$$

and $f_{X,Y}(x,y) = 0$ otherwise.

27.3.2 Extensions

Exercise 27.11. Consider a pair of random variables X, Y with constant joint density on the triangle with vertices at $(0,0)$, $(3,0)$, and $(0,3)$.

a. For $0 \le y \le 3$, find the conditional density $f_{X|Y}(x \mid y)$ of X, given $Y = y$.

b. Find the conditional probability that $X \le 1$, given $Y = 1$.

c. Find the conditional probability that $X \le 1$, given $Y \le 1$.

Exercise 27.12. Consider a pair of random variables X, Y with constant joint density on the quadrilateral with vertices $(0,0)$, $(2,0)$, $(2,6)$, $(0,12)$.

a. For $0 \le y \le 6$, find the conditional density $f_{X|Y}(x \mid y)$ of X, given $Y = y$.

b. For $6 \le y \le 12$, find the conditional density $f_{X|Y}(x \mid y)$ of X, given $Y = y$.

c. Find the conditional probability that $X \le 1$, given $3 \le Y \le 9$.

Exercise 27.13. Let X, Y have joint density $f_{X,Y}(x, y) = 14e^{-2x-7y}$ for $x > 0$ and $y > 0$; and $f_{X,Y}(x, y) = 0$ otherwise.

a. For $y > 0$, find the conditional density $f_{X|Y}(x \mid y)$ of X, given $Y = y$.

b. Find the conditional probability that $X \ge 1$, given $Y = 3$.

c. Find the conditional probability that $Y \le 1/5$, given $X = 2.7$.

Exercise 27.14. Let X, Y have joint density $f_{X,Y}(x, y) = 18e^{-2x-7y}$ for $0 < y < x$; and $f_{X,Y}(x, y) = 0$ otherwise.

a. For $y > 0$, find the conditional density $f_{X|Y}(x \mid y)$ of X, given $Y = y$.

b. For $x > 0$, find the conditional density $f_{Y|X}(y \mid x)$ of Y, given $X = x$.

Exercise 27.15. Suppose X, Y have joint density

$$f_{X,Y}(x, y) = \begin{cases} \frac{1}{9}(3 - x)(2 - y) & \text{if } 0 \le x \le 3 \text{ and } 0 \le y \le 2, \\ 0 & \text{otherwise.} \end{cases}$$

a. For $0 \le y \le 2$, find the conditional density $f_{X|Y}(x \mid y)$ of X, given $Y = y$.

b. Find the conditional probability that $X \le 2$, given $Y = 3/2$.

c. Find the conditional probability that $Y \ge 1$, given $X = 5/4$.

Exercise 27.16. Every day, a student calls his mother and then (afterwards) calls his girlfriend. Let X be the time (in hours) until he calls his mother, and let Y be the time (in hours) until he calls his girlfriend. Since he always calls his mother first, then $X < Y$. So let the joint density of the time be

$$f_{X,Y}(x, y) = 10e^{-3x-2y} \qquad \text{for } 0 < x < y,$$

and $f_{X,Y}(x, y) = 0$ otherwise. Given that $X = 1/2$, find the conditional probability that $Y > 2/3$. In other words, find $P(Y > 2/3 \mid X = 1/2)$.

Exercise 27.17. Assume that Wyoming is shaded exactly like a rectangle, and that a person's location in Wyoming is $0 \le X \le 350$ and $0 \le Y \le 276$. Assume that the joint density of a person's location is constant on this region.

Assume that I-80 runs perfectly straight along the east-to-west line $Y = 30$, as in Figure 27.5. Given that the person is located on I-80, what is the conditional probability that he is within 10 miles of a State border? (He could be either west or east; please take both into account.)

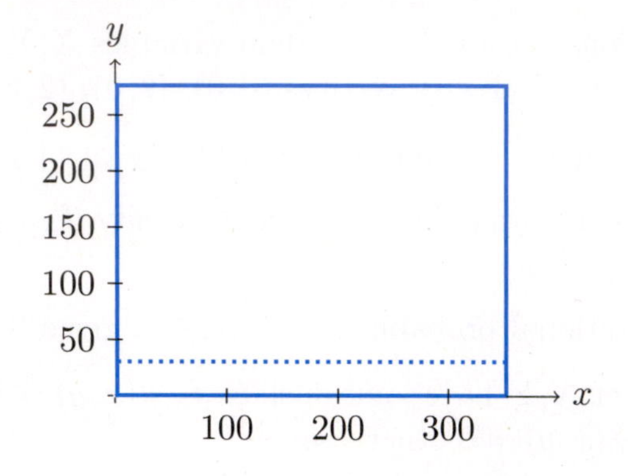

FIGURE 27.5: State of Wyoming.

Exercise 27.18. Consider the semicircle with radius 2 seen in Figure 27.6. A dancer is randomly located in the semicircle, with the joint density of their location to be

$$f_{X,Y}(x, y) = \frac{2}{9\pi} \qquad \text{if } x, y \text{ is in the semicircle,}$$

and $f_{X,Y}(x, y) = 0$ otherwise. Find the probability that the person is in the region where $-1 \le X \le 1$ and $0 \le Y \le 1$.

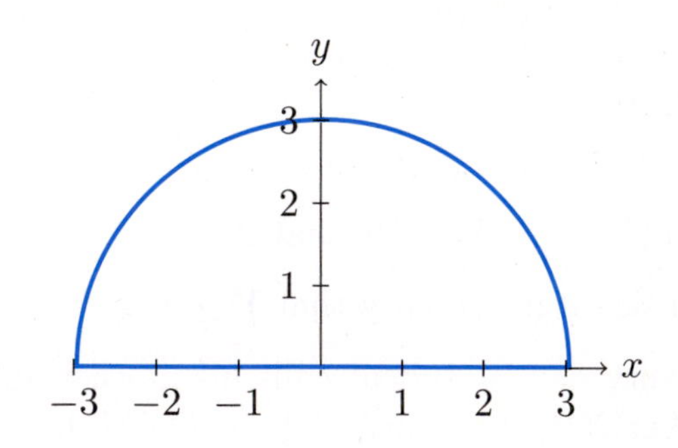

FIGURE 27.6: Picture of a semicircle-shaped stage.

Exercise 27.19. Show that a conditional PDF satisfies the conditions of a PDF.

Chapter 28

Expected Values of Continuous Random Variables

A thing long expected takes the form of the unexpected when at last it comes.
—Mark Twain

How long should we expect to have to wait at a traffic light? What is the expected length of a randomly chosen song on our playlist? What is the expected distance from the center, when a dart is thrown at a dartboard?

28.1 Introduction

To a great extent, the way that expected values are calculated for continuous random variables should not be surprising: the method is analogous to how expected values are calculated for discrete random variables. For continuous random variables, we cannot use the probabilities of values of random variables as a method of scaling, because the probability of any particular value is 0, i.e., $P(X = x) = 0$ for each x. Instead, we use the density $f_X(x)$ for scaling. In other words, we integrate x over all possible x's (just as we summed over all possible x's with discrete random variables), and we give a weight of $f_X(x)$ to each x (just as we gave a weight of $p_X(x)$ to each x with discrete random variables).

Definition 28.1. Expected value of a continuous random variable
A continuous random variable X with density $f_X(x)$ has expected value

$$\mathbb{E}(X) = \int_{-\infty}^{\infty} x f_X(x)\, dx.$$

(This should remind us of the formula for the expected value of a discrete random variable X, i.e., $\mathbb{E}(X) = \sum_x x p_X(x)\,dx$.)

We revisit some earlier examples and calculate the expected values of the associated random variables:

Example 28.2. If X is a continuous random variable with density

$$f_X(x) = \begin{cases} \frac{1}{26}(4x+1) & \text{if } 2 \leq x \leq 4, \\ 0 & \text{otherwise,} \end{cases}$$

find the expected value of X.

The expected value of X is

$$\mathbb{E}(X) = \int_{-\infty}^{\infty} x f_X(x)\,dx,$$

but $f_X(x) = 0$ for x's outside the range $2 \leq x \leq 4$. Thus, we have

$$\begin{aligned}
\mathbb{E}(X) &= \int_2^4 (x)\frac{1}{26}(4x+1)\,dx \\
&= \int_2^4 \frac{1}{26}(4x^2 + x)\,dx \\
&= \frac{1}{26}\left(\frac{4}{3}x^3 + \frac{x^2}{2}\right)\Big|_{x=2}^4 \\
&= \frac{1}{26}\left(\left(\frac{4}{3}4^3 + \frac{4^2}{2}\right) - \left(\frac{4}{3}2^3 + \frac{2^2}{2}\right)\right) \\
&= 121/39 \\
&= 3.1026
\end{aligned}$$

Note that, if a continuous random variable is always found in a finite-length interval, then the expected value of X must be found in this interval too:

Theorem 28.3. Bounds on the expected value of a continuous random variable
If $a \leq X \leq b$, then we must also have $a \leq \mathbb{E}(X) \leq b$ too.

To see this:

$$\begin{aligned}
\mathbb{E}(X) &= \int_a^b x f_X(x)\,dx \\
&\le \int_a^b b f_X(x)\,dx \qquad \text{since } x \le b \text{ on the interval } [a,b] \\
&= b\int_a^b f_X(x)\,dx \\
&= (b)(1)\,dx \qquad \text{since the density must integrate to 1} \\
&= b,
\end{aligned}$$

and similarly

$$\mathbb{E}(X) = \int_a^b x f_X(x)\,dx \ge \int_a^b a f_X(x)\,dx = a\int_a^b f_X(x)\,dx = (a)(1)\,dx = a.$$

So, as we claimed, if we have $a \le X \le b$, we must have $a \le \mathbb{E}(X) \le b$ too.

Example 28.4. Suppose X has density

$$f_X(x) = \frac{3}{8}(x)(2-x)(3-x) \qquad \text{for } 0 \le x \le 2,$$

and $f_X(x) = 0$ otherwise.

The expected value of X is

$$\begin{aligned}
\mathbb{E}(X) &= \int_{-\infty}^{\infty} x f_X(x)\,dx \\
&= \int_0^2 (x)\left(\frac{3}{8}\right)(x)(2-x)(3-x)\,dx \\
&= \frac{3}{8}\int_0^2 (6x^2 - 5x^3 + x^4)\,dx \\
&= \frac{3}{8}\left(6x^3/3 - 5x^4/4 + x^5/5\right)\Big|_{x=0}^{2} \\
&= \frac{3}{8}(6(2)^3/3 - 5(2)^4/4 + 2^5/5) \\
&= \frac{3}{8}(960 - 1200 + 384)/60 \\
&= 9/10
\end{aligned}$$

Example 28.5. Let X be a random variable with

$$f_X(x) = 3e^{-3x} \qquad \text{for } x > 0,$$

and $f_X(x) = 0$ otherwise.

The expected value of X is

$$\mathbb{E}(X) = \int_{-\infty}^{\infty} x f_X(x)\, dx = \int_0^{\infty} (x)(3)e^{-3x}\, dx$$

Now we use integration by parts with $u = 3x$ and $dv = e^{-3x}\, dx$, so $du = 3\, dx$ and $v = \frac{e^{-3x}}{-3}$. So we get

$$\mathbb{E}(X) = (3x)\left(\frac{e^{-3x}}{-3}\right)\bigg|_{x=0}^{\infty} - \int_0^{\infty} (3)\frac{e^{-3x}}{-3}\, dx = -x\, e^{-3x}\big|_{x=0}^{\infty} + \frac{e^{-3x}}{-3}\bigg|_{x=0}^{\infty}$$

We note that $-xe^{-3x} \to -0e^{-(3)(0)} = 0$ as $x \to 0$. Also $-xe^{-3x} = -\frac{x}{e^{3x}} \to -\frac{\infty}{\infty}$ as $x \to \infty$, so, using L'Hospital's Rule, $-\frac{x}{e^{3x}} \to -\frac{1}{3e^{3x}} \to 0$ as $x \to \infty$. Also $\frac{e^{-3x}}{-3} \to 0$ as $x \to \infty$ and $\frac{e^{-3x}}{-3} \to -1/3$ as $x \to 0$. Thus

$$\mathbb{E}(X) = 1/3.$$

28.2 Some Generalizations about Expected Values

Suppose that a and b are any constants, with $a < b$. Suppose that $f_X(x)$ is constant on the interval $[a, b]$ and is 0 otherwise. Then the "constant" must be $1/(b-a)$ so that the integral of the density yields 1. Thus

$$f_X(x) = \frac{1}{b-a} \qquad \text{for } a \le x \le b,$$

and $f_X(x) = 0$ otherwise.

Theorem 28.6. Expected value of a continuous random variable that is constant on an interval and zero otherwise

If X is a continuous random variable with density

$$f_X(x) = \frac{1}{b-a} \qquad \text{for } a \le x \le b,$$

and $f_X(x) = 0$ otherwise, then

$$\mathbb{E}(X) = \frac{a+b}{2}.$$

To derive the expected value in the previous box, we compute

$$\mathbb{E}(X) = \int_{-\infty}^{\infty} x f_X(x)\,dx = \int_a^b (x)\left(\frac{1}{b-a}\right)\,dx = (x^2/2)\left(\frac{1}{b-a}\right)\Big|_{x=a}^{b}$$

Substituting in the values a and b for x, we conclude

$$\mathbb{E}(X) = ((b^2-a^2)/2)\left(\frac{1}{b-a}\right) = ((a+b)(b-a)/2)\left(\frac{1}{b-a}\right) = (a+b)/2.$$

Theorem 28.7. Expected value of a continuous random variable with exponentially decreasing density

If, for some constant $\lambda > 0$, the random variable X has density

$$f_X(x) = \lambda e^{-\lambda x} \qquad \text{for } x > 0,$$

and $f_X(x) = 0$ otherwise, then

$$\mathbb{E}(X) = 1/\lambda.$$

To derive this expected value, we follow the earlier example, in which λ was equal to 3. The same argument works here. The expected value of X is

$$\mathbb{E}(X) = \int_{-\infty}^{\infty} x f_X(x)\,dx = \int_0^{\infty} x\lambda e^{-\lambda x}\,dx$$

Now we use integration by parts with $u = \lambda x$ and $dv = e^{-\lambda x}\,dx$, so $du = \lambda\,dx$ and $v = \frac{e^{-\lambda x}}{-\lambda}$. So we get

$$\mathbb{E}(X) = (\lambda x)\left(\frac{e^{-\lambda x}}{-\lambda}\right)\Big|_{x=0}^{\infty} - \int_0^{\infty} (\lambda)\frac{e^{-\lambda x}}{-\lambda}\,dx = -x\,e^{-\lambda x}\Big|_{x=0}^{\infty} + \frac{e^{-\lambda x}}{-\lambda}\Big|_{x=0}^{\infty}$$

We note that $-xe^{-\lambda x} \to -0e^{-(\lambda)(0)} = 0$ as $x \to 0$. Also $-xe^{-\lambda x} = -\frac{x}{e^{\lambda x}} \to -\frac{\infty}{\infty}$ as $x \to \infty$, so, using L'Hospital's Rule, $-\frac{x}{e^{\lambda x}} \to -\frac{1}{\lambda e^{\lambda x}} \to 0$ as $x \to \infty$. Also $\frac{e^{-\lambda x}}{-\lambda} \to 0$ as $x \to \infty$ and $\frac{e^{-\lambda x}}{-\lambda} \to -1/\lambda$ as $x \to 0$. Thus

$$\mathbb{E}(X) = 1/\lambda.$$

28.3 Some Applied Problems with Expected Values

Example 28.8. Let X be the time (in minutes) that Maxine waits for a traffic light to turn green, and let Y be the time (in minutes, at a different intersection) that Daniella waits for a traffic light to turn green. Suppose that X and Y have joint density

$$f_{X,Y}(x,y) = 15e^{-3x-5y}, \qquad \text{for } x > 0 \text{ and } y > 0,$$

and $f_{X,Y}(x,y) = 0$ otherwise. Find the expected time that each of them wait at their lights.

In this example, $f_X(x) = 0$ for $x \le 0$, and $f_Y(y) = 0$ for $y \le 0$. We could notice that $15e^{-3x-5y}$ factors into the product of two densities, $(3e^{-3x})(5e^{-5y})$, or we could get the density of X by integrating Y out of the picture, as follows:

$$\begin{aligned} f_X(x) &= \int_{-\infty}^{\infty} f_{X,Y}(x,y)\,dy \\ &= \int_0^{\infty} 15e^{-3x-5y}\,dy \\ &= -3e^{-3x-5y}\Big|_{y=0}^{\infty} \\ &= 3e^{-3x} \end{aligned}$$

Thus

$$f_X(x) = 3e^{-3x} \qquad \text{for } x > 0,$$

and $f_X(x) = 0$ otherwise.

Using the boxed result in the previous section, it follows that

$$\mathbb{E}(X) = 1/3.$$

We could also have integrated from the start:

$$\mathbb{E}(X) = \int_0^{\infty}\int_0^{\infty} (x)(15e^{-3x-5y})\,dy\,dx.$$

Similarly, for $y > 0$,

$$f_Y(y) = 5e^{-5y} \qquad \text{for } y > 0,$$

and $f_Y(y) = 0$ otherwise. Again using the boxed result in the previous section, it follows that

$$\mathbb{E}(Y) = 1/5.$$

Example 28.9. Suppose that a person throws a dart at a square dart board. Let X and Y denote, respectively, the x- and y-coordinates (in feet) of the location where the dart lands; the middle of the dart board is at $(0,0)$. Suppose that the dart board is two feet wide and two feet tall, so that the dart only lands on the dart board if $-1 \le X \le 1$, $-1 \le Y \le 1$. Also suppose that the person always hits the dart board, and moreover, X and Y have joint density:

$$f_{X,Y}(x,y) = \frac{9}{16}(1-x^2)(1-y^2) \qquad \text{for } -1 \le x \le 1,\ -1 \le y \le 1,$$

and $f_{X,Y}(x,y) = 0$ otherwise. Find the expected values of X and Y.

In this example, $f_X(x) = 0$ for $x > 1$ and for $x < -1$.

For $-1 \leq x \leq 1$, we can get the density of X by integrating Y out of the picture, as follows:

$$\begin{aligned} f_X(x) &= \int_{-\infty}^{\infty} f_{X,Y}(x,y)\,dy \\ &= \int_{-1}^{1} \frac{9}{16}(1-x^2)(1-y^2)\,dy \\ &= \frac{9}{16}(1-x^2)(y-y^3/3)\Big|_{y=-1}^{1} \\ &= \frac{9}{16}(1-x^2)((2/3)-(-2/3))\Big|_{y=-1}^{1} \\ &= \frac{3}{4}(1-x^2) \end{aligned}$$

Thus

$$\mathbb{E}(X) = \int_{-1}^{1}(x)\left(\frac{3}{4}\right)(1-x^2)\,dx = \frac{3}{4}\int_{-1}^{1}(x-x^3)\,dx = \frac{3}{4}\left(x^2/2 - x^4/4\right)\Big|_{x=-1}^{1} = 0.$$

Since the entire problem is symmetric in X and Y, it follows also that $\mathbb{E}(Y) = 0$.

28.4 Exercises

28.4.1 Practice

For the densities in Exercises 28.1 to 28.8, find the expected value of the random variable.

Exercise 28.1. A student models the number of revolutions X on a pencil sharpener that are needed to fully sharpen his pencil, by using density

$$f_X(x) = \begin{cases} \frac{2}{3}(x-19) & \text{for } 19 \leq x \leq 20; \\ -\frac{1}{3}(x-22) & \text{for } 20 \leq x \leq 22; \end{cases}$$

and $f_X(x) = 0$ otherwise.

Exercise 28.2. The time (in minutes) it takes until Julie's boyfriend calls is a random variable X with density

$$f_X(x) = \frac{1}{10}e^{-x/10}, \qquad \text{for } 0 < x,$$

and $f_X(x) = 0$ otherwise.

Exercise 28.3. Let X have density

$$f_X(x) = \frac{1}{3}e^{-x/3}, \qquad \text{for } 0 < x,$$

and $f_X(x) = 0$ otherwise.

Exercise 28.4. Let X have density

$$f_X(x) = x, \qquad \text{for } 0 \le x \le \sqrt{2},$$

and $f_X(x) = 0$ otherwise.

Exercise 28.5. Suppose that the length X of a tungsten wire is

$$f_X(x) = \frac{3}{1000}x^2, \qquad \text{for } 0 \le x \le 10,$$

and $f_X(x) = 0$ otherwise.

Exercise 28.6. Edward's trip to school is three miles long, down a straight highway. Let X denote his location at a randomly chosen point on his drive to school, with density

$$f_X(x) = \frac{1}{3}, \qquad \text{for } 0 \le x \le 3,$$

and $f_X(x) = 0$ otherwise.

Exercise 28.7. Bruno goes fishing every day in the summer for 5 hours a day. Let X be the amount of time (in hours) that goes by until he notices that he catches his first fish (there is always a fish on his line by the end of the day, because he fishes in a very densely populated pond). So X has density

$$f_X(x) = \frac{3}{245}(x^2 + 8), \qquad \text{for } 0 \le x \le 5,$$

and $f_X(x) = 0$ otherwise.

Exercise 28.8. Let X have density

$$f_X(x) = \frac{2}{25}x, \qquad \text{for } 0 \le x \le 5,$$

and $f_X(x) = 0$ otherwise.

Exercise 28.9. Consider a pair of random variables X, Y with constant joint density on the triangle with vertices at $(0, 0)$, $(3, 0)$, and $(0, 3)$. Find the expected value $\mathbb{E}(X)$. (Notice that, by symmetry, $\mathbb{E}(Y)$ is just the same!)

Exercise 28.10. Consider a pair of random variables X, Y with constant joint density on the quadrilateral with vertices $(0, 0)$, $(2, 0)$, $(2, 6)$, $(0, 12)$.

a. Find the expected value $\mathbb{E}(X)$.

b. Find the expected value $\mathbb{E}(Y)$.

Exercise 28.11. Let X, Y have joint density $f_{X,Y}(x, y) = 14e^{-2x-7y}$ for $x > 0$ and $y > 0$; and $f_{X,Y}(x, y) = 0$ otherwise.

a. Find the expected value $\mathbb{E}(X)$.

b. Find the expected value $\mathbb{E}(Y)$.

Exercise 28.12. Let X, Y have joint density $f_{X,Y}(x, y) = 18e^{-2x-7y}$ for $0 < y < x$; and $f_{X,Y}(x, y) = 0$ otherwise.

a. Find the expected value $\mathbb{E}(X)$.

b. Find the expected value $\mathbb{E}(Y)$.

Exercise 28.13. Suppose X, Y have joint density

$$f_{X,Y}(x, y) = \begin{cases} \frac{1}{9}(3 - x)(2 - y) & \text{if } 0 \le x \le 3 \text{ and } 0 \le y \le 2, \\ 0 & \text{otherwise.} \end{cases}$$

a. Find the expected value $\mathbb{E}(X)$.

b. Find the expected value $\mathbb{E}(Y)$.

Exercise 28.14. Suppose X, Y have joint density

$$f_{X,Y}(x, y) = \begin{cases} \frac{4}{3}(x)(1 - x)(3 - y) & \text{if } 0 \le x \le 1 \text{ and } 0 \le y \le 3, \\ 0 & \text{otherwise.} \end{cases}$$

a. Find the expected value $\mathbb{E}(X)$.

b. Find the expected value $\mathbb{E}(Y)$.

Exercise 28.15. Let X have density

$$f_X(x) = \begin{cases} \frac{3}{4} & \text{for } 0 \le x \le 1, \\ \frac{1}{4} & \text{for } 3 \le x \le 4, \\ 0 & \text{otherwise.} \end{cases}$$

Find $\mathbb{E}(X)$.

Exercise 28.16. Let X have density

$$f_X(x) = \begin{cases} \frac{7}{8} & \text{for } 0 \le x \le 1, \\ \frac{1}{8} & \text{for } 7 \le x \le 8, \\ 0 & \text{otherwise.} \end{cases}$$

Find $\mathbb{E}(X)$.

Exercise 28.17. Consider the joint density of X and Y from Example 26.2:

$$f_{X,Y}(x,y) = \frac{1}{8}(1-x^2)(3-y) \qquad \text{for } -1 \le x \le 1 \text{ and } -1 \le y \le 1,$$

and $f_{X,Y}(x,y) = 0$ otherwise.

a. Find $\mathbb{E}(X)$.

b. Find $\mathbb{E}(Y)$.

Exercise 28.18. Consider the joint density of X and Y from Example 26.4:

$$f_{X,Y}(x,y) = \frac{3}{2}xy \qquad \text{for } 0 \le x \text{ and } 0 \le y \text{ and } x+y \le 2,$$

and $f_{X,Y}(x,y) = 0$ otherwise. Find $\mathbb{E}(Y)$.

Exercise 28.19. As in Example 27.2, a bird lands in a grassy region described as follows: $0 \le x$, and $0 \le y$, and $x + y \le 10$. Let X and Y be the coordinates of the bird's landing. Assume that X and Y have the joint density

$$f_{X,Y}(x,y) = 1/50 \qquad \text{for } 0 \le x \text{ and } 0 \le y \text{ and } x+y \le 10,$$

and $f_{X,Y}(x,y) = 0$ otherwise. Find $\mathbb{E}(Y)$. (Hint: The answer is *not* "$\mathbb{E}(Y) = 5$.")

28.4.2 Extensions

Exercise 28.20. Consider two points that are independently placed on a line of length 10, at locations X and Y. Thus the joint density of X and Y is

$$f_{X,Y}(x,y) = 1/100 \qquad \text{for } 0 \le x \le 10 \text{ and } 0 \le y \le 10,$$

and $f_{X,Y}(x,y) = 0$ otherwise.

a. First, try to show that, if Z denotes the distance between X and Y (so $0 \le Z \le 10$), then the cumulative distribution function $F_Z(z)$ of Z is

$$F_Z(z) = \frac{1}{5}z - \frac{1}{100}z^2 \qquad \text{for } 0 \le z \le 10$$

and $F_Z(z) = 0$ for $z \le 0$ and $F_Z(z) = 1$ for $z \ge 10$. (Hint: Use the complement; try to find $P(Z > z)$, i.e., the probability that the distance between X and Y is more than z. A picture should help.)

b. Regardless of whether you can accomplish the part above, just find $f_Z(z)$, by differentiating $F_Z(z)$.

c. Use the density $f_Z(z)$ to find $\mathbb{E}(Z)$, i.e., the expected distance between X and Y.

Exercise 28.21. Let X and Y have joint density

$$f_{X,Y}(x,y) = \frac{3}{80}(x^2 + y), \qquad \text{for } 0 \le x \le 2 \text{ and } 0 \le y \le 4,$$

and $f_{X,Y}(x,y) = 0$ otherwise. Find $\mathbb{E}(X)$.

28.4.3 Advanced

Exercise 28.22. Three children are skating around the perimeter of a circle in an ice rink of radius 50 feet. Assume that their locations around the perimeter of the circle are independent, continuous random variables, each with constant density. Their mother comes to the door of the ice rink (located at a fixed position on perimeter of the circle). When she appears, what is the expected distance around the perimeter from her to the closest of the three children?

Chapter 29

Variance of Continuous Random Variables

If you do not expect the unexpected, you will not find it; for it is hard to be sought out, and difficult.
—Heraclitus

To what extent are the students' grades in a course near the average? How can we quantify the spread of randomly thrown darts from the center of the board?

29.1 Variance of a Continuous Random Variable

In the previous chapter, we extensively studied the fact that the expected value $\mathbb{E}(X)$ of a continuous random variable X with density $f_X(x)$ is

$$\mathbb{E}(X) = \int_{-\infty}^{\infty} x f_X(x)\, dx.$$

The variance of a continuous random variable measures the spread of the random variable's distribution around the mean (just as the variance does with a discrete random variable as well). We first point out two different random variables that have the same expected value but different variances.

Example 29.1. For our first example, we define two random variables that each have expected value 10/3. The random variable Y is from Exercise 28.19, and the random variable X will be defined below. Since these two random variables have the same expected value, we will use their variances to compare them.

In Exercise 28.19, a bird lands in a grassy region described as follows: $x \geq 0$, and $y \geq 0$, and $x + y \leq 10$. Let X and Y be the coordinates of the bird's landing. Assume that X and Y have the joint density

$$f_{X,Y}(x,y) = 1/50 \qquad \text{for } 0 \leq x \text{ and } 0 \leq y \text{ and } x + y \leq 10,$$

and $f_{X,Y}(x,y) = 0$ otherwise. We already found $f_Y(y)$, for $0 \leq y \leq 10$:

$$f_Y(y) = \int_0^{10-y} 1/50\,dx = (10-y)/50$$

and $f_Y(y) = 0$ otherwise, and then we computed the expected value of Y:

$$\mathbb{E}(Y) = \int_0^{10} (y)(10-y)/50\,dy = 10/3.$$

Now we compare to a different random variable that also has expected value 10/3: Suppose that X is the waiting time until the next bus arrives, given in minutes. If X has density

$$f_X(x) = \frac{3}{10}e^{-3x/10} \qquad \text{for } x > 0,$$

and $f_X(x) = 0$ otherwise, then the density of X is exponentially decreasing, with $\lambda = 3/10$, so we can conclude, by Theorem 28.7 that $\mathbb{E}(X) = 1/\lambda = 10/3$. Now a question immediately arises: Both random variables, X and Y, have expected value 10/3; can we say something more about the distributions of X and Y to better understand how X and Y behave? With discrete random variables, we sometimes looked at the masses; for continuous random variables, we can look at the densities. The densities of Y and X are given, respectively, on the left and right sides of Figure 29.1.

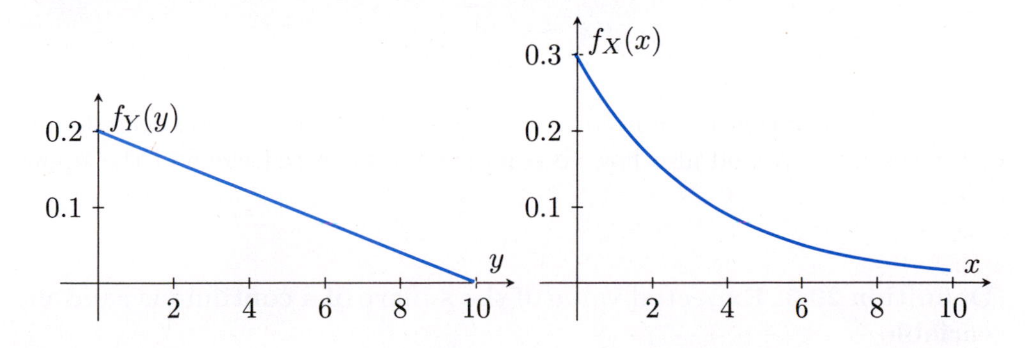

FIGURE 29.1: Left: The density $f_Y(y) = (10-y)/50$ of Y. Right: The density $f_X(x) = \frac{3}{10}e^{-3x/10}$ of X.

It is still difficult to quantify which distribution in Example 29.1 is more "spread out" from the expected value. To quantify this—as we quantified such a spread with discrete variables—we use the concept of the variance.

Definition 29.2. Variance
For a random variable X with expected value $\mu_X = \mathbb{E}(X)$, the variance of X is

$$\mathrm{Var}(X) = \mathbb{E}((X - \mu_X)^2).$$

We notice that

$$\mathrm{Var}(X) = \mathbb{E}((X - \mu_X)^2) = \mathbb{E}(X^2 - 2\mu_X X + \mu_X^2) = \mathbb{E}(X^2) - 2\mu_X \mathbb{E}(X) + \mu_X^2,$$

but μ_X and $\mathbb{E}(X)$ are equal, so this yields:

Remark 29.3. Variance
For X with expected value $\mu_X = \mathbb{E}(X)$, the variance of X is always

$$\mathrm{Var}(X) = \mathbb{E}(X^2) - (\mathbb{E}(X))^2.$$

The spread of a random variable is sometimes quantified in terms of the square root of the variance instead of the variance itself:

Definition 29.4. Standard deviation
For a random variable X with variance $\mathrm{Var}(X)$, the standard deviation of X is

$$\sigma_X = \sqrt{\mathrm{Var}(X)}.$$

The definition of **variance** is the same for continuous and discrete random variables.

To find the variance, we need to know how to compute the expected value of the square of X and also how to compute the expected value of the square of Y.

Definition 29.5. Expected value of the square of a continuous random variable
If X is a continuous random variable X with density $f_X(x)$, then the expected value of X^2 is

$$\mathbb{E}(X^2) = \int_{-\infty}^{\infty} x^2 f_X(x)\,dx.$$

Example 29.1 (continued) In Example 29.1, we introduced a random variable Y such that

$$\begin{aligned}\mathbb{E}(Y^2) &= \int_{-\infty}^{\infty} y^2 f_Y(y)\,dy \\ &= \int_0^{10} y^2(10-y)/50\,dy \\ &= \int_0^{10} (10y^2 - y^3)/50\,dy \\ &= (10y^3/3 - y^4/4)/50\Big|_{y=0}^{10} \\ &= (10000/3 - 10000/4)/50 \\ &= 50/3\end{aligned}$$

and thus

$$\text{Var}(Y) = \mathbb{E}(Y^2) - (\mathbb{E}(Y))^2 = 50/3 - (10/3)^2 = 50/9 = 5.555556.$$

So the standard deviation of Y is

$$\sigma_Y = \sqrt{50/9} = 2.357.$$

We have to work a little harder to get the expected value of X^2:

$$\begin{aligned}\mathbb{E}(X^2) &= \int_{-\infty}^{\infty} x^2 f_X(x)\,dx \\ &= \int_0^{\infty} (x^2)\left(\frac{3}{10}\right) e^{-3x/10}\,dx\end{aligned}$$

and we use integration by parts, with $u = x^2$ and $dv = \frac{3}{10}e^{-3x/10}\,dx$, and thus $du = 2x\,dx$ and $v = -e^{-3x/10}$, so

$$\mathbb{E}(X^2) = (x^2)(-e^{-3x/10})\Big|_{x=0}^{\infty} - \int_0^{\infty} (2x)(-e^{-3x/10})\,dx$$

The first part, $(x^2)(-e^{-3x/10})\big|_{x=0}^{\infty}$, evaluates to 0. We can manipulate the second part by factoring the "2" out, and by multiplying and dividing by 3/10, so that we have

$$\mathbb{E}(X^2) = \frac{2}{3/10}\int_0^{\infty} (x)\left(\frac{3}{10}\right) e^{-3x/10}\,dx = \frac{2}{3/10}\mathbb{E}(X) = \left(\frac{20}{3}\right)\left(\frac{10}{3}\right) = \frac{200}{9}.$$

and finally we obtain

$$\text{Var}(X) = \mathbb{E}(X^2) - (\mathbb{E}(X))^2 = 200/9 - (10/3)^2 = 100/9 = 11.111111,$$

and

$$\sigma_X = \sqrt{100/9} = 3.333.$$

Since the standard deviation σ_X of X is greater than the standard deviation σ_Y of Y, then X is more "spread out" around its expected value, as compared to how much Y is spread around its own expected value.

Example 29.6. If X is a continuous random variable with density

$$f_X(x) = \begin{cases} \frac{1}{26}(4x+1) & \text{if } 2 \le x \le 4, \\ 0 & \text{otherwise,} \end{cases}$$

find the expected value of X^2, and also the variance and standard deviation of X.

The expected value of X^2 is

$$\mathbb{E}(X^2) = \int_{-\infty}^{\infty} x^2 f_X(x)\,dx,$$

but $f_X(x) = 0$ for x's outside the range $2 \le x \le 4$. Thus, we have

$$\begin{aligned}
\mathbb{E}(X^2) &= \int_2^4 (x^2)\frac{1}{26}(4x+1)\,dx \\
&= \int_2^4 \frac{1}{26}(4x^3 + x^2)\,dx \\
&= \frac{1}{26}\left(\frac{4}{4}x^4 + \frac{x^3}{3}\right)\Bigg|_{x=2}^4 \\
&= \frac{1}{26}\left(\left(4^4 + \frac{4^3}{3}\right) - \left(2^4 + \frac{2^3}{3}\right)\right) \\
&= 388/39 \\
&= 9.9487
\end{aligned}$$

From Example 28.2, we know $\mathbb{E}(X) = 121/39$. Thus

$$\begin{aligned}
\text{Var}(X) &= \mathbb{E}(X^2) - (\mathbb{E}(X))^2 \\
&= 388/39 - (121/39)^2 \\
&= 491/1521 \\
&= 0.3228.
\end{aligned}$$

The standard deviation of X is

$$\sigma_X = \sqrt{491/1521} = \frac{1}{39}\sqrt{491} = 0.5682.$$

29.2 Expected Values of Functions of One Random Variable

To compute the variance of a continuous random variable X, we need to calculate $\mathbb{E}(X^2)$, the expected value of the square of X. Sometimes we need the expected value of other functions of a random variables. Such expected values are calculated in an analogous way:

Definition 29.7. Expected value of a function of a continuous random variable
If X is a continuous random variable X with density $f_X(x)$, and if g is an arbitrary function, then the expected value of $g(X)$ is

$$\mathbb{E}(g(X)) = \int_{-\infty}^{\infty} g(x) f_X(x)\,dx.$$

Example 29.8. Returning to the above Example, in which X is a continuous random variable with density

$$f_X(x) = \begin{cases} \frac{1}{26}(4x+1) & \text{if } 2 \le x \le 4, \\ 0 & \text{otherwise,} \end{cases}$$

find the expected value of $1/X$.

The expected value of $1/X$ is

$$\mathbb{E}(1/X) = \int_{-\infty}^{\infty} (1/x) f_X(x)\,dx,$$

but $f_X(x) = 0$ for x's outside the range $2 \le x \le 4$. Thus, we have

$$\begin{aligned}
\mathbb{E}(1/X) &= \int_2^4 (1/x)\frac{1}{26}(4x+1)\,dx \\
&= \int_2^4 \frac{1}{26}\left(4 + \frac{1}{x}\right)\,dx \\
&= \frac{1}{26}\,(4x + \ln x)\Big|_{x=2}^{4} \\
&= \frac{1}{26}\,(((4)(4) + \ln 4) - ((4)(2) + \ln 2)) \\
&= \frac{4}{13} + \frac{1}{26}\ln 2 \\
&= 0.3344
\end{aligned}$$

In particular, we emphasize that $\mathbb{E}(1/X) \neq 1/\mathbb{E}(X)$. In this case, referring back to Example 28.2, we have $1/\mathbb{E}(X) = 39/121 = 0.3223$, which is not equal to $\mathbb{E}(1/X) = 0.3344$.

Example 29.9. When Joyce reaches in her refrigerator for a drink from a 2-liter bottle of soda, her roommates have usually already helped themselves to some of the soda. So the amount of soda that remains is a random variable X. If X has density $f_X(x) = 1/2$ for $0 \le x \le 2$, and $f_X(x) = 0$ otherwise, calculate the cost of the portion that has already been drunk by the roommates. Suppose that the cost of soda is \$1.38 for 2 liters, i.e., \$0.79 per liter.

The quantity that has been drunk is $2 - X$ liters, and its cost is $(0.79)(2 - X)$. So the desired expected value is

$$\begin{aligned}\mathbb{E}((0.79)(2-X)) &= \int_0^2 (0.79)(2-x)(1/2)\,dx \\ &= 0.395 \int_0^2 (2-x)\,dx \\ &= 0.395\,(2x - x^2/2)\big|_{x=0}^2 \\ &= (0.395)(2) \\ &= 0.79\end{aligned}$$

Theorem 29.10. Expected value of a linear function of a continuous random variable
If X is a continuous random variable X with density $f_X(x)$, and if a and b are any constants, then the expected value of $aX + b$ is

$$\mathbb{E}(aX + b) = a\mathbb{E}(X) + b.$$

To see that the theorem is true, just write

$$\begin{aligned}\mathbb{E}(aX+b) &= \int_{-\infty}^{\infty} (ax+b) f_X(x)\,dx \\ &= a\int_{-\infty}^{\infty} x f_X(x)\,dx + b\int_{-\infty}^{\infty} f_X(x)\,dx \\ &= a\mathbb{E}(X) + b.\end{aligned}$$

In Section 28.2, we learned that if the density of a random variable is constant on the interval $[a, b]$, then the expected value is $(a + b)/2$. In this case, the density of X is the constant $1/2$ on the interval $[0, 2]$, so $\mathbb{E}(X) = 1$. Thus, using the boxed result above, we can decompose the soda example above, and we again see that the expected cost of soda already drunk by the roommates is

$$\mathbb{E}((0.79)(2-X)) = \mathbb{E}(1.58 - 0.79X) = 1.58 - 0.79\mathbb{E}(X) = 0.79.$$

29.3 Expected Values of Functions of Two Random Variables

Example 29.11. In Exercise 28.17 and Example 26.2, we considered:

$$f_{X,Y}(x,y) = \frac{1}{8}(1-x^2)(3-y) \qquad \text{for } -1 \le x \le 1 \text{ and } -1 \le y \le 1,$$

and $f_{X,Y}(x, y) = 0$ otherwise. Then we asked to find $\mathbb{E}(X)$ and to find $\mathbb{E}(Y)$.

Now, revisiting this question, and considering the random variable $X + Y$, we ask what is the expected value of $X + Y$, i.e., what is $\mathbb{E}(X + Y)$?

One might guess (correctly, by the way!) that $\mathbb{E}(X+Y)$ is just equal to $\mathbb{E}(X)+\mathbb{E}(Y)$, and then use the answers from Exercise 28.17 to conclude that

$$\mathbb{E}(X+Y)=\mathbb{E}(X)+\mathbb{E}(Y)=0-1/9=-1/9.$$

(Another response would be to let $Z=X+Y$, and then observe $-2 \leq Z \leq 2$, and then find $F_Z(z)$, and then differentiate to find $f_Z(z)$, and finally to integrate $zf_Z(z)$ over the range $-2 \leq z \leq 2$. This method of solution would be quite complicated, so we do not use it.)

Fortunately, there is a straightforward method of computing the answer (without relying on Exercise 28.17) that will prove even more useful. We just integrate $x+y$ over all possible x's and y's (i.e., using a double integral), weighting each possible value $x+y$ by the joint density, i.e., by $f_{X,Y}(x,y)$. Using this method, we get:

$$\begin{aligned}\mathbb{E}(X+Y)&=\int_{-\infty}^{\infty}\int_{-\infty}^{\infty}(x+y)f_{X,Y}(x,y)\,dy\,dx\\&=\int_{-1}^{1}\int_{-1}^{1}(x+y)\frac{1}{8}(1-x^2)(3-y)\,dy\,dx\end{aligned}$$

One nice feature of this method is that it prevents us from having to take the intermediate step to compute the density of either of the variables. The downside is that it makes the integration a little more complicated, but we handle the whole computation in one fell swoop! We get the following:

$$\begin{aligned}\mathbb{E}(X+Y)&=\int_{-1}^{1}\int_{-1}^{1}(x+y)\frac{1}{8}(1-x^2)(3-y)\,dy\,dx\\&=\int_{-1}^{1}\frac{1}{8}(1-x^2)\int_{-1}^{1}(3x-xy+3y-y^2)\,dy\,dx\\&=\int_{-1}^{1}\frac{1}{8}(1-x^2)\Big[\,(3xy-xy^2/2+3y^2/2-y^3/3)\big|_{y=-1}^{1}\Big]\,dx\\&=\int_{-1}^{1}\frac{1}{8}(1-x^2)(6x-2/3)\,dx\\&=\int_{-1}^{1}\frac{1}{8}(6x-6x^3-2/3+2x^2/3)\,dx\\&=\frac{1}{8}\,(6x^2/2-6x^4/4-2x/3+2x^3/9)\big|_{x=-1}^{1}\\&=-1/9\end{aligned}$$

The ability to compute $\mathbb{E}(X+Y)$ in one calculation is very nice..., but even nicer is the fact that this method works much more generally. For any function of X and Y, we can calculate the expected value of the function of X and Y using this method.

Definition 29.12. Expected values of functions of continuous random variables
If X and Y are continuous random variables with joint density $f_{X,Y}(x,y)$, and if g is any function of two variables, then

$$\mathbb{E}(g(X,Y)) = \int_{-\infty}^{\infty}\int_{-\infty}^{\infty} g(x,y)f_{X,Y}(x,y)\,dy\,dx$$

Now, by using the new method for expected values, and letting $g(x,y) = x$, we have a very succinct way to calculate the expected value of X directly, using a joint density:

Remark 29.13. Expected values of a continuous random variable, using a joint density
If X and Y are continuous random variables with joint density $f_{X,Y}(x,y)$, then the expected value of X is

$$\mathbb{E}(X) = \int_{-\infty}^{\infty}\int_{-\infty}^{\infty} xf_{X,Y}(x,y)\,dy\,dx \qquad .$$

Generalizing the example above, we can always get the expected value of the sum of two random variables, by just letting $g(x,y) = x + y$.

Remark 29.14. Expected value of the sum of two continuous random variables
If X and Y are continuous random variables with joint density $f_{X,Y}(x,y)$, then the expected value of $X + Y$ is

$$\begin{aligned}\mathbb{E}(X+Y) &= \int_{-\infty}^{\infty}\int_{-\infty}^{\infty} (x+y)f_{X,Y}(x,y)\,dy\,dx \\ &= \int_{-\infty}^{\infty}\int_{-\infty}^{\infty} xf_{X,Y}(x,y)\,dy\,dx + \int_{-\infty}^{\infty}\int_{-\infty}^{\infty} yf_{X,Y}(x,y)\,dy\,dx \\ &= \mathbb{E}(X) + \mathbb{E}(Y)\end{aligned}$$

Example 29.15. In Example 25.9, a person throws a dart at a square dart board. Let X and Y denote, respectively, the x- and y-coordinates (in feet) of the location where the dart lands; the middle of the dart board is at $(0,0)$. Suppose that the dart board is two feet wide and two feet tall, so that the dart only lands on the dart board if $-1 \le X \le 1$, $-1 \le Y \le 1$. Also suppose that the person always hits the dart board, and moreover, X and Y have joint density:

$$f_{X,Y}(x,y) = \frac{9}{16}(1-x^2)(1-y^2) \qquad \text{for } -1 \le x \le 1,\ -1 \le y \le 1,$$

and $f_{X,Y}(x,y) = 0$ otherwise.

Back in Example 25.9, we already computed $\mathbb{E}(X) = 0$, but we stopped in the middle of that problem to compute $f_X(x)$, which was somewhat unnecessary. Instead, with our new method, we can just compute directly, without stopping in the middle.

We compute $\mathbb{E}(X)$ directly:

$$\begin{aligned}
\mathbb{E}(X) &= \int_{-\infty}^{\infty}\int_{-\infty}^{\infty} x f_{X,Y}(x,y)\,dy\,dx \\
&= \int_{-1}^{1}\int_{-1}^{1} (x)\left(\frac{9}{16}\right)(1-x^2)(1-y^2)\,dy\,dx \\
&= \frac{9}{16}\int_{-1}^{1}(x-x^3)\,dx \int_{-1}^{1}(1-y^2)\,dy \\
&= \frac{9}{16}\left(x^2/2 - x^4/4\right)\Big|_{x=-1}^{1}\left(y-y^3/3\right)\Big|_{y=-1}^{1} \\
&= \frac{9}{16}(0)(4/3) \\
&= 0
\end{aligned}$$

Example 29.16. As a crowning achievement, we recompute Exercise 28.20: Consider two points that are independently placed on a line of length 10, at locations X and Y. Thus the joint density of X and Y is

$$f_{X,Y}(x,y) = 1/100 \qquad \text{for } 0 \le x \le 10 \text{ and } 0 \le y \le 10,$$

and $f_{X,Y}(x,y) = 0$ otherwise.

Let $Z = |X-Y|$ be the absolute value of the distance between X and Y. Find the expected distance $\mathbb{E}(Z) = \mathbb{E}(|X-Y|)$ between X and Y.

We compute

$$\begin{aligned}
\mathbb{E}(Z) &= \mathbb{E}(|X-Y|) \\
&= \int_{-\infty}^{\infty}\int_{-\infty}^{\infty} |x-y| f_{X,Y}(x,y)\,dy\,dx \\
&= \int_{0}^{10}\int_{0}^{10} |x-y|\frac{1}{100}\,dy\,dx
\end{aligned}$$

The only thing we need to do is split the integral into two parts, according to whether $x \ge y$, in which case $|x-y| = x-y$, or whether $x \le y$, in which case

$|x - y| = y - x$. So we continue

$$\begin{aligned}
\mathbb{E}(Z) &= \int_0^{10}\int_0^{x}(x-y)\frac{1}{100}\,dy\,dx + \int_0^{10}\int_x^{10}(y-x)\frac{1}{100}\,dy\,dx \\
&= \int_0^{10}(xy - y^2/2)\frac{1}{100}\bigg|_{y=0}^{x}\,dx + \int_0^{10}(y^2/2 - xy)\frac{1}{100}\bigg|_{y=x}^{10}\,dx \\
&= \int_0^{10}(x^2 - x^2/2)\frac{1}{100}\,dx + \int_0^{10}((10^2/2 - 10x) - (x^2/2 - x^2))\frac{1}{100}\,dx \\
&= \int_0^{10}(x^2 - 10x + 50)\frac{1}{100}\,dx \\
&= (x^3/3 - 10x^2/2 + 50x)\frac{1}{100}\bigg|_{x=0}^{10} \\
&= (1000/3 - 1000/2 + 500)\frac{1}{100} \\
&= 10/3
\end{aligned}$$

Notice that we did not need to compute $F_Z(z)$ or $f_Z(z)$ en route to computing $\mathbb{E}(Z)$ with this method. So there are some advantages to just computing expected values using the powerful formula

$$\mathbb{E}(g(X,Y)) = \int_{-\infty}^{\infty}\int_{-\infty}^{\infty} g(x,y) f_{X,Y}(x,y)\,dy\,dx.$$

In the case of this example, we were using $g(X,Y) = |X - Y|$.

29.4 More Friendly Facts about Continuous Random Variables

(Several of these facts have analogous discrete versions; see Section 12.3.)

We already showed that $\mathbb{E}(aX + b) = a\mathbb{E}(X) + b$ for any random variable X and any constants a and b. More is true:

Theorem 29.17. Expected value of a (linear) sum of two continuous random variables

If X and Y are continuous random variables with joint density $f_{X,Y}(x,y)$, and if a and b are any constants, then the expected value of $aX + bY$ is

$$\mathbb{E}(aX + bY) = a\mathbb{E}(X) + b\mathbb{E}(Y).$$

To see that $\mathbb{E}(aX + bY) = a\mathbb{E}(X) + b\mathbb{E}(Y)$, just write

$$\begin{aligned}\mathbb{E}(aX + bY) &= \int_{-\infty}^{\infty}\int_{-\infty}^{\infty}(ax + by)f_{X,Y}(x, y)\,dy\,dx\\ &= a\int_{-\infty}^{\infty}\int_{-\infty}^{\infty}xf_{X,Y}(x, y)\,dy\,dx + b\int_{-\infty}^{\infty}\int_{-\infty}^{\infty}yf_{X,Y}(x, y)\,dy\,dx\\ &= a\mathbb{E}(X) + b\mathbb{E}(Y)\end{aligned}$$

Applying the above fact over and over shows that:

Corollary 29.18. Expected value of a (linear) sum of two or more continuous random variables
If $X_1, X_2, \ldots, X_n$ are continuous random variables and if $a_1, a_2, \ldots, a_n$ are any constants, then the expected value of $a_1X_1 + a_2X_2 + \cdots + a_nX_n$ is

$$\mathbb{E}(a_1X_1 + a_2X_2 + \cdots + a_nX_n) = a_1\mathbb{E}(X_1) + a_2\mathbb{E}(X_2) + \cdots + a_n\mathbb{E}(X_n).$$

Caution: The X and Y must be independent in order to use the formulas in the box below!!!

Theorem 29.19. The expected value of the product of the functions of two *independent* random variables equals the product of the expected values of the functions of the two random variables
If X and Y are independent random variables, and g and h are any two functions, then

$$\mathbb{E}(g(X)h(Y)) = \mathbb{E}(g(X))\mathbb{E}(h(Y)).$$

To see this, we compute

$$\begin{aligned}\mathbb{E}(g(X)h(Y)) &= \int_{-\infty}^{\infty}\int_{-\infty}^{\infty}g(x)h(y)f_{X,Y}(x, y)\,dy\,dx\\ &= \int_{-\infty}^{\infty}\int_{-\infty}^{\infty}g(x)h(y)f_X(x)f_Y(y)\,dy\,dx \qquad \text{since } X, Y \text{ are indep.}\\ &= \int_{-\infty}^{\infty}g(x)f_X(x)\,dx\int_{-\infty}^{\infty}h(y)f_Y(y)\,dy\\ &= \mathbb{E}(g(X))\mathbb{E}(h(Y))\end{aligned}$$

Using identity functions for f and g, i.e., $f(X) = X$ and $g(Y) = Y$, the following nice fact follows immediately:

Corollary 29.20. The expected value of the product of two *independent* random variables equals the product of the expected values of the two random variables
If X and Y are independent random variables, then

$$\mathbb{E}(XY) = \mathbb{E}(X)\mathbb{E}(Y).$$

Again, caution: The X and Y must be independent, to utilize these results!!

The argument for the following is given as Exercise 29.32:

Theorem 29.21. If $X_1, \ldots, X_n$ are independent random variables, and $a_1, \ldots, a_n$ are constants, then

$$\mathrm{Var}(a_1X_1 + \cdots + a_nX_n) = a_1^2\,\mathrm{Var}(X_1) + \cdots + a_n^2\,\mathrm{Var}(X_n).$$

When $n = 2$, $a_1 = a$, $a_2 = 1$, $X_1 = X$, and $X_2 = b$ (a constant), then $\mathrm{Var}(X_2) = \mathbb{E}(X_2^2) - (\mathbb{E}(X_2))^2 = \mathbb{E}(b^2) - (\mathbb{E}(b))^2 = b^2 - b^2 = 0$, so this gives us immediately:

Corollary 29.22. If X is any random variable, and if a and b are any constants, then

$$\mathrm{Var}(aX + b) = a^2\,\mathrm{Var}(X).$$

If we use $a_i = 1$ for all i, then Theorem 29.21 yields:

Corollary 29.23. The variance of the sum of *independent* random variables equals the sum of the variances
If $X_1, \ldots, X_n$ are independent random variables, then

$$\mathrm{Var}(X_1 + \cdots + X_n) = \mathrm{Var}(X_1) + \cdots + \mathrm{Var}(X_n).$$

(We emphasize that the $X_1, \ldots, X_n$ are n independent values, not just n copies of the same value.) For instance, if the random variables represent n independent waiting times (e.g., at traffic lights), or heights of people, or other kinds of independent measurements, we get the variance of the sum by adding the variances of the individual random variables.

29.5 Exercises

29.5.1 Practice

For the densities in Exercises 29.1 to 29.8, find the variance of the given random variable.

Exercise 29.1. A child has lost her ring somewhere along a 20-foot line in a narrow strip of grass. Let X denote the location (lengthwise) along the grass where it was dropped, so

$$f_X(x) = \frac{1}{20}, \qquad \text{for } 0 \le x \le 20,$$

and $f_X(x) = 0$ otherwise.

Exercise 29.2. Let X have density

$$f_X(x) = \frac{2}{25}x, \qquad \text{for } 0 \leq x \leq 5,$$

and $f_X(x) = 0$ otherwise.

Exercise 29.3. When meeting a friend at a coffee house, Donald sees that his friend has not yet arrived, so he assumes that his friend's arrival time, X, in minutes, will have density

$$f_X(x) = \frac{1}{5}, \qquad \text{for } 0 \leq x \leq 5,$$

and $f_X(x) = 0$ otherwise.

Exercise 29.4. The waiting time X, in minutes, until the next green car passes has density

$$f_X(x) = \frac{1}{2}e^{-x/2}, \qquad \text{for } 0 < x,$$

and $f_X(x) = 0$ otherwise.

Exercise 29.5. Darius always arrives at Wiley dining hall 50 minutes before it closes. He estimates that the time X necessary to stand in line, in minutes, has density

$$f_X(x) = \frac{x}{50}, \qquad \text{for } 0 \leq x \leq 10,$$

and $f_X(x) = 0$ otherwise.

Exercise 29.6. Let X have density

$$f_X(x) = \frac{1}{4}, \qquad \text{for } 2 \leq x \leq 6,$$

and $f_X(x) = 0$ otherwise.

Exercise 29.7. Upon arriving at a restaurant and finding that no table is available, Diana and Markus have a waiting time X (in hours) with density

$$f_X(x) = 2e^{-2x}, \qquad \text{for } 0 < x,$$

and $f_X(x) = 0$ otherwise.

Exercise 29.8. Let X be the time, in hours, that Christine spends talking on the telephone. Then X has density

$$f_X(x) = 3e^{-3x}, \qquad \text{for } 0 < x,$$

and $f_X(x) = 0$ otherwise.

Exercise 29.9. Consider a pair of random variables X, Y with constant joint density on the triangle with vertices at $(0,0)$, $(3,0)$, and $(0,3)$.

a. Find the expected value of the sum of X and Y, i.e., find $\mathbb{E}(X+Y)$.

b. Find the variance of X, i.e., find $\text{Var}(X)$.

Exercise 29.10. Consider a pair of random variables X, Y with constant joint density on the quadrilateral with vertices $(0,0)$, $(2,0)$, $(2,6)$, $(0,12)$.

a. Find the variance of X, i.e., find $\text{Var}\,X$.

b. Find the variance of Y, i.e., find $\text{Var}\,Y$.

Exercise 29.11. Let X, Y have joint density $f_{X,Y}(x,y) = 14e^{-2x-7y}$ for $x > 0$ and $y > 0$; and $f_{X,Y}(x,y) = 0$ otherwise. Find the variance of the sum of X and Y, i.e., find $\text{Var}(X+Y)$.

Exercise 29.12. Let X, Y have joint density $f_{X,Y}(x,y) = 18e^{-2x-7y}$ for $0 < y < x$; and $f_{X,Y}(x,y) = 0$ otherwise. Find the variance of Y.

Exercise 29.13. Suppose X, Y have joint density

$$f_{X,Y}(x,y) = \begin{cases} \frac{1}{9}(3-x)(2-y) & \text{if } 0 \le x \le 3 \text{ and } 0 \le y \le 2, \\ 0 & \text{otherwise.} \end{cases}$$

Find the expected value of $X^2 + Y^3$, i.e., find $\mathbb{E}(X^2 + Y^3)$.

Exercise 29.14. Let X have density

$$f_X(x) = \frac{1}{5}, \qquad \text{for } 2 \le x \le 7,$$

and $f_X(x) = 0$ otherwise.

a. Find the expected value of X^2, i.e., $\mathbb{E}(X^2)$.

b. Find the expected value of $1/X^2$, i.e., $\mathbb{E}(1/X^2)$.

c. Find the variance $\text{Var}(X)$.

d. Find the standard deviation σ_X.

Exercise 29.15. Two convicts are running away from prison. One man, John, runs east X miles, and another man, Jeff, runs north Y miles. John cannot run more than 4 miles, and Jeff cannot run more than 6 miles. At a random point in time, their locations are spotted by a helicopter. Assume that the joint density of their location is

$$f_{X,Y}(x,y) = \frac{1}{24}, \qquad \text{for } 0 \le x \le 6, \text{ and } 0 \le y \le 4,$$

and $f_{X,Y}(x,y) = 0$ otherwise. Let $g(x,y) = x + y$ be the total distance that the two men have run so far.

a. Find the expected value of the sum of these two distances.

b. Find the variance of the sum of these two distances.

Exercise 29.16. The distance X, in yards, that a small person can throw a 50-pound weight, has density

$$f_X(x) = -0.0375x^2 + 0.075x + 0.3, \qquad \text{for } 0 \le x \le 4,$$

and $f_X(x) = 0$ otherwise. Find the variance of X.

For the densities in Exercises 29.17 to 29.22, and the given function $g(x, y)$ in each exercise, find $\mathbb{E}(g(X, Y))$.

Exercise 29.17. Let X, Y have joint density

$$f_{X,Y}(x, y) = \frac{1}{36}, \qquad \text{for } 0 \le x \le 6,\ 0 \le y \le 6,$$

and $f_{X,Y}(x, y) = 0$ otherwise. Let $g(x, y) = 2x + y$.

Exercise 29.18. Let X, Y have joint density

$$f_{X,Y}(x, y) = 1, \qquad \text{for } 0 \le x \le 1,\ 0 \le y \le 1,$$

and $f_{X,Y}(x, y) = 0$ otherwise. Let $g(x, y) = x^2 + y^2$.

Exercise 29.19. Let X, Y have joint density

$$f_{X,Y}(x, y) = \frac{1}{36}, \qquad \text{for } 0 \le x \le 4,\ 0 \le y \le 9,$$

and $f_{X,Y}(x, y) = 0$ otherwise. Let $g(x, y) = xy$.

Exercise 29.20. Let X, Y have joint density

$$f_{X,Y}(x, y) = 1, \qquad \text{for } 0 \le x \le 1,\ 0 \le y \le 1,$$

and $f_{X,Y}(x, y) = 0$ otherwise. Let $g(x, y) = x + y$.

Exercise 29.21. Let X, Y have joint density

$$f_{X,Y}(x, y) = \frac{1}{6}, \qquad \text{for } 8 \le x \le 10, \text{ and } 0 \le y \le 3,$$

and $f_{X,Y}(x, y) = 0$ otherwise. Let $g(x, y) = x^2y^3$.

Exercise 29.22. Let X, Y have joint density

$$f_{X,Y}(x, y) = \frac{xy^2}{72}, \qquad \text{for } 0 \le x \le 4,\ 0 \le y \le 3,$$

and $f_{X,Y}(x, y) = 0$ otherwise. Let $g(x, y) = \frac{x-y}{2}$.

29.5.2 Extensions

Exercise 29.23. Geoffrey does not like to be low on gas, so he randomly stops to fill up his tank. He has a 14-gallon tank, and the current price of gas is \$3.75 per gallon. Whenever he stops to buy gas, he always buys a candy bar for \$1.30. If X is the amount of gas (in gallons) in his tank when he stops for a purchase, then $f_X(x) = 1/14$ for $0 \le x \le 14$, and $f_X(x) = 0$ otherwise. He always fills the tank, so he will always buy $14 - X$ gallons. Find the expected amount of money Geoffrey spends on a purchase of gas and a candy bar.

Exercise 29.24. Let X have density

$$f_X(x) = \frac{1}{3}e^{-x/3}, \qquad \text{for } 0 < x,$$

and $f_X(x) = 0$ otherwise.

a. Find the expected value of X^2, i.e., $\mathbb{E}(X^2)$.

b. Find the variance $\text{Var}(X)$.

c. Find the standard deviation σ_X.

Exercise 29.25. Let X and Y correspond to the horizontal and vertical coordinates in the triangle with corners at $(2,0)$, $(0,2)$, and the origin. Let $f_{X,Y}(x,y) = \frac{15}{28}(xy^2 + y)$ for (x,y) inside the triangle, and $f_{X,Y}(x,y) = 0$ otherwise. Find $\mathbb{E}(XY)$.

Exercise 29.26. Let X and Y correspond to the horizontal and vertical coordinates in the rectangle with corners at $(15,0)$, $(15,10)$, $(0,10)$, and the origin. Let $f_{X,Y}(x,y) = \frac{1}{150}$ for (x,y) inside the rectangle, and $f_{X,Y}(x,y) = 0$ otherwise. Find $\mathbb{E}(XY^2)$.

Exercise 29.27. Let X have density

$$f_X(x) = \frac{2}{75},$$

for x in the quadrilateral with vertices at $(0,0)$, $(10,0)$, $(5,5)$, $(0,5)$, and $f_X(x) = 0$ otherwise.

a. Find $\mathbb{E}(X)$.

b. Find $\text{Var}(X)$.

Exercise 29.28. Let X have density

$$f_X(x) = 25xe^{-5x}, \qquad \text{for } 0 < x,$$

and $f_X(x) = 0$ otherwise. Find $\text{Var}(X)$.

Exercise 29.29. Let X have density

$$f_X(x) = \frac{2\ln x}{x}, \qquad \text{for } 1 \le x \le e,$$

and $f_X(x) = 0$ otherwise. Find $\text{Var}(X)$.

29.5.3 Advanced

Exercise 29.30. Every day, a student calls his mother and then (afterwards) calls his girlfriend. Let X be the time (in hours) until he calls his mother, and let Y be the time (in hours) until he calls his girlfriend. Since he always calls his mother first, then $X < Y$. So let the joint density of the time be

$$f_{X,Y}(x, y) = 10e^{-3x-2y} \qquad \text{for } 0 < x < y,$$

and $f_{X,Y}(x, y) = 0$ otherwise.

Let $Z = |Y - X|$ be time in between the beginning of the two calls.

a. Find $\mathbb{E}(Z)$.

b. Find $\text{Var}(Z)$.

Exercise 29.31. Let X have density

$$f_X(x) = \frac{1}{3}e^{-x/3}, \qquad \text{for } 0 < x,$$

and $f_X(x) = 0$ otherwise.

a. Find the expected value of X^3, i.e., $\mathbb{E}(X^3)$.

b. Find a general form, which handles all positive integers n, for $\mathbb{E}(X^n)$, often called the nth moment of X.

Exercise 29.32. Show that, if $X_1, \ldots, X_n$ are independent random variables, and $a_1, \ldots, a_n$ are constants, then

$$\text{Var}(a_1X_1 + \cdots + a_nX_n) = a_1^2 \text{Var}(X_1) + \cdots + a_n^2 \text{Var}(X_n).$$

Exercise 29.33. Suppose that X and Y have joint probability density function $f_{X,Y}(x, y) = 16e^{-4x-4y}$ for $x > 0$ and $y > 0$, and $f_{X,Y}(x, y) = 0$ otherwise. Find $\mathbb{E}(X + Y)$.

Exercise 29.34. If the joint density of X and Y is $\frac{8}{27}xy$ on the interior of the triangle with corners at $(0, 0)$, $(3, 0)$, $(0, 3)$ (and the joint density is 0 elsewhere), find $\mathbb{E}(X)$.

Chapter 30

Review of Continuous Random Variables

30.1 Summary of Continuous Random Variables

What can you do with a density?

- Remember that it is important to check that you have a valid density first.
- Find probabilities, e.g., $P(1.7 \le X \le 3.2)$.
- Graph the density $f_X(x)$.
- Calculate expected value (average), variance, and standard deviation of X and of functions of X.
- Calculate the CDF by integrating the density.

What can you do with a CDF?

- Find cumulative probabilities, e.g., $F_X(5.77) = P(X \le 5.77)$.
- Graph the CDF $F_X(x)$.
- Find the median and other percentiles.
- Calculate the density by taking the derivative of the CDF.

The formulas in this chapter were for continuous random variables only. Earlier in the text, we discussed discrete random variable formulas which are similar, but they did not use calculus.

How do you know if your density is valid?

- The density must account for every possible outcome. It must describe everything that could happen from $-\infty$ to ∞, although the density is sometimes zero on large portions of the real line.

- If you integrate the density from $-\infty$ to ∞, the result is 1. (The area under the density is 1.)

$$\int_{-\infty}^{\infty} f_X(x)\,dx = 1$$

How do you find the cumulative distribution function (CDF)?

$$F_X(a) = P(X \le a) = \int_{-\infty}^{a} f_X(x)\,dx$$

How do you find the median?

Look at the CDF graph. The first x-value where the CDF is equal to 0.5 is your median.

How do you find other percentiles?

To find the pth percentile, just find the x-value where $F_X(x) = P(X \le x) = p$. This works for the median: the median is the value of x for which $F_X(x) = P(X \le x) = 1/2$. It works for other percentiles too. For instance, the 95th percentile is the x-value for which $F_X(x) = P(X \le x) = 0.95$. (You find the first x-value where the CDF is equal to 0.95.)

How do you get the mean/(weighted) average/expected value?

$$\mathbb{E}(X) = \int_{-\infty}^{\infty} x f_X(x)\,dx$$

$$\mathbb{E}(g(X)) = \int_{-\infty}^{\infty} g(x) f_X(x)\,dx$$

How do you get the variance and standard deviation?

$$\mathrm{Var}(X) = \mathbb{E}(X^2) - (\mathbb{E}(X))^2$$

$$\sigma_X = \sqrt{\mathrm{Var}(X)}$$

What do you do if you have a linear combination of X? Here a and b are constants.

$$\begin{aligned}
\mathbb{E}(aX + b) &= a\mathbb{E}(X) + b \\
\mathrm{Var}(aX + b) &= a^2\,\mathrm{Var}(X) \\
\sigma_{aX+b} &= \sqrt{a^2\,\mathrm{Var}(X)} = |a|\sqrt{\mathrm{Var}(X)}
\end{aligned}$$

What do you do if you have a sum of random variables and need an expectation? (This works regardless of whether the variables are independent or dependent.)

$$\mathbb{E}\left(\sum_{i=1}^{n} X_i\right) = \sum_{i=1}^{n} \mathbb{E}(X_i)$$

What if all of these random variables have the same distribution? (Again, this works regardless of whether the variables are independent or dependent.)

$$\mathbb{E}\left(\sum_{i=1}^{n} X_i\right) = n\mathbb{E}(X_1)$$

What do you do if you need the variance of n independent random variables?

$$\text{Var}\left(\sum_{i=1}^{n} X_i\right) = \sum_{i=1}^{n} \text{Var}(X)$$

$$\sigma_{\sum_{i=1}^{n} X_i} = \sqrt{\sum_{i=1}^{n} \text{Var}(X)}$$

What if all of these random variables are independent and also have the same distribution?

$$\text{Var}\left(\sum_{i=1}^{n} X_i\right) = n\,\text{Var}(X)$$

$$\sigma_{\sum_{i=1}^{n} X_i} = \sqrt{n\,\text{Var}(X)}$$

30.2 Exercises

Exercise 30.1. Using the following probability density function

$$f_X(x) = \begin{cases} 2 & \text{if } 0 \le x \le 1/2, \\ 0 & \text{otherwise,} \end{cases}$$

a. Find the cumulative distribution function $F_X(x)$.

b. Find $P(0.25 \le X \le 0.45)$.

c. Find $P(X > 0.45)$.

d. Find $P(X = 0.25)$.

e. Find the mean $\mathbb{E}(X)$.

f. Find the standard deviation σ_X.

g. Graph the density $f_X(x)$.

h. Graph the CDF $F_X(x)$. Identify the median on the graph of the CDF.

Exercise 30.2. Using the following probability density function,

$$f_X(x) = \begin{cases} \frac{1}{8}e^{-x/8} & \text{if } x > 0, \\ 0 & \text{otherwise,} \end{cases}$$

a. Find the cumulative distribution function $F_X(x)$.

b. Find $P(X > 4)$.

c. Find $P(-2 \leq X \leq 12)$.

d. Find $P(X < 240)$.

e. Find the mean $\mathbb{E}(X)$.

f. Find the standard deviation σ_X.

g. Graph the density $f_X(x)$.

h. Graph the CDF $F_X(x)$. Identify the median on the graph of the CDF.

Exercise 30.3. Using the following probability density function,

$$f_X(x) = \begin{cases} \frac{1}{x^2} & \text{if } x \geq 1, \\ 0 & \text{otherwise,} \end{cases}$$

a. Find the cumulative distribution function $F_X(x)$.

b. Find $P(X > 3)$.

c. Find $P(-0.5 \leq X < 5)$.

d. Find $P(X < 4.2)$.

e. Find the mean $\mathbb{E}(X)$.

f. Graph the density $f_X(x)$.

g. Graph the CDF $F_X(x)$. Identify the 25th, 50th (median), and 75th percentiles on the graph of the CDF.

Exercise 30.4. Using the following CDF,

$$F_X(x) = \begin{cases} 0 & \text{if } x < 2, \\ \frac{x-2}{2} & \text{if } 2 \leq x \leq 4, \\ 1 & \text{if } 4 < x, \end{cases}$$

find the corresponding density $f_X(x)$.

Exercise 30.5. What is the constant k that makes the following function a valid density?

$$f_X(x) = \begin{cases} k(e^{-x} + e^{-4x}) & \text{if } 0 \leq x, \\ 0 & \text{otherwise,} \end{cases}$$

Exercise 30.6. What is the constant k that makes the following function a valid density?

$$f_X(x) = \begin{cases} kx^2(1-x)^7 & \text{if } 0 \leq x \leq 1, \\ 0 & \text{otherwise,} \end{cases}$$

Exercise 30.7. Which of the following can be true? If an answer is false, state why it is false.

a. A CDF $F_X(x)$ can have the value 4.3.

b. A density $f_X(x)$ can have the value 4.3.

c. A mass $p_X(x)$ can have the value 4.3.

Exercise 30.8. Which of the following can be true? If an answer is false, state why it is false.

a. A CDF $F_X(x)$ can be 1 for two or more values of x.

b. A density $f_X(x)$ can be 1 for two or more values of x.

c. A mass $p_X(x)$ can be 1 for two or more values of x.

Exercise 30.9. Which of the following can be true? If an answer is false, state why it is false.

a. The (indefinite) integral of $F_X(x)$ is $f_X(x)$.

b. The (indefinite) integral of $f_X(x)$ is $F_X(x)$.

c. The area under the curve of $f_X(x)$ between $-\infty$ and x is $F_X(x)$.

d. The area under the curve of $F_X(x)$ between $-\infty$ and x is $f_X(x)$.

e. The derivative of $F_X(x)$ is $f_X(x)$.

f. The derivative of $f_X(x)$ is $F_X(x)$.

Exercise 30.10. Which of the following can be true? If an answer is false, state why it is false.

a. The area under the $F_X(x)$ curve from $-\infty$ to ∞ is 1.

b. The area under the $f_X(x)$ curve from $-\infty$ to ∞ is 1.

Exercise 30.11. Which of the following can be true? If an answer is false, state why it is false.

a. The CDF $F_X(x)$ can be negative.

b. The density $f_X(x)$ can be negative.

c. The mass $p_X(x)$ can be negative.

Exercise 30.12. Which of the following can be true? If an answer is false, state why it is false.

a. The graph of $f_X(x)$ can have jumps (i.e., can be discontinuous).

b. The graph of $F_X(x)$ for discrete random variables can have jumps.

c. The graph of $F_X(x)$ for continuous random variables can have jumps.

d. The graph of $p_X(x)$ can have jumps.

Part VI

Named Continuous Random Variables

For discrete distributions, you learned the rules for general distributions and then about some of the most widely used types of random variables, to make your life easier in scenarios that match specific situations (Binomial, Geometric, etc.). For continuous distributions, you have already learned the rules for general distributions during the previous part of the book. Now in the chapters that follow, we will show you some formulas for scenarios matching very common types of situations, and the named continuous random variables that are appropriate in these situations. For instance, we dedicate several chapters to the Normal distribution, perhaps the most important continuous distribution for probability and statistics. At the end of this part of the book, we will review some strategies for comparing and contrasting all of these distributions, to help you pick the correct distribution for your situation.

There are many other continuous distributions than the five that we cover here, but these are the most common ones.

By the end of this part of the book, you should be able to:

1. Distinguish between Continuous Uniform, Exponential, Gamma, Beta, and Normal distributions when reading a story.

2. Identify the variable and the parameters in a story, and state in English what the variable and its parameters mean.

3. Use the formulas for the density, CDF, expected value, and variance to answer questions and find probabilities.

4. Utilize joint densities to study problems about two or more random variables at the same time.

5. Use all discrete and continuous named distributions where appropriate.

Math skills you will need: e^x, derivatives, integrals (in particular, integration by parts), and the Gamma function.

Additional resources: Computer programs (such as Excel, Maple, Mathematica, Matlab, Minitab, R, SPSS, etc.) and calculators may be used to assist in the calculations. You will also need the standard Normal table that appears in Chapter 35 and also appears in the back of the book.

Chapter 31

Continuous Uniform Random Variables

We make chance a matter of absolute calculation. We subject the unlooked for and unimagined, to the mathematical formulæ of the schools.

—"The Mystery of Marie Rogêt" by Edgar Allan Poe (Snowden's Ladies' Companion, 1842 and 1843)

If you want to board a bus at a random time, and the bus circles campus once every 30 minutes, how long do you expect to wait until the bus arrives? What is the probability that the bus will come within the next 5 minutes? What is the variance of the time until the next bus comes? What is the probability that you just missed the bus by 1 minute or less?

31.1 Introduction

For the next few chapters, we focus on some very specific continuous random variables. The most simple type of continuous random variable is called a *Continuous Uniform random variable*, often just called a *Uniform*. We say that X is *Uniform* anytime that the density of X is constant on some interval. In such a case, the density of a Uniform random variable must be equal to 1 divided by the length of the interval. The reason is that, when the density (which is constant) is integrated over the length (or area, or volume, in the case of two or three dimensions) of the region, the integral needs to evaluate to 1.

One of the nicest properties of Uniform random variables is that, since the density is constant, the integrals for finding various probabilities are especially easy, because we can essentially avoid having to perform an integration. We can just multiply the constant (i.e., the density) by the length over which we are "integrating," and then we obtain the desired probability, without actually having to use any complicated integrating techniques.

The name "Continuous Uniform" (often abbreviated simply as "Uniform") is reminiscent of the Discrete Uniform random variables from Chapter 20. A Discrete Uniform random variable has a mass that is constant on a collection of points, just as a continuous random variable has a density that is constant on some interval. (Similarly, a pair of jointly distributed Uniform random variables has a density that is constant on some two-dimensional region.)

Continuous Uniform random variables

The common thread: Constant density. If a random variable has a density that is constant on a region, the variable is Uniform. This means that, for instance, the random variable is equally likely to be located in any regions that have the same size. If the location X of a hidden object is is Uniform on $[0, 10]$, then X is equally likely to be found in $[2.2, 3.4]$, or $[0, 1.2]$, or $[8.74, 9.94]$, etc., etc., because all of these regions have length 1.2.

Things to look for: Constant density on an interval (or a pair of random variables with constant joint density on a two-dimensional region).

The variable:

$$X = \text{an exact position or arrival time}$$

(for example: your exact arrival time, in minutes, if we know that you will arrive sometime between 1:00 and 1:15 pm). There are no general limits to what values X can take, but a specific range will be defined for each problem.

The parameters: a and b, the endpoints of the interval $[a, b]$ or (a, b) where the density of X is nonzero. Sometimes a pair of Continuous Uniform random variables is defined on a two-dimensional region.

Density:

$$f_X(x) = \begin{cases} \frac{1}{b-a}, & a \le x \le b, \\ 0 & \text{otherwise} \end{cases}$$

CDF:

$$F_X(x) = \begin{cases} 0 & x < a, \\ \frac{x-a}{b-a}, & a \le x \le b, \\ 1 & b < x \end{cases}$$

Expected value formula:

$$\mathbb{E}(X) = (a + b)/2$$

Variance formula:

$$\text{Var}(X) = (b - a)^2/12$$

The notation for a Uniform random variable on the interval (a, b) looks like:

$$X \sim \text{C.Uniform}(a, b).$$

Where does the cumulative distribution function formula come from?

If we integrate the probability density function formula from $-\infty$ to c, where $a \leq c \leq b$, we get

$$F_X(c) = \int_{-\infty}^{c} f_X(x)\,dx = \int_a^c \frac{1}{b-a}\,dx = \frac{c-a}{b-a}.$$

Thus $F_X(x) = \frac{x-a}{b-a}$ for $a \leq x \leq b$.

However, it is also useful to think about the cumulative distribution function as the area under the curve for x in the range $[a, c]$.

Where does the expected value formula come from?

The expected value $\mathbb{E}(X)$ for the Uniform random variable is just the midpoint of the range where the density $f_X(x)$ is not 0. (This is not true for other types of continuous random variables.) Since the Continuous Uniform distribution is symmetric, the median (center) and the mean (expected value) are the same.

31.2 Examples

Example 31.1. If X is a Continuous Uniform random variable on the interval $[a, b]$, then the expected value of X is $\mathbb{E}(X) = (a + b)/2$, as shown in Theorem 28.6. The expected value of X^2 is:

$$\begin{aligned}
\mathbb{E}(X^2) &= \int_a^b x^2 \frac{1}{b-a}\,dx \\
&= \frac{x^3}{3}\frac{1}{b-a}\bigg|_{x=a}^{b} \\
&= \frac{b^3 - a^3}{3}\frac{1}{b-a} \\
&= \frac{(b-a)(a^2+ab+b^2)}{3}\frac{1}{b-a} \\
&= \frac{a^2+ab+b^2}{3}
\end{aligned}$$

So the variance of X is:

$$\begin{aligned}
\frac{a^2+ab+b^2}{3} - \left(\frac{a+b}{2}\right)^2 &= \frac{4a^2+4ab+4b^2}{12} - \frac{3a^2+6ab+3b^2}{12} \\
&= \frac{a^2-2ab+b^2}{12} \\
&= \frac{(b-a)^2}{12}
\end{aligned}$$

Example 31.2. Sasha is a clerk in a convenience store. She needs to go to the bathroom and leaves her counter unattended at 1:02 PM. When she returns at 1:08 PM, there is an impatient customer waiting at the counter.

a. Why is this a Continuous Uniform distribution situation? What are the parameters? What is X?

The customer was equally likely to have arrived anytime in the continuous range between 1:02 and 1:08. The parameters are $a = 2$ and $b = 8$. The random variable X is the customer's actual arrival time.

b. Show the graph of the probability density function for the impatient customer's arrival time.

The formula for the density is

$$f_X(x) = \begin{cases} 1/6 & \text{if } 2 \leq x \leq 8, \\ 0 & \text{otherwise} \end{cases}$$

See the left side of Figure 31.1.

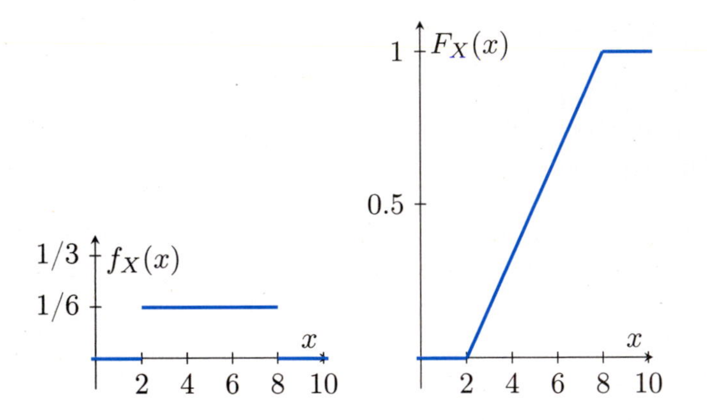

FIGURE 31.1: Left: The density $f_X(x) = 1/6$. Right: The CDF $F_X(x) = \frac{x-2}{6}$, of Uniform random variable on $[2, 8]$.

c. What is the expected time the customer arrived?

$$\mathbb{E}(X) = \frac{2+8}{2} = 5, \qquad \text{which is 1:05 PM.}$$

d. What is the standard deviation for the time the customer arrived?

The variance is

$$\text{Var}(X) = \frac{(8-2)^2}{12} = 3,$$

so the standard deviation is

$$\sigma_X = \sqrt{3} = 1.7321 \text{ minutes.}$$

e. What is the probability the customer arrived within the last 2 minutes (i.e., between 1:06 PM and 1:08 PM)?

We are looking for $P(6 \le X \le 8)$. You could do this problem three different ways.

Method #1. *For the Continuous Uniform distribution X, the probability of X being in some range is equal to the length of the range divided by the entire length of the interval where X is defined.* This is true in general, for Continuous random variables, because the density we are integrating (to obtain the probabilities) is a *constant function.* Thus $P(6 \le X \le 8) = \frac{\text{length of } [6,8]}{\text{length of } [2,8]} = 1/3$.

(Note: If some of the desired interval falls outside the range where X is defined, that part of the range must be ignored. For example, consider $P(6 \le X \le 10)$. We know that X is between 2 and 8, so the portion from 8 to 10 can be ignored. So, $P(6 \le X \le 10) = P(6 \le X \le 8) = 2/6 = 1/3$.)

Method #2. If you use the probability density function, you could integrate

$$P(6 \le X \le 8) = \int_6^8 \frac{1}{6} = 2/6 = 1/3.$$

Method #3. If you know the cumulative distribution function, you could just use

$$P(6 \le X \le 8) = P(X \le 8) - P(X \le 6) = F_X(8) - F_X(6).$$

See the right side of Figure 31.1 for a plot of the cumulative distribution function.

f. What is the cumulative distribution function? Also show the graph.

The CDF is

$$F_X(x) = \begin{cases} 0 & x \le 2, \\ \frac{x-2}{8-2} & 2 \le x \le 8, \\ 1 & 8 \le x, \end{cases}$$

The CDF is shown on the right in Figure 31.1.

g. Given that the impatient customer did not arrive in the first 2 minutes, what is the probability that he arrived in the last 2 minutes of this interval?

This is a conditional probability. When working with Uniform random variables, if a condition removes some of the interval, all of the remaining conditional probabilities are still Uniform on the remaining interval. E.g., in this example, we have the condition $X \ge 4$. So the portion $[2, 4]$ of the original interval is removed. What remains is the rest of the interval, i.e., $[4, 8]$. Thus, the conditional density is constant on the interval $[4, 8]$. So

$$P(X > 6 \mid X > 4) = \frac{\text{length of } [6, 8]}{\text{length of } [4, 8]} = 2/4 = 1/2.$$

Remark 31.3. Conditional probabilities and Continuous Uniform Random Variable

Consider X which is Uniform on $[a, b]$. If we are given that $X > c$, for some c in the interval $[a, b]$, then all of the resulting conditional probabilities are Uniform on $[c, b]$.

Similarly, if we are given $X < c$ (for c in $[a, b]$), then all of the resulting conditional probabilities are Uniform on $[a, c]$.

This calculation can also be written as

$$P(X > 6 \mid X > 4) = \frac{P(X > 6 \text{ and } X > 4)}{P(X > 4)} = \frac{P(X > 6)}{P(X > 4)} = \frac{1 - F_X(6)}{1 - F_X(4)}.$$

Therefore, $P(X > 6 \mid X > 4) == \frac{1-(6-2)/6}{1-(4-2)/6} = 1/2$. As another way to compute this probability, we can consider a new variable Y that is Uniform on $[4, 8]$. The density of Y is

$$f_Y(y) = \begin{cases} 1/4 & 4 \le y \le 8 \\ 0 & \text{otherwise} \end{cases}$$

This density is shown in Figure 31.2. The desired probability is $P(Y > 6) = 2/4 = 1/2$.

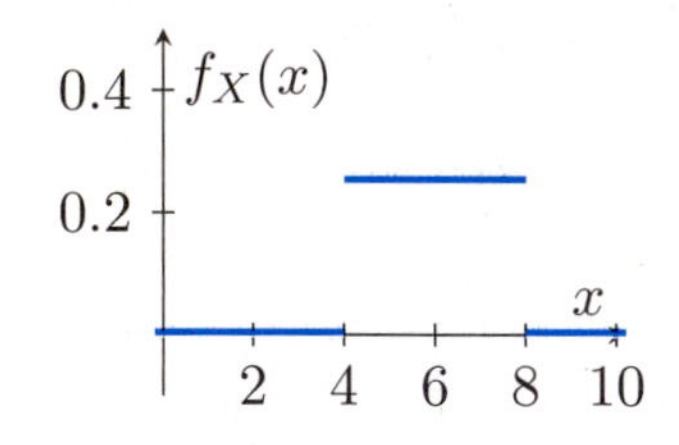

FIGURE 31.2: Density $f_Y(y) = 1/4$ of Uniform random variable Y on $[4, 8]$.

Example 31.4. If a burglary happens in an apartment, the costs to the renter from the theft, T, can be modeled by a Continuous Uniform random variable ranging from no charge up to \$4000.

a. What is the probability distribution function for the cost of damage?

$$f_T(x) = \begin{cases} \frac{1}{4000} & \text{if } 0 \le x \le 4000, \\ 0 & \text{otherwise} \end{cases}$$

b. What is the expected cost of damage?

$$\mathbb{E}(T) = \frac{0 + 4000}{2} = \$2000.$$

c. What is the standard deviation in the cost of damage?

$$\text{Var}(T) = \frac{(4000 - 0)^2}{12} = 1{,}333{,}333,$$

$$\sigma_T = \sqrt{1{,}333{,}333} = \$1154.70.$$

d. An insurance company issues a renter's policy with a deductible of $800. This means that if there is a claim the insurance company will reimburse the renter all damage costs between $800 and $4000, but the renter is responsible for the first $800 in costs.

What is the expected out-of-pocket cost C to the renter if damage occurs?

We observe that $T = C + M$, where M is the amount that the insurance company pays. The out-of-pocket cost C is just a function of T:

$$C = \begin{cases} T & \text{if } 0 \le T \le 800, \\ 800 & \text{if } 800 < T \le 4000 \end{cases}$$

Since C is just a function of T, then we can compute $\mathbb{E}(C)$ by integrating the density of T times the cost:

$$\begin{aligned} \mathbb{E}(C) &= \int_0^{800} x \frac{1}{4000}\, dx + \int_{800}^{4000} 800 \frac{1}{4000}\, dx \\ &= \frac{x^2}{2} \frac{1}{4000} \bigg|_{x=0}^{800} + (3200)800 \frac{1}{4000} \\ &= 80 + 640 \\ &= \$720.00 \end{aligned}$$

e. How much should the insurance company expect to pay the renter if damage occurs?

Remember that the insurance company only pays if the damage is above $800. Again, use M to denote the amount that the insurance company pays. Then M is just a function of T:

$$M = \begin{cases} 0 & \text{if } 0 \le T \le 800, \\ T - 800 & \text{if } 800 \le T \le 4000 \end{cases}$$

As before, since M is just a function of T, then we can compute $\mathbb{E}(M)$ by integrating the density of T times the amount M that the insurance company pays:

$$\begin{aligned}\mathbb{E}(M) &= \int_0^{800} 0\frac{1}{4000}\,dx + \int_{800}^{4000}(x-800)\frac{1}{4000}\,dx \\ &= \left(\frac{x^2}{2} - 800x\right)\frac{1}{4000}\Big|_{x=800}^{4000} \\ &= \$1280.00\end{aligned}$$

Notice that the amount the insurance company expects to pay,

$$\mathbb{E}(M) = \$1280.00,$$

added to the amount the renter expects to pay

$$\mathbb{E}(C) = \$720.00,$$

is equal to the expected total amount of damage

$$\mathbb{E}(T) = \$2000.00$$

Now we return to the method of using the lengths of intervals, as discussed in Example 31.2.

If X is Uniformly distributed on the interval $[a, b]$, then

$$f_X(x) = \frac{1}{b-a} \qquad \text{for } a \le x \le b,$$

and $f_X(x) = 0$ otherwise. So we can use the lengths of intervals, instead of having to integrate.

Remark 31.5. Using lengths of intervals to calculate probabilities, with Continuous Uniform random variables

If C is a subinterval of $[a, b]$, then

$$P(X \in C) = \int_C \frac{1}{b-a}\,dx = \frac{\text{length of } C}{b-a} = \frac{\text{length of } C}{\text{length of } [a,b]}.$$

Similar concepts work in higher dimensions. When X and Y have a joint density that is constant, we say that X and Y have a jointly Uniform distribution on a two-dimensional region. The joint density must be 1 divided by the area of the region, so that the integral over the region is 1.

Example 31.6. Let X and Y denote the latitude and longitude of a person who is thought to be lost in a certain part of a forest. If we assume that the densities of X and Y are each constant, say, $2 \le X \le 8.2$ and $3.01 \le Y \le 6.3$, then

$$f_X(x) = \frac{1}{8.2 - 2} \qquad \text{for } 2 \le x \le 8.2,$$

and $f_X(x) = 0$ otherwise, and

$$f_Y(y) = \frac{1}{6.3 - 3.01} \qquad \text{for } 3.01 \le y \le 6.3,$$

and $f_Y(y) = 0$ otherwise. If, moreover, we assume that X and Y are independent, then the joint density of X and Y is constant too:

$$\begin{aligned} f_{X,Y}(x,y) &= f_X(x) f_Y(y) \\ &= \left(\frac{1}{8.2-2}\right)\left(\frac{1}{6.3-3.01}\right) \qquad \text{for } 2 \le x \le 8.2 \text{ and } 3.01 \le y \le 6.3, \end{aligned}$$

and $f_{X,Y}(x,y) = 0$ otherwise.

More generally, if X is Uniform on $[a,b]$ and Y is Uniform on $[c,d]$, and if X and Y are independent, then

$$f_{X,Y}(x,y) = \left(\frac{1}{b-a}\right)\left(\frac{1}{d-c}\right) \qquad \text{for } a \le x \le b \text{ and } c \le y \le d,$$

and $f_{X,Y}(x,y) = 0$ otherwise.

We can also have X and Y Uniform on non-rectangular regions too.

Example 31.7. For instance, if X and Y are Uniform on a circular dartboard of radius 2, then

$$f_{X,Y}(x,y) = 1/(4\pi) \qquad \text{for } x, y \text{ on the dartboard,}$$

and $f_{X,Y}(x,y) = 0$ otherwise.

The reason that $f_{X,Y}(x,y) = 1/(4\pi)$ on the dartboard is that—in general—the density of a Uniform random variable defined on a region is equal to 1 divided by the length, area, volume, etc., of the region. That way, when integrating the density (which is constant) over the length, area, volume, of the region, the integral will evaluate to 1. In this case, the circular-shaped dartboard has area 4π, so the density must be $f_{X,Y}(x,y) = 1/(4\pi)$ on the circle.

If, for instance, D represents the distance from the location of the dart to the center of the board, then

$$F_D(a) = P(D \le a) = \frac{\text{area of circle of radius } a}{\text{area of circle of radius } 2} = \frac{\pi a^2}{\pi 2^2} = a^2/4,$$

for $0 \le D \le 2$. We can use such a cumulative distribution function to find out, for instance, about the density of D, or the expected value of D, etc.

Example 31.8. When a joint density is constant, then the conditional densities (e.g., if the value of one random variable is given) are uniform too. For instance, in Example 31.7, if $X = 0$ is given, then the conditional density of Y is uniform on the interval $[-2, 2]$, i.e.,

$$f_{Y|X}(y \mid 0) = 1/4 \qquad \text{for } -2 \leq y \leq 2.$$

As another example with the dartboard, if $Y = 1$ is given, then the conditional density of X is uniform on the interval $[-\sqrt{3}, \sqrt{3}]$, i.e.,

$$f_{X|Y}(x \mid 1) = \frac{1}{2\sqrt{3}} \qquad \text{for } -\sqrt{3} \leq y \leq \sqrt{3}.$$

For some other examples of this phenomenon, refer to Example 27.2 (where $f_{X|Y}(x \mid 2) = 1/8$ for $0 \leq x \leq 8$) and Example 27.3 (where $f_{X|Y}(x \mid 2) = \frac{1}{2\sqrt{5}}$ for $-\sqrt{5} \leq x \leq \sqrt{5}$).

Remark 31.9. Constant joint densities and conditional distributions

If X, Y have a constant joint density (i.e., a joint Uniform distribution), then the conditional densities are constant (i.e., Uniform) too.

Example 31.10. Now consider two students who each choose—independently, and Uniformly—numbers between 0 and 10.

Let X the be first student's choice, and let Y be the second student's choice. Since X and Y are independent, we can multiply to get the joint density of X and Y:

$$f_{X,Y}(x, y) = 1/100 \qquad \text{for } 0 \leq x \leq 10,\ 0 \leq y \leq 10,$$

and $f_{X,Y}(x, y) = 0$ otherwise.

Suppose that we want to find the expected value of the minimum of the two numbers.

Method #1: To find the expected value, we can integrate

$$\mathbb{E}(\min(X, Y)) = \int_0^{10} \int_0^{10} \min(x, y) \frac{1}{100}\, dy\, dx$$

It is difficult to know whether x or y is the minimum when we integrate over all x's and y's at once, so we suggest breaking the integral into two pieces, depending on whether Y or X is actually the minimum. In Figure 31.3, we show these two regions.

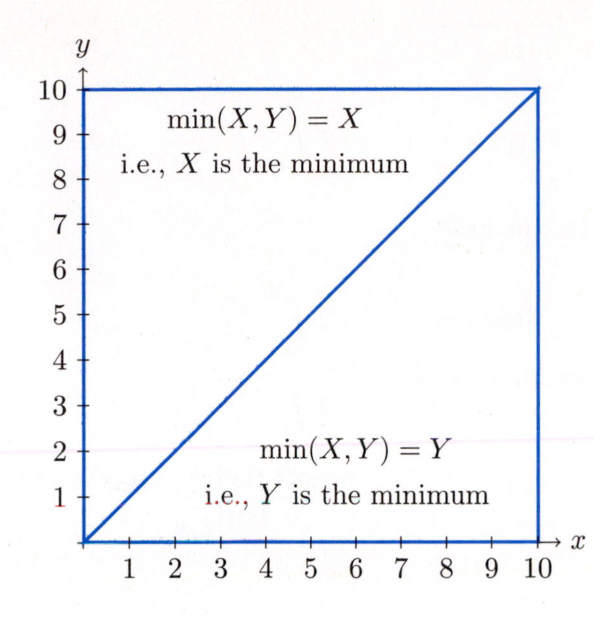

FIGURE 31.3: The regions where $\min(X, Y) = X$, and where $\min(X, Y) = Y$.

Now we integrate $\min(X, Y)f_{X,Y}(x, y)$ over the entire square, shown in Figure 31.3. When we are in the lower triangle, where $\min(X, Y) = Y$, then

$$\min(X, Y)f_{X,Y}(x, y) = (y)(1/100).$$

When we are in the upper triangle, where $\min(X, Y) = X$, then

$$\min(X, Y)f_{X,Y}(x, y) = (x)(1/100).$$

We integrate over the range where Y is the min and then over the range where X is the min:

$$\mathbb{E}(\min(X, Y)) = \int_0^{10}\int_0^x \min(x, y)\frac{1}{100}\,dy\,dx + \int_0^{10}\int_x^{10} \min(x, y)\frac{1}{100}\,dy\,dx$$

and we get

$$\mathbb{E}(\min(X, Y)) = \int_0^{10}\int_0^x y\frac{1}{100}\,dy\,dx + \int_0^{10}\int_x^{10} x\frac{1}{100}\,dy\,dx$$

If we wanted to, we could switch the bounds of integration on the second integral, so that we integrate over x on the inside and over y on the outside. The reasons will be clear in a brief moment:

$$\mathbb{E}(\min(X, Y)) = \int_0^{10}\int_0^x y\frac{1}{100}\,dy\,dx + \int_0^{10}\int_0^y x\frac{1}{100}\,dx\,dy$$

Finally, we see that we have essentially just written the same integral twice, just swapping letters the second time, so we can save ourselves some time by just writing, more simply:

$$\mathbb{E}(\min(X,Y)) = 2\int_0^{10}\int_0^x y\frac{1}{100}\,dy\,dx$$

Finally, we solve the integral

$$\begin{aligned}\mathbb{E}(\min(X,Y)) &= 2\int_0^{10} \frac{y^2}{2}\frac{1}{100}\Big|_{y=0}^{x}\,dx\\ &= 2\int_0^{10}\frac{x^2}{2}\frac{1}{100}\,dx\\ &= 2\,\frac{x^3}{6}\frac{1}{100}\Big|_{x=0}^{10}\\ &= 10/3\end{aligned}$$

Method #2: Another possible method is to always write U as the smaller of the two values X and Y, and write V as the larger of the two values X and Y. In other words,

$$U = \min(X,Y),$$

and

$$V = \max(X,Y).$$

Then U and V must have constant density on the triangle where $0 \le U \le V \le 10$, since this joint density is induced by the constant joint density for X and Y. So the density must be $1/50$ because the density must integrate to 1 over the whole region (which has area 50). In other words,

$$f_{U,V}(u,v) = 1/50 \qquad \text{for } 0 \le u \le v \le 10,$$

and $f_{U,V}(u,v) = 0$ otherwise. This triangle is shown in Figure 31.4. So the

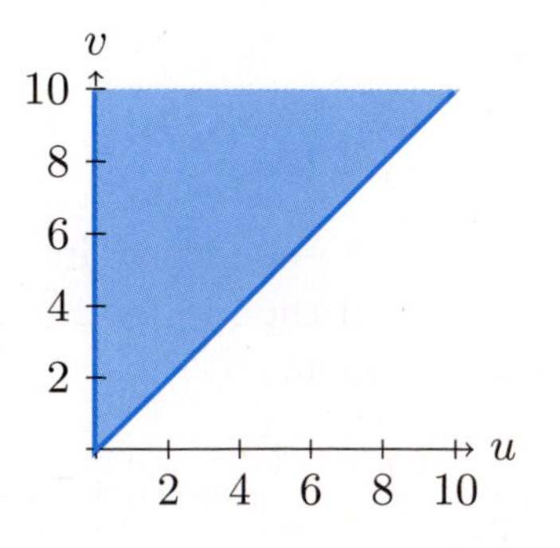

FIGURE 31.4: The region where $0 \le U \le V \le 10$.

integral for the expected value becomes:

$$\begin{aligned}
\mathbb{E}(\min(X, Y)) &= \mathbb{E}(U) \\
&= \int_0^{10} \int_0^{v} (u) \left(\frac{1}{50}\right) du\, dv \\
&= \int_0^{10} \left(\frac{u^2}{2}\right) \left(\frac{1}{50}\right)\Big|_{u=0}^{v} dv \\
&= \int_0^{10} \left(\frac{v^2}{2}\right) \left(\frac{1}{50}\right) dv \\
&= \left(\frac{v^3}{6}\right) \left(\frac{1}{50}\right)\Big|_{v=0}^{10} \\
&= 10/3,
\end{aligned}$$

which agrees with our result from the first method.

31.3 Linear Scaling of a Uniform Random Variable

Theorem 31.11. Linear scaling of a Uniform random variable

If X is a Uniform random variable on the interval $[s, t]$, and if a, b are constants, then $aX + b$ is a Uniform random variable too.

If $a > 0$, then $aX + b$ is Uniform on the interval $[as + b, at + b]$.

If $a < 0$, then the interval is reversed, i.e., $aX + b$ is Uniform on the interval $[at + b, as + b]$.

To see that this is true, we show the result for $a > 0$. Since $s < X < t$, then $as + b < aX + b < at + b$. So if we write $Y = aX + b$, we see that $f_Y(y) = 0$ for y outside the interval $[as + b, at + b]$. For y inside the interval $[as + b, at + b]$, we compute

$$F_Y(y) = P(Y \le y) = P(aX + b \le y) = P\left(X \le \frac{y - b}{a}\right) = \frac{\frac{y-b}{a} - s}{t - s},$$

and thus

$$f_Y(y) = \frac{d}{dy} F_Y(y) = \frac{1/a}{t - s} = \frac{1}{a(t - s)}.$$

So the density of Y is constant on the interval $[as+b, at+b]$ and is 0 otherwise. Thus Y is a Uniform random variable on the interval $[as + b, at + b]$.

Example 31.12. With this information in hand, we return to Exercise 29.23 and calculate much more than just the expected value of the purchase price:

Geoffrey does not like to be low on gas, so he randomly stops to fill up his tank. He has a 14-gallon tank, and the current price of gas is $2.75 per gallon. Whenever he stops to buy gas, he also buy a candy bar for $1.30. If X is the amount of gas (in gallons) in his tank when he stops for a purchase, then $f_X(x) = 1/14$ for $0 \leq x \leq 14$, and $f_X(x) = 0$ otherwise. He always fills the tank, so he will always buy $14 - X$ gallons.

The price of the purchase is

$$Y = (2.75)(14 - X) + 1.30 = 39.80 - 2.75X.$$

Since X is Uniform on $[0, 14]$, it follows that Y is Uniform on the interval

$$[39.80 - (2.75)(14),\ 39.80 - (2.75)(0)] = [1.30, 39.80].$$

Thus, we confirm the fact that the expected value of the purchase price is

$$\mathbb{E}(Y) = \frac{39.80 + 1.30}{2} = 20.55.$$

We know much more now; for instance, we know that the variance of Y is

$$\text{Var}(Y) = \frac{(39.80 - 1.30)^2}{12} = 123.5208333.$$

and we know that the density of Y is

$$f_Y(y) = \frac{1}{39.80 - 1.30} = \frac{1}{38.50} \qquad \text{for } 1.30 \leq y \leq 39.80,$$

and $f_Y(y) = 0$ otherwise.

31.4 Exercises

31.4.1 Practice

Exercise 31.1. Rope checks. In a certain manufacturing process, an automated quality control computer checks 10 yards of rope at a time. If no defects are detected in that 10-yard section, that portion of the rope is passed on. However, if there is a defect detected, a person will have to check the rope over more carefully to determine where (measured from the left side in yards) the defect is. If exactly 1 defect is detected in a rope section, we would like to find the probabilities for its location.

a. Why is this a Continuous Uniform problem?

b. What does X represent in this scenario?

c. What are the parameters in this scenario?

d. What is the expected value for the location of the defect?

e. What is the standard deviation?

f. What is the probability density function? Write it in function notation, and also graph it.

g. What is the cumulative distribution function? Write it in function notation, and also graph it.

h. Find $P(X > 8)$.

i. Find $P(2.3 \leq X \leq 5.2)$

j. Find $P(X < 2|X < 5)$

Exercise 31.2. **Harry Potter and the Missed Exit.** You are driving down an interstate when you suddenly realize you have missed your exit because you were busy listening to an exciting chapter of a Harry Potter book on CD. According to the CD case, the chapter lasts for 14 minutes. Assume you are driving 60 miles per hour (in other words, 1 mile per minute). You could have passed your exit anytime over the past 14 minutes. You want to know about the probabilities for where it was exactly (how far back in miles) as you turn around to head back to find it.

a. Why is this a Continuous Uniform problem?

b. What does X represent in this scenario?

c. What are the parameters in this scenario?

d. What is the expected time it will take for you to find the missed exit?

e. What is the standard deviation?

f. What is the probability density function? Write it in function notation, and also graph it.

g. What is the cumulative distribution function? Write it in function notation, and also graph it.

h. Find the probability that the exit was passed within a mile of the half-way point.

i. Find the probability that the exit was passed in the first minute of the chapter.

j. You are pretty sure that you were paying attention to the road until the last 4 minutes of the chapter when Voldemort tried to attack Harry, so within that 4-minute interval, what is the probability that you missed the exit within the last 30 seconds?

k. How many total miles do you expect to have to drive out of your way to get back to the missed exit (miles past the exit $\times 2$)?

l. Also find the standard deviation in the total miles out of your way that you had to drive.

m. If you get reimbursed by your company 43 cents a mile for your trip plus the $3.20 you spent on a grande mocha at Starbucks (conveniently located at the turn-around spot) to wake yourself up as you get turned around, how much is the expected amount of money this detour will cost the company?

n. What is the standard deviation in this amount of money that you get reimbursed in part m?

Exercise 31.3. **Network latency.** The network latency per request from your computer to a server is Uniform between 30 ms and 150 ms.

a. What is the probability that a particular request is 60 ms or less?

b. If your computer makes 320 requests to the server, how many are expected to have a latency that is 60 ms or less?

Exercise 31.4. **A fly is flying around.** You believe that there is a fly somewhere less than 6 feet away from you. If you believe that he is located Uniformly in a circle of radius 6 feet away from you, what is the probability that he is more than 2 feet away from you?

Exercise 31.5. **Bird on a wire.** A bird lands at a location that is Uniformly distributed along an electrical wire of length 150 feet. The wire is stretched tightly between two poles. What is the probability that the bird is 20 feet or less from one or the other of the poles?

Exercise 31.6. **Touchdown.** A quarterback needs to throw the ball quickly, and in his haste, the location it lands is Uniformly distributed within a 30×120 sq. ft. area in the endzone. The receiver can only catch the ball the ball lands within 7 feet of him, i.e., within a circle of radius 7 feet, completely contained in the endzone. What is the probability that the receiver catches the ball? (Assume that the receiver is positioned so that a circle of 7 feet around him is completely contained within the endzone.)

Exercise 31.7. **Target art.** A wall in a room is 108 inches tall and 132 inches wide. There is a painting on the wall that is 18 inches by 24 inches. If a tennis ball is accidentally flung at the wall, and the location where it lands is Uniformly distributed on the wall, what is the probability that the tennis ball hits the painting?

Exercise 31.8. **Solar cell.** A solar cell measures 10 inches by 10 inches, and it has 4 non-overlapping regions that generate electricity. Each of these regions is

4 inches by 4 inches in size (there is a border between the electricity-generating regions). What is the probability that a photon that hits the solar cell at a location which is Uniformly distributed on the solar cell will actually hit one of these electricity-generating regions?

31.4.2 Extensions

Exercise 31.9. Cab ride. Suppose that the length X of a randomly selected passenger's trip in a cab is Uniformly distributed between 5 and 30 miles. The charge induced for such a trip, in dollars is $Y = 2.50X + 3.00$.

a. How much should a randomly selected customer expect to pay?

b. What is the standard deviation of the amount that a randomly selected customer pays?

Exercise 31.10. There's a fly in the house! Consider the two-dimensional house in Figure 31.5. Suppose that a fly is found somewhere in the house, with a location that is Uniformly distributed.

a. Find the conditional density of Y given that $X = 2$.

b. Find the conditional probability that the fly is upstairs, i.e., $Y > 10$, given that $X = 2$, i.e., find $P(Y > 10 \mid X = 2)$.

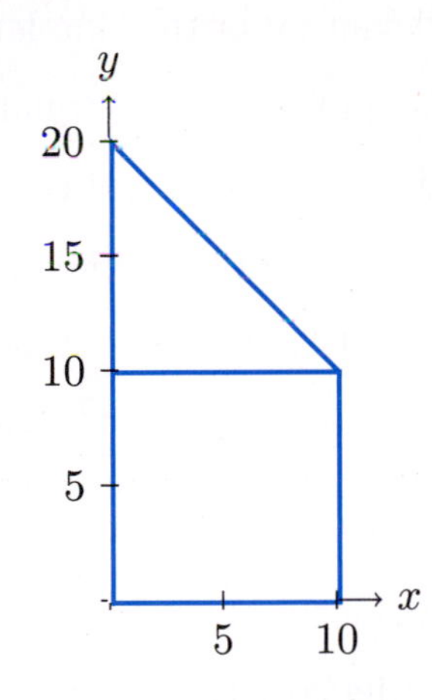

FIGURE 31.5: Picture of the house.

Exercise 31.11. There's still a fly in the house! Consider the two-dimensional house in Figure 31.5. Suppose that a fly is still found somewhere in the house, with joint density

$$f_{X,Y}(x, y) = 1/150 \qquad \text{for } x, y \text{ in the house,}$$

and $f_{X,Y}(x, y) = 0$ otherwise. (We no longer assume $X = 2$.)

a. Find $\mathbb{E}(X)$.

b. Find $\text{Var}(X)$.

c. Find $\mathbb{E}(Y)$.

d. Find $\text{Var}(Y)$.

Exercise 31.12. **Long jump.** Suppose that a particular long jumper assumes that each of his jumps are Uniformly distributed between 6.5 and 7.2 meters. He is happy whenever he jumps 7 meters or more. If he makes 10 such jumps, what is the probability that he is happy with exactly 4 of his jumps?

Exercise 31.13. **Student arrivals.** Five students will arrive to the classroom during the next 2 minutes. Assume that their arrival times are independent, and each arrival time is Uniformly distributed between 0 and 2 minutes. What is the probability that exactly two of the students arrive during the next 30 seconds?

Exercise 31.14. **Manufactured cubes.** A machine manufactures cubes with a side length which varies Uniformly over the interval $[0.2, 0.3]$ in millimeters. For the following problems, make sure you use the correct units. (Assume the sides of the base and the height are all the same.)

a. What is the expected side length?

b. What is the standard deviation of the side length?

c. What is the expected area of one of the square bases?

d. What is the standard deviation of one of the square bases?

e. What is the expected volume of one of the cubes?

f. What is the standard deviation of the volume of one of the cubes?

In the rest of the problem, assume that the cost to make the cubes is 12 cents per cubic millimeter and 6 cents for the general cost (labor, electricity, etc.) per cube.

f. What is the expected cost for making 1 cube?

g. What is the variance in the cost for making 1 cube?

h. What is the expected cost for making 10 cubes?

i. What is the variance in the cost for making 10 cubes?

Exercise 31.15. **Dartboard.** Kelly throws a dart at a circular dartboard of radius 3 feet. Let X and Y denote the location where the dart lands. Assume that $-3 \leq X \leq 3$ and $-3 \leq Y \leq 3$ and $X^2 + Y^2 \leq 9$, i.e., the dart lands on the dartboard. Moreover, assume that the dart's location is Uniform on the dartboard, i.e.,

$$f_{X,Y}(x, y) = 1/(9\pi) \qquad \text{if } x, y \text{ are on the dartboard, i.e., } x^2 + y^2 \leq 9,$$

and $f_{X,Y} = 0$ otherwise.

Let $D = \sqrt{X^2 + Y^2}$ be the distance from the dart to the center of the dartboard. Find $\mathbb{E}(D)$.

Exercise 31.16. Bugs on a wire. An automatic camera is taking pictures of insects that land on a wire. The distance between each insect and the camera (at the time of the picture) is random, with constant density on the interval $[1, 6]$, measured in feet. (Assume that the locations of the insects are independent.) A "very good" picture is taken when an insect is less than 3 feet away. What is the probability that none of the camera's 3 attempted pictures are "very good"?

Exercise 31.17. More bugs on a wire. For the scenario in Exercise 31.16, if X, Y, Z are the distances of the camera from the insect for his first, second, and third pictures, respectively, then find the expected value of the smallest of these three distances, i.e., find $\mathbb{E}(\min(X, Y, Z))$.

Exercise 31.18. Laundry cycles. You are doing two loads of laundry: one was just put in the washer, the other was simultaneously put in the dryer. You know it takes the dryer between 25 and 40 minutes to completely dry your clothes, and the washer takes between 27 and 37 minutes to complete its cycle. Those distributions are independent and Uniform on their respective time intervals. What is the probability the dryer is already done if the washer just finished?

Exercise 31.19. Pencil darts. In a game of chance, a circle of radius 3 inches is drawn on a piece of paper that is 8.5×11 sq. in. While blindfolded, the student tries to place her pencil tip inside the circle. She wins \$3 if she is successful, or loses \$1 if unsuccessful. (If she misses the paper altogether, she can try again until she hits the paper.) Assume that when she hits the paper, the location of her pencil tip is Uniformly distributed on the paper's surface. What are her expected winnings/loses in this game?

Exercise 31.20. Clay pots. Emmanuel makes pots out of clay. The clay costs \$5.00 per pound. Each pot is made from a random amount X of clay (in pounds), Uniformly distributed between 1.9 and 2.7 pounds. There is also a charge to bake the pot in a kiln (a kiln is a big oven for making clay get very hard). The charge to use the kiln to harden a pot is \$0.40. What is the probability that he spends more than \$12.00 altogether (cost of clay plus cost of using the kiln) to make a pot of clay?

Exercise 31.21. Lost toy. A child is playing in a sandbox and loses a very small toy. If the sand box is a square the measures 5 feet by 5 feet, what is the probability that the toy is actually located in a right triangle at the southwest corner of the sandbox, with sides measuring 1.5 feet by 2.5 feet?

Exercise 31.22. Active child. Suppose that a piece of chalk is broken into two pieces, with the breaking point Uniformly distributed along the length of the chalk, which is 8 cm. What is the probability that, as a result, one or the other of the pieces is shorter than 1 cm?

Exercise 31.23. Mole rat. The location of a mole rat is Uniform inside a circular enclosure that has diameter 40 feet. What is the probability that the mole rat is within 2 feet from the edge of the enclosure?

Exercise 31.24. Broken chalk. Assume a mother's child has a random location that is Uniformly distributed across the 80 foot by 120 foot playground shown in Figure 31.6 below.

a. What is the probability that the child is in the grass?

b. On the swings?

c. On the slides?

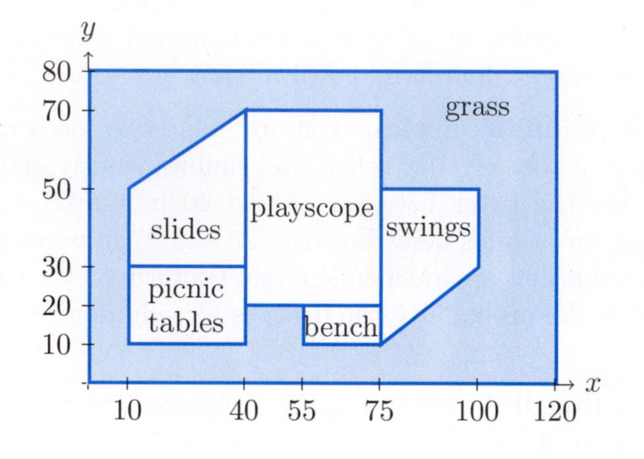

Figure 31.6: A playground.

Exercise 31.25. Crayon factory. In a crayon factory, wax is rolled into cylinders, each of which are exactly 3.6 inches long, but the radius (in inches) is a Uniform random variable X on the interval $[0.15, 0.17]$. Find the probability that the volume of a crayon exceeds 0.30 cubic inches. Hint: The volume of a cylinder is $\pi r^2 \ell$ where r is the radius and ℓ is the length.

Exercise 31.26. Let X, Y, Z be independent and uniformly distributed on the interval $[0, 10]$. Find the probability that Y is the middle value, i.e., find $P(X < Y < Z \text{ or } Z < Y < X)$.

Exercise 31.27. Suppose X, Y have constant joint density on the triangle with vertices at $(0, 0)$, $(3, 0)$, and $(0, 3)$. Find $\mathbb{E}(X)$.

31.4.3 Advanced

Exercise 31.28. A fly near the wall. There is a fly randomly located Uniformly in a room that is 10 feet high, 14 feet long, and 13 feet wide. What is the probability that the fly is within 1 foot of the walls, ceiling, or floor?

Exercise 31.29. Let X_1, X_2, X_3 be independent continuous random variables, each Uniformly distributed in the interval $[0, 10]$. Let Y denote the middle of the three values. Find the cumulative distribution function $F_Y(a) = P(Y \leq a)$ of the random variable Y.

Exercise 31.30. If X has a Continuous Uniform distribution on the interval $(1, 10)$, find $\mathbb{E}(\ln X)$.

Chapter 32

Exponential Random Variables

Waiting is painful. Forgetting is painful. But not knowing which to do is the worst kind of suffering.

—*By the River Piedra I Sat Down and Wept* by Paulo Coelho (Harper, 1994)

What is the expected time until your best friend sends you a text message? When a mother is waiting for her three children to call her, what is the probability that the first call will arrive within the next 5 minutes?

32.1 Introduction

Exponential random variables are often waiting times for the next event to happen. For instance, the time until the telephone rings, or a guest arrives, or black car passes, etc., are all modeled by Exponential random variables. We saw examples of Exponential random variables, for example, in Example 25.5 and Example 28.8. Exponential random variables are always positive. (It would not make sense, for instance, for a waiting time to be negative.) Exponential random variables, as we will see, also have the memoryless property, as did Geometric random variables. In many ways, Exponential random variables are a continuous version of Geometric random variables. Besides being memoryless, the density of an Exponential random variable decreases exponentially in x, just as the mass of a Geometric random variable decreased exponentially.

The notation for an Exponential random variable looks like:

$$X \sim \text{Exponential}(\lambda).$$

Exponential random variables

The common thread: Time until the 1st success occurs. For example, the time until the 1st customer enters a shop, or the telephone rings, or one's girlfriend/boyfriend arrives at the door, or until the 1st black car drives by.

Things to look for: The waiting time until the first event occurs, or gaps between events.

The variable:

$$X = \text{time until the next event occurs, } X \geq 0$$

The parameters:

$$\lambda = \text{the average rate, e.g., number of arrivals per time period.}$$

When the time between arrivals is exponential, the number of arrivals in a fixed time interval is Poisson with the mean λ.

Density:

$$f_X(x) = \begin{cases} \lambda e^{-\lambda x}, & x > 0, \\ 0 & \text{otherwise} \end{cases}$$

CDF:

$$F_X(x) = \begin{cases} 1 - e^{-\lambda x}, & x > 0, \\ 0 & \text{otherwise} \end{cases}$$

Expected value formula:

$$\mathbb{E}(X) = 1/\lambda$$

Variance formula:

$$\text{Var}(X) = 1/\lambda^2$$

32.2 Average and Variance

We use integration by parts to show why Exponential random variables have expected value $1/\lambda$. We use $x = u$ and $dx = du$, as well as $dv = \lambda e^{-\lambda x}$ and $v = -e^{-\lambda x}$, to get

$$\mathbb{E}(X) = \int_{-\infty}^{\infty} x f_X(x)\, dx = \int_0^{\infty} x \lambda e^{-\lambda x}\, dx = (-xe^{-\lambda x})\Big|_0^{\infty} - \int_0^{\infty} -e^{-\lambda x}\, dx.$$

We observe that $-xe^{-\lambda x} = 0$ for $x = 0$, and $-xe^{-\lambda x} \to 0$ as $x \to \infty$. Thus

$$\mathbb{E}(X) = \int_0^{\infty} e^{-\lambda x}\, dx = -\left.\frac{e^{-\lambda x}}{\lambda}\right|_0^{\infty} = 1/\lambda.$$

The second moment is calculated similarly. We again use integration by parts, with $x^2 = u$ and $2x\,dx = du$, as well as $dv = \lambda e^{-\lambda x}$ and $v = -e^{-\lambda x}$, to get

$$\begin{aligned}\mathbb{E}(X^2) &= \int_{-\infty}^{\infty} x^2 f_X(x)\,dx \\ &= \int_0^{\infty} (x^2)(\lambda)e^{-\lambda x}\,dx \\ &= (-x^2 e^{-\lambda x})\Big|_0^{\infty} - \int_0^{\infty} -2xe^{-\lambda x}\,dx.\end{aligned}$$

Since $-x^2e^{-\lambda x} = 0$ for $x = 0$, and since $-x^2e^{-\lambda x} \to 0$ as $x \to \infty$, then $\mathbb{E}(X^2) = \int_0^\infty 2xe^{-\lambda x}\,dx$. We can multiply and divide by 2 and λ, to get what remains into the form of the first moment, i.e.,

$$\mathbb{E}(X^2) = \int_0^{\infty} 2xe^{-\lambda x}\,dx = \frac{2}{\lambda}\int_0^{\infty} x\lambda e^{-\lambda x}\,dx = \frac{2}{\lambda}\mathbb{E}(X) = \frac{2}{\lambda^2}.$$

So the variance is:

$$\text{Var}(X) = \frac{2}{\lambda^2} - \frac{1}{\lambda^2} = 1/\lambda^2.$$

Some examples of the densities and CDFs for Exponential random variables with various values of λ are given in Figure 32.1. These examples show that no

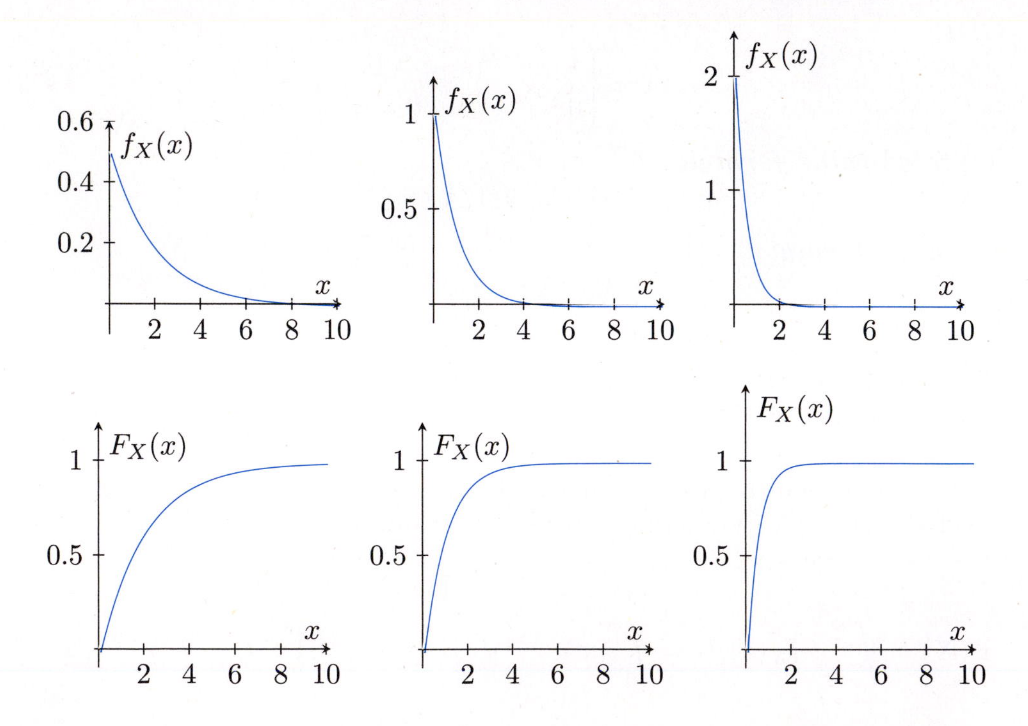

Figure 32.1: Upper row: The density $f_X(x)$ of an Exponential random variable with parameter λ equal to 1/2, 1, or 2 (respectively). Lower row: The analogous CDF $F_X(x)$ for λ equal to 1/2, 1, or 2 (respectively).

matter what the λ parameter is, the density starts at λ when $x = 0$ and then quickly moves closer to 0 as $x \to \infty$. The CDF starts at 0 but quickly climbs close to 1 as $x \to \infty$. The density and CDF curves are steeper for larger values of λ.

Example 32.1. The time between fatal car accidents on a stretch of desolate highway between two cities was found to follow an Exponential distribution with a mean of one accident every 44 days. If an accident has occurred today (the count starts over), the sheriff's office is interested in when the next accident may occur.

a. What does X represent in this story? What values can X take?

The random variable X is the time until the next fatal accident. Since you can't have a negative waiting time, then $X > 0$.

b. Why is this an example of the Exponential distribution?

We know the rate these accidents occur, and we are waiting for the *next* accident (one event). We are measuring our wait in days (continuous), not in specific trials (discrete).

c. What is the parameter for this distribution?

The parameter is λ, which is the number of accidents/time, i.e., 1 accident/44 days or 0.02273 accidents/day.

d. How long should the sheriff's office expect to wait for the next accident to occur?

$$\mathbb{E}(X) = \frac{1}{\lambda} = \frac{1}{1 \text{ accident}/44 \text{ days}} = 44 \text{ days/accident}.$$

e. What is the standard deviation in the time the sheriff's office will wait for the next accident to occur?

$$\text{Var}(X) = \frac{1}{(0.02273)^2} = 1936 \text{ days}^2$$

$$\sigma_X = \sqrt{1936} = 44 \text{ days}$$

f. Find the probability that the next accident occurs within the next 31 days.

$$P(X \leq 31) = F_X(31) = 1 - e^{-(0.02273)(31)} = 0.5057.$$

g. Today is August 1st. What is the probability that no accident will have occurred by August 9th?

This means that we will have to wait more than 8 days for the next accident to occur. So the probability is

$$P(X > 8) = 1 - F_X(8) = 1 - (1 - e^{-(0.02273)(8)}) = 1 - (1 - 0.8337) = 0.8337$$

h. You need an estimate of how long you would wait so that there is a 95% chance that the next accident would happen before that day. What is that cut-off length of time?

We want the time a such that

$$P(X \leq a) = 0.95,$$

or equivalently

$$F_X(a) = 0.95 = 1 - e^{-0.02273x}.$$

Thus $e^{-0.02273x} = 0.05$. Taking the natural logarithm of both sides, we have

$$-0.02273x = -2.996,$$

so $x = 131.7964$. In other words, we are 95% certain that the next accident will happen within the next 131.7964 days.

We saw an example of Exponential random variables in Example 26.7. The "lifetime" of a music player might not seem like a "waiting time," but it can be interpreted that way, i.e., the time we wait until the music player dies.

We saw another example of waiting times near the start of section 29.1, where X is the waiting time until the next bus arrives, given in minutes. That waiting time was exponential with $\lambda = 3/10$. So now we can immediately conclude that $\mathbb{E}(X) = 1/\lambda = 10/3$.

We also saw waiting times in Exercise 26.6, about the response time for a police car to an emergency call.

Example 32.2. Let X be the time (in minutes) that Maxine waits for a traffic light to turn green, and let Y be the time (in minutes, at a different intersection) that Daniella waits for a traffic light to turn green. Suppose that X and Y have joint density

$$f_{X,Y}(x, y) = 15e^{-3x-5y}, \qquad \text{for } x > 0 \text{ and } y > 0,$$

and $f_{X,Y}(x, y) = 0$ otherwise.

In this case, we notice that the joint density can be factored, so there are some constants c and d so that

$$f_X(x) = ce^{-3x} \qquad \text{for } x > 0,$$

and $f_X(x) = 0$ otherwise, and

$$f_Y(y) = de^{-5y} \qquad \text{for } y > 0,$$

and $f_Y(y) = 0$ otherwise. Since each of these densities must integrate to 1, we see that $c = 3$ and $d = 5$, and X and Y are independent Exponential random variables, with parameters 3 and 5, respectively.

32.3 Properties of Exponential Random Variables

32.3.1 Complement of the CDF

If X is an Exponential random variable with parameter λ, then the CDF of X is

$$F_X(x) = 1 - e^{-\lambda x} \qquad \text{for } x > 0,$$

and $F_X(x) = 0$ otherwise. Thus, for $a > 0$, we see that $P(X > a)$ has an especially nice form:

$$P(X > a) = 1 - P(X \le a) = 1 - (1 - e^{-\lambda a}) = e^{-\lambda a}.$$

32.3.2 Memoryless Property of Exponential Random Variables

Just like the Geometric distribution, the Exponential distribution has the memoryless property. The memoryless property helps streamline our ability to calculate conditional distributions for Exponential random variables. (Recall from Chapter 16 that the Geometric distribution also has a memoryless property.) For example, if X is an Exponential random variable, then $P(X > 5 \mid X > 2) = P(X > 3)$. Intuitively, if we have already waited at least two minutes for an event to occur, then the conditional probability that we wait at least five minutes altogether is equal to the probability that, starting from right now, we will wait at least three additional minutes until the event occurs. One important difference is that, since the Exponential distribution is continuous, then $P(X \ge 5)$ and $P(X > 5)$ are exactly the same value, because $P(X = 5) = 0$.

As another example, if we know that a waiting time X is bigger than 7, and we want the probability that X is bigger than 11, then we only need to know that—after 7 units of time have passed—we have to wait at least 4 additional units of time. So, for example, it would make sense if

$$P(X > 11 \mid X > 7) = P(X > 4).$$

We compute

$$\begin{aligned} P(X > 11 \mid X > 7) &= \frac{P(X > 11 \text{ and } X > 7)}{P(X > 7)} \\ &= \frac{P(X > 11)}{P(X > 7)} \\ &= \frac{e^{-11\lambda}}{e^{-7\lambda}} \\ &= e^{-4\lambda} \\ &= P(X > 4) \end{aligned}$$

This phenomenon is true in general. If a and b are positive constants, and we are given $X > a$, then the probability that $X > a + b$ is just equal to the (unconditional) probability that $X > b$:

$$\begin{aligned} P(X > a + b \mid X > a) &= \frac{P(X > a + b \text{ and } X > a)}{P(X > a)} \\ &= \frac{P(X > a + b)}{P(X > a)} \\ &= \frac{e^{-(a+b)\lambda}}{e^{-a\lambda}} \\ &= e^{-b\lambda} \\ &= P(X > b) \end{aligned}$$

This is called the "memoryless property," which we also saw was true for Geometric random variables. If we already know that X is bigger than a, then the conditional probability (given X is bigger than a) that X is bigger than $a + b$ is just equal to the unconditional probability that X is bigger than b. This is true with waiting times in life too, in many cases. For instance, if we have already waited 10 minutes for the phone to ring, then the probability that we wait a total of 12 minutes or less is just equal to the unconditional probability of waiting 2 minutes or less from the outset—the previous 10 minutes do not affect the future waiting time. (We are assuming that we do not know when the phone will ring, i.e., that there is not a planned calling time in advance. So this example works best, for example, if we imagine that we are answering a telephone in an office, for a telethon, etc., when the timing of the next phone call is variable and is not planned ahead of time.)

Theorem 32.3. Memoryless property of an Exponential Random Variable

If X is an Exponential random variable with parameter λ, then for any positive constants a and b,

$$P(X > a + b \mid X > a) = P(X > b).$$

Example 32.4. (continued from Example 32.1) The time between accidents for all fatal car accidents on a stretch of highway between two cities was found to follow an Exponential distribution with a mean of one accident every 44 days. If an accident has occurred today (the count starts over), the sheriff's office is interested in when the next accident may occur.

Today is August 1st. Given that no accident has occurred by August 25th, what is the probability that no accident will have occurred by September 2nd?

No accident occurring by August 25th means that there is no accident in the next 24 days, so our wait time will be more than 24 days, so $X > 24$.

No accident occurring by September 2nd means that there is no accident in the next 32 days so our wait time will be more than 32 days.

So the conditional probability is

$$\begin{aligned} P(X > 32 \mid X > 24) &= \frac{P(X > 32 \text{ and } X > 24)}{P(X > 24)} \\ &= \frac{P(X > 32)}{P(X > 24)} \\ &= \frac{1 - F_X(32)}{1 - F_X(24)} \\ &= \frac{1 - (1 - e^{-(0.02273)(32)})}{1 - (1 - e^{-(0.02273)(24)})} \\ &= \frac{e^{-(0.02273)(32)}}{e^{-(0.02273)(24)}} \\ &= e^{-(0.02273)(8)} \\ &= 0.8337 \end{aligned}$$

That was a lot of math to go through. However, thanks to the memoryless property, the result is the same as if you rename the start time as August 25th instead of August 1st.

$$P(X > 32 \mid X > 24) = P(X > 8) \qquad \text{by the memoryless property,}$$

so by the calculation from Example 32.1g, we have

$$P(X > 32 \mid X > 24) = P(X > 8) = 1 - F_X(8) = 0.8337.$$

32.3.3 Minimum of Independent Exponential Random Variables

We already saw, in Exercise 25.17, that the *maximum* of two independent Exponential random variables does not have any type of form that we have seen before... but the *minimum* of two independent Exponential random variables is an Exponential random variable too! In other words, if X and Y are independent Exponential random variables with parameters λ_1, λ_2, respectively, and we define $Z = \min(X, Y)$, then Z is Exponential random variable as well. To see this, we notice that Z must be positive since X and Y are both positive; thus $F_Z(z) = 0$ for $z \leq 0$. Now consider $z > 0$. We see that

$$\begin{aligned} F_Z(z) &= P(Z \leq z) \\ &= 1 - P(Z > z) \\ &= 1 - P(X > z \text{ and } Y > z) \\ &= 1 - P(X > z)P(Y > z) \\ &= 1 - e^{-\lambda_1 z}e^{-\lambda_2 z} \\ &= 1 - e^{-(\lambda_1+\lambda_2)z} \end{aligned}$$

So Z is an Exponential random variable with parameter $\lambda_1 + \lambda_2$. Applying this rule over and over, we get the following result:

Theorem 32.5. Minimum of a collection of Exponential Random Variables

If $X_1, X_2, \ldots, X_n$ are independent Exponential random variables with parameters $\lambda_1, \lambda_2, \ldots, \lambda_n$, respectively, and we define

$$Z = \min(X_1, X_2, \ldots, X_n),$$

then Z is also an Exponential random variable, with parameter $\lambda_1 + \cdots + \lambda_n$.

Example 32.6. Now consider two random variables, X and Y, with joint distribution

$$f_{X,Y}(x, y) = 36e^{-5x-4y} \qquad \text{for } 0 < x < y,$$

and $f_{X,Y}(x, y) = 0$ otherwise. Then X has an exponential distribution but Y does not.

We compute, for $y > 0$, the density of Y:

$$f_Y(y) = \int_0^y 36e^{-5x-4y}\,dx = \frac{36}{5}(e^{-4y} - e^{-9y}).$$

The density of Y tells us that Y is not exponentially distributed. On the other hand, for $x > 0$, when we compute the density of X,

$$f_X(x) = \int_x^\infty 36e^{-5x-4y}\,dy = 9e^{-9x},$$

so X is an Exponential random variable with parameter $\lambda = 9$. Finally, since $0 < X < Y$, we know that X and Y are dependent.

The expected value of Y is

$$\mathbb{E}(Y) = \int_0^\infty (y)\left(\frac{36}{5}(e^{-4y} - e^{-9y})\right) dy = \frac{13}{36}.$$

Since X is Exponential with parameter 9, we do not need to evaluate an integral to find the expected value. We know that the expected value is the inverse of the parameter, i.e., $\mathbb{E}(X) = 1/9$.

32.3.4 Poisson Process

There is an important link between the Poisson distribution (Chapter 18) and the Exponential distribution in this chapter. There is a name for the process of combining the Poisson, Exponential, and Gamma (next chapter) distributions. (A sequence of events that have independent Exponential waiting times between consecutive events is called a Poisson process.) Here we discuss how these distributions are related.

Consider a process in which events occur, such as the times when cars pass by an observer, when the telephone rings, when a text message arrives, etc. If the times between consecutive occurrences are independent Exponential random variables, each with parameter λ, then such a process is called a Poisson process. This concept is fundamental in courses on Stochastic Processes, but we mention here a basic property of such processes: The number of occurrences in a fixed time interval of length t is a Poisson random variable with parameter λt.

Example 32.7. Suppose that the times between the arrival of consecutive emails are independent Exponential random variables, each of which has an average of 1/2 a minute, i.e., the parameter is $\lambda = 2$. So we expect 2 emails to arrive per minute. The expected number of emails to arrive in a fixed 10 minute time interval (say, between 1:35 PM and 1:45 PM) is a Poisson random variable with mean $(2)(10) = 20$.

Example 32.8. A traffic engineer has studied a lonesome desert highway, and she believes that the average time between consecutive cars at an observation point is 12 minutes, and these waiting times are independent Exponential random variables. Thus, on average, 5 cars pass the observation point per hour. Moreover, the number of cars that pass the observation point per hour is Poisson with mean 5. The number of cars that pass during a two-hour period is Poisson with mean 10. The number of cars that pass during a 30-minute period is Poisson with mean 2.5 (even though the number of cars that pass is an integer-valued random variable, the parameter λ is allowed to be a non-integer).

The probability that exactly 4 cars pass between 1:10 PM and 2:10 PM is

$$\frac{e^{-5}5^4}{4!} = 0.1755.$$

The probability that exactly 13 cars pass between 1:35 PM and 3:35 PM is

$$\frac{e^{-10}10^{13}}{13!} = 0.0729.$$

The probability that exactly 2 cars pass between 4:30 PM and 5:00 PM is

$$\frac{e^{-2.5}2.5^2}{2!} = 0.2565.$$

32.3.5 Moments of an Exponential Random Variable (Optional)

We refer to $\mathbb{E}(X^k)$, for a fixed integer k, as the **kth moment** of X. For instance, for an Exponential random variable, the kth moment is:

$$\begin{aligned}
\mathbb{E}(X^k) &= \int_0^\infty (x^k)(\lambda e^{-\lambda x})\,dx \\
&= \Big(-(x^k)(e^{-\lambda x}) \\
&\quad -(kx^{k-1})(e^{-\lambda x})/\lambda \\
&\quad -((k)(k-1)x^{k-2})(e^{-\lambda x})/\lambda^2 \\
&\quad -((k)(k-1)(k-2)x^{k-3})(e^{-\lambda x})/\lambda^3 \\
&\quad -\cdots - k!(e^{-\lambda x})/\lambda^k \Big)\Big|_{x=0}^{\infty} \\
&= k!/\lambda^k.
\end{aligned}$$

Theorem 32.9. Moments of an Exponential Random Variable

If X is an Exponential random variable with parameter λ, then the kth moment of X is

$$\mathbb{E}(X^k) = k!/\lambda^k.$$

In particular, the expected value of X is

$$\mathbb{E}(X) = 1/\lambda,$$

and the second moment of X is

$$\mathbb{E}(X^2) = 2/\lambda^2,$$

so the variance of X is

$$\text{Var}(X) = \frac{2}{\lambda^2} - \left(\frac{1}{\lambda}\right)^2 = 1/\lambda^2.$$

32.4 Exercises

32.4.1 Practice

Exercise 32.1. Egg laying. Chickens at Rolling Meadows Farm lay an average of 18 eggs per day. The farmer has rigged a fancy monitoring device to the nesting boxes so that he can monitor exactly when the hens lay their eggs. Assume that no 2 eggs will be laid at exactly the same time and that the eggs (and chickens) are independent from each other. The farmer wants to know how long he will have to wait (in minutes) for the next egg to be laid if he starts monitoring at the first rooster crow in the morning.

a. Why is this an Exponential problem?

b. What does X represent in this scenario?

c. What is the parameter in units matching this specific question?

d. What is the expected length of time (in minutes) the farmer will have to wait for the first egg to be laid?

e. What is the standard deviation?

f. What is the probability density function for the wait time (in minutes) for the first egg? Write it in function form and also graph it.

g. What is the cumulative distribution function for the wait time (in minutes) for the first egg? Write it in function form and also graph it.

h. What is the probability the farmer will have to wait longer than it takes him to make his cup of coffee (the first 10 minutes after the rooster crows)?

i. What is the probability the first egg will be laid while he is milking the cow (between 30 minutes and an hour after the rooster crows)?

j. Given that the first egg did not come while he was making his coffee, what is the probability that the first egg will come while he is milking the cow?

k. The farmer is out doing chores and can't check the fancy egg monitor. If the farmer wants to wait to go to the henhouse to feed the hens until he is 90% sure the first egg has been laid, how long should he wait?

Exercise 32.2. Hurricanes. On the average, hurricane of category 4 or stronger (on the Stafford/Simpson scale) strikes the United States once every 6 years. A hurricane of this strength has winds of at least 131 miles per hour and can cause extreme damage. (`http://www.aoml.noaa.gov/hrd/Landsea/deadly/index.html`) An insurance agency is considering whether it might want to stop insuring oceanfront homes and wants to assess the risk involved. The president of the company wants to know how long (in years) before the next hurricane that is category 4 or stronger.

a. Why is this an Exponential scenario?

b. What does X represent in this scenario?

c. What is the parameter?

d. What is the expected length of time (in years) between now and when the next hurricane that is category 4 or stronger?

e. What is the variance in this length of time?

f. What is the probability density function for the length of time before the next hurricane that is category 4 or stronger? Write your answer in function form and show a graph.

g. What is the CDF for the length of time before the next hurricane that is category 4 or stronger? Write your answer in function form and show a graph.

h. What is the probability that there will not be any hurricane that is category 4 or stronger, during the next 3 years?

i. What is the probability that there will be a hurricane that is category 4 or stronger, during the period that is between 5 to 10 years from now?

j. Given that there are no hurricanes that are category 4 or stronger during the next 3 years, what is the probability that there will not be any during the next 10 years?

k. How long a waiting time do we need, if we want to be 75% sure that there is a hurricane that is category 4 or stronger during the waiting time?

Exercise 32.3. **Waiting by the phone.** Bob is waiting for his girlfriend Alice to call. His waiting time X is Exponentially distributed, with expected waiting time $\mathbb{E}(X) = 0.20$ hours, i.e., 12 minutes.

a. What is the probability that he must wait more than 12 minutes for her to call?

b. What is the probability that he must wait more than 12 minutes (altogether) for her to call, given that he has already waited 3 minutes?

c. What is the probability that he must wait at most 10 minutes (altogether) for her to call, given that he has already waited 3 minutes?

d. Given that he waited less than 10 minutes, what is the probability that he waited less than 3 minutes?

e. Find the unique time "a" in hours such that $P(X \leq a) = 1/2$ and $P(X > a) = 1/2$. This is the *median* waiting time; also see Exercise 32.13.

Exercise 32.4. **Waiting for a bus.** A student waits for a bus. Let X be the number of hours that the student waits. Assume that the waiting time is Exponential with average 20 minutes.

a. What is the probability that the student waits more than 30 minutes?

b. What is the probability that the student waits more than 45 minutes (total), given that she has already waited for 20 minutes?

c. Given that someone waits less than 45 minutes, what is the probability that they waited less than 20 minutes?

d. What is the standard deviation of the student's waiting time?

Exercise 32.5. Waiting for a ride. The waiting time for rides at an amusement park has an Exponential distribution with average waiting time of 1/2 an hour. Find the time "t" such that 80% of the people have waiting time t or less.

Exercise 32.6. Happy birthday. It is your birthday and you are waiting for someone to write a "Happy Birthday" message on your Facebook wall. Your waiting time is approximately Exponential with average waiting time of 10 minutes between such postings; assume that the times of the postings are independent.

a. What is the probability that the next posting takes 15 minutes or longer to appear?

b. What is the standard deviation of the time in between consecutive Happy Birthday messages?

c. Suppose that the most recent posting was done at 1:40 PM, and it is now 1:45 PM (i.e., no postings have been made during the last five minutes). What is the expected time for the next message to appear?

Exercise 32.7. Falling asleep. Lily estimates that her time to fall asleep each night is approximately Exponential, with an average time of 30 minutes until she falls asleep.

a. What is the probability that it takes her less than 10 minutes to fall asleep?

b. What is the probability that it takes her more than 1 hour to fall asleep?

c. If she has already laid awake for 1 hour, what is the probability that it will take her more than 90 minutes (altogether) to fall asleep?

32.4.2 Extensions

Exercise 32.8. Pizza delivery. Suppose that the times until Hector, Ivan, and Jacob's pizza arrives are independent exponential random variables, each with average of 20 minutes. Find the probability that none of the waiting times exceed 20 minutes, i.e., find $P(\max(X, Y, Z) \leq 20)$.

Exercise 32.9. Flight delays. Suppose that, when an airplane waits on the runway, the company must pay each customer a fee if the waiting time exceeds 3 hours. Suppose that an airplane with 72 passengers waits an exponential amount of time on the runway, with average 1.5 hours. If the waiting time X, in hours, is bigger than 3, then the company pays each customer $(100)(X-3)$ dollars (otherwise, the company pays nothing). What is the amount that the company expects to pay for the 72 customers on the airplane altogether? (Of course their waiting times are all the same.)

Exercise 32.10. Let X be uniform on $[0, 10]$. Let Y be exponential with $\mathbb{E}(Y) = 5$. Find $P(X < Y)$.

Exercise 32.11. Let X be exponential with expected value 3. Let Y be another random variable that depends on X as follows: if $X > 5$, then $Y = X - 5$; otherwise, $Y = 0$.

a. Find the expected value of Y.

b. Find the variance of Y.

Exercise 32.12. Vending machine. The time between consecutive uses of a vending machine is Exponential with average 15 minutes.

a. Given that the machine has not been used in the previous 5 minutes, what is the probability that the machine will not be used during the next 10 minutes?

b. How many purchases are expected within the next hour?

c. What is the distribution of the number of purchases to be made within the next hour?

d. Suppose that a person pays 75 cents for each beverage in the vending machine. The supplier pays 40 cents per beverage in the machine, and thus makes a profit of 35 cents per beverage. What is the expected profit made during a given 8 hour period?

Exercise 32.13. Median. Suppose Y is an Exponential random variable with parameter $\lambda > 0$. What is the unique value "a" such that $P(Y \leq a) = 1/2$ and $P(Y > a) = 1/2$? (This value "a" is called the "median.")

Exercise 32.14. If Y is an Exponential random variable with parameter $\lambda > 0$, the expected value of Y is $\mathbb{E}(Y) = 1/\lambda$. What is the probability that Y is bigger than its expected value, i.e., what is $P(Y > \frac{1}{\lambda})$?

Exercise 32.15. Still waiting by the phone. Dan, Dominic, and Doug are waiting together in the living room for their girlfriends, Sally, Shellie, and Susanne, to call. Their waiting times (in hours) are independent Exponential random variables, with parameters 2.1, 3.7, and 5.5, respectively. What is the probability that the phone will ring (i.e., the first call will arrive) within the next 30 minutes (i.e., within the next 1/2 an hour)?

Exercise 32.16. Air traffic control. Air traffic control stations often have insufficient numbers of air traffic controllers, sometimes just one person on duty. In a recent study, a lone air traffic controller is managing an airstrip in which the expected time between arrivals of airplanes is 15 minutes. Assume that the times between consecutive arrivals are independent Exponential random variables. If he falls asleep for a period of 5 minutes, what is the probability that one or more airplanes arrive during this period (but cannot land because the air traffic controller is not awake to guide them)?

32.4.3 Advanced

Exercise 32.17. Consider an Exponential random variable X with parameter $\lambda > 0$. Let $Y = \lfloor X \rfloor$, which means we get Y by rounding X down to the nearest integer (in particular, Y itself is a discrete random variable, because Y is always an integer). For example, if $X = 7.2$, then $Y = 7$. If $X = 12.9999$, then $Y = 12$. If $X = 5.01$, then $Y = 5$, etc.

So the mass of Y is exactly

$$p_Y(y) = P(Y = y) = P(y \leq X < y + 1).$$

a. Find an expression for the mass of Y. (Your expression will have λ in it; i.e., just integrate to find the value of $P(y \leq X < y + 1)$, and then simplify.)

b. Do you recognize the mass of Y? (Yes, you should!) What type of random variable is Y? What are the parameters of Y? (Hint: Y is one of the types of named discrete random variables.)

Exercise 32.18. Let X be an Exponential random variable with $\mathbb{E}(X) = 1/\lambda$. Define $Y = \lceil X \rceil$, i.e., Y is the least integer that is greater than or equal to Y. For instance, if $X = 5.3$, then $Y = 6$. If $X = 4.99$, then $Y = 5$. If $X = 5.00001$ then $Y = 6$. If $X = 5.99999$ then $Y = 6$. What is the distribution of Y?

Exercise 32.19. A store manager is very impatient at the start of the day. Let X be the time (in minutes) until his first customer arrives. The workers at the store decided to find a way to measure his impatience level as a function of X:

$$g(X) = aX^2 + bX + c,$$

where a, b, c are fixed constants.

a. Calculate the manager's expected impatience level.

b. Find the variance of the manager's impatience level.

Exercise 32.20. Suppose that the time in between customers at a store are independent Exponential random variables, with an average of 2 minutes between consecutive customers. Let X be the time until the 3rd customer arrives.

a. Find $\mathbb{E}(X)$.

b. Find $\text{Var}(X)$.

c. Find the density of X.

Chapter 33

Gamma Random Variables

I have noticed that people who are late are often so much jollier than the people who have to wait for them.

—E. V. (Edward Verrall) Lucas

An employee wonders how long it will take until the fourth customer arrives at the store. How is this related to the length of time until the first arrival? Is the time to the fourth arrival always four times as long as the time to the first arrival? Why or why not?

33.1 Introduction

A Negative Binomial random variable is the sum of r independent Geometric random variables, i.e., the number of (discrete) trials until the rth success occurs. The Gamma random variable has a similar motivation, but in a continuous setting. A Gamma random variable is the sum of r independent Exponential random variables. If an Exponential random variable is viewed as the waiting time until the first event occurs, then a Gamma random variable is the waiting time until the rth event occurs. Thus, if $X_1, X_2, \ldots, X_r$ are independent Exponential random variables that each have parameter λ, and if we define

$$X = X_1 + X_2 + \cdots + X_r,$$

then we say that X is a Gamma random variable with parameters λ and r.

The density of a Gamma random variable X with parameters λ and r is

$$f_X(x) = \frac{\lambda^r}{\Gamma(r)} x^{r-1} e^{-\lambda x} \qquad \text{for } x > 0,$$

and $f_X(x) = 0$ otherwise.

Gamma random variables

The common thread: Time until the rth event occurs. For example, the time until the 10th customer enters a shop, or the phone rings for the 3rd time, or the time until the 10th black car drives by.

Things to look for: Repeated independent trials until rth event; same parameter for each waiting time.

The variable:

$$X = \text{time until the } r\text{th event occurs, } X \geq 0$$

The parameters:

$$r = \text{total number of arrivals/events that you are waiting for}$$
$$\lambda = \text{the average rate, e.g., number of arrivals per time period}$$

The λ for the Gamma distribution is the same as the one for the Exponential distribution. The r parameter is often called the "shape" parameter, and λ is the "scale" parameter.

Density:

$$f_X(x) = \begin{cases} \frac{\lambda^r}{\Gamma(r)} x^{r-1} e^{-\lambda x}, & x > 0, \\ 0 & \text{otherwise} \end{cases}$$

where $\Gamma(r) = (r-1)!$ since r is a positive integer.
When $r = 1$, a Gamma random variable is just an Exponential random variable.

CDF:

$$F_X(x) = \begin{cases} 1 - e^{-\lambda x} \sum_{j=0}^{r-1} \frac{(\lambda x)^j}{j!}, & x > 0, \\ 0 & \text{otherwise} \end{cases}$$

Expected value formula:

$$\mathbb{E}(X) = r/\lambda$$

Variance formula:

$$\text{Var}(X) = r/\lambda^2$$

The notation for the Gamma looks like:

$$X \sim \text{Gamma}(r, \lambda).$$

Since a Gamma random variable is the sum of r Exponential random variables, then the expected value of a Gamma random variable is just equal to r times the expected value of an Exponential random variable:

$$\mathbb{E}(X) = \mathbb{E}(X_1 + \cdots + X_n) = \mathbb{E}(X_1) + \cdots + \mathbb{E}(X_r) = \frac{1}{\lambda} + \cdots + \frac{1}{\lambda} = \frac{r}{\lambda}.$$

Since the X_j's are independent, the variance of a Gamma random variable is just equal to r times the variance of an Exponential random variable:

$$\mathrm{Var}(X) = \mathrm{Var}(X_1 + \cdots + X_n) = \mathrm{Var}(X_1) + \cdots + \mathrm{Var}(X_n) = \frac{1}{\lambda^2} + \cdots + \frac{1}{\lambda^2} = \frac{r}{\lambda^2}.$$

Some examples of densities for Gamma random variables with various values of r and λ are given in Figures 33.1, 33.2, and 33.3.

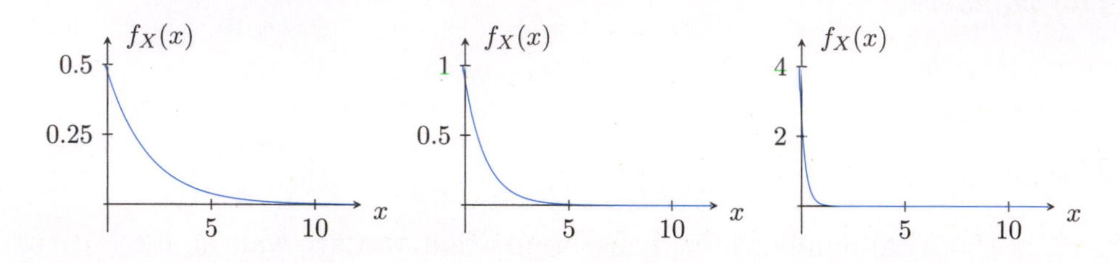

FIGURE 33.1: The density $f_X(x)$ of a Gamma random variable with $r = 1$ and (left) $\lambda = 1/2$, (middle) $\lambda = 1$, and (right) $\lambda = 4$. Notice that the scale of the y-axis changes is changing as the values of λ change.

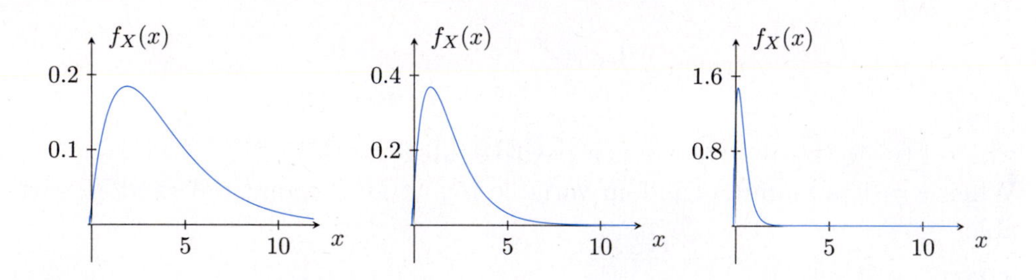

FIGURE 33.2: The density $f_X(x)$ of a Gamma random variable with $r = 2$ and (left) $\lambda = 1/2$, (middle) $\lambda = 1$, and (right) $\lambda = 4$.

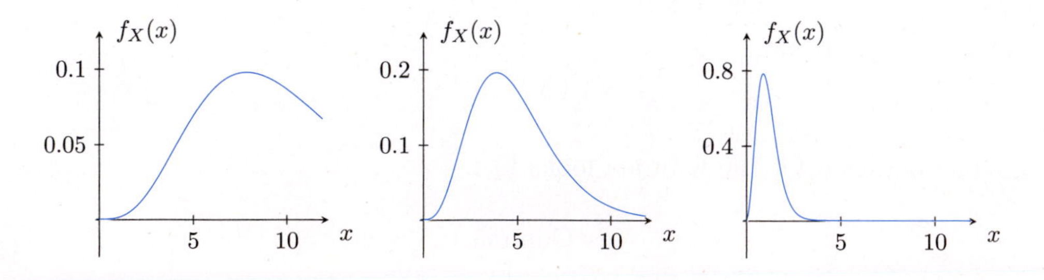

FIGURE 33.3: The density $f_X(x)$ of a Gamma random variable with $r = 5$ and (left) $\lambda = 1/2$, (middle) $\lambda = 1$, and (right) $\lambda = 4$.

33.2 Examples

Example 33.1. Customers arrive at a shoe store at an average rate of 1 every 5 minutes. The inter-arrival times from customer to customer are independent, and each inter-arrival time is Exponentially distributed. Since the clerks work on commission, arguments over which customers belong to which clerk can get very heated. Bill has made a deal with his coworker Shirley that he will wait on the first 3 customers, she will wait on the next 3 customers after that, and then he takes over again. Shirley wants to grab a cup of coffee while she is waiting for her turn, but she wants to estimate how much time she will have for her break. In particular, Shirley wants to make sure that she is ready when the 4th customer arrives.

a. What does X represent in this story? What values can X take?

The random variable X is the time Shirley waits for the 4th customer to arrive. In particular, X is nonnegative. (Theoretically all 4 customers could arrive near the very beginning of Bill's shift, or the 4th customer might not arrive for days.)

b. Why is this an example of the Gamma distribution?

Correct answer: The random variable X has a Gamma distribution because we know the average customer arrival rate, and we are measuring the waiting time until the 4th customer.

Other answers you might have considered (and why they are wrong):

If we were measuring the waiting time for just the 1st customer, this would be an Exponential random variable.

If we were counting up how many customers arrive in the next 30 minutes, this would be a Poisson random variable.

If we were counting the number of customers Shirley had to help until she had her first customer who purchased red stiletto high heels (and if we know the probability that customers in general want to purchase red stiletto high heels), then this would be a Geometric random variable.

If we wanted to count the number of customers until we found the 4th customer who purchased red stiletto high heels, then this would be a Negative Binomial random variable.

c. What are the parameters for this distribution?

The parameter λ is the average waiting time between customer arrivals, e.g., 2 customers every 10 minutes (or 1 customer every 5 minutes if we want to simplify).

The parameter r is the number of customers Shirley is waiting for. She is interested in the arrival time of the 4th customer, so $r = 4$.

d. How long should Shirley expect to have for her coffee break before it is her turn to wait on the customers?

$$\mathbb{E}(X) = \frac{r}{\lambda} = \frac{4 \text{ customers}}{1 \text{ customer } / \text{ } 5 \text{ minutes}} = 20 \text{ minutes}.$$

e. What is the standard deviation in the amount of time Shirley will have for her coffee break?

$$\text{Var}(X) = \frac{r}{\lambda^2} = \frac{4}{(1/5)^2} = 100 \text{ minutes}^2$$

$$\sigma_X = \sqrt{100} = 10 \text{ minutes}$$

f. What is the density of the time until the 4th customer of the day arrives?

$$f_X(x) = \begin{cases} \frac{(1/5)^4}{\Gamma(4)} x^{4-1} e^{-(1/5)x}, & x > 0, \\ 0 & \text{otherwise} \end{cases}$$

which, since $\Gamma(4) = 3! = 6$, simplifies to

$$f_X(x) = \begin{cases} \frac{1}{3750} x^3 e^{-(1/5)x}, & x > 0, \\ 0 & \text{otherwise} \end{cases}$$

g. What is the probability that she will have at least 30 minutes before she has her first customer (the fourth customer of the day)?

The probability is

$$\begin{aligned} P(X > 30) &= 1 - P(X \leq 30) \\ &= 1 - F_X(30) \\ &= 1 - \left(1 - e^{-(1/5)(30)} \sum_{j=0}^{4-1} \frac{((1/5)(30))^j}{j!}\right) \\ &= e^{-6} \sum_{j=0}^{3} \frac{6^j}{j!} \\ &= e^{-6}(1 + 6 + 18 + 36) \\ &= 61e^{-6} \\ &= 0.1512 \end{aligned}$$

Example 33.2. David works at a customer call center. He talks to customers on the telephone. The length (in hours) of each conversation is Exponential with average 1/3, and the lengths of calls are independent. As soon as one conversation is finished, he hangs up the phone, and immediately picks up the phone again to start another call (i.e., there are no gaps in between the calls). Thus, if he conducts r phone calls in a row, *the total amount of time* he spends on the telephone is $X_1 + \cdots + X_r$, where the X_j's are independent, and each X_j has the same density.

We see that *the total time* it takes David to make r calls is exactly a Gamma random variable $X = X_1 + \cdots + X_r$ with parameters $\lambda = 3$ and r.

The expected time that it takes him to make r calls is

$$\mathbb{E}(X) = r/3.$$

The variance of the time that it takes him to make r calls is

$$\text{Var}(X) = r/9,$$

and thus the standard deviation is $\sqrt{r/9}$.

We can also compute the probability that he completes two calls within the first 1 hour, by using the density of X given above. When $r = 2$ and $\lambda = 3$, then the density of X becomes

$$f_X(x) = x^{2-1}3^2e^{-3x}/(2-1)! = 9xe^{-3x} \qquad \text{for } x > 0,$$

and $f_X(x) = 0$ otherwise. Completing two calls in the first hour means $X < 1$; thus

Here we are using integration by parts.

$$\begin{aligned} P(X < 1) &= \int_0^1 9xe^{-3x}\,dx \\ &= \left(9xe^{-3x}/(-3) - 9e^{-3x}/(9)\right)\Big|_{x=0}^{1} \\ &= (-3e^{-3} - e^{-3}) - (-1) \\ &= 1 - 4e^{-3} \\ &= 0.8009 \end{aligned}$$

Example 33.3. Consider 300 students who are waiting for service at the registrar's office. Assume that their waiting times are independent exponential random variables, and each waiting time (in minutes) has density $f_X(x) = 2e^{-2x}$ for $x > 0$, and $f_X(x) = 0$ otherwise. Find the probability that the 300 students collectively (i.e., altogether) spend between 145 and 152 hours waiting for their appointments.

This is the same probability as $145 \le Y \le 152$ where Y is a Gamma random variable Y with parameters $r = 300$ and $\lambda = 2$. So the probability is

$$P(145 \le Y \le 152) = \int_{145}^{152} \frac{2^{300}}{299!}x^{299}e^{-2x}\,dx = 0.3122,$$

but this requires a high powered calculator or symbolic computing environment (such as Maple or Mathematica) to calculate.

33.3 Exercises

33.3.1 Practice

Exercise 33.1. Egg laying. (See Exercise 32.1.) Chickens at Rolling Meadows Farm lay an average of 18 eggs per day. The farmer has rigged a fancy monitoring device to the nesting boxes so that he can monitor exactly when the hens lay their eggs. Assume that no 2 eggs will be laid at exactly the same time and that the eggs (and chickens) are independent from each other. The farmer wants to know how long he will have to wait (in minutes) for the next half-dozen (6) eggs to be laid so that he can bake a chocolate cake if he starts monitoring at the first rooster crow in the morning.

a. Why is this a Gamma problem? What makes this a Gamma situation instead of an Exponential situation?

b. What does X represent in this scenario?

c. What are the parameters in units matching this specific question?

d. What is the expected length of time (in minutes) the farmer will have to wait for the half-dozen eggs to be laid?

e. What is the standard deviation?

f. What is the probability density function for the wait time (in minutes) for the half-dozen eggs? Write it in function form and also graph it.

g. What is the cumulative distribution function for the wait time (in minutes) for the half-dozen eggs? Write it in function form and also graph it.

h. Find the probability the farmer will have to wait for his half-dozen eggs longer than it takes him to do his morning chores and run his errands in town (6 hours after the rooster crows).

i. Find the probability that the sixth egg will be laid while he is eating dinner (between 12 and 13 hours after the rooster crows)?

Exercise 33.2. Hurricanes. (See Exercise 32.2.) On the average, a category 4 (on the Stafford/Simpson scale) or stronger hurricane strikes the United States once every 6 years. A hurricane of this strength has winds of at least 131 miles per hour and can cause extreme damage. (`http://www.aoml.noaa.gov/hrd/Landsea/deadly/index.html`) An insurance agency is considering whether they might want to stop insuring oceanfront homes and wants to assess the risk involved. The president of the company wants to know how long (in years) before the next 3 hurricanes that are category 4 or stronger.

a. Why is this a Gamma scenario?

b. What does X represent in this scenario?

c. What are the parameters?

d. What is the expected length of time (in years) between now and when the third hurricane that is category 4 or stronger will come?

e. What is the variance in this length of time?

f. What is the probability density function for the length of time before the third hurricane that is category 4 or stronger will come? Write your answer in function form and show a graph.

g. What is the CDF for the length of time before the third hurricane that is category 4 or stronger? Write your answer in function form and show a graph.

h. What is the probability that the third hurricane that is category 4 or stronger will arrive in the next 10 years?

Exercise 33.3. Flight delays. The time (in minutes) until a person's flight departs at an airport has density

$$f_X(x) = \frac{1}{45}e^{-x/45}, \qquad \text{for } x > 0,$$

and $f_X(x) = 0$ otherwise.

a. What is the expected total waiting time of seven passengers on seven different flights?

b. What is the standard deviation of the total waiting time for seven passengers on seven different flights?

Exercise 33.4. Waiting for a ride. The waiting time for rides at an amusement park has an Exponential distribution with average waiting time of 1/2 an hour (assume that the waiting times are independent).

a. If a person rides 5 rides, what is the expected amount of time that the person spends waiting in line?

b. If a person rides 5 rides, what is the standard deviation of the time that the person spends waiting in line?

c. Find the probability that the person spends more than 1 hour altogether while waiting for two rides (i.e., that their actual waiting time is longer than the expected waiting time).

Exercise 33.5. Waiting for a bus. A student waits for a bus 10 times during a week. Let X be the number of hours that the student waits. Assume that the waiting times are independent Exponential random variables, each with average 30 minutes.

a. What is his expected time spent waiting for the 10 buses altogether?

b. What is the standard deviation of his waiting time for the 10 buses altogether?

Exercise 33.6. Homework. A student estimates that the time needed to solve each homework problem is Exponentially distributed and is independent of all the other homework problems. Each problem takes, on average, 15 minutes to solve.

a. What is the expected time spent on a 6-question homework assignment?

b. What is the variance of the time spent on a 6-question homework assignment?

Exercise 33.7. Waiting at the store. The time that each customer spends in a grocery store line is Exponential with average waiting time 3 minutes. If I am the 7th customer in line, how long do I expect to wait until all 7 of us have had our groceries processed?

Exercise 33.8. Taking an exam. Thirteen students take an exam. Each student spends an Exponential amount of time on problem #2, with average of 10 minutes per student. What is the density of the sum of the time that the 13 students spend on problem #2 altogether?

Exercise 33.9. Happy birthday. It is your birthday and you are waiting for someone to write a "Happy Birthday" message on your Facebook wall. Your waiting time is approximately Exponentially distributed with average waiting time of 10 minutes between such postings; assume that the times of the postings are independent.

What is the probability that you have to wait 24 minutes or less until you get your second Happy Birthday message?

Exercise 33.10. Pinball. While listening to "Pinball Wizard," you decide to conduct an experiment in which you play pinball over and over again. Each game takes an Exponential amount of time to finish, with expected value of 3 minutes.

a. What is the density of the time that you spend, if you play two games in a row?

b. What is the probability that you will be finished with two games in a total of five minutes or less?

Exercise 33.11. Meteor shower. An astronomer watches a meteor shower. He believes that the density of the time in between each meteor is

$$f_X(x) = 10e^{-10x}, \qquad \text{for } x > 0;$$

otherwise, $f_X(x) = 0$. He has also estimates that there are 500 meteors in this shower. How many minutes does he expect to be watching this meteor shower?

Exercise 33.12. Starting a band. Five friends decide to start a band. They each start to practice their own instruments until they can perform their first song. Unfortunately, they practice independently of each other, so their waiting times until they are ready to perform are independent Exponential random variables, each with an average of 3 days. How much time do they expect to spend practicing altogether (i.e., the sum of their practice times) until they are all ready to get together and rehearse?

33.3.2 Extensions

Exercise 33.13. Lots of homework. Kelsey gets back to her apartment and immediately starts all her homework assigned that day for her three classes. The time to finish each is an Exponential random variable, with average 30 minutes; these three times are independent. What is the probability that it takes her more than 90 minutes (altogether) to complete the three assignments?

Exercise 33.14. Passing cars. A student procrastinates, watching cars pass his house. After 10 black cars have passed, he will (finally) start his homework. He notices that the time between consecutive cars that pass is Exponentially distributed, and the times are independent, each with expected time of $1/5$ of a minute. He also assumes that each car has a 30% chance of being black, and the colors of the cars are independent. How long should he expect to wait until the 10th black car arrives?

Exercise 33.15. Waiting for a therapist. The waiting time to see a therapist is Exponential with average of 10 minutes during each visit. What is the probability that, during 3 separate visits to the therapist, a patient spends a total of 40 minutes or more waiting?

33.3.3 Advanced

Exercise 33.16. If X is a Gamma random variable with $r = 2$, show that X does not have the memoryless property.

Exercise 33.17. If X is a Gamma random variable with $r = 2$, find the probability that X exceeds $\mathbb{E}(X)$.

Chapter 34

Beta Random Variables

Life shrinks or expands in proportion to one's courage.

—*The Diary of Anais Nin, Volume 3* by Anaïs Nin (Swallow Press, 1971)

What is the percent of students who pass the SOA/CAS P/1 actuarial exam?

34.1 Introduction

The Beta distribution deals with percents, proportions, or fractions. The notation for the Beta distribution looks like:

$$X \sim \text{Beta}(\alpha, \beta)$$

When α and β are both 1, a Beta random variable is simply a Uniform random variable.

Unfortunately, in general, there is no nice shortcut for the Beta cumulative distribution function. Since the power of x depends on the parameters α and β, the density (and, therefore, its integral) depends essentially on those parameters. Thus, a concise form for the CDF is not available.

The density has a familiar shape for many pairs of parameters. We give some examples. The plots are only drawn for the values $0 \leq x \leq 1$ because the density equals 0 elsewhere. We give four density and CDF example plots in Figures 34.1, 34.2, 34.3, and 34.4.

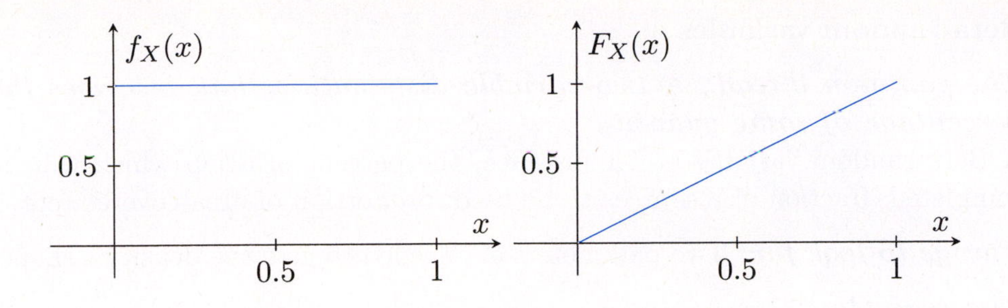

FIGURE 34.1: When $\alpha = 1$ and $\beta = 1$, (left) the density $f_X(x) = 1$, and (right) the CDF $F_X(x) = x$ of a Beta$(1, 1)$ random variable.

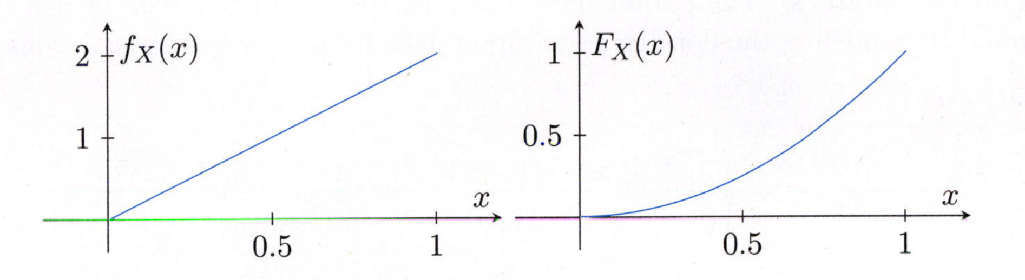

FIGURE 34.2: When $\alpha = 2$ and $\beta = 1$, (left) the density $f_X(x) = 2x$, and (right) the CDF $F_X(x) = x^2$ of a Beta$(2, 1)$ random variable.

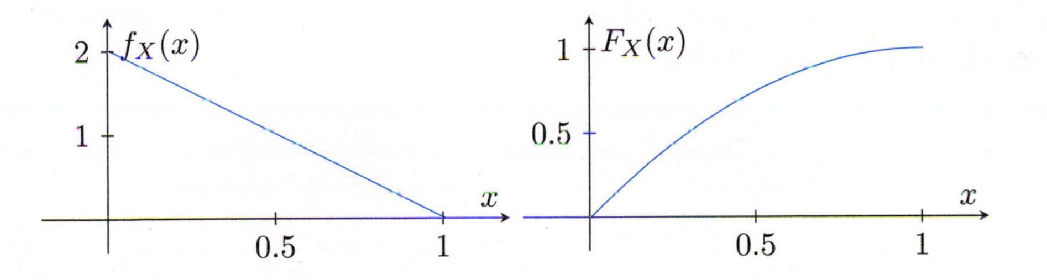

FIGURE 34.3: When $\alpha = 1$ and $\beta = 2$, (left) the density $f_X(x) = 2(1 - x)$, and (right) the CDF $F_X(x) = x(2 - x)$ of a Beta$(1, 2)$ random variable.

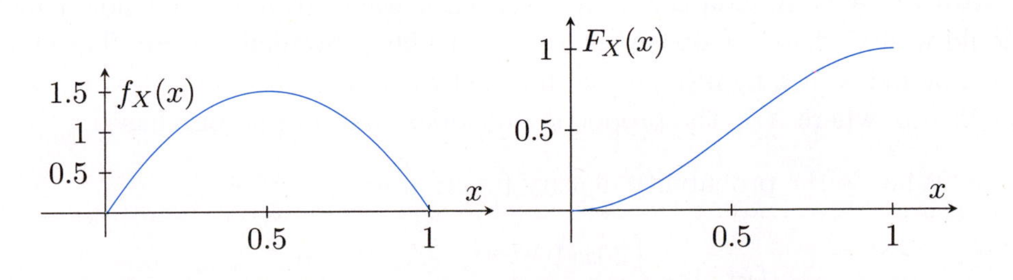

FIGURE 34.4: When $\alpha = 2$ and $\beta = 2$, (left) the density $f_X(x) = 6x(1 - x)$, and (right) the CDF $F_X(x) = x^2(3 - 2x)$ of a Beta$(2, 2)$ random variable.

Beta random variables

The common thread: A two-variable distribution that describes the percentage of some quantity.
A Beta random variable is, for instance, the percent of a job which will be completed, fraction of resources to be used, proportion of area covered, etc.

Things to look for: Two parameters are needed to give the density's shape.

The variable:

$$X = \text{the proportion, fraction, or percent of a quantity of interest.}$$

The parameters: The parameters, α and β, will either be given or can be found by modeling the density based upon data from previous observations.

Density:

$$f_X(x) = \begin{cases} \frac{\Gamma(\alpha+\beta)}{\Gamma(\alpha)\Gamma(\beta)} x^{\alpha-1}(1-x)^{\beta-1}, & \text{for } 0 \leq x \leq 1, \\ 0 & \text{otherwise} \end{cases}$$

When α or β is a positive integer, it is helpful to remember that $\Gamma(n) = (n-1)!$.

Expected value formula:

$$\mathbb{E}(X) = \frac{\alpha}{\alpha + \beta}$$

Variance formula:

$$\text{Var}(X) = \frac{\alpha\beta}{(\alpha+\beta)^2(\alpha+\beta+1)}$$

34.2 Examples

Example 34.1. A soda company distributor wants to figure out how long he should wait to deliver more sodas to a particular convenience store. The store's stock of sodas in a month can be modeled by a Beta distribution with $\alpha = 2$ and $\beta = 5$, where X is the proportion of sodas that will be purchased.

a. What is the probability density function for X?

$$f_X(x) = \begin{cases} 30x(1-x)^4, & \text{for } 0 \leq x \leq 1, \\ 0 & \text{otherwise} \end{cases}$$

This density is depicted in Figure 34.5.

b. What is the cumulative distribution function for X?

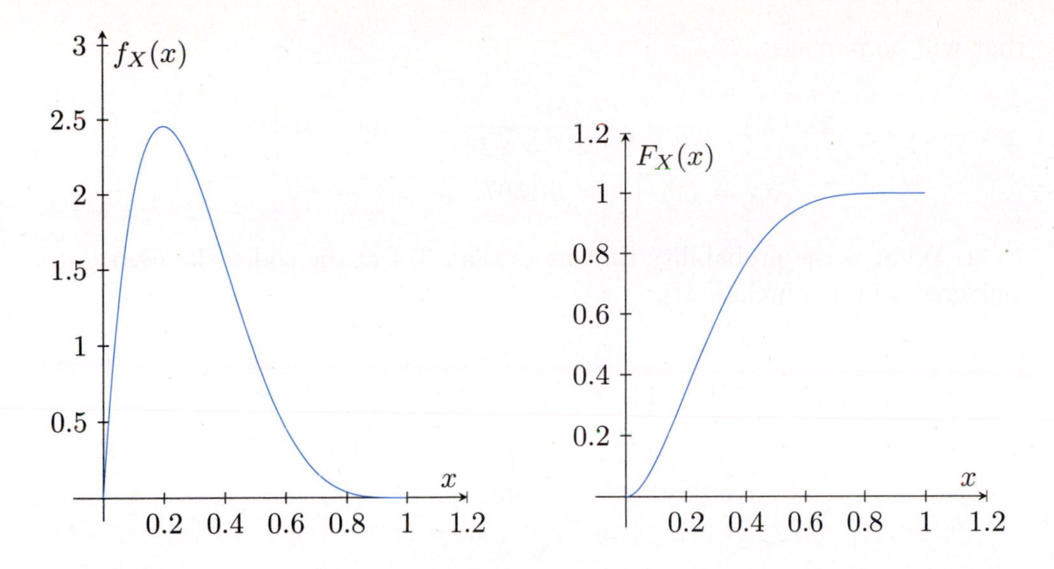

FIGURE 34.5: When $\alpha = 2$ and $\beta = 5$, a Beta(2, 5) random variable has density $f_X(x) = 30x(1-x)^4$ (left) and CDF $F_X(x) = 1 - 6(1-x)^5 + 5(1-x)^6$ (right).

The CDF $F_X(x)$ is 0 for $x < 0$ and is 1 for $x > 1$. The interesting portion of $F_X(x)$ is for $0 \le x \le 1$. In this region, it is profitable to use u-substitution, because the $(1-x)$ portion of the density has a higher power (namely, 4) than the x portion of the density (namely, 1). So we calculate, for $0 \le a \le 1$,

$$F_X(a) = \int_0^a 30x(1-x)^4\,dx.$$

Using $u = 1 - x$ and $du = -dx$, we obtain

$$\begin{aligned} F_X(a) &= \int_1^{1-a} -30(1-u)u^4\,du \\ &= \int_{1-a}^{1} 30(u^4 - u^5)\,du \\ &= 30\left(\frac{1}{5}u^5 - \frac{1}{6}u^6\right)\Big|_{u=1-a}^{1} \\ &= 1 - 6(1-a)^5 + 5(1-a)^6. \end{aligned}$$

This CDF is depicted in Figure 34.5.

c. What is the expected proportion of sodas that will be purchased?

$$\mathbb{E}(X) = \frac{2}{2+5} = 2/7 = 0.2857.$$

d. What is the variance and standard deviation of the proportion of sodas that will be purchased?

$$\text{Var}(X) = \frac{(2)(5)}{(2+5)^2(2+5+1)} = 5/196 = 0.0255,$$
$$\sigma_X = \sqrt{5/196} = 0.1597.$$

e. What is the probability that more than 3/4 of the sodas the distributor delivered will be purchased?

$$\begin{aligned} P(X > 3/4) &= 1 - F_X(3/4) \\ &= 1 - \left(1 - 6(1-3/4)^5 + 5(1-3/4)^6\right) \\ &= 6(1/4)^5 - 5(1/4)^6 \\ &= 19/4096 \\ &= 0.0046 \end{aligned}$$

f. What is the probability that between 1/2 and 3/4 of the sodas the distributor delivered will be purchased?

The probability is $P(1/2 \le X \le 3/4) = F_X(3/4) - F_X(1/2)$. Using the results from part e, we know $F_X(3/4) = 1 - P(X > 3/4) = 1 - 19/4096 = 4077/4096$. Also, we have

$$P(X \le 1/2) = F_X(1/2) = 1 - 6(1-1/2)^5 + 5(1-1/2)^6 = 57/64,$$

so it follows that

$$\begin{aligned} P(1/2 \le X \le 3/4) &= P(X \le 3/4) - P(X \le 1/2) \\ &= 4077/4096 - 57/64 \\ &= 0.1047 \end{aligned}$$

g. The distributor was already told that more than half of the sodas delivered earlier this month have been purchased, so what is the conditional probability that less than 3/4 of the sodas will be purchased?

$$\begin{aligned} P(X < 3/4 \mid X > 1/2) &= \frac{P(1/2 < X < 3/4)}{P(X > 1/2)} \\ &= \frac{429/4096}{1 - F_X(1/2)} \\ &= \frac{429/4096}{1 - 57/64} \\ &= 429/448 \\ &= 0.9576 \end{aligned}$$

34.3 Exercises

34.3.1 Practice

Exercise 34.1. Qualifying exam. The proportion of people who pass a professional qualifying exam on the first try has a Beta distribution with $\alpha = 3$ and $\beta = 4$.

a. What is the expected proportion of people who will pass on the first try at the next exam?

b. What is the standard deviation in the proportion of people who will pass on the first try at the next exam?

Exercise 34.2. Shelf space. A grocery store chain is trying to decide how much shelf space to devote to organic produce. If they don't stock enough, their more upscale customers will shop at the competing grocery store. If they stock too much, then much of the expensive organic produce will have to be thrown out when it expires. The percentage of organic produce which is purchased during the week after delivery can be modeled with a Beta distribution with $\alpha = 4$ and $\beta = 3$.

a. What is the expected proportion of organic produce is purchased??

b. What is the standard deviation of the proportion of organic produce that is purchased?

34.3.2 Extensions

Exercise 34.3. Horse coloration. The proportion of horses with a particular coloration is modeled by the following function:

$$f_X(x) = \begin{cases} k(-2x^3 + x^4 + x^2) & \text{if } 0 \leq x \leq 1 \\ 0 & \text{otherwise} \end{cases}$$

a. Why is this a Beta distribution?

b. What are the parameters?

c. What is the value of k?

d. What is the expected value?

Exercise 34.4. Email checking. The percent of time on a workday that employees at a company spend checking email can be modeled by the following function:

$$f_X(x) = \begin{cases} k(1 - 3x + 3x^2 - x^3) & \text{if } 0 \leq x \leq 1 \\ 0 & \text{otherwise} \end{cases}$$

a. Why is this a Beta distribution?

b. What are the parameters?

c. What is the value of k?

d. What is the expected value?

Chapter 35

Normal Random Variables

Sir,—It has been wittily remarked that there are three kinds of falsehood: the first is a 'fib,' the second is a downright lie, and the third and most aggravated is statistics.

—"Letter to the Editor" of *The National Observer*, written by T. Mackay, dated June 8, 1891, published June 13, 1891

What is the probability that a randomly chosen student is more than 6 feet tall? What is the probability the randomly selected student will be between 5 and 6 feet tall? What is the cutoff length that separates the tallest 5% of people from the rest of the population?

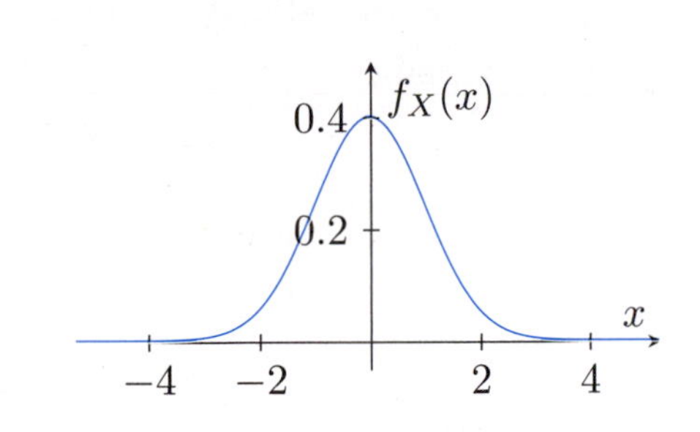

FIGURE 35.1: The density of a standard Normal random variable.

35.1 Introduction

We encounter Normal random variables in many situations. The shape of the density of a standard Normal random variable, seen in Figure 35.1, is the familiar bell curve that is so often associated with randomly distributed quantities

The density of a Normal random variable is often referred to as a bell curve.

and measurements. Many things that are clustered near their expected value are Normally distributed. In fact, the bell curve looks like a bell because most of the density is concentrated in a bell shape near the expected value. Also, as with a bell, only a relatively small amount of the density is concentrated far from the center.

Weights and heights of many animals, plants, as well as manufactured products of all shapes and sizes, are often Normally distributed. The growth of plants, the volumes of liquids in bottles, and many other biological, physical, and financial random variables are Normally distributed.

Normal random variables

The common thread: Concentration (often of some kind of measurement) near the expected value, so the distribution is like a bell curve. For instance, the weight, height, lifespan, and intelligence scores of living things are often Normally distributed. The actual volume of soda in a can filled by a machine in a factory, the high temperatures for June 17th in a given community, and many other variables in the biological, physical, and financial world, are all Normally distributed.

Things to look for: Random quantities that are "close" to their expected values a relatively large portion of the time.

The variable:

$$X = \text{the actual height, volume, weight, score, etc.}$$

The parameters:

$$\mu_X = \mathbb{E}(X) = \text{the expected value}$$
$$\sigma_X^2 = \text{the variance}$$

Sometimes the standard deviation σ_X is given instead.

Density:

$$f_X(x) = \frac{e^{-(x-\mu_X)^2/(2\sigma^2)}}{\sqrt{2\pi\sigma^2}}, \qquad -\infty < x < \infty$$

Expected value formula:

$$\mathbb{E}(X) = \mu_X$$

Variance formula:

$$\text{Var}(X) = \sigma_X^2$$

Perhaps Normal random variables are the most prevalent kind of random variables in everyday applications. The Normal distribution is often referred to as the *Gaussian* distribution. (It is named for the 19th century mathematician Gauss; `http://en.wikipedia.org/wiki/Carl_Friedrich_Gauss`.) Since we experience Normal random variables every day, why have we not practiced working with the density of the Normal distribution yet? The reason is this: The Normal distribution has a density that we can easily write down, but we cannot easily integrate the density. In fact, we do not have *any* closed-formed way to express probabilities that arise from the Normal distribution. Similarly, we cannot even write down the cumulative distribution function of the Normal distribution with a closed-formula.

For Normal random variables, with parameters μ_X and σ_X^2, as we will see in Section 35.2, the expected value is exactly $\mathbb{E}(X) = \mu_X$ and the variance is exactly $\text{Var}(X) = \sigma_X^2$. As we will see in later chapters, the Normal distribution plays a crucial role in describing the asymptotic behavior of the sum and of the average of a large collection of random variables that are either independent or loosely dependent. Many limiting theorems have been discovered that are associated with the Normal distribution.

Another nice property of Normal random variables is that the sum of independent Normal random variables is also a Normal random variable (this will be established in the next chapter), and also if X is Normal, then $aX + b$ is Normal too (to be seen in Section 35.2). These properties of Normal random variables make them very desirable to work with, even though we must consult a chart to look up the probabilities associated with Normal random variables (since we cannot integrate the density by hand).

The notation for a random variable X with Normal distribution is:

$$X \sim \mathcal{N}(\mu_X, \sigma_X^2).$$

Since there is not a closed formula for the CDF of the Normal distribution, we are forced to read all of the probabilities for the Normal distribution from a table. This is not because of any inadequacy in our mathematical ability or understanding; nobody has such a closed formula. So mathematicians and statisticians and practitioners do exactly the same thing—or one can use a calculator (if it has a button for the Normal distribution), or a program on a computer.

A Normal random variable with parameters 0 and 1 (respectively) is used so frequently that we have a special name for it: a *standard* Normal random variable. When working with Normal random variables, whenever we write Z, we are referring to a standard Normal random variable.

Definition 35.1. Density of a Standard Normal Random Variable
We say Z is a *standard* Normal random variable if it has parameters 0 and 1 respectively. A standard Normal random variable Z has density

$$f_Z(z) = \frac{1}{\sqrt{2\pi}} e^{-z^2/2} \qquad \text{for } -\infty < z < \infty.$$

One of the very nice things about Normal random variables—especially from the perspective of using the Normal distribution—is that we *do not need* to

The authors assume that a significant portion of our readers will be taking the Society of Actuaries P exam/ Casualty Actuarial Society exam 1; only $P(Z \leq z)$ for $z \geq 0$ are given.

calculate integrals in order to get probabilities from the density. We are *unable* to integrate the density in closed form. Instead, we provide a table that gives (cumulative) probabilities of the form $P(Z \leq z)$, for z's in the range 0.00 to 3.09. The table is included in this chapter and also at the back of the book.

It might appear that half of the values we need, $P(Z \leq z)$ for $z < 0$, are missing from the table, because we often need to calculate $P(Z \leq z)$ for negative values of z too! Fortunately, the density of Z is symmetric around $\mu_Z = 0$ (i.e., about the origin), so we can calculate the missing probabilities by using the complementary probabilities. We will practice this technique many times.

We also emphasize that only the probabilities for *standard* Normal random variables are given in the table. Of course, for every possible combination of μ_X and σ_X^2, there is a unique Normal distribution. We cannot provide a separate Normal table for every possible Normal distribution, so in Section 35.2 we will learn how to *standardize* or *normalize* whatever Normal curve we need, by doing a transformation of any Normal random variable X to the standard Normal random variable Z.

It is helpful to see how changing μ_X and σ_X will affect the shape of $f_X(x)$. The shape is always a bell curve, with area 1 below the curve. The curves always stretch from negative infinity to positive infinity, but there is never much area once you get far beyond the middle of the curve. For smaller standard deviations, the curve will be taller and skinnier. For larger standard deviations, the curve will be shorter and wider. In Figure 35.2, we plot the density $f_X(x)$ when $\sigma_X^2 = 1$ is fixed but μ_X takes on values -2, 0, and 1, respectively. Changing μ_X just shifts the density to the left or right. The curve with $\mu_X = -2$ is on the left; the curve with $\mu_X = 0$ is in the middle; and the curve with $\mu_X = 1$ is on the right.

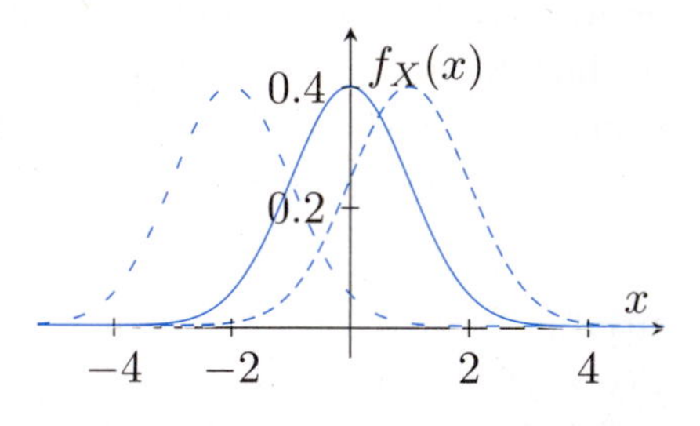

FIGURE 35.2: Three examples of the density of Normal random variables with $\mu_X = -2$ (loosely dashed), $\mu_X = 0$ (solid line), and $\mu_X = 1$ (densely dashed), respectively. In each case, $\sigma_X^2 = 1$.

In Figure 35.3, we see that smaller values of σ_X give thin, tall densities, because such densities are very tightly concentrated about μ_X. In contrast, large values of σ_X give wide, short densities, because such densities are not very concentrated around μ_X at all.

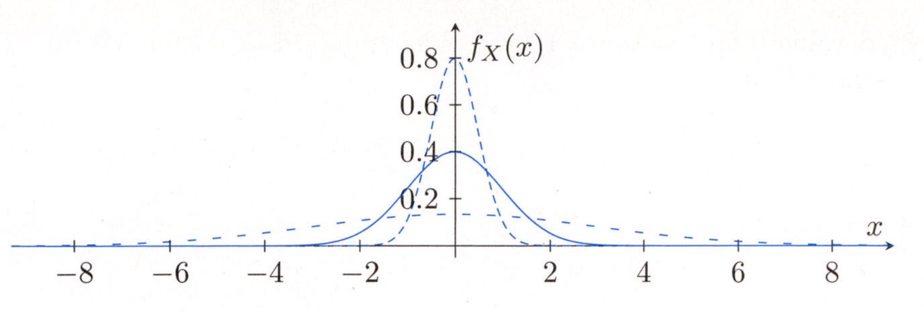

FIGURE 35.3: Three examples of the density of Normal random variables with $\mu_X = 0$ in each case, and $\sigma_X = 0.5$ (the tall, well concentrated one, drawn with densely dashed lines), $\sigma_X = 1$ (the standard Normal, drawn with a solid line), and $\sigma_X = 3$ (the short, very widely distributed one, drawn with loosely dashed lines), respectively.

35.2 Transforming Normal Random Variables

If b, c are any fixed real numbers, and Z is standard Normal, then $bZ + c$ is a Normal random variable. To see this, first assume $b > 0$. We compute

$$P(bZ + c \le a) = P(Z \le (a - c)/b) = \int_{-\infty}^{(a-c)/b} \frac{e^{-z^2/2}}{\sqrt{2\pi}}\,dz$$

Now we substitute $x = bz + c$, and thus $dx = b\,dz$, to get

$$P(bZ + c \le a) = \int_{-\infty}^{a} \frac{e^{-(x-c)^2/(2b^2)}}{\sqrt{2\pi b^2}}\,dx.$$

Thus, $bZ + c$ has density $\frac{e^{-(x-c)^2/(2b^2)}}{\sqrt{2\pi b^2}}$, so $bZ + c$ is Normal with parameters $\mu_{bZ+c} = c$ and $\sigma^2_{bZ+c} = b^2$. Since Z is symmetric about 0, then $-Z$ has the same distribution as Z, i.e., both Z and $-Z$ are standard Normal random variables. Thus $b(-Z) + c = -bZ + c$ also is Normal with the same parameters as $bZ+c$. So every random variable of the form $X = bZ+c$ is Normal (regardless of whether b is positive or negative), with $\mu_X = c$ and $\sigma^2_X = b^2$.

Theorem 35.2. If Z is a standard Normal random variable, and b, c are any constants, then $bZ + c$ is a Normal random variable with mean c and variance c^2.

We claimed that if X has the density

$$f_X(x) = \frac{e^{-(x-\mu)^2/(2\sigma^2)}}{\sqrt{2\pi\sigma^2}}, \qquad -\infty < x < \infty,$$

then $\mathbb{E}(X) = \mu$ and $\text{Var}(X) = \sigma^2_X$, but we did not prove it yet. We prove it first for the standard Normal random variable Z, i.e., we now prove that Z has expected value 0 and variance 1 (and thus standard deviation 1 too). To see this, we note that

$$\mathbb{E}(Z) = \int_{-\infty}^{\infty} (z)\Big(\frac{e^{-z^2/2}}{\sqrt{2\pi}}\Big)\,dz, \tag{35.1}$$

but

$$\int (z)\Big(\frac{e^{-z^2/2}}{\sqrt{2\pi}}\Big)\,dz = -\frac{e^{-z^2/2}}{\sqrt{2\pi}}.$$

Taking limits as $z \to -\infty$ and $z \to \infty$, we verify that $\mathbb{E}(Z) = 0$.

To see $\text{Var}(Z) = 1$, we write $\text{Var}(Z) = \mathbb{E}(Z^2) - (\mathbb{E}(Z))^2$; since $\mathbb{E}(Z) = 0$, then

$$\text{Var}(Z) = \mathbb{E}(Z^2) = \int_{-\infty}^{\infty} (z^2)\Big(\frac{e^{-z^2/2}}{\sqrt{2\pi}}\Big)\,dz.$$

Using integration by parts, with $u = z$ and $du = dz$, and $dv = (z)\Big(\frac{e^{-z^2/2}}{\sqrt{2\pi}}\Big)\,dz$ and $v = -\frac{e^{-z^2/2}}{\sqrt{2\pi}}$, we obtain

$$\text{Var}(Z) = \int_{-\infty}^{\infty} (z^2)\Big(\frac{e^{-z^2/2}}{\sqrt{2\pi}}\Big)\,dz = -z\frac{e^{-z^2/2}}{\sqrt{2\pi}}\Big|_{z=-\infty}^{\infty} - \int_{-\infty}^{\infty} -\frac{e^{-z^2/2}}{\sqrt{2\pi}}\,dz.$$

The first term is 0 because $-z\frac{e^{-z^2/2}}{\sqrt{2\pi}} \to 0$ as $z \to -\infty$ and as $z \to \infty$. The second term is 1 because it is equal to $\int_{-\infty}^{\infty} f_Z(z)\,dz$, i.e., the integral of the density of Z over all values of z. Thus $\text{Var}(Z) = 1$. (See Exercise 35.25 for one additional note.)

More generally, consider a Normal random variable X with parameters μ_X and σ_X^2. By Theorem 35.2, $\sigma_X Z + \mu_X$ is Normal too. A Normal random variable's distribution is completely and uniquely specified by the values of its two parameters. Since X and $\sigma_X Z + \mu_X$ have the same expected value and same variance and are both Normal random variables, then X and $\sigma_X Z + \mu_X$ have the same distribution. Also, $\sigma_X Z + \mu_X$ has mean μ_X and variance σ_X^2, so it follows that the parameters of X are actually the mean and variance of X, respectively. So we have proved:

Theorem 35.3. If X is any Normal random variable with parameters μ_X and σ_X^2, then X has expected value μ_X and variance σ_X^2, and thus standard deviation σ_X. In particular, a standard Normal random variable has expected value $\mu_Z = 0$ and variance $\sigma_Z^2 = 1$, and thus standard deviation $\sigma_Z = 1$.

Within our proof, the following useful fact was also established (since the density of a Normal random variable is completely determined by its two parameters).

Theorem 35.4. If Z is a standard Normal random variable, and if X is Normal with expected value μ_X and variance σ_X^2, then X has the same distribution as $\sigma_X Z + \mu_X$.

Thus, every Normal random variable can be written in a standardized form, also called a normalized form:

Corollary 35.5. Standardizing a Normal random variable

If X is any Normal random variable, with parameters μ_X and σ_X^2, and if Z is a standard Normal random variable, then X and $\sigma_X Z + \mu_X$ have the same distribution. So X can be "scaled" to a standard Normal random variable, i.e.,

$$\frac{X - \mu_X}{\sigma_X} \quad \text{is a standard Normal random variable.}$$

As another corollary of Theorem 35.4, suppose X is a Normal random variable and r, s are any two constants. We know X and $\sigma_X Z + \mu_X$ have the same distribution. Thus $rX + s$ and $r(\sigma_X Z + \mu_X) + s = r\sigma_X Z + (r\mu_X + s)$ have the same distribution too. So $rX + s$ is a Normal random variable with standard deviation $r\sigma_X$ and expected value $r\mu_X + s$.

Theorem 35.6. If X is a Normal random variable and r, s are any two constants, then $rX + s$ is a Normal random variable with standard deviation $r\sigma_X$ and expected value $r\mu_X + s$.

As an analogy about scaling a Normal random variable: Think about measuring a table in centimeters versus inches. The table is the same length, no matter how we measure it. We just change the units of measurement. So, for instance, if X is given in inches, then $Y = 2.54X$ is the same measurement, given in centimeters, because 1 inch equals 2.54 centimeters. This is comparable to the situation above, with $\mu = 0$ and $\sigma = 2.54$. See Figure 35.4 for another example of scaling a Normal random variable.

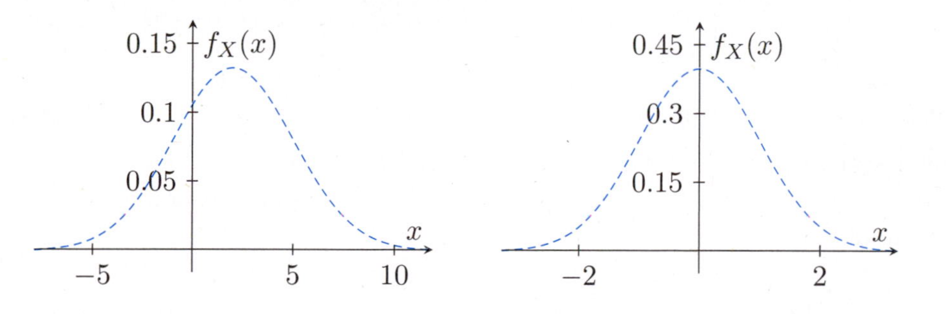

Figure 35.4: Scaling a Normal random variable essentially just changes the labeling in both the x- and y-dimensions. Left: The density of a Normal random variable with $\mu_X = 2$ and $\sigma_X = 3$. Right: The density of a standard Normal random variable, i.e., with $\mu_X = 0$ and $\sigma_X = 1$. Notice that the shape stays the same, but the center, spread, and height have changed.

Standard Normal Table

z	_._0	_._1	_._2	_._3	_._4	_._5	_._6	_._7	_._8	_._9
0.0	0.5000	0.5040	0.5080	0.5120	0.5160	0.5199	0.5239	0.5279	0.5319	0.5359
0.1	0.5398	0.5438	0.5478	0.5517	0.5557	0.5596	0.5636	0.5675	0.5714	0.5753
0.2	0.5793	0.5832	0.5871	0.5910	0.5948	0.5987	0.6026	0.6064	0.6103	0.6141
0.3	0.6179	0.6217	0.6255	0.6293	0.6331	0.6368	0.6406	0.6443	0.6480	0.6517
0.4	0.6554	0.6591	0.6628	0.6664	0.6700	0.6736	0.6772	0.6808	0.6844	0.6879
0.5	0.6915	0.6950	0.6985	0.7019	0.7054	0.7088	0.7123	0.7157	0.7190	0.7224
0.6	0.7257	0.7291	0.7324	0.7357	0.7389	0.7422	0.7454	0.7486	0.7517	0.7549
0.7	0.7580	0.7611	0.7642	0.7673	0.7704	0.7734	0.7764	0.7794	0.7823	0.7852
0.8	0.7881	0.7910	0.7939	0.7967	0.7995	0.8023	0.8051	0.8078	0.8106	0.8133
0.9	0.8159	0.8186	0.8212	0.8238	0.8264	0.8289	0.8315	0.8340	0.8365	0.8389
1.0	0.8413	0.8438	0.8461	0.8485	0.8508	0.8531	0.8554	0.8577	0.8599	0.8621
1.1	0.8643	0.8665	0.8686	0.8708	0.8729	0.8749	0.8770	0.8790	0.8810	0.8830
1.2	0.8849	0.8869	0.8888	0.8907	0.8925	0.8944	0.8962	0.8980	0.8997	0.9015
1.3	0.9032	0.9049	0.9066	0.9082	0.9099	0.9115	0.9131	0.9147	0.9162	0.9177
1.4	0.9192	0.9207	0.9222	0.9236	0.9251	0.9265	0.9279	0.9292	0.9306	0.9319
1.5	0.9332	0.9345	0.9357	0.9370	0.9382	0.9394	0.9406	0.9418	0.9429	0.9441
1.6	0.9452	0.9463	0.9474	0.9484	0.9495	0.9505	0.9515	0.9525	0.9535	0.9545
1.7	0.9554	0.9564	0.9573	0.9582	0.9591	0.9599	0.9608	0.9616	0.9625	0.9633
1.8	0.9641	0.9649	0.9656	0.9664	0.9671	0.9678	0.9686	0.9693	0.9699	0.9706
1.9	0.9713	0.9719	0.9726	0.9732	0.9738	0.9744	0.9750	0.9756	0.9761	0.9767
2.0	0.9772	0.9778	0.9783	0.9788	0.9793	0.9798	0.9803	0.9808	0.9812	0.9817
2.1	0.9821	0.9826	0.9830	0.9834	0.9838	0.9842	0.9846	0.9850	0.9854	0.9857
2.2	0.9861	0.9864	0.9868	0.9871	0.9875	0.9878	0.9881	0.9884	0.9887	0.9890
2.3	0.9893	0.9896	0.9898	0.9901	0.9904	0.9906	0.9909	0.9911	0.9913	0.9916
2.4	0.9918	0.9920	0.9922	0.9925	0.9927	0.9929	0.9931	0.9932	0.9934	0.9936
2.5	0.9938	0.9940	0.9941	0.9943	0.9945	0.9946	0.9948	0.9949	0.9951	0.9952
2.6	0.9953	0.9955	0.9956	0.9957	0.9959	0.9960	0.9961	0.9962	0.9963	0.9964
2.7	0.9965	0.9966	0.9967	0.9968	0.9969	0.9970	0.9971	0.9972	0.9973	0.9974
2.8	0.9974	0.9975	0.9976	0.9977	0.9977	0.9978	0.9979	0.9979	0.9980	0.9981
2.9	0.9981	0.9982	0.9982	0.9983	0.9984	0.9984	0.9985	0.9985	0.9986	0.9986
3.0	0.9987	0.9987	0.9987	0.9988	0.9988	0.9989	0.9989	0.9989	0.9990	0.9990
3.1	0.9990	0.9991	0.9991	0.9991	0.9992	0.9992	0.9992	0.9992	0.9993	0.9993
3.2	0.9993	0.9993	0.9994	0.9994	0.9994	0.9994	0.9994	0.9995	0.9995	0.9995
3.3	0.9995	0.9995	0.9995	0.9996	0.9996	0.9996	0.9996	0.9996	0.9996	0.9997
3.4	0.9997	0.9997	0.9997	0.9997	0.9997	0.9997	0.9997	0.9997	0.9997	0.9998
3.5	0.9998	0.9998	0.9998	0.9998	0.9998	0.9998	0.9998	0.9998	0.9998	0.9998
3.6	0.9998	0.9998	0.9999	0.9999	0.9999	0.9999	0.9999	0.9999	0.9999	0.9999

For a standard Normal random variable Z, these are the values of the cumulative distribution function $F_Z(z) = P(Z \le z)$, also corresponding to the area under the density f_Z to the left of z.

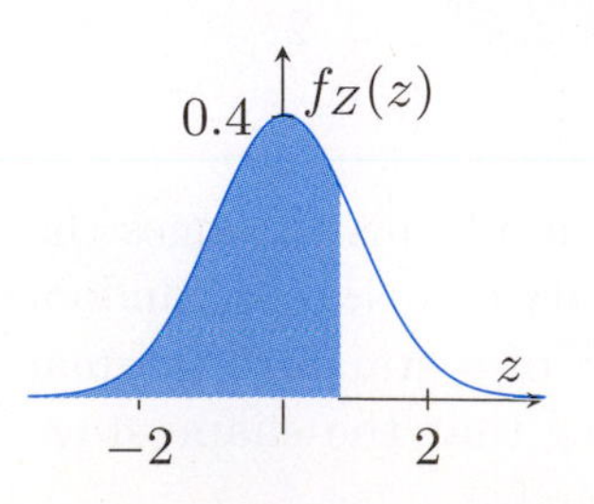

For example, in this graph, the area under the curve is $P(Z \le 0.75) = 0.7734$.

Now we practice using the standard Normal table. We have four types of ways to use the table. In every case, we encourage you to quickly sketch a Normal curve—accuracy is not required at all—because we just want you to see whether the probability that results is bigger or smaller than 1/2. This will be easy to tell when looking at the picture, because either more or less than 1/2 of the area under the curve is shaded. Remember that the probability is just the area under the curve of the density function.

In the next several examples, we will demonstrate how to use the standard Normal table to calculate probabilities in several different situations.

Example 35.7. If $z \geq 0$ and we want $P(Z \leq z)$, we just directly look up $P(Z \leq z)$ in the table. For instance, in Figure 35.5 $P(Z \leq 1.24) = 0.8925$ is the area under the curve.

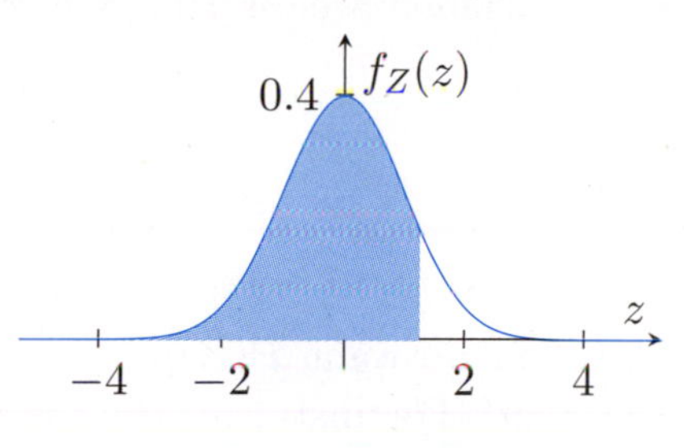

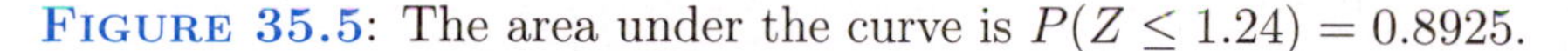

FIGURE 35.5: The area under the curve is $P(Z \leq 1.24) = 0.8925$.

Example 35.8. If $z \geq 0$ and we want $P(Z > z)$, we first look up the complement, $P(Z \leq z)$, in the table, and then we use $P(Z > z) = 1 - P(Z \leq z)$. For instance, see Figure 35.6.

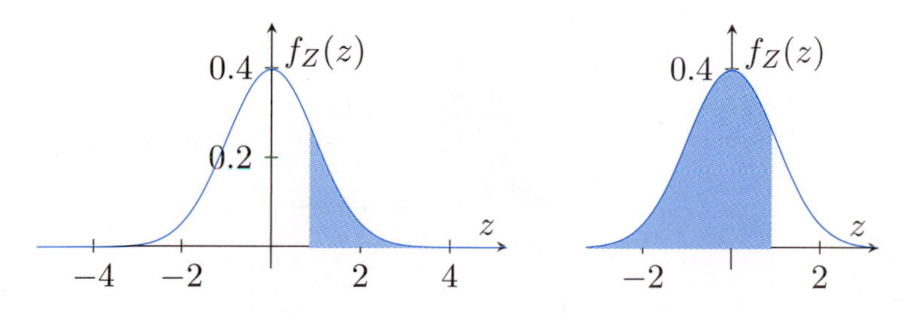

FIGURE 35.6: Left: We want the probability $P(Z > 0.88)$, but it is not in the table. Right: We find $P(Z \leq 0.88) = 0.8106$ in the table. So the desired probability is $P(Z > 0.88) = 1 - P(Z \leq 0.88) = 1 - 0.8106 = 0.1894$.

Example 35.9. If $z \leq 0$ and we want $P(Z \geq z)$ (not on the table) we use symmetry! We just use symmetry and then look up the value in the table. For instance, both plots in Figure 35.7 have the same shaded area.

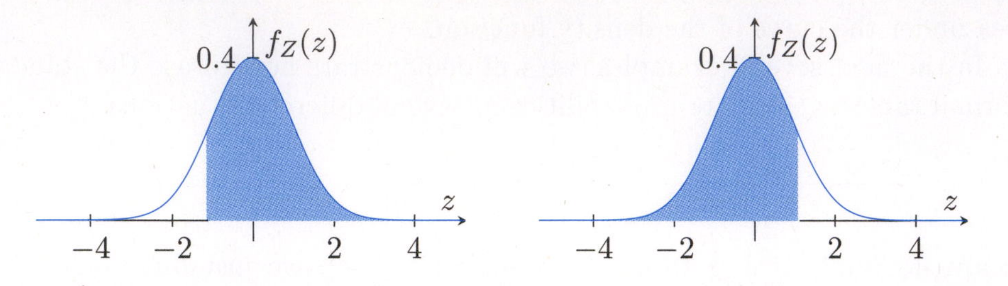

FIGURE 35.7: Both plots have the same shaded area. Left: The shaded area is $P(Z \geq -1.10)$. Right: The shaded area is $P(Z \leq 1.10)$. Each one is 0.8643.

Example 35.10. If $z \leq 0$ and we want $P(Z \leq z)$ (not on the table) we use symmetry and the complement! For instance, the upper left and upper right plots of Figure 35.8 (same shaded area) are mirror images. The complementary probability is in the Normal table; it is displayed at the bottom of Figure 35.8.

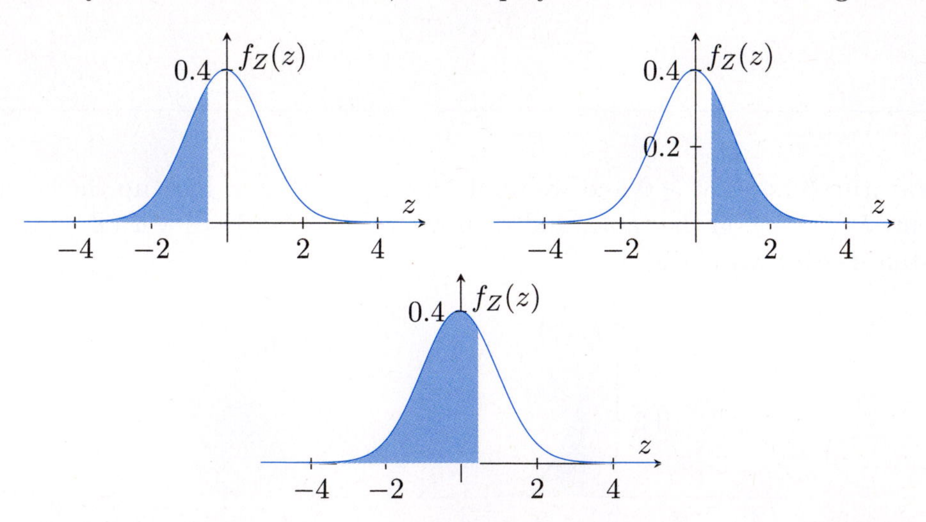

FIGURE 35.8: Upper left: The shaded area is $P(Z \leq -0.48)$. Upper right: The shaded area is $P(Z \geq 0.48)$ (a mirror image). Bottom: The shaded area is $P(Z \leq 0.48)$, which is 0.6844 (see the Normal table). So the upper left and upper right ones are each $P(Z \geq 0.48) = 1 - P(Z \leq 0.48) = 1 - 0.6844 = 0.3156$.

Finally, for the probability of Z occurring in some given range, we need two separate probabilities. For $P(a \leq Z \leq b) = P(Z \leq b) - P(Z \leq a)$, we compute $P(Z \leq b)$ and $P(Z \leq a)$ separately.

Example 35.11. As an example of computing the probability of Z in some range, consider the computation of

$$P(1.22 \leq Z \leq 1.92) = P(Z \leq 1.92) - P(Z \leq 1.22) = 0.9726 - 0.8888 = 0.0838.$$

The analogous computation is given in Figure 35.9.

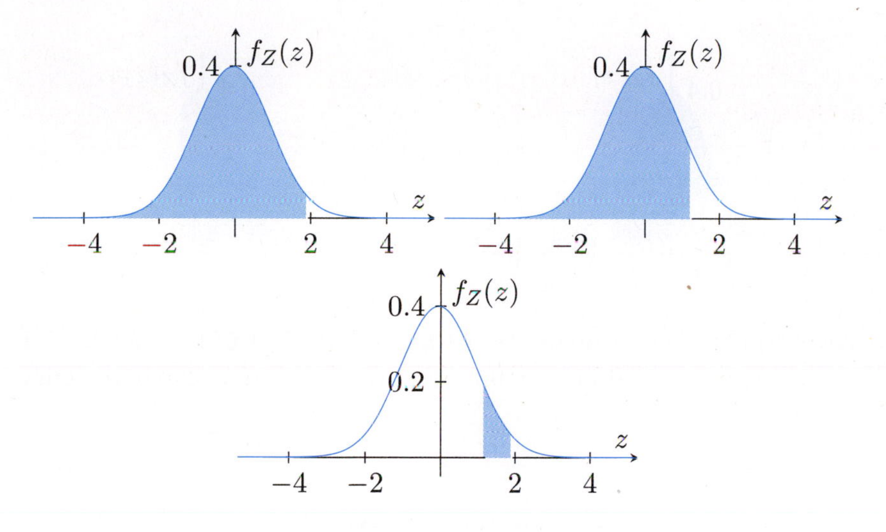

FIGURE 35.9: Step-by-step method for computing $P(1.22 \leq Z \leq 1.92) = P(Z \leq 1.92) - P(Z \leq 1.22) = 0.9726 - 0.8888 = 0.0838$ in stages. Upper left: The shaded area is $P(Z \leq 1.92) = 0.9726$. Upper right: The shaded area is $P(Z \leq 1.22) = 0.8888$. Bottom: The shaded area is $P(1.22 \leq Z \leq 1.92) = P(Z \leq 1.92) - P(Z \leq 1.22) = 0.9726 - 0.8888 = 0.0838$.

Example 35.12. Consider $P(-1.51 \leq Z \leq -0.57)$, in which neither value seems to be on the table. By symmetry, this is equal to $P(0.57 \leq Z \leq 1.51)$. We first turn the range of negative numbers into a range of positive numbers, and then use the Normal table to get the relevant values.

$$\begin{aligned} P(-1.51 \leq Z \leq -0.57) &= P(0.57 \leq Z \leq 1.51) \\ &= P(Z \leq 1.51) - P(Z \leq 0.57) \\ &= 0.9345 - 0.7157 \\ &= 0.2188. \end{aligned}$$

The plots of the relevant Normal curves are given in Figures 35.10 and 35.11.

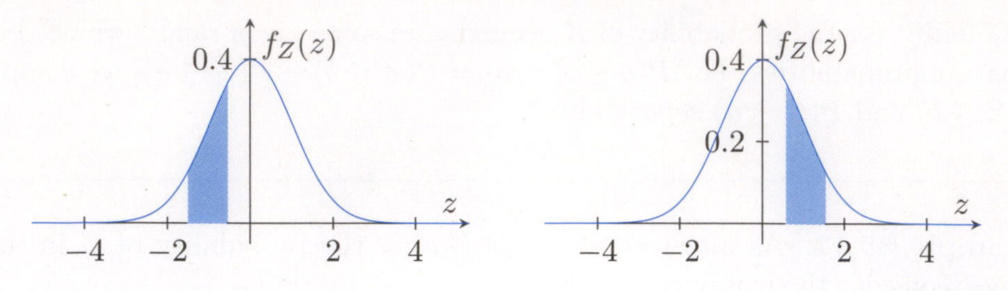

FIGURE 35.10: Left: The shaded area is $P(-1.51 \le Z \le -0.57)$. Right: The shaded area is $P(0.57 \le Z \le 1.51)$ (mirror image).

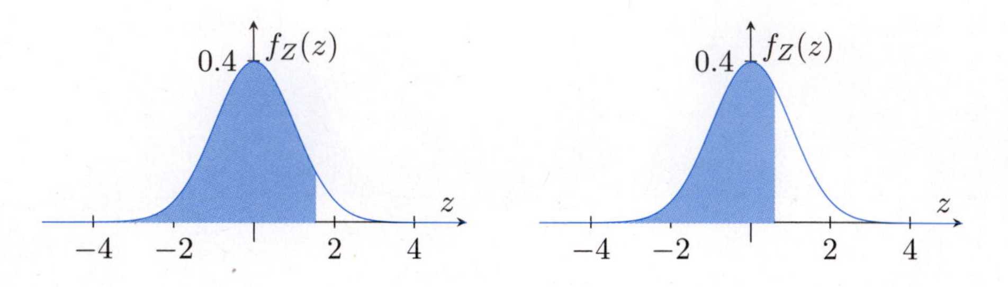

FIGURE 35.11: The probability $P(0.57 \le Z \le 1.51) = P(Z \le 1.51) - P(Z \le 0.57) = 0.9345 - 0.7157 = 0.2188$ is the same area under the curve as in Figure 35.10.

Example 35.13. Consider $P(-0.50 \le Z \le 1.50)$, in which only one value is on the table. The value that is not on the table will have to be handled with symmetry. We decompose the desired probability as:

$$P(-0.50 \le Z \le 1.50) = P(Z \le 1.50) - P(Z \le -0.50).$$

The latter probability, $P(Z \le -0.50)$, is not in the Normal table. So we use a mirror image and then a complementary probability:

$$\begin{aligned} P(Z \le -0.50) &= P(Z \ge 0.50) \\ &= 1 - P(Z \le 0.50) \\ &= 1 - 0.6915 \end{aligned}$$

So the entire calculation is:

$$\begin{aligned} P(-0.50 \le Z \le 1.50) &= P(Z \le 1.50) - P(Z \ge 0.50) \\ &= P(Z \le 1.50) - (1 - P(Z \le 0.50)) \\ &= 0.9332 - (1 - 0.6915) \\ &= 0.6247 \end{aligned}$$

The shaded area corresponding to the probability $P(-0.50 \leq Z \leq 1.50) = 0.6247$ is given in Figures 35.12.

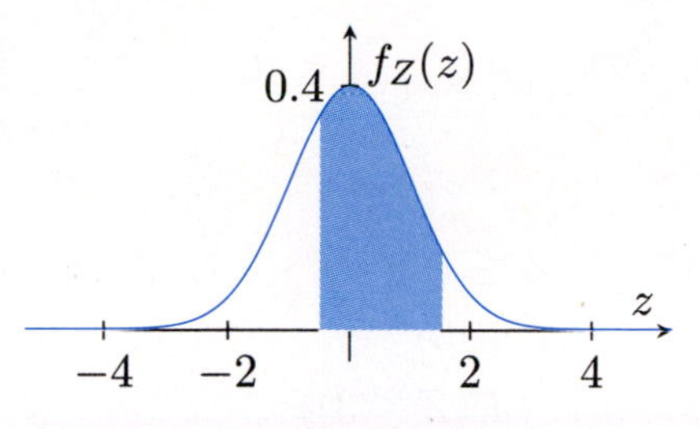

FIGURE 35.12: The area under the curve is $P(-1.51 \leq Z \leq -0.57) = P(0.57 \leq Z \leq 1.51) = P(Z \leq 1.51) - P(Z \leq 0.57) = 0.9345 - 0.7157 = 0.2188$.

Example 35.14. It is helpful to know the probability that X is within one or two or three standard deviations of its expected value, μ_X. The probability that X is within one standard deviation of its μ_X is

$$\begin{aligned}
&P(\mu_X - \sigma_X \leq X \leq \mu_X + \sigma_X) \\
&\quad = P\left(\frac{\mu_X - \sigma_X - \mu_X}{\sigma_X} \leq \frac{X - \mu_X}{\sigma_X} \leq \frac{\mu_X + \sigma_X - \mu_X}{\sigma_X}\right) \\
&\quad = P(-1 \leq Z \leq 1) \\
&\quad = P(Z \leq 1) - P(Z \leq -1) \\
&\quad = P(Z \leq 1) - P(Z \geq 1) \\
&\quad = P(Z \leq 1) - (1 - P(Z \leq 1)) \\
&\quad = 0.8413 - (1 - 0.8413) \\
&\quad = 0.6826
\end{aligned}$$

The probability that X is within two standard deviations of μ_X is

If X is between $\mu_X - 2\sigma_X$ and $\mu_X + 2\sigma_X$ with probability 95%, how often is X above $\mu_X + 2\sigma_X$?

$$\begin{aligned}
&P(\mu_X - 2\sigma_X \leq X \leq \mu_X + 2\sigma_X) \\
&\quad = P\left(\frac{\mu_X - 2\sigma_X - \mu_X}{\sigma_X} \leq \frac{X - \mu_X}{\sigma_X} \leq \frac{\mu_X + 2\sigma_X - \mu_X}{\sigma_X}\right) \\
&\quad = P(-2 \leq Z \leq 2) \\
&\quad = P(Z \leq 2) - P(Z \leq -2) \\
&\quad = P(Z \leq 2) - P(Z \geq 2) \\
&\quad = P(Z \leq 2) - (1 - P(Z \leq 2)) \\
&\quad = 0.9772 - (1 - 0.9772) \\
&\quad = 0.9544
\end{aligned}$$

The probability that X is within three standard deviations of μ_X is

$$\begin{aligned}
P(\mu_X - 3\sigma_X \leq X \leq \mu_X + 3\sigma_X) \\
&= P\left(\frac{\mu_X - 3\sigma_X - \mu_X}{\sigma_X} \leq \frac{X - \mu_X}{\sigma_X} \leq \frac{\mu_X + 3\sigma_X - \mu_X}{\sigma_X}\right) \\
&= P(-3 \leq Z \leq 3) \\
&= P(Z \leq 3) - P(Z \leq -3) \\
&= P(Z \leq 3) - P(Z \geq 3) \\
&= P(Z \leq 3) - (1 - P(Z \leq 3)) \\
&= 0.9987 - (1 - 0.9987) \\
&= 0.9974
\end{aligned}$$

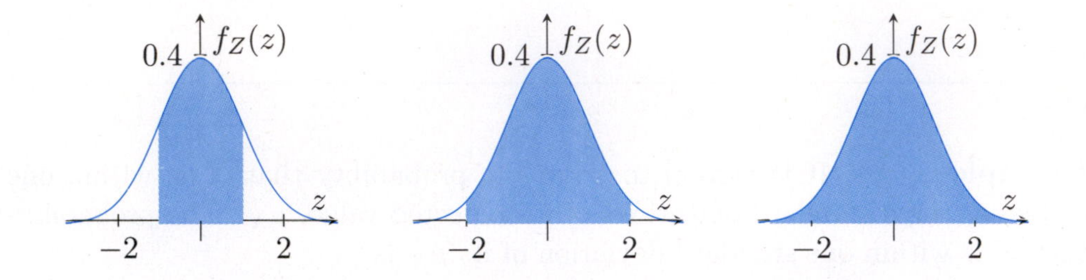

FIGURE 35.13: Left: The shaded area shows $P(\mu_X - \sigma_X \leq X \leq \mu_X + \sigma_X) = P(-1 \leq Z \leq 1) = 0.6826$. Middle: The shaded area shows $P(\mu_X - 2\sigma_X \leq X \leq \mu_X + 2\sigma_X) = P(-2 \leq Z \leq 2) = 0.9544$. Right: The shaded area shows $P(\mu_X - 3\sigma_X \leq X \leq \mu_X + 3\sigma_X) = P(-3 \leq Z \leq 3) = 0.9974$.

Remark 35.15. For quick calculations—the Empirical Rule
People sometimes use the "68–95–99.7% rule" for Normal curves. This means that if your distribution is Normal-shaped, then approximately:
68% of the distribution will fall between $\mu_X - \sigma_X$ and $\mu_X + \sigma_X$. 95% of the distribution will fall between $\mu_X - 2\sigma_X$ and $\mu_X + 2\sigma_X$. 99.7% of the distribution will fall between $\mu_X - 3\sigma_X$ and $\mu_X + 3\sigma_X$.

Now we consider some examples that require us to convert Normal random variables (which are not standard) into standard Normal random variables. We will first convert the Normal random variable X to a standard Normal random variable Z by using $Z = \frac{X-\mu_X}{\sigma_X}$. Then we will use the appropriate inequality changes to make the form match the Normal table. Finally, we will look up the probability from the table for the standard Normal distribution.

Example 35.16. Checking account balances at a bank are approximately Normally distributed with expected value \$1325 and standard deviation \$250. Bill has a balance of \$775. For each question below, we we give the related curve for the standard Normal distribution with the relevant area shaded.

a. What is the probability that an account will have less money than Bill's account?

We let $X \sim \mathcal{N}(\mu_X = 1325, \sigma_X^2 = 62500)$ be the amount in the selected account. We compute

$$\begin{aligned} P(X < 775) &= P\left(\frac{X - \mu_X}{\sigma_X} < \frac{775 - 1325}{250}\right) \\ &= P(Z < -2.20) \\ &= P(Z > 2.20) \\ &= 1 - P(Z \leq 2.20) \\ &= 1 - 0.9861 \\ &= 0.0139. \end{aligned}$$

See Figure 35.14 (left side) to visualize this probability.

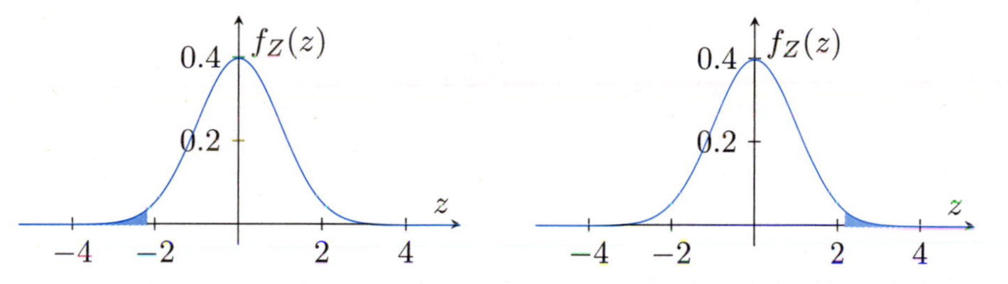

FIGURE 35.14: Left: The shaded area is $P(Z \leq -2.20) = 0.0139$. Right: The shaded area is $P(Z > 2.20) = 0.0139$.

b. What is the probability that an account will have more than \$1875?

Again let $X \sim \mathcal{N}(1325, 62500)$. Then

$$\begin{aligned} P(X > 1875) &= P\left(\frac{X - \mu_X}{\sigma_X} > \frac{1875 - 1325}{250}\right) \\ &= P(Z > 2.20) \\ &= 1 - P(Z \leq 2.20) \\ &= 1 - 0.9861 \\ &= 0.0139. \end{aligned}$$

See Figure 35.14 (right side) to visualize this as the shaded area under the Normal curve.

c. What is the probability that an account will have exactly \$1875?

Again let $X \sim \mathcal{N}(1325, 62500)$. Then $P(X = 1875) = 0$, because the probability that *any* continuous random variable equals a particular value is always zero (this holds true in general, not just for Normal random variables).

d. What is the probability that an account will have less than \$1325, which is the expected value of the money in an account?

Since the density of Normal random variables are always symmetric about the expected value, we do not need a calculator for this value. We know that a Normal random variable is less than its expected value exactly half the time, so the desired probability is $1/2$. If we want to calculate this quantity formally, we can again let $X \sim \mathcal{N}(1325, 62500)$, and then:

$$\begin{aligned} P(X < 1325) &= P\left(\frac{X - \mu_X}{\sigma_X} < \frac{1325 - 1325}{250}\right) \\ &= P(Z < 0) \\ &= 0.5000 \end{aligned}$$

See Figure 35.15 to visualize the probability.

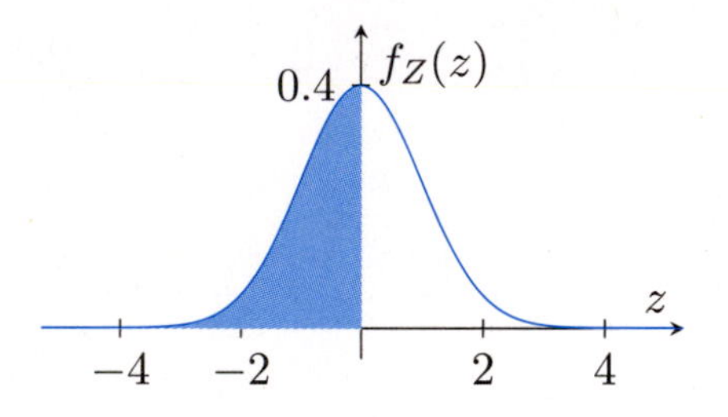

FIGURE 35.15: The probability $P(Z < 0) = 0.5000$ is the area under the curve.

e. What is the probability that an account will have less than \$10?

Again let $X \sim \mathcal{N}(1325, 62500)$. Then

$$\begin{aligned} P(X < 10) &= P\left(\frac{X - \mu_X}{\sigma_X} < \frac{10 - 1325}{250}\right) \\ &= P(Z < -5.26) \\ &= 0 \end{aligned}$$

The reason that $P(Z < -5.26) \approx 0$ is that Z is very well concentrated about the mean; the probability $P(Z < z)$ is practically 0 for relatively small values of z, e.g., for $z < -3$.

f. What is the probability that an account will have between \$1075 and \$1825?

Again let $X \sim \mathcal{N}(1325, 250)$. Then

$$\begin{aligned} P(1075 < X < 1825) &= P\left(\frac{1075 - 1325}{250} < \frac{X - \mu_X}{\sigma_X} < \frac{1825 - 1325}{250}\right) \\ &= P(-1 < Z < 2) \\ &= P(Z < 2) - P(Z < -1) \end{aligned}$$

The probability $P(-1 < Z < 2)$ is given in Figure 35.16.

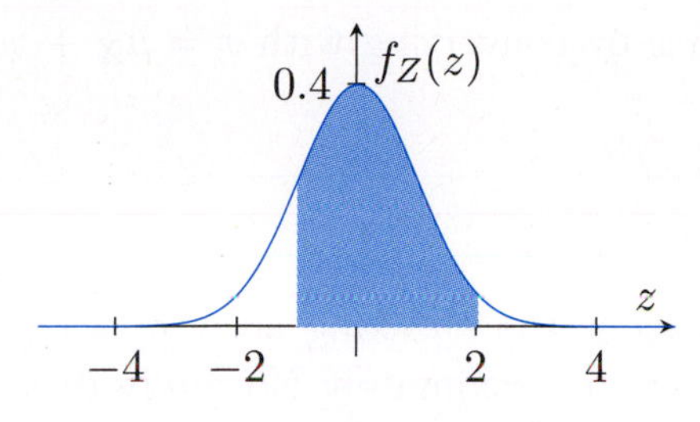

FIGURE 35.16: The probability $P(-1 < Z < 2) = 0.8185$ is the area under the curve.

We can look up $P(Z < 2)$ in the table directly, i.e., $P(Z < 2) = 0.9772$. On the other hand, we must manipulate $P(Z < -1)$ learning the methods we learning earlier:

$$P(Z < -1) = P(Z > 1) = 1 - P(Z < 1) = 1 - 0.8413 = 0.1587.$$

So we get

$$P(1075 < X < 1825) = 0.9772 - 0.1587 = 0.8185.$$

So an account has between \$1075 and \$1825 with probability 0.8185.

35.3 "Backward" Normal Problems

In the previous problems, we knew the x-value and we wanted to find the probability $P(X \leq x)$ or $P(X \geq x)$.

What if you know the probability or percentile (for example, "top quarter" or "bottom tenth" or "middle 50%"), but you don't know the cut-off x-value that will give you this probability? We will call this situation a "backward" Normal problem because you solve for the x-value in a procedure that is backward from what you did in the previous types of problems in this chapter.

This is like working from the inside to the outside.

1. Start with $P(Z \leq z)$ as the given probability. Work backward from the probability in Normal table to get the desired corresponding z value.

2. Use the complement if necessary, e.g., $P(Z > z) = 0.2$ is the same as $P(Z < z) = 1 - 0.2 = 0.8$.

   If you have a two-sided probability, use

$$\begin{aligned} P(-z < Z < z) &= P(Z < z) - P(Z < -z) \\ &= P(Z < z) - P(Z > z) \\ &= P(Z < z) - (1 - P(Z < z)) \\ &= 2P(Z < z) - 1 \end{aligned} \tag{35.2}$$

3. Convert the z to x by converting with $x = \mu_X + z\sigma_X$.

We rearrange $Z = (X - \mu_X)/\sigma_X$ to solve for X.

Example 35.17. Consider the checking account scenario in Example 35.16, in which the balances are approximately Normally distributed with a mean of \$1325 and a standard deviation of \$250.

a. Find the range for the top 15% of balances.

We begin with $P(Z > z) = 0.15$, but again this is not in the table for the Normal distribution, so we must use the complement. We have

$$0.15 = P(Z > z) = 1 - P(Z \leq z),$$

as shown on the left side of Figure 35.17. Therefore, we also have

$$P(Z \leq z) = 1 - 0.15 = 0.85,$$

as shown on the right side of Figure 35.17. Looking in the chart, this means $z = 1.04$. We still need to convert from Z, the number of standard deviations

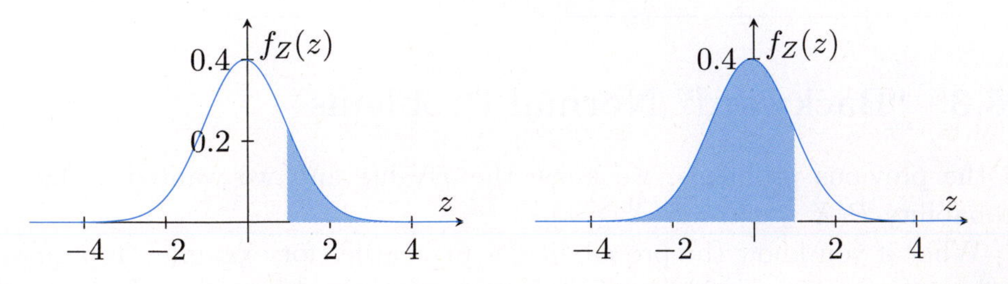

FIGURE 35.17: Left: The shaded area is $P(Z > z) = 0.15$. Right: The shaded area is $P(Z \leq z) = 0.85$.

away from the mean our account balance is, to actual account balances.

$$x_0 = \mu_X + z\sigma_X$$
$$x_0 = 1325 + (1.04)(250)$$
$$x_0 = 1585$$

Now we can put this back into our inequality expression:

$$0.15 = P(Z > 1.04) = P(X > 1585).$$

b. Find the range for the bottom 20% of balances.

We start with $P(Z \leq z) = 0.20$, but none of the probabilities on the table for the Normal distribution (at the end of this chapter) are smaller than 0.5, so we must use the complement. The technique is the same as in Example 35.10. We describe this in a series of three plots, in Figure 35.18.

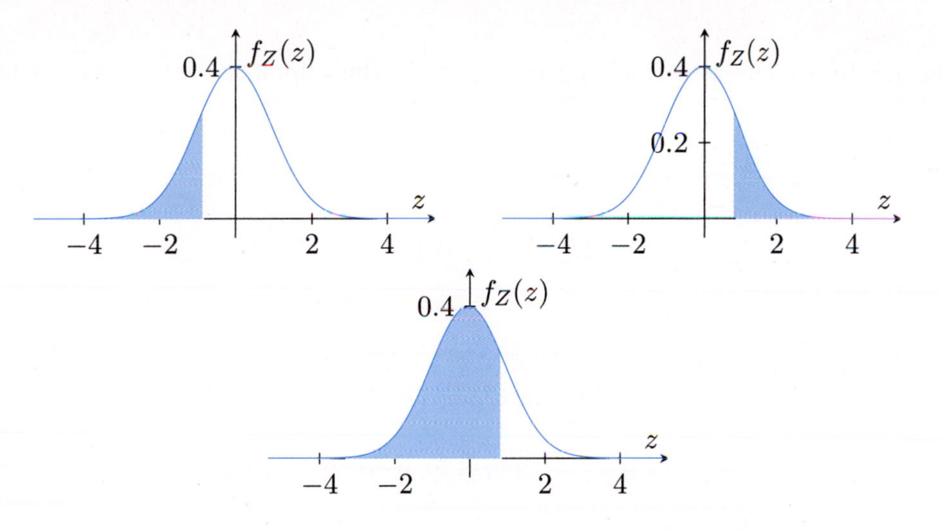

FIGURE 35.18: Upper left: The shaded area is $P(Z \leq z) = 0.20$. Upper right: The shaded area is $P(Z \geq -z) = 0.20$ (a mirror image). Bottom: The shaded area is $P(Z \leq -z) = 0.80$, so $-z = 0.84$ (see the Normal table). So the upper left area is $P(Z \leq -0.84) = 0.20$, and the upper right area is $P(Z \geq 0.84) = 0.20$.

We know that z will be negative because the desired probability $P(Z \leq z) = 0.20$ is less than 0.5 (i.e., is not found on the chart for the standard Normal table); see Figure 35.18 (left side). The probability is the same if we look at the mirror image, but now our cut-off value is $-z$, i.e., $P(Z \geq -z) = 0.20$; see Figure 35.18 (middle). Finally, we need to take the complement of this probability so that we can look up the value on the table, i.e.,

When we use the Normal table to work backward, we go from the inside (probabilities) to outside (z-values).

$$0.20 = P(Z \leq z) = P(Z \geq -z) = 1 - P(Z \leq -z).$$

See Figure 35.18 (right side).

Thus

$$P(Z \leq -z) = 1 - 0.20 = 0.80$$

So the Normal table gives us $-z = 0.84$, i.e., $z = -0.84$. Now we use this information about the standard Normal random variable Z to convert the cutoff from z's into x's. We can relate the cutoffs x and z such that $P(Z \leq z) = 0.20 = P(X \leq x)$ as follows: $x = \mu_X + z\sigma_X$, so we get

$$x = 1325 + (-0.84)(250) = 1115.$$

Thus, 20% of the balances are less than \$1115, i.e.,

$$P(Z \leq -0.84) = 0.20 = P(X \leq 1115).$$

Example 35.18. Between what two central values do 40% of the balances fall?

We have

$$\begin{aligned} 0.40 &= P(-z \leq Z \leq z) \\ &= 2P(Z < z) - 1 \qquad \text{by (35.2)} \end{aligned}$$

So $2P(Z < z) = 1.40$, and thus $P(Z < z) = 0.70$. So the table tells us $z = 0.52$, or perhaps $z = 0.53$. We could compromise and use $z = 0.525$. Thus, the upper balance is

$$X = \mu_X + \sigma_X z = 1325 + (250)(0.525) = 1456.25,$$

and the lower balance is

$$X = \mu_X + \sigma_X z = 1325 - (250)(0.525) = 1193.75.$$

35.4 Summary: "Forward" Vs. "Backward" Normal

Forward:

You know x and you are looking for a probability of percentage, e.g., find $P(X \geq 2)$. Convert X to Z at beginning, using $Z = (X - \mu_X)/\sigma_X$. Use the Normal table from the outside (z-values) to the inside (probabilities).

Backward:

You know the probability or percentage, and you want x, e.g., such that $P(X \geq x) = 0.25$. Use the Normal table from the inside (probabilities) to the outside (z-values). Convert Z to X at the end, using $X = \mu_X + \sigma_X Z$.

35.5 Exercises

35.5.1 Practice

Exercise 35.1. Use the Normal table to find the following probabilities starting from Z. Also sketch a standard Normal curve, and shade the region corresponding to the given probability.

a. $P(Z < 1.47)$

b. $P(Z > 1.47)$

c. $P(Z < -1.47)$

d. $P(Z > -1.47)$

e. $P(Z = -1.47)$

Exercise 35.2. Use the Normal table to find the following probabilities starting from Z. Also sketch a standard Normal curve, and shade the region corresponding to the given probability.

a. $P(Z \leq 0.19)$

b. $P(Z \leq 1.90)$

c. $P(Z \geq 9.10)$

d. $P(0.19 < Z < 1.90)$

e. $P(Z \geq -1.90)$

f. $P(Z = -1.90)$

Exercise 35.3. Assume that X has a Normal distribution with a mean of 2 and a standard deviation of 3. Use the Normal table to find the following probabilities starting from X. Also sketch a standard Normal curve, and shade the region corresponding to the given probability.

a. $P(X < 1.62)$

b. $P(X > -8.49)$

c. $P(-4 < X < 1)$

Exercise 35.4. Assume that X has a Normal distribution with a mean of -1.32 and a standard deviation of 0.34. Use the Normal table to find the following probabilities starting from X. Also sketch a standard Normal curve, and shade the region corresponding to the given probability.

a. $P(X > -2)$

b. $P(X < 2.56)$

c. $P(1.47 < X < 4.12)$

Exercise 35.5. Use the standard Normal table to find the following cut-off values for Z. Also sketch a standard Normal curve, and shade the region corresponding to the given probability.

a. $P(Z < z) = 0.95$

b. $P(Z > z) = 0.15$

c. $P(-z < Z < z) = 0.65$

Exercise 35.6. Use the standard Normal table to find the following cut-off values for Z. Also sketch a standard Normal curve, and shade the region corresponding to the given probability.

a. $P(Z < z) = 0.5$

b. $P(Z > z) = 0.87$

c. $P(-z < Z < z) = 0.25$

Exercise 35.7. Assume that X has a Normal distribution with a mean of 2 and a standard deviation of 3. Use the standard Normal table to find the following cut-off values for X. Also sketch a standard Normal curve, and shade the region corresponding to the given probability.

a. $P(X < x) = 0.78$

b. $P(X > x) = 0.21$

c. What are the two central values such that 90% of the X values are in the range between these two numbers?

Exercise 35.8. Assume that X has a Normal distribution with a mean of -1.32 and a standard deviation of 0.34. Use the standard Normal table to find the following cut-off values for X. Also sketch a standard Normal curve, and shade the region corresponding to the given probability.

a. What is the cut-off for the top 5% of X values?

b. What is the cut-off for the bottom 10% of X values?

c. What are the boundaries for the middle 50% of X values?

Exercise 35.9. Find the value of a so that, if Z is a standard Normal random variable, then

$$P(a \leq Z \leq 0.54) = 0.3898.$$

Exercise 35.10. **Exam scores.** The students in my class have Exam 1 scores which are Normally distributed with a mean of 75 and a standard deviation of 9. If a student is selected at random,

a. What is the probability the student will have a score of more than 90 (an A)?

b. What is the probability the student will have a score of less than 60 (an F)?

c. What is the probability a student will have a score between 80 and 89 (the B range)?

d. What is the range of scores for the middle 50% of the student scores?

e. What is the range for the central 99.7% of the scores?

f. What is the lower cut-off for the top 3/4 of the students?

Exercise 35.11. Sugary candy. The quantity of sugar X (measured in grams) in a randomly selected piece of candy is Normally distributed, with expected value $\mathbb{E}(X) = \mu_X = 22$ and variance $\text{Var}(X) = \sigma_X^2 = 8$. Find the probability that a randomly selected piece of candy has less than 20 grams of sugar.

Exercise 35.12. Annual precipitation. Assume that the annual precipitation in a student's hometown is Normally distributed, with expected value $\mu_X = 36.3$ inches and variance $\sigma_X^2 = 8.41$. A rare species of frog lives in the town. This rare species of frog is known to reproduce during the year only if the annual precipitation is between 35 and 39 inches. What is the probability that the species of frog is able to reproduce this year?

Exercise 35.13. Getting to class. The distance a student lives (in miles) from their probability classroom is approximately Normally distributed with a mean of 3 miles and a standard deviation of 1.2 miles.

a. How far away do the closest 10% of students live?

b. What is the probability that a student will live too close to get a parking permit (less than 1 mile away)?

c. What is the probability that a student will live further away than 5 miles or less than 1 mile away?

Exercise 35.14. Movie length. Children's movies run an average of 98 minutes with a standard deviation of 10 minutes. You check out a movie, selected at random without reading the running time on the box, from the library to entertain your kids so you can study for your probability test. Assume that your kids will be occupied for the entire length of the movie.

a. What is the probability that your kids will be occupied for at least the 2 hours you would like to study?

b. What is range for the bottom quartile (lowest 25%) of time they will be occupied?

c. What are the limits for the central 95% of times your kids will be occupied?

Exercise 35.15. Weighing beagles. Let X be the weight (in pounds) of a beagle. Then X is Normally distributed with $\mu_X = 17.2$ pounds and $\sigma_X^2 = 3.2$. Find the probability that a beagle weighs 20 pounds or less.

Exercise 35.16. Weighing elephants. A certain type of elephant has average weight 11,000 pounds, with standard deviation of 1,000 pounds. Find the probability that such an elephant's weight exceeds 13,000 pounds.

Exercise 35.17. Can of soda. The quantity of liquid in a can of soda is Normally distributed with average 11.92 ounces and standard deviation 0.05 ounces. Find the probability that the can of soda contains 12 ounces of liquid or more.

Exercise 35.18. Blonde hairs. Suppose that a randomly chosen blonde-haired woman has 140,000 hairs on their head, with standard deviation 20,000. Find the probability that such a person has fewer than 150,000 hairs on her head.

Exercise 35.19. Sunflower heights. A full-grown sunflower stands 12 feet tall, on average, with standard deviation of 1 foot. Find the probability that a sunflower is between 11 to 13 feet tall.

Exercise 35.20. Weighing pumpkins. In a certain field, the weight of pumpkins is 14 pounds, with standard deviation of 3.7 pounds. What is the probability of finding a pumpkin that weighs 15 pounds or more?

Exercise 35.21. Physics exams. In a large physics course, students have average score of 82 percent, with standard deviation of 5 percent. Find the probability that a randomly chosen student's score exceeds 90 percent.

Exercise 35.22. Drag race. In a quarter-mile drag race, the average time of completion is 13.2 seconds, with standard deviation of 0.11 seconds. Find the probability that a car completes the race in 13 seconds or less.

Exercise 35.23. Weighing cereal. Chocolate-coated Sugar Fun Blast cereal is filled into boxes by weight, with the weight approximately Normally distributed with an average of 16 ounces and a standard deviation of 0.2 ounces.

a. What is the probability that the box you buy (chosen at random) will weigh less than 15.5 ounces?

b. What is the probability that the box you buy will weigh between 16 and 16.25 ounces?

c. What is the probability that the box you buy will weigh more than 17 ounces?

d. What is the upper cut-off for the 90th percentile (bottom 90%) of weights?

e. What is the range for the middle 60% of weights?

f. What is the range for the upper 2.5%?

35.5.2 Extensions

Exercise 35.24. Female heights. Assume that the height of an American female is Normal with expected value $\mu = 64$ and standard deviation $\sigma = 2.5$.

a. What is the probability that an American female's height is 66 inches or taller?

b. The heights of 10 American females are measured (in inches). Let Y be the number of the 10 females whose height is 66 inches or taller. Find $P(Y = 7)$. What is the distribution of Y?

35.5.3 Advanced

Exercise 35.25. We have not actually verified here that the density of a Normal random variable integrates to 1, because the usual argument to do this requires a perhaps surprising conversion to polar coordinates and a double integral; see, for instance, Section 5.4 of Ross [5] or Section 5.3 of Pitman [4]. Construct such an argument that $\int_{-\infty}^{\infty} f_Z(z)\,dz = 1$.

Chapter 36

Sums of Independent Normal Random Variables

Probability is a mathematical discipline whose aims are akin to those, for example, of geometry of analytical mechanics. In each field we must carefully distinguish three aspects of the theory:

(a) the formal logical content,

(b) the intuitive background,

(c) the applications.

The character, and the charm, of the whole structure cannot be appreciated without considering all three aspects in their proper relation.

—*An Introduction to Probability Theory and its Applications, Volume 1, 3rd edition* by William Feller (Wiley, 1971)

What is the total amount of food eaten by all of the students in a dining hall on a given day? In a given month? In a given year? How can we scale our results to compare the average, variance, and standard deviation of these food totals?

36.1 Sums of Independent Normal Random Variables

We already demonstrated in the last chapter that the Normal random variable is really useful in practice because many random variables encountered in real life are Normal random variables. Many quantities associated with physical life are Normally distributed (as we mentioned, lengths, volumes, widths, heights, etc., related to all kinds of animals and plants and other physical beings, are often Normally distributed). On these grounds alone, Normal random variables are beautiful and important enough to study in a course of probability.

We emphasize that Normal random variables are also extremely important for another reason! Informally stated:

When many independent random variables (not necessarily Normal) are added together, the sum of the random variables is approximately Normal. We will begin to formalize this idea in Chapter 37. Also, the average of many independent random variables is approximately Normal too. The study of these "limit laws" is a major area of probability theory, in which people investigate the limiting properties of random variables. One of the most important parts of the theory of limit laws is devoted to the study of Normal random variables and their usefulness in approximating the sum of random variables and the averages of random variables. In Chapter 37, we will state some of the most commonly used limit laws and discuss their applications. Some of the results are useful and pleasantly surprising! Before studying limit laws, however, we discuss sums of independent Normal random variables.

In the previous chapter, Theorem 35.6 gave us the remarkable fact:

If X is a Normal random variable and r, s are any two constants, then $rX + s$ is a Normal random variable with standard deviation $r\sigma_X$ and expected value $r\mu_X + s$.

The following theorem, to be proved in Section 36.2, is also very useful:

Theorem 36.1. The sum of independent Normal random variables is a Normal random variable too (First Version)

If $X_1, X_2, \ldots, X_n$ are *independent Normal* random variables, with expected values $\mu_1, \mu_2, \ldots, \mu_n$, respectively, and with variances $\sigma_1^2, \sigma_2^2, \ldots, \sigma_n^2$, respectively, then

$$X_1 + \cdots + X_n \text{ is also a Normal random variable}$$

with expected value $\mu_1 + \cdots + \mu_n$ and variance $\sigma_1^2 + \cdots + \sigma_n^2$.

The amazing part of the fact above is that if $X_1, \ldots, X_n$ are independent Normal random variables, then $X_1 + \cdots + X_n$ is a *Normal random variable too.* The expected value must be

$$\mathbb{E}(X_1 + \cdots + X_n) = \mathbb{E}(X_1) + \cdots + \mathbb{E}(X_n) = \mu_1 + \cdots + \mu_n;$$

(this is not surprising at all; this works for *all sums* of all random variables, with or without the independent assumption and with or without the assumption that the random variables are Normal). The variance must be

$$\text{Var}(X_1 + \cdots + X_n) = \text{Var}(X_1) + \cdots + \text{Var}(X_n) = \sigma_1^2 + \cdots + \sigma_n^2;$$

(this is not surprising either; this works for all sums of *independent* random variables, with or without the assumption that the random variables are Normal). Once again, we emphasize that the *delightful part* of the boxed statement above is that the sum of independent Normals is *Normal* too.

In the case in which the Normal random variables are not only independent, but also all have the same expected values and the same variances, the boxed formula above has the following nice form:

Corollary 36.2. The sum of independent Normal random variables is a Normal random variable too (Identically Distributed Version)

If $X_1, X_2, \ldots, X_n$ are *independent Normal* random variables, which each have expected value μ and variance σ^2, then

$$X_1 + \cdots + X_n \text{ is also a Normal random variable}$$

with expected value $n\mu$ and variance $n\sigma^2$.

These two nice, boxed results have analogous versions, if we subtract the expected value from the sum of the random variables and divide by the variance. In the case of the first version, we get:

Corollary 36.3. The sum of independent Normal random variables, properly scaled, is a standard Normal random variable (First Version)

If $X_1, X_2, \ldots, X_n$ are *independent Normal* random variables, with expected values $\mu_1, \mu_2, \ldots, \mu_n$, respectively, and with variances $\sigma_1^2, \sigma_2^2, \ldots, \sigma_n^2$, respectively, then

$$\frac{X_1 + \cdots + X_n - (\mu_1 + \cdots + \mu_n)}{\sqrt{\sigma_1^2 + \cdots + \sigma_n^2}} \text{ is a } \textit{standard} \text{ Normal random variable.}$$

In the case of the second version, we get:

Corollary 36.4. The sum of independent Normal random variables, properly scaled, is a standard Normal random variable (Alternate Version)

If $X_1, X_2, \ldots, X_n$ are *independent Normal* random variables, which each have expected value μ and variance σ^2, then

$$\frac{X_1 + \cdots + X_n - n\mu}{\sqrt{n\sigma^2}} \text{ is a } \textit{standard} \text{ Normal random variable.}$$

Example 36.5. In Exercise 35.11, we encountered a type of candy such that the quantity of sugar X (measured in grams) in a randomly selected piece of candy is Normally distributed, with expected value $\mathbb{E}(X) = \mu_X = 22$ and variance $\text{Var}(X) = \sigma_X^2 = 8$.

Suppose that Hector eats ten pieces of this candy on Halloween night. What is the probability that he consumed more than 200 grams of sugar?

If $X_1, \ldots, X_{10}$ denote the amount of sugar in the first, second, ..., tenth pieces of candy, respectively, then the total amount of sugar that he ate is $X_1 + \cdots + X_{10}$. We know that the amount of sugar in the jth piece of candy is a Normal random variable $X_j \sim \mathcal{N}(\mu = 22, \sigma^2 = 8)$, so the total amount of sugar is also a Normal random variable with expected value $10\mu = (10)(22) = 220$ and variance $10\sigma^2 = (10)(8) = 80$. So

$$Z = \frac{X_1 + \cdots + X_{10} - 220}{\sqrt{80}}$$

is a standard Normal random variable. With this in mind, we compute the probability that Hector consumed more than 200 grams of sugar. The desired probability is:

$$\begin{aligned} P(X_1 + \cdots + X_{10} > 200) &= P\left(\frac{X_1 + \cdots + X_{10} - 220}{\sqrt{80}} > \frac{200 - 220}{\sqrt{80}}\right) \\ &= P(Z > -2.24) \\ &= P(Z < 2.24) \\ &= 0.9875 \end{aligned}$$

So there is a 98.75% chance that Hector ate more than 200 grams of sugar on Halloween night.

Example 36.6. Consider a species of ant whose body weight (in milligrams) is Normally distributed with mean $\mu = 5$ and variance $\sigma^2 = 1.3$.

Suppose that an ant colony contains 100,000 of these ants. What is the probability that the ant colony weighs more than 500,500 milligrams?

Let $X_1, \ldots, X_{100{,}000}$ denote the weights of the 100,000 ants. So the total weight of the colony is $X_1 + \cdots + X_{100{,}000}$. The weight of each ant is Normally distributed, with $X_j \sim \mathcal{N}(\mu = 5, \sigma^2 = 1.3)$ for each j. Thus the total weight of the colony is also a Normal random variable with expected value $100{,}000\mu = (100{,}000)(5) = 500{,}000$ and variance $100{,}000\sigma^2 = (100{,}000)(1.3) = 130{,}000$. We use this information to shift and scale the sum of the weights. The sum of the weights, after this adjustment, is

$$Z = \frac{X_1 + \cdots + X_{100{,}000} - 500{,}000}{\sqrt{130{,}000}},$$

a standard Normal random variable.

Now we calculate the probability that the colony weighs more than 500,500 milligrams:

$$P\left(\sum_{j=1}^{100,000} X_j > 500{,}500\right) = P\left(\frac{\sum_{j=1}^{100,000} X_j - 500{,}000}{\sqrt{130{,}000}} > \frac{500{,}500 - 500{,}000}{\sqrt{130{,}000}}\right)$$
$$= P(Z > 1.39)$$
$$= 1 - P(Z \leq 1.39)$$
$$= 1 - 0.9177$$
$$= 0.0823$$

Thus, there is an 8.23% chance that the weight of the ant colony exceeds 500,500 milligrams.

Example 36.7. In planning for a new bookstore, a bookseller estimates that the thickness of each book to be sold is Normally distributed, with expected value $\mu = 1.125$ inches and standard deviation $\sigma = 0.25$ inches.

The bookseller is thinking about the thickness of a book of the month to be on a feature display at the front of her store. (If the book is too thick, and the shelf is too thin, the book will fall off the shelf.) She wants to be 95 percent (or more) sure that the book will fit onto her feature display. How thick must her feature display be, to guarantee that this month's book of the month will fit on it? (The feature display must be as deep—or deeper—than the thickness of the book itself.)

If X is the thickness of the book of the month, then

$$Z = \frac{X - 1.125}{0.25}$$

is a standard Normal random variable. Let a be the required thickness of the feature display so that $P(X \leq a) \geq 0.95$. Then we have

$$0.95 \leq P(X \leq a)$$
$$= P\left(\frac{X - 1.125}{0.25} \leq \frac{a - 1.125}{0.25}\right)$$
$$= P\left(Z \leq \frac{a - 1.125}{0.25}\right)$$

Looking at the Normal table, we must have $\frac{a-1.125}{0.25} \geq 1.65$, in order for the probability above to exceed 95%. So we get

$$\frac{a - 1.125}{0.25} \geq 1.65,$$

which simplifies to

$$a \geq (0.25)(1.65) + 1.125 = 1.5375.$$

So the feature display must be at least 1.5375 inches deep, in order to accommodate the book of the month at least 95% of the time.

Example 36.8. In the scenario from the previous example, within the store itself, the bookseller wants to be able to place n books on a regular shelf that is 72 inches long, and still be 95 percent (or more) sure that all of these n books will fit. What is the largest value of n that she can plan to place on each shelf?

Let $X_1, X_2, \ldots$ denote the thickness of the books. The total thickness of n books is $X_1 + \cdots + X_n$. We are given that $X_j \sim \mathcal{N}(\mu = 1.125, \sigma^2 = 0.0625)$, for each j. Thus the total width of the n books on the shelf is also a Normal random variable with expected value $n\mu = 1.125n$ and variance $n\sigma^2 = 0.0625n$. So

$$Z = \frac{X_1 + \cdots + X_n - 1.125n}{\sqrt{0.0625n}}$$

is a standard Normal random variable. The bookseller wants $P(X_1 + \cdots + X_n \leq 72) \geq 0.95$. So we calculate:

$$\begin{aligned}
0.95 &\leq P(X_1 + \cdots + X_n \leq 72) \\
&= P\left(\frac{X_1 + \cdots + X_n - 1.125n}{\sqrt{0.0625n}} \leq \frac{72 - 1.125n}{\sqrt{0.0625n}}\right) \\
&= P\left(Z \leq \frac{72 - 1.125n}{\sqrt{0.0625n}}\right)
\end{aligned}$$

Thus, looking at the chart of the Normal distribution, we see that we must have

$$1.645 \leq \frac{72 - 1.125n}{\sqrt{0.0625n}},$$

or equivalently (rearranging the terms),

$$0 \geq 1.125n + 1.645\sqrt{0.0625n} - 72.$$

To find the value of n where equality is reached (i.e., where 0 is obtained) in the previous equation, we can write: $x = \sqrt{n}$ and $a = 1.125$, $b = 1.645\sqrt{0.0625}$, and $c = -72$. Then we just need to remember that the quadratic equation $0 = ax^2 + bx + c$ has solution $x = \frac{-b \pm \sqrt{b^2 - 4ac}}{2a}$. In this case, we know x (i.e., $\sqrt{n}$) is positive, so we only need the larger of the two solutions. So we get $\sqrt{n} = 7.8193$ and $n = 61.1416$. Thus, the maximum value of n that the bookseller can use, and still maintain the desired conditions, is $n = 61$. In other words, if the bookseller puts 61 books onto a shelf, she can still be at least 95% confident that they will all fit onto the 72 inch shelf.

If X, Y are independent Normal random variables, then $-1X = -X$ is Normal, and if we add Y, Theorem 36.1 implies that $Y - X$ is Normal too.

Corollary 36.9. The difference of independent Normal random variables is a Normal random variable too

If X, Y are independent Normal random variables, then $Y - X$ is a Normal random variable too. We have $\mathbb{E}(Y - X) = \mathbb{E}(Y) - \mathbb{E}(X)$, and $\text{Var}(Y - X) = \text{Var}(Y) + \text{Var}(-X) = \text{Var}(Y) + \text{Var}(X)$.

Example 36.10. As in Exercise 35.24, assume that the height (in inches) of an American female is Normal with expected value $\mu_1 = 64$ and standard deviation $\sigma_1 = 2.5$. Also assume that the height of an American male is Normal with expected value $\mu_2 = 69$ and standard deviation $\sigma_2 = 3.0$. Let X denote the female's height and Y denote the male's height. What is the probability that a randomly selected male is taller than a randomly selected female?

The expected value of $Y - X$ is

$$\mathbb{E}(Y - X) = \mathbb{E}(Y) - \mathbb{E}(X) = 69 - 64 = 5.$$

The variance of $Y - X$ is

$$\text{Var}(Y - X) = \text{Var}(Y) + \text{Var}(X) = 3.0^2 + 2.5^2 = 15.25.$$

The standard deviation of $Y - X$ is $\sqrt{\text{Var}(Y - X)} = \sqrt{15.25} = 3.91$.

The probability that a randomly selected male is taller than a randomly selected female is

$$\begin{aligned} P(Y > X) &= P(Y - X > 0) \\ &= P\left(\frac{Y - X - 5}{3.91} > \frac{0 - 5}{3.91}\right) \\ &= P(Z > -1.28) \\ &= P(Z < 1.28) \\ &= 0.8997 \end{aligned}$$

36.2 Why a Sum of Independent Normals Is Normal (Optional)

Near the start of this Chapter, we claimed that: If $X_1, X_2, \ldots, X_n$ are *independent Normal* random variables, with expected values $\mu_1, \mu_2, \ldots, \mu_n$, respectively, and with variances $\sigma_1^2, \sigma_2^2, \ldots, \sigma_n^2$, respectively, then

$X_1 + \cdots + X_n$ is also a Normal random variable

with expected value $\mu_1 + \cdots + \mu_n$ and variance $\sigma_1^2 + \cdots + \sigma_n^2$.

Here we present all of the details to support that claim.

To see that $X_1 + \cdots + X_n$ is a Normal random variable, *we will show that* $X_1 + X_2$ *is Normal.* This implies that

$$(X_1 + X_2) + X_3 \text{ is Normal,}$$

which implies that

$$(X_1 + X_2 + X_3) + X_4 \text{ is Normal}$$

etc., and by similar reasoning, the sum of n independent Normals is seen to be Normal. To see that $X_1 + X_2$ is Normal, we note that X_1 and X_2 have the same distribution as $\sigma_1 Z_1 + \mu_1$ and $\sigma_2 Z_2 + \mu_2$, where Z_1, Z_2 are independent standard Normal random variables. So it suffices to show that $\sigma_1 Z_1 + \sigma_2 Z_2 + \mu_1 + \mu_2$ is Normal. In fact, using the rule "X is Normal implies $\alpha X + \beta$ is Normal too," with $\alpha = \sigma_1$ and $\beta = \mu_1 + \mu_2$, it is enough to show the simpler statement that $Z_1 + \frac{\sigma_2}{\sigma_1} Z_2$ is Normal. For simplicity, we write $c = \sigma_2/\sigma_1$, and we prove that $Z_1 + cZ_2$ is Normal with expected value 0 and variance $1 + c^2$. We work with the CDF of $Z_1 + cZ_2$ to accomplish this. For any real-valued a, we show that the CDF of $Z_1 + cZ_2$ is

Note: if $\sigma_1 = 0$, do not divide by 0; instead, $\sigma_1 Z_1 + \sigma_2 Z_2 + \mu_1 + \mu_2$ would simplify to $\sigma_2 Z_2 + \mu_1 + \mu_2$ and is thus already known to be Normal, so we'd be done.

$$F_{Z_1+cZ_2}(a) = \int_{-\infty}^{a} \frac{1}{\sqrt{2\pi(1+c^2)}} \exp\left(-\frac{t^2}{2(1+c^2)}\right) dt.$$

The rest of the argument is dedicated to show that $F_{Z_1+cZ_2}(a)$ has this form. We compute

$$\begin{aligned} F_{Z_1+cZ_2}(a) &= P(Z_1 + cZ_2 \le a) \\ &= \int_{-\infty}^{\infty} \int_{-\infty}^{(a-z_1)/c} \frac{1}{\sqrt{2\pi}} \exp\left(-\frac{z_1^2}{2}\right) \frac{1}{\sqrt{2\pi}} \exp\left(-\frac{z_2^2}{2}\right) dz_2\, dz_1. \end{aligned}$$

We want to reverse the order of integration, so we must get rid of the z_1's in the range of z_2. Thus, we use the transformation $t = z_1 + cz_2$ (i.e., $z_2 = (t - z_1)/c$), so $dt = c\, dz_2$, and in particular, when $z_2 = (a - z_1)/c$, we have $t = a$. This yields

$$F_{Z_1+cZ_2}(a) = \int_{-\infty}^{\infty} \int_{-\infty}^{a} \frac{1}{\sqrt{2\pi}} \exp\left(-\frac{z_1^2}{2}\right) \frac{1}{\sqrt{2\pi}} \exp\left(-\frac{(t-z_1)^2}{2c^2}\right) \frac{1}{c}\, dt\, dz_1.$$

Now we expand the $(t - z_1)^2$ and extract the terms that have t's but no z_1's. We also switch the order of integration. We get

$$\begin{aligned} F_{Z_1+cZ_2}(a) &= \int_{-\infty}^{\infty} \int_{-\infty}^{a} \frac{1}{\sqrt{2\pi}} \exp\left(-\frac{z_1^2}{2}\right) \frac{1}{\sqrt{2\pi}} \exp\left(-\frac{t^2 - 2tz_1 + z_1^2}{2c^2}\right) \frac{1}{c}\, dt\, dz_1 \\ &= \int_{-\infty}^{a} \frac{1}{\sqrt{2\pi}} \exp\left(\frac{-t^2}{2c^2}\right) \frac{1}{c} \int_{-\infty}^{\infty} \frac{1}{\sqrt{2\pi}} \exp\left(-\frac{z_1^2}{2} + \frac{2tz_1 - z_1^2}{2c^2}\right) dz_1\, dt. \end{aligned}$$

We simplify the z_1 expression:

$$-\frac{z_1^2}{2}+\frac{2tz_1-z_1^2}{2c^2} = -\frac{c^2z_1^2-2tz_1+z_1^2}{2c^2}$$
$$= -\frac{(1+c^2)z_1^2-2tz_1}{2c^2}$$
$$= -\frac{(1+c^2)}{2c^2}\left(z_1^2-\frac{2tz_1}{1+c^2}\right).$$

Then we "complete the square" on the z_1 expression:

$$-\frac{z_1^2}{2}+\frac{2tz_1-z_1^2}{2c^2} = -\frac{(1+c^2)}{2c^2}\left(z_1^2-\frac{2tz_1}{1+c^2}+\frac{t^2}{(1+c^2)^2}\right)+\frac{(1+c^2)}{2c^2}\frac{t^2}{(1+c^2)^2},$$

which simplifies to

$$-\frac{z_1^2}{2}+\frac{2tz_1-z_1^2}{2c^2} = -\frac{(1+c^2)}{2c^2}\left(z_1-\frac{t}{1+c^2}\right)^2+\frac{t^2}{2c^2(1+c^2)}.$$

Now we substitute back into the expression for the CDF to get

$$F_{Z_1+cZ_2}(a) = \int_{-\infty}^{a}\frac{1}{\sqrt{2\pi}}\exp\left(\frac{-t^2}{2c^2}\right)\frac{1}{c}$$
$$\times\int_{-\infty}^{\infty}\frac{\exp\left(-\frac{(1+c^2)}{2c^2}\left(z_1-\frac{t}{1+c^2}\right)^2+\frac{t^2}{2c^2(1+c^2)}\right)}{\sqrt{2\pi}}\,dz_1\,dt.$$

Combining the exponential expressions with t's, using

$$\frac{-t^2}{2c^2}+\frac{t^2}{2c^2(1+c^2)} = -\frac{t^2}{2(1+c^2)},$$

the CDF expression simplifies to

$$F_{Z_1+cZ_2}(a) = \int_{-\infty}^{a}\frac{1}{\sqrt{2\pi}}\exp\left(-\frac{t^2}{2(1+c^2)}\right)\frac{1}{c}$$
$$\times\int_{-\infty}^{\infty}\frac{1}{\sqrt{2\pi}}\exp\left(-\frac{(1+c^2)}{2c^2}\left(z_1-\frac{t}{1+c^2}\right)^2\right)\,dz_1\,dt.$$

The expression for the z_1 terms almost looks like the integral for the density of a Normal random variable. To see this more clearly, we write $\mu_t = \frac{t}{1+c^2}$ and $\sigma^2 = \frac{c^2}{1+c^2}$, so that we have

$$F_{Z_1+cZ_2}(a) = \int_{-\infty}^{a}\frac{1}{\sqrt{2\pi}}\exp\left(-\frac{t^2}{2(1+c^2)}\right)\frac{1}{c}$$
$$\times\int_{-\infty}^{\infty}\frac{1}{\sqrt{2\pi}}\exp\left(-\frac{1}{2\sigma^2}(z_1-\mu_t)^2\right)\,dz_1\,dt.$$

Then we multiply and divide by a factor of $\frac{\sqrt{1+c^2}}{c} = \frac{1}{\sigma} = \frac{1}{\sqrt{\sigma^2}}$, to get

$$F_{Z_1+cZ_2}(a) = \int_{-\infty}^{a} \frac{1}{\sqrt{2\pi}} \exp\left(-\frac{t^2}{2(1+c^2)}\right) \frac{1}{c} \frac{c}{\sqrt{1+c^2}}$$
$$\times \int_{-\infty}^{\infty} \frac{1}{\sqrt{2\pi\sigma^2}} \exp\left(-\frac{1}{2\sigma^2}(z_1-\mu_t)^2\right) dz_1\, dt.$$

When evaluating the inner integral, the value of t (and thus of μ_t) is fixed, so that the inner integrand, $\frac{1}{\sqrt{2\pi\sigma^2}} \exp\left(-\frac{(z_1-\mu_t)^2}{2\sigma^2}\right)$ is exactly the density of a Normal random variable with expected value μ_t and variance σ^2. So the inner integral evaluates to 1, because we are integrating over all z_1. Thus, the entire expression simplifies to

$$F_{Z_1+cZ_2}(a) = \int_{-\infty}^{a} \frac{1}{\sqrt{2\pi(1+c^2)}} \exp\left(-\frac{t^2}{2(1+c^2)}\right) dt.$$

So, the CDF of $Z_1 + cZ_2$ shows $Z_1 + cZ_2$ is Normal with $\mathbb{E}(Z_1 + cZ_2) = 0$ and $\text{Var}(Z_1 + cZ_2) = 1 + c^2$.

36.3 Exercises

36.3.1 Practice

Exercise 36.1. **Haircuts.** The time that it takes a random person to get a haircut is Normally distributed, with an average of 23.8 minutes and a standard deviation of 5 minutes. Assume that different people have independent times of getting their hair cut. Find the probability that, if there are four customers in a row (with no gaps in between), they will all be finished getting their hair cut in 1.5 hours (altogether) or less.

Exercise 36.2. **Annual precipitation.** As in Exercise 35.12, assume that the annual precipitation in a student's hometown is Normally distributed, with expected value $\mu = 36.3$ inches and variance $\sigma^2 = 8.41$. Also assume that the amount of precipitation is independent in distinct years.

Let $X_1, \ldots, X_{10}$ denote the precipitation in the 10 distinct years of a decade. Find the probability that the total rainfall during the decade exceeds 380 inches.

Exercise 36.3. **Printer pages.** A printer can produce, on average, 30 pages per minute, i.e., one page every 2 seconds. Each page's printing time has standard deviation of 0.3 seconds. If the pages run times are independent, find the probability that the total print time for a 30 page job is 62 seconds or less.

Exercise 36.4. **Counting calories.** Let X denote the number of calories that a person eats in a single day. Suppose that X is Normally distributed with $\mu_X = 2000$ and $\sigma_X^2 = 10{,}000$. If the person's eating habits are independent from day to day, find the probability that the person eats 735,000 calories or more during a 365-day year.

Exercise 36.5. Customers' purchases. At a local store, there are 14 customers during a given evening. Each customer has an expected purchase of \$4.90, with a standard deviation of \$1.50. If the customers' purchases are assumed to be independent and approximately Normal, what is the approximate probability that the revenue exceeds \$74 on a given evening?

Exercise 36.6. Weighing books. At the library, a student checks out 10 books. She estimates that each book has a Normally distributed weight, with mean 12 ounces and standard deviation of 3 ounces. What is the probability that the total weight of the books exceeds 125 ounces altogether?

Exercise 36.7. Candy sticks. A student has 23 candy sticks in a bag, with lengths that are Normally distributed. Each stick is, on average 1.8 cm long, with standard deviation 0.5 cm. What is the probability that the total length of the candy is less than 40 cm?

Exercise 36.8. Waiting for a bus. Dylan waits at the bus stop an average of 6 minutes per day, with a standard deviation of 1.5 minutes. He believes that his waiting times are approximately Normally distributed.

a. If he waits for the bus on 20 mornings, what is the probability that he spends more than 115 minutes at the bus stop altogether in the morning?

b. If he takes into account the evening waiting times too, and he believes that the mornings and evenings both have about the same kind of distributions (e.g., independent, Normally distributed, etc.), approximate the probability that he spends at least 230 minutes at the bus stop altogether in the mornings and the evenings during the month.

Exercise 36.9. Weighing apples. According to `http://www.fowlerfarms.com/apple_a_day.htm`, a certain kind of apple weighs 150 grams, on average. Suppose that the standard deviation of this type of apple is 20 grams. When picking 66 such apples, what is the probability that they weigh (altogether) 9966 grams or more?

Exercise 36.10. Waiting for girls. Preparing for their dates on a Friday night, it takes each girl (approximately) a Normally distributed amount of time to prepare, with an average of 40 minutes (per girl) to get ready, and a standard deviation of 15 minutes. The 50 men who are waiting for them in the lobby are mostly science majors, and they start to wonder: What is the probability that the average waiting time exceeds 36 minutes (where the average is taken over the 50 preparation times of the women)?

36.3.2 Extensions

Exercise 36.11. Tree heights. After planting 20 evergreen tree saplings on his farm, Don wonders about their heights. If the heights are independent and Normally distributed, with average height 3 feet, and standard deviation of 1.2 feet, what is the probability that none of the 20 trees' heights exceeds 5 feet?

Exercise 36.12. Distances between cars. On a busy interstate highway, there are 21 cars in a particular lane. Let $X_1, \ldots, X_{20}$ denote the 20 distances between these 21 cars (i.e., X_1 is the distance between the first and second cars; X_2 is the distance between the second and third cars; etc.). At a particular moment, the X_j's are judged to be approximately Normal, with an average of 500 feet between consecutive cars and standard deviation of 75 feet. Find the probability that the row of 21 cars is less than two miles long (each mile contains 5280 feet) if:

a. The length of each car is assumed to be negligible, i.e., we do not take into account the lengths of the 21 cars themselves.

b. The length of each car is assumed to be fixed, i.e., 13.5 feet long. Thus, we want the probability of

$$X_1 + \cdots + X_{20} + (13.5)(21) < (5280)(2).$$

c. The lengths of the cars are assumed to be independent and Normally distributed too, each with average length 13.5 feet and standard deviation 1 foot. In this case, if $Y_1, \ldots, Y_{21}$ are the length of the cars Thus, we want the probability of

$$X_1 + \cdots + X_{20} + Y_1 + \cdots + Y_{21} < (5280)(2).$$

Exercise 36.13. Long jumps. An athletic director is recording long-jump scores for a group of students. The expected value of each of their long jumps is 7 meters, and the standard deviation is 0.2 meters. If 20 such jumps are independent and approximately Normally distributed, find the probability that the average of these 20 jumps is between 6.95 and 7.05.

Exercise 36.14. Baking cakes. Roberto and Sally are busy making cakes for a bake sale. They need to make 35 cakes for the sale. The cooking times of the cakes are independent and Normally distributed. Sally makes 19 of the cakes and Roberto makes 16 of them. Sally's have average cook time of 45 minutes each, with standard deviation of 3 minutes. Roberto's cook, on average, 42 minutes each, with standard deviation of 4 minutes. What is the probability that the average of their 35 baking times is actually between 43 to 44 minutes altogether?

Exercise 36.15. Losing weight. A new low-carb diet author claims that people following his plan will lose an average of 3 pounds with a standard deviation of 1.4 pounds each week. An author of a low-fat diet claims that people following her plan will lose an average of 2 pounds with a standard deviation of 1.8 pounds each week. What is the probability that a low-carb dieter will lose at least a pound more than a low-fat dieter in a week?

36.3.3 Advanced

Exercise 36.16. Consider a group of n independent, identically distributed Normal-loss contracts, $X_1, \ldots, X_n$, each with average μ and standard deviation σ. A company plans to keep an amount of funds R on reserve so that $X_1 + \cdots + X_n \leq R$ with probability 95%, i.e., so that the company is 95% sure that all of its losses can be covered.

a. If $n = 25$, find the needed reserve R. What is the value of R/n, i.e., the reserve needed per contract?

b. Same question, with $n = 100$.

c. Same question, with $n = 10{,}000$.

d. Same question, for general n. (Notice that, for a bigger aggregation of contracts, the amount of reserve, per contract, gets smaller; in other words, R/n is a decreasing function of n, which gets closer to the average loss per contract.)

Chapter 37

Central Limit Theorem

If people do not believe that mathematics is simple, it is only because they do not realize how complicated life is.

—John von Neumann (remark at the first national meeting of the Association for Computing Machinery, 1947)

A basketball player plans to practice shooting hoops until she gets at least 10 successful baskets. She's meeting up with some friends after practice, so she wants to plan how long that will take, given that she usually makes about 60% of the baskets she shoots. If she wants to be 95% sure of her planning, how many baskets should she plan on attempting, to be sure that she has at least 10 successful shots?

37.1 Introduction

Gerolamo Cardano (1501–1576) studied games of chance using methods that were among the earliest predecessors to the modern discipline of probability. Five hundred years after Cardano, many others have contributed to the study of probability as it related to the limiting properties of randomness; these scholars include: Bernoulli (actually several Bernoullis), Laplace, Poisson, Chebyshev, Markov, Borel, Cantelli, Kolmogorov, and Khinchin. Probability is a major area of study that transcends mathematics and has applications in every discipline that involves randomness or uncertainty or chance. Within the study of probability theory, the subtopic of limiting properties receives a significant amount of attention. Many articles and books have been devoted to the study of limiting properties of random phenomena. We will only present some of the most commonly referenced laws about limits in probability theory. This is just the first stone off a large mountain of known results about the limiting properties about random variables.

We do not prove the Weak or Strong Law of Large Numbers or the Central Limit Theorem. More advanced techniques (for instance, characteristic functions) are needed. See, for instance, Billingsley [1] or Durrett [2] for details.

37.2 Laws of Large Numbers

Theorem 37.1. The Weak Law Of Large Numbers

Consider a sequence of independent random variables $X_1, X_2, X_3, \ldots$ that each have finite expected value μ. Also consider any positive number $\epsilon > 0$. The *weak law of large numbers* states that the average of the first n of the X_j's will not be too far from μ. More specifically, the *probability* that the average of $X_1, \ldots, X_n$ is more than ϵ away from μ will *converge to 0*, as $n \to \infty$. In other words,

$$\lim_{n\to\infty} P\left(\left|\frac{X_1 + \cdots + X_n}{n} - \mu\right| < \epsilon\right) = 1,$$

or equivalently,

$$\lim_{n\to\infty} P\left(\left|\frac{X_1 + \cdots + X_n}{n} - \mu\right| \geq \epsilon\right) = 0,$$

So if we fix a small ϵ (for instance, $\epsilon = 1/1000$), then the probability of the average $\frac{X_1+\cdots+X_n}{n}$ of the X_j's being more than ϵ away from μ will converge to 0 as $n \to \infty$.

Theorem 37.2. The Strong Law Of Large Numbers

Consider a sequence of independent random variables $X_1, X_2, X_3, \ldots$ that each have finite expected value μ. The *strong law of large numbers* states that the average of the first n of the X_j's will converge as $n \to \infty$ to μ with probability 1. In other words, the *probability* that the average of $X_1, \ldots, X_n$ converges to μ as $n \to \infty$ is 100%. In other words,

$$P\left(\lim_{n\to\infty} \frac{X_1 + \cdots + X_n}{n} = \mu\right) = 1.$$

(Note: It is possible for a sequence of random variables to converge weakly (i.e., the average converges in probability) without the sequence of random variables converging strongly. The opposite is impossible, since strong convergence implies weak convergence. The Strong Law of Large Numbers is a stronger statement than the Weak Law of Large Numbers.)

37.3 Central Limit Theorem

The name of the next theorem is a bit of a misnomer, because there are many versions of central limit theorems. The following is one of the most simple and widely used:

Theorem 37.3. The Central Limit Theorem

Consider a sequence of independent random variables $X_1, X_2, X_3, \ldots$ that each have finite expected value μ and finite variance σ^2. Let $Z \sim \mathcal{N}(0, 1)$ be a standard Normal random variable. The *central limit theorem* states that the probability of the average of $X_1, \ldots, X_n$, properly scaled (i.e., with subtraction of $n\mu$ and then division by $\sqrt{n\sigma^2}$), being less than "a," will converge to the cumulative distribution function of a standard Normal random variable, evaluated at a. In other words,

$$\lim_{n\to\infty} P\left(\frac{X_1 + \cdots + X_n - n\mu}{\sqrt{n\sigma^2}} \leq a\right) = \int_{-\infty}^{a} \frac{1}{\sqrt{2\pi}} e^{-z^2/2}\, dz,$$

or equivalently,

$$\lim_{n\to\infty} P\left(\frac{X_1 + \cdots + X_n - n\mu}{\sqrt{n\sigma^2}} \leq a\right) = F_Z(a).$$

The Central Limit Theorem, often affectionately called the CLT, is perhaps surprising because we do not need the random variables $X_1, X_2, X_3, \ldots$ to be Normal. This theorem basically says that sums of n independent random variables (of *any type*) are distributed similarly to a Normal random variable when n is large. (There is no minimum n necessary before the CLT applies, but the CLT is more effective, the larger n is.) This is a truly powerful concept that is used throughout the sciences and beyond. We give some applications in the next several examples.

37.4 CLT for Sums of Continuous Random Variables

Example 37.4. Consider the volumes of soda remaining in 100 cans of soda that are nearly empty. Let $X_1, \ldots, X_{100}$, denote the volumes (in ounces) of cans one through one hundred, respectively. Suppose that the volumes X_j are independent, and that each X_j is Uniformly distributed between 0 and 2.

Find the probability that the 100 cans of soda contain less than 90 ounces of soda *total.*

The expected value of X_j is $\mathbb{E}(X_j) = (0 + 2)/2 = 1$. The variance of X_j is $\text{Var}(X_j) = (2 - 0)^2/12 = 1/3$. Thus, the expected value of $X_1 + \cdots + X_{100}$ is

$$\mathbb{E}(X_1 + \cdots + X_{100}) = (100)(1),$$

and the variance of $X_1 + \cdots + X_{100}$ (since the X_j's are independent) is

$$\text{Var}(X_1 + \cdots + X_{100}) = (100)(1/3).$$

The content of the Central Limit Theorem, applied to this scenario, is that

$$\frac{X_1 + \cdots + X_{100} - (100)(1)}{\sqrt{(100)(1/3)}}$$

is approximately Normally distributed, i.e.,

$$P\left(\frac{X_1 + \cdots + X_{100} - (100)(1)}{\sqrt{(100)(1/3)}} \leq a\right) \approx F_Z(a).$$

Suppose we want to find the probability that the total amount is no more than 90 ounces. We compute

$$\begin{aligned}
P(X_1 + \cdots + X_{100} \leq 90) &= P\left(\frac{X_1 + \cdots + X_{100} - (100)(1)}{\sqrt{(100)(1/3)}} \leq \frac{90 - (100)(1)}{\sqrt{(100)(1/3)}}\right) \\
&\approx P(Z \leq -1.73) \\
&= P(Z \geq 1.73) \\
&= 1 - P(Z \leq 1.73) \\
&= 1 - 0.9582 \\
&= 0.0418
\end{aligned}$$

In other words, the probability is approximately 4.18% that the 100 cans of soda contain a *total* of less than 90 ounces of soda.

Example 37.4 (continued) Now we find the probability that the total volume of soda remaining is between 97 and 103 ounces.

We compute

$$\begin{aligned}
&P(97 \leq X_1 + \cdots + X_{100} \leq 103) \\
&\quad = P\left(\frac{97 - (100)(1)}{\sqrt{(100)(1/3)}} \leq \frac{X_1 + \cdots + X_{100} - (100)(1)}{\sqrt{(100)(1/3)}} \leq \frac{103 - (100)(1)}{\sqrt{(100)(1/3)}}\right) \\
&\quad \approx P(-0.52 \leq Z \leq 0.52) \\
&\quad = P(Z \leq 0.52) - P(Z \leq -0.52) \\
&\quad = P(Z \leq 0.52) - P(Z \geq 0.52) \\
&\quad = P(Z \leq 0.52) - 1 + P(Z \leq 0.52) \\
&\quad = 2P(Z \leq 0.52) - 1 \\
&\quad = (2)(0.6985) - 1 \\
&\quad = 0.3970
\end{aligned}$$

So the probability is approximately 39.70% that the 100 cans of soda contain between 97 and 103 ounces of soda.

Example 37.5. We return to the setup in Exercise 33.2: David works at a customer call center. He talks to customers on the telephone. The length (in hours) of each conversation has density $f_X(x) = 3e^{-3x}$ for $x > 0$ and $f_X(x) = 0$ otherwise. The lengths of calls are independent. As soon as one conversation is finished, he hangs up the phone, and immediately picks up the phone again to handle another call (i.e., there are no gaps in between the calls). Thus, if he conducts n phone calls in a row, the total amount of time he spends on the telephone is $X_1 + \cdots + X_n$, where the X_j's are independent, and each X_j has the density above.

What is the probability that David can handle 135 calls within a 40 hour period? In other words, what is $P(X_1 + \cdots + X_{135} \leq 40)$?

If we only use the fact that $X_1 + \cdots + X_{135}$ is a Gamma random variable, we would need to compute 135 nested integrals. Yuck! Computing 135 nested integrals would definitely be unreasonable, but such a computation would be necessary to get the *exact* probability.

Fortunately, we can use the Central Limit Theorem to get an excellent *approximation* to the answer, and we will not need to solve any integrals! Since $\mathbb{E}(X_j) = 1/3$ for each j, so

$$\mathbb{E}(X_1 + \cdots + X_{135}) = (135)(1/3).$$

Since the X_j's are independent, we can add the variances to obtain

$$\text{Var}(X_1 + \cdots + X_{135}) = (135)(1/9).$$

Thus, we compute $P(X_1 + \cdots + X_{135} \leq 40)$ as follows:

$$\begin{aligned} P\left(\frac{X_1 + \cdots + X_{135} - (135)(1/3)}{\sqrt{(135)(1/9)}} \leq \frac{40 - (135)(1/3)}{\sqrt{(135)(1/9)}}\right) &\approx P(Z \leq -1.29) \\ &= P(Z \geq 1.29) \\ &= 1 - P(Z \leq 1.29) \\ &= 1 - 0.9015 \\ &= 0.0985 \end{aligned}$$

Thus, we conclude that the probability is approximately 38.59% that David can handle 135 calls within 40 hours.

Example 37.6. As in Example 37.5, consider the lengths of calls handled by David in a call center. The calls are independent Exponential random variables, and each call lasts, on average, 1/3 of an hour. On a particular day, David records the lengths of 24 consecutive calls. What is the probability that the average of these 24 calls exceeds 1/4 of an hour?

Let $X_1, \ldots, X_{24}$ denote the lengths of the calls. The average length of the 24 calls is $(X_1 + \cdots + X_{24})/24$. So the average exceeds 1/4 of an hour if

$$P\left(\frac{X_1 + \cdots + X_{24}}{24} > \frac{1}{4}\right).$$

Multiplying by 24 on both sides of the inequality, we have, equivalently,

$$P(X_1 + \cdots + X_{24} > 6).$$

Now we normalize the sum of the random variables, to get

$$\begin{aligned} P\left(\frac{X_1 + \cdots + X_{24} - (24)(1/3)}{\sqrt{(24)(1/9)}} > \frac{6 - (24)(1/3)}{\sqrt{(24)(1/9)}}\right) &\approx P(Z > -1.22) \\ &= P(Z < 1.22) \\ &= 0.8888 \end{aligned}$$

37.5 CLT for Sums of Discrete Random Variables

When we apply the Central Limit Theorem to sums of discrete random variables, the method of setup and solution is almost completely the same. The only change is a concept called *continuity correction*. We only use *continuity correction* when working with sums of discrete random variables that are integer valued.

To motivate the need for continuity correction, consider a sum of random variables $X_1, X_2, \ldots, X_n$ that are integer-valued. Then $X_1 + \cdots + X_n$ is also integer valued. In other words, $X_1 + \cdots + X_n$ cannot take on fractional values that are not integers. So, for instance, if we are trying to compute

$$P(X_1 + \cdots + X_n \leq 40),$$

this is exactly the same as

$$P(X_1 + \cdots + X_n < 41).$$

Which number should we use in our computation, 40 or 41??? The answer will not be greatly affected either way, but the rule of *continuity correction* tells us to use the number halfway in between the two candidate values. So we must first determine which two numbers are candidates (one value will correspond to a less-than-or-equal or a greater-than-or-equal, and the other value will correspond to a strictly-less-than or a strictly-greater-than). Afterwards, we take the average of the two numbers, in this case, 40.5. So we would compute

$$P\left(\frac{X_1 + \cdots + X_n - n\mu}{\sqrt{n\sigma^2}} \leq \frac{40.5 - n\mu}{\sqrt{n\sigma^2}}\right).$$

Remark 37.7. Caution: Continuity correction is only used for applications of the Central Limit Theorem to sums of discrete random variables. We do not need continuity correction when applying the Central Limit Theorem to sums of continuous random variables.

Example 37.8. One thousand students participate in a survey to see how many breakfasts that they each eat, within a two week period. Let $X_1, X_2, \ldots, X_{1000}$ denote the numbers of breakfasts that student 1, 2, ..., 1000 eats, respectively. Since X_j is the number of breakfasts consumed by the jth student, then $0 \leq X_j \leq 14$ for each j. Assume that X_j is Binomial, with $n = 14$ and $p = 0.6$, and assume that the students behave independently.

Find the probability that the students eat strictly more than 8350 breakfasts altogether during the two week period.

The expected value of X_j is $\mathbb{E}(X_j) = np = (14)(0.6) = 8.4$. The variance of X_j is $\text{Var}(X_j) = np(1-p) = (14)(0.6)(0.4) = 3.36$. Thus, the expected value of $X_1 + \cdots + X_{1000}$ is

$$\mathbb{E}(X_1 + \cdots + X_{1000}) = (1000)(8.4) = 8400,$$

and the variance of $X_1 + \cdots + X_{1000}$ (since the X_j's are independent) is

$$\text{Var}(X_1 + \cdots + X_{1000}) = (1000)(3.36).$$

The content of the Central Limit Theorem, applied to this scenario, is that

$$\frac{X_1 + \cdots + X_{1000} - (1000)(8.4)}{\sqrt{(1000)(3.36)}}$$

is approximately Normally distributed, i.e.,

$$P\left(\frac{X_1 + \cdots + X_{1000} - (1000)(8.4)}{\sqrt{(1000)(3.36)}} \leq a\right) \approx F_Z(a).$$

The only way that the continuity correction affects our calculation is in deciding whether to compute

$$P(X_1 + \cdots + X_{1000} > 8350)$$

or

$$P(X_1 + \cdots + X_{1000} \geq 8351).$$

These are the same probabilities, but they will generate slightly different results in our approximation below. So we pick the average value of 8350 and 8351 (which is 8350.5), and we proceed as follows: So we compute

$$P(X_1 + \cdots + X_{1000} > 8350.5) = P\left(\frac{X_1 + \cdots + X_{1000} - (1000)(8.4)}{\sqrt{(1000)(3.36)}} > \frac{8350.5 - (1000)(8.4)}{\sqrt{(1000)(3.36)}}\right)$$
$$\approx P(Z > -0.85)$$
$$= P(Z < 0.85)$$
$$= 0.8023$$

In other words, the probability is approximately 80.23% that the students eat strictly more than 8350 breakfasts during the two week period.

Example 37.8 (continued) Suppose that we now want to find the probability that between 8380 and 8435 breakfasts (inclusive) are served, i.e., find

$$P(8380 \leq X_1 + \cdots + X_{1000} \leq 8435).$$

We could equivalently compute

$$P(8379 < X_1 + \cdots + X_{1000} < 8436),$$

because that would be the exact same probability. So we settle for the average of each endpoint. In other words, we compute

$$P(8379.5 < X_1 + \cdots + X_{1000} < 8435.5).$$

The derivation is the following:

$$P(8379.5 \leq X_1 + \cdots + X_{1000} \leq 8435.5)$$
$$= P\left(\frac{8379.5 - 8400}{\sqrt{(1000)(3.36)}} \leq \frac{X_1 + \cdots + X_{1000} - 8400}{\sqrt{(1000)(3.36)}} \leq \frac{8435.5 - 8400}{\sqrt{(1000)(3.36)}}\right)$$
$$\approx P(-0.35 \leq Z \leq 0.61)$$
$$= P(Z \leq 0.61) - P(Z \leq -0.35)$$
$$= P(Z \leq 0.61) - P(Z \geq 0.35)$$
$$= P(Z \leq 0.61) - 1 + P(Z \leq 0.35)$$
$$= 0.7291 - 1 + 0.6368$$
$$= 0.3659$$

So the probability is approximately 36.59% that the students eat between 8380 and 8435 breakfasts (inclusive) during the two week period.

Example 37.9. We modify the scenario from Example 17.3 of Chapter 17 as follows: Suppose that 220 basketball players are under consideration for a position on a basketball team. Suppose that each player makes 80% of her free throws successfully. Each player attempts to make a basket as many times as necessary, until scoring the first basket. Let X_j denote the number of attempts that the jth player needs. So $X_1 + \cdots + X_{220}$ is the total number of attempts that all 220 players need (collectively) until they have each made one basket successfully. Also let the X_j's be independent, i.e., the number of attempts by one player does not affect the number of attempts from any of the other players.

What is the probability that strictly between 260 and 280 attempts are needed? In other words, what is

$$P(260 < X_1 + \cdots + X_{220} < 280)?$$

Equivalently, what is

$$P(261 \leq X_1 + \cdots + X_{220} \leq 279)?$$

Since we are approximating an integer-valued sum of random variables, we again need to use continuity correction. So we compute

$$P(260.5 \leq X_1 + \cdots + X_{220} \leq 279.5).$$

We know from Example 17.3 that each player has expected number of shots $\mathbb{E}(X_j) = 1/0.80 = 1.25$ and variance $\mathrm{Var}(X_j) = 0.20/0.80^2 = 0.3125$.

We compute:

$$\begin{aligned}
&P(260.5 \leq X_1 + \cdots + X_{220} \leq 279.5) \\
&= P\left(\frac{260.5 - (220)(1.25)}{\sqrt{(220)(0.3125)}} \leq \frac{\sum_{i=1}^{220} X_i - (220)(1.25)}{\sqrt{(220)(0.3125)}} \leq \frac{279.5 - (220)(1.25)}{\sqrt{(220)(0.3125)}}\right) \\
&\approx P(-1.75 \leq Z \leq 0.54) \\
&= P(Z \leq 0.54) - P(Z \leq -1.75) \\
&= P(Z \leq 0.54) - P(Z \geq 1.75) \\
&= P(Z \leq 0.54) - 1 + P(Z \leq 1.75) \\
&= 0.7054 - 1 + 0.9599 \\
&= 0.6653
\end{aligned}$$

So the probability is approximately 66.53% that (strictly) between 260 and 280 attempts are needed altogether.

37.6 Approximations of Binomial Random Variables

The Central Limit Theorem states that the *sum* of a large number of independent random variables behaves approximately like a Normal random variable. The CLT can also be applied to *one* Binomial(n, p) random variable (not a sum of Binomial random variables) to show that a Binomial has some similarities to a Normal random variable if $np(1-p)$ is significantly large. In practice, $np(1-p) \geq 10$ is sometimes used a rule of thumb for this application of the CLT. Notice that, if $np(1-p)$ is large, then n is large too, because $n \geq np(1-p)$ always. There is no definitive, comprehensive rule that states "n is large enough" for all such approximations.

Consider such a Binomial(n, p) random variable X, with $np(1-p)$ large. The main idea is that we can view X as the sum of a large number of independent Bernoulli random variables $X_1, X_2, \ldots, X_n$, each with parameter p. For instance, if X is a Binomial random variable with parameters $n = 300$ and $p = 1/4$ then $np(1-p) = 56.25$, so $np(1-p)$ is sufficiently large that X will roughly behave like a Normal random variable. In such a case, we let $X_1, \ldots, X_{300}$ be 300 independent Bernoulli random variables, each with parameter $p = 1/4$, and then both X and $X_1 + \cdots + X_{300}$ have the same mass:

$$\begin{aligned} p_X(j) &= P(X = j) \\ &= \binom{300}{j}(1/4)^j(3/4)^{300-j} \\ &= P(X_1 + \cdots + X_{300} = j) \\ &= p_{X_1+\cdots+X_{300}}(j). \end{aligned}$$

Since Binomial random variables are discrete, continuity correction is needed when doing a Normal approximation to a Binomial random variable.

As another example, if X is a Binomial random variable with parameters $n = 1000$ and $p = 9/10$ then $np(1-p) = 90$, so $np(1-p)$ is large enough to make X approximately Normal. We let $X_1, \ldots, X_{1000}$ be independent Bernoulli random variables, each with parameter $p = 9/10$, and then

$$p_X(j) = \binom{1000}{j}(9/10)^j(1/10)^{1000-j} = p_{X_1+\cdots+X_{1000}}(j).$$

So, in other words, if X is a Binomial random variable with parameters n and p such that $np(1-p)$ is large, then X can be treated as the sum of n independent Bernoulli random variables, each with parameter p, and thus, *by the Central Limit Theorem, X is approximately a Normal random variable*, with $\mathbb{E}(X) = np$ and $\text{Var}(X) = np(1-p)$.

Another way to see that Binomial random variables with parameters n and p (such that $np(1-p)$ is sufficiently large) are approximately Normal random variables is to graph the mass of such random variables. The mass of a Binomial random variable looks like the density of a Normal random variable, i.e., like a bell curve. In contrast to this idea, the mass of a Geometric random variable does not look like a bell curve at all, and a Geometric random variable is not approximated by a Normal random variable.

Note that the random variables are only approximately Normal, not completely Normal. For instance, the mass of a Binomial random variable is still discrete (although it looks like a bell curve) and the mass $p_X(j)$ is only defined for integers $0 \leq j \leq n$. See Figures 37.1 and 37.2. We will still use continuity correction with such Binomial random variables, since they are discrete.

Remark 37.10. Approximating a Binomial random variable by a Normal random variable

If X is a Binomial random variable with parameters n and p such that $np(1-p)$ is large (e.g., $np(1-p) \geq 10$), then X behaves approximately like a Normal random variable with expected value np and variance $np(1-p)$ and standard deviation $\sqrt{np(1-p)}$. Thus

$$Z \approx \frac{X - \mu_X}{\sigma_X} = \frac{X - np}{\sqrt{np(1-p)}}$$

i.e., we approximately have a standard Normal random variable, when we standardize the X. In other words,

$$P\left(\frac{X - np}{\sqrt{np(1-p)}} \leq a\right) \approx P(Z \leq a) = F_Z(a).$$

Such approximations should also include continuity correction, which makes a slight adjustment to the value of a, as discussed earlier in this chapter.

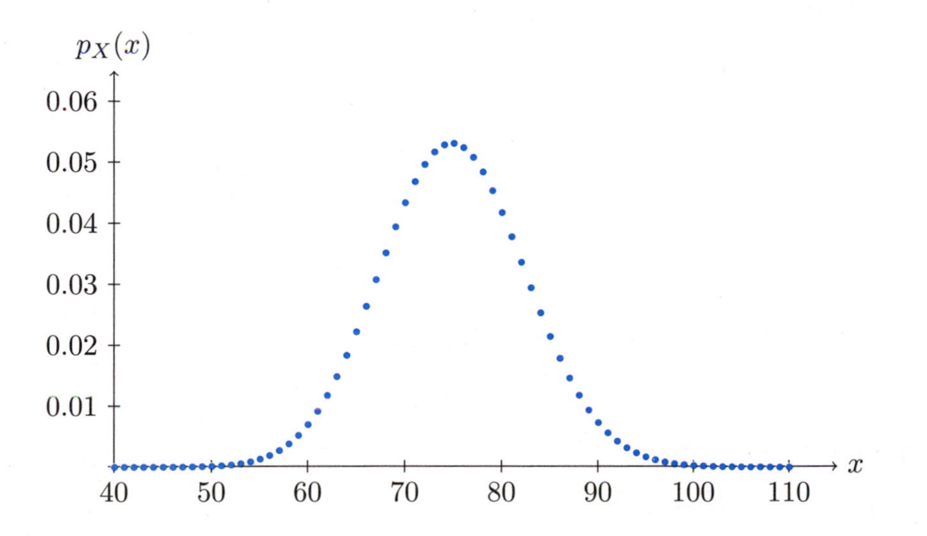

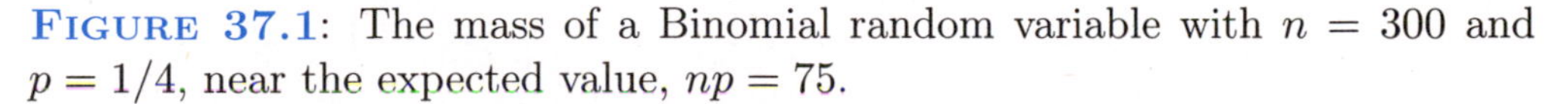

FIGURE 37.1: The mass of a Binomial random variable with $n = 300$ and $p = 1/4$, near the expected value, $np = 75$.

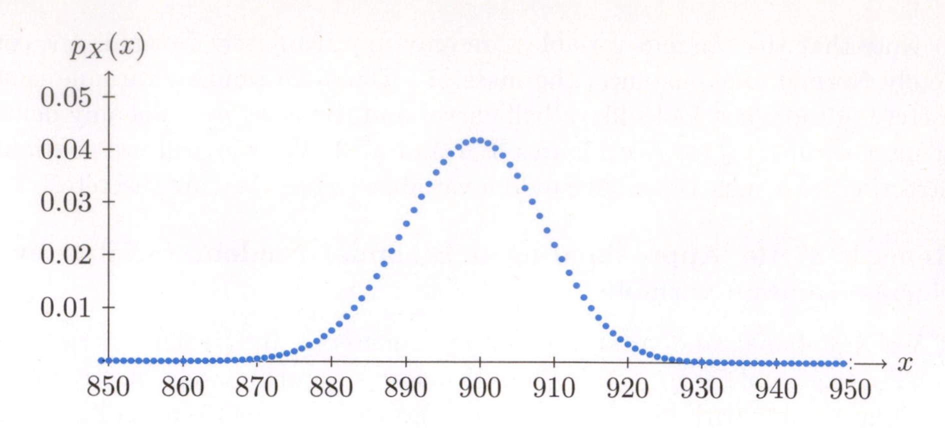

FIGURE 37.2: The mass of a Binomial random variable with $n = 1000$ and $p = 9/10$, near the expected value, $np = 900$.

Example 37.11. Consider the independent single-baby births, in which the probability of a boy or girl in each birth is assumed to be equally likely. Let X be the number of girls.

The number X of girls is Binomial with expected value $np = (1000)(1/2) = 500$ and variance $np(1-p) = (1000)(1/2)(1/2) = 250$. Thus, X is approximately Normal with expected value 500 and variance 250.

The probability that strictly more than 470 of the babies are girls is exactly

$$\begin{aligned} P(X > 470) &= P(X = 471) + P(X = 472) + \cdots + P(X = 1000) \\ &= \binom{1000}{471}(1/2)^{471}(1/2)^{529} + \binom{1000}{472}(1/2)^{472}(1/2)^{528} \\ &\quad + \cdots + \binom{1000}{1000}(1/2)^{1000}(1/2)^{0} \end{aligned}$$

but this is very unwieldy to calculate, and many calculators could not handle such a computation anyway! So we use a Normal approximation to the Binomial, with continuity correction, noting that

$$P(X > 470) = P(X \geq 471),$$

so we use $P(X > 470.5)$, and we get

$$\begin{aligned} P(X > 470) &= P(X > 470.5) \\ &= P\left(\frac{X - 500}{\sqrt{250}} > \frac{470.5 - 500}{\sqrt{250}}\right) \\ &\approx P(Z > -1.87) \\ &= P(Z < 1.87) \\ &= 0.9693 \end{aligned}$$

Example 37.11 (continued) Now, what if we want to approximate a specific probability? For instance, what if we want to know the probability that exactly 500 of the 1000 babies are girls?

We know that

$$P(X = 500) = \binom{1000}{500}(1/2)^{500}(1/2)^{500},$$

but again, this is very difficult for many calculators to handle. So we reformulate this as

$$P(500 \le X \le 500),$$

or equivalently

$$P(499 < X < 501),$$

and with continuity correction, this becomes

$$P(499.5 \le X \le 500.5),$$

which will give a pretty good approximation. We calculate:

$$\begin{aligned} P(499.5 \le X \le 500.5) &= P\left(\frac{499.5 - 500}{\sqrt{250}} \le \frac{X - 500}{\sqrt{250}} \le \frac{500.5 - 500}{\sqrt{250}}\right) \\ &\approx P(-0.03 \le Z \le 0.03) \\ &= P(Z \le 0.03) - P(Z \le -0.03) \\ &= P(Z \le 0.03) - P(Z \ge 0.03) \\ &= P(Z \le 0.03) - 1 + P(Z \le 0.03) \\ &= 0.5120 - 1 + 0.5120 \\ &= 0.0240 \end{aligned}$$

By the way, the exact answer, using a computer, is 0.02522501818..., so our approximation is very comparable to the actual value.

Example 37.12. At a certain local restaurant, students are known to prefer Japanese pan noodles 40% of the time (it is a very popular and tasty dish!). Consider 2000 randomly chosen students; let X denote the number of students that order Japanese pan noodles.

We know that X is Binomial with $n = 2000$ and $p = 0.40$, so $\mathbb{E}(X) = np = (2000)(0.40) = 800$ and $\text{Var}(X) = np(1-p) = (2000)(0.40)(0.60) = 480$, so the standard deviation of X is $\sqrt{np(1-p)} = \sqrt{480} = 21.9089$. So X is approximately Normal with expected value 800 and variance 480.

The probability that at most 840 of the students eat Japanese pan noodles is $P(X \leq 840)$, i.e., $P(X < 841)$, so we use $P(X \leq 840.5)$ for the continuity correction, and we calculate

$$\begin{aligned} P(X \leq 840.5) &= P\left(\frac{X - 800}{\sqrt{480}} \leq \frac{840.5 - 800}{\sqrt{480}}\right) \\ &\approx P(Z \leq 1.85) \\ &= 0.9678 \end{aligned}$$

37.7 Approximations of Poisson Random Variables

A Poisson random variable with large parameter λ will be distributed like a Normal random variable. One way to see this is to graph the mass of such a Poisson random variable, which also looks like a bell curve; see Figure 37.3.

Since Poisson random variables are discrete, continuity correction is needed when doing a Normal approximation to a Poisson random variable.

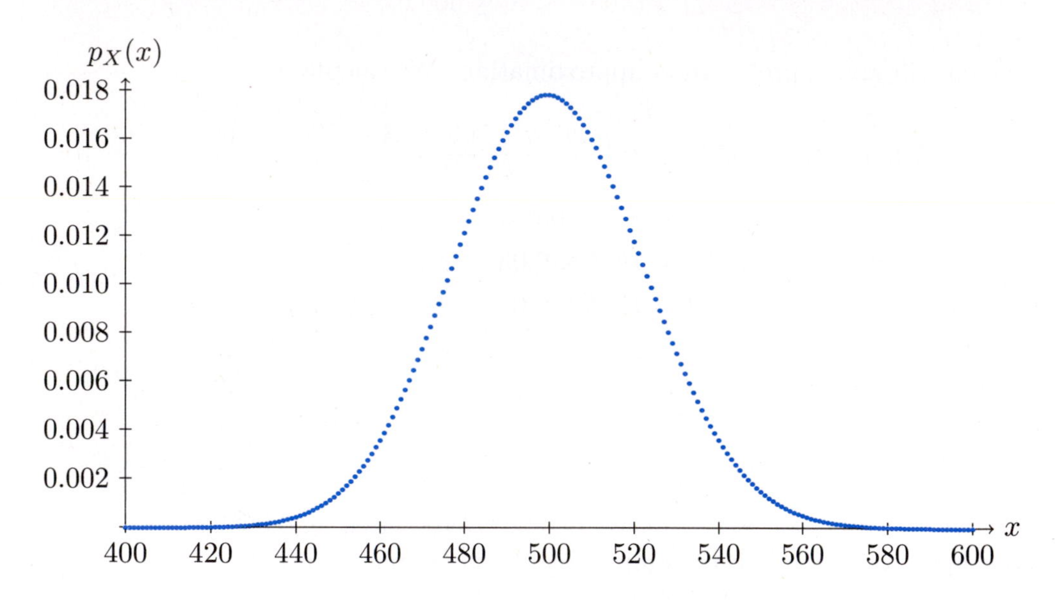

FIGURE 37.3: The mass of a Poisson random variable with $\lambda = 500$, near the expected value, $\lambda = 500$.

Of course, Poisson random variables are discrete, and moreover the mass $p_X(j)$ is defined only for the nonnegative integers $0, 1, 2, \ldots$. Thus, we must use continuity correction with approximations of Poisson random variables too.

Having the parameter $\lambda \geq 10$ is one possible rule of thumb for getting relatively good approximations of a Normal random variable to a Poisson random variable. We can view X as the sum of a large number of Poisson random variables. As an example, if X is a Poisson random variable with parameter $\lambda = 500$, we can let $X_1, \ldots, X_{500}$ be a collection of independent Poisson random variables, each with parameter 1, so both X and $X_1 + \cdots + X_{500}$ have the same mass:

$$p_X(j) = e^{-500} 500^j / j! = p_{X_1 + \cdots + X_{500}}(j).$$

This comes from the fact that the sum of independent Poisson random variables is also a Poisson random variable, and that the parameter of the sum is equal to the sum of the parameters. In this case, $\overbrace{1 + 1 + \cdots + 1}^{500} = 500$. So we have just decomposed a Poisson random variable X with parameter 500 into a sum of 500 independent Poisson random variables $X_1, \ldots, X_{500}$ that each have parameter 1. Thus, *by the Central Limit Theorem, X is approximately a Normal random variable.* We note that X has expected value $\lambda = 500$ and variance 500, and therefore standard deviation $\sqrt{500}$.

We can even decompose the very same Poisson random variable more finely. For instance, if X is Poisson with parameter $\lambda = 500$, we can let $X_1, \ldots, X_{3000}$ be a collection of independent Poisson random variables, each with parameter 1/6 (we use 1/6 here, so that the mean is $500 = (3000)(1/6)$), then both X and $X_1 + \cdots + X_{3000}$ have the same mass:

$$p_X(j) = e^{-500} 500^j / j! = p_{X_1 + \cdots + X_{3000}}(j).$$

This works because $\overbrace{1/6 + 1/6 + \cdots + 1/6}^{3000} = 500$.

Methods like this allow us to even decompose Poisson random variables with parameters that are not integers! For instance, if X is Poisson with parameter 372.9, we can let $X_1, \ldots, X_{1000}$ be a collection of independent Poisson random variables, each with parameter 0.3729, and then both X and $X_1, \ldots, X_{1000}$ have the same mass:

$$p_X(j) = e^{-372.9} 372.9^j / j! = p_{X_1 + \cdots + X_{1000}}(j).$$

This works because $\overbrace{0.3729 + 0.3729 + \cdots + 0.3729}^{1000} = 372.9$.

Remark 37.13. Approximating a Poisson random variable using a Normal random variable

If X is a Poisson random variable with parameter λ sufficiently large, e.g., $\lambda \geq 10$, then X behaves approximately like a Normal random variable with expected value λ, variance λ, and standard deviation $\sqrt{\lambda}$. Thus

$$\frac{X - \mu_X}{\sigma_X} = \frac{X - \lambda}{\sqrt{\lambda}}$$

behaves approximately like a standard Normal random variable Z. In other words,

$$P\left(\frac{X - \lambda}{\sqrt{\lambda}} \leq a\right) \approx P(Z \leq a) = F_Z(a).$$

As with Binomial random variables, these approximations of Poisson random variables should also include continuity correction, resulting in a slight adjustment to the value of a, as discussed earlier.

Example 37.14. A certain grocery store is believed to have a Poisson number of customers each day, with parameter $\lambda = 1200$.

What is the probability that the number of customers on a certain day is strictly between 1150 and 1250?

We let X denote the number of customers. So $\mathbb{E}(X) = \lambda = 1200$ and $\text{Var}(X) = \lambda = 1200$ too. We want to compute

$$P(1150 < X < 1250),$$

or equivalently,

$$P(1151 \leq X \leq 1249),$$

so with continuity correction, we use

$$P(1150.5 \leq X \leq 1249.5).$$

Now we compute

$$\begin{aligned}
P(1150.5 \leq X \leq 1249.5) &= P\left(\frac{1150.5 - 1200}{\sqrt{1200}} \leq \frac{X - 1200}{\sqrt{1200}} \leq \frac{1249.5 - 1200}{\sqrt{1200}}\right) \\
&\approx P(-1.43 \leq Z \leq 1.43) \\
&= P(Z \leq 1.43) - P(Z \leq -1.43) \\
&= P(Z \leq 1.43) - P(Z \geq 1.43) \\
&= P(Z \leq 1.43) - 1 + P(Z \leq 1.43) \\
&= 0.9236 - 1 + 0.9236 \\
&= 0.8472
\end{aligned}$$

For comparison, the actual answer, using a computer, is

$$
\begin{aligned}
P(1150 < X < 1250) &= P(X = 1151) + P(X = 1152) + \cdots + P(X = 1249) \\
&= \sum_{j=1151}^{1249} \frac{e^{-1200}1200^j}{j!} \\
&= 0.8470129469\ldots
\end{aligned}
$$

We could not perform such a calculation very easily by hand, because it has nearly 100 terms that would need to be calculated, and each term includes the use of very large numbers, e.g., 1200^j and $j!$ for $1151 \le j \le 1249$, and also very small numbers, e.g., e^{-1200} for $1151 \le j \le 1249$.

Example 37.15. A publisher estimates that a 500 page book has a Poisson number of errors with average $\lambda = 100$ errors/book. Estimate the probability that there are strictly more than 80 errors in such a book.

If we let X denote the number of errors in a newly published 500 page book, the probability that there are strictly more than 80 errors is $P(X > 80)$, which is equivalent to $P(X \ge 81)$.

Why can we not calculate this manually, without Normal approximation? If we try a manual calculation, we first realize that there are infinitely many terms:

$$
P(X \ge 81) = \frac{e^{-100}100^{81}}{81!} + \frac{e^{-100}100^{82}}{82!} + \frac{e^{-100}100^{83}}{83!} + \frac{e^{-100}100^{84}}{84!} + \cdots,
$$

which is hopeless. So we rearrange and use the complement:

$$
\begin{aligned}
P(X \ge 81) &= 1 - P(X \le 80) \\
&= \frac{e^{-100}100^{0}}{0!} + \frac{e^{-100}100^{1}}{1!} + \frac{e^{-100}100^{2}}{2!} + \cdots + \frac{e^{-100}100^{80}}{80!},
\end{aligned}
$$

but this is still hopeless, because there are 81 terms in the summation. So the value of a Normal approximation—which can be evaluated in one fell swoop—should now seem apparent.

Also remember that we use $P(X \ge 80.5)$, for the purposes of continuity correction. This gives us

$$
\begin{aligned}
P(80.5 \le X) &= P\left(\frac{80.5 - 100}{\sqrt{100}} \le \frac{X - 100}{\sqrt{100}}\right) \\
&\approx P(-1.95 \le Z) \\
&= P(1.95 \ge Z) \\
&= P(Z \le 1.95) \\
&= 0.9744
\end{aligned}
$$

Example 37.16. (Continuation of Example 37.15)

If we buy 10 different books from this publisher, each containing 500 pages, the number of our books that have strictly more than 80 errors each is Binomial with $n = 10$ and $p = 0.9744$. The probability that exactly nine of the books have 80 or more errors each (and the remaining book has 79 or fewer errors) is

$$\binom{10}{9}p^9(1-p)^1 = \binom{10}{9}(0.9744)^9(0.0256)^1 = 0.2027.$$

37.8 Exercises

37.8.1 Practice

Exercise 37.1. Concert tickets. At a certain university, each student who tries to purchase concert tickets for an upcoming show is successful at connecting to the concert ticket website with probability 0.85. If unsuccessful in logging on, he or she tries again a few minutes later, over and over, until finally getting tickets. The concert is large (tens of thousands of seats), so assume that such attempts are independent. Within a group of 300 students, find the probability that *strictly more than 360 attempts* are necessary for these 300 students to successfully get tickets.

Exercise 37.2. Blind auction bids. At an auction, exactly 282 people place requests for an item. The bids are placed "blindly," which means that they are placed independently, without knowledge of the actions of any other bidders. Assume that each bid (measured in dollars) is a Continuous Uniform random variable on the interval $[10.50, 19.30]$. Find the probability that the sum of all the bids exceeds \$4150.

Exercise 37.3. Baseball statistics. A certain baseball player has, on average, 0.7 RBI's (runs batted in) per game, with standard deviation of 0.2. What is the approximate probability that the player has at most 110 RBI's during a given season, which contains 162 games?

Exercise 37.4. Tootsie Pops. Back in 1970, a television commercial for Tootsie Pops (a candy) asked how many licks would be needed to get to the Tootsie Roll Center of a Tootsie Pop. This question has become somewhat famous and well-known among children. High school, undergraduate, and Ph.D. students have all performed experiments to uncover the answer (some have even built licking machines to test this question). Suppose that an average 364 licks are required, with a standard deviation of 40 licks. If a group of 50 children test their licking abilities, what is the approximate probability that, among the 50 children, the average number of licks per student is 380 or more?

Exercise 37.5. Strawberry milkshakes. The number of strawberry milkshakes that are sold at a local diner in a given day has an average of 97.5 per day and standard deviation of 7.8. In a given 30-day month, what is the approximate probability that they sell more than 3000 milkshakes?

Exercise 37.6. Spelling. A student is learning to spell some difficult words. He estimates that it takes him 6 attempts to learn each word, with a variance of 3.5 per word. Find the approximate probability that he will need 125 or more attempts to learn 20 words.

Exercise 37.7. Water bottles. Water is sold in bottles that are advertised to contain 1-liter each. Barbara is an inspector for the company and notices that the amount of water in each bottle, unfortunately, has an average of only 0.99 liters, and surprisingly high standard deviation of 0.03 liters. Estimate the probability that a 12-pack of water will actually contain 12 liters or more of water.

Exercise 37.8. Memory chip. Rafael is pursuing a major in computer science. He notices that a memory chip containing $2^{12} = 4096$ bits is full of data that seems to have been generated, bit-by-bit, at random, with 0's and 1's equally likely, and the bits are stored independently. If the each bit is equally likely to be a 0 or 1, estimate the probability that there are actually 4120 or more 1's stored in the memory chip?

Exercise 37.9. Compost worms. Josephine estimates that the worms in her composter can eat, on average, 3.05 pounds of waste per week, with standard deviation of 0.3 pounds. She tests this conjecture with her worms over a 50-week period. Estimate the probability that the average over the 50-week period exceeds 3 pounds per week.

Exercise 37.10. Shower length. Each morning a student takes a shower that lasts 15 minutes, with a standard deviation of 4 minutes. Find an estimate of the probability that the student spends between 11 and 12 hours in the shower during a 45-day period.

Exercise 37.11. Student enrollment. The registrar estimates that each student enrolls for 15.3 credits on average, with a standard deviation of 1.2 credits per student. Estimate the probability that, on a campus with 4000 enrolled students, the total number of enrolled credit hours is strictly between 61,100 and 61,300.

Exercise 37.12. Toy geyser. A student builds a toy geyser for her engineering class that has height (in inches) of density

$$f_X(x) = \frac{1}{8}e^{-x/8} \qquad \text{for } x > 0,$$

and $f_X(x) = 0$ otherwise. If she makes 20 such geysers, what is the rough probability that the average geyser height, among these 20 geysers, is 7 inches or more?

Exercise 37.13. Die rolls. A group of students in a probability course is skeptical about Geometric random variables, so one day, they all bring a die to class. Each of the 40 students rolls her/his die repeatedly. Each student stops upon their first 5 that appears, independent of the other students. What is the estimated probability that it takes the students a total of 250 or more die rolls to accomplish their experiment?

Exercise 37.14. Phone calls. Suppose that each call a student makes to his girlfriend has an average length of 12 minutes, and the standard deviation of the length of each call is also 3 minutes. Estimate the probability that he completely uses up a 300-minute calling card while talking to his girlfriend during 24 conversations.

Exercise 37.15. Journal pages. Suppose that the number of pages in an issue of a journal is Uniformly distributed between 256 and 384. During 40 years of publication, the journal publishes a total of 160 issues. Give an estimate for the probability that between 51,000 and 52,000 pages (inclusive) were used for these 160 issues.

Exercise 37.16. Car sales. A car dealership expects to sell 6 cars per day, with standard deviation of 2.1 cars per day. During a 100-day period, estimate the probability that they sell strictly more than 625 cars.

Exercise 37.17. Rainfall. If the amount of rainfall in a given region is assumed to be Uniform on the interval $[0.2, 4.0]$ (in inches) each month, what is the approximate probability that there are 53 or more inches of rain during a 24-month period?

Exercise 37.18. Class attendance. In a math class with 200 students, suppose that the students' decisions to attend the class are independent, and each student attends with probability 93%. On a given day, find the approximate probability that strictly fewer than 179 students attend.

Exercise 37.19. Student workers. Consider a group of students whose are assigned to work a random number of hours. Their hours per student, per week, are modeled by a Binomial random variable with $n = 20$ and $p = 0.8$. (Each hour assigned to work will count as a "success" in the Binomial model.) If there are 100 students in the fraternity, and the numbers of hours spent working are independent, find an estimate for the probability that they work between 1580 and 1620 hours altogether during a given week.

Exercise 37.20. Weighing burgers. A restaurant claims that each of their burgers has a 1/4 pound of meat after it is prepared. Of course this is not exact. The students in a local high school measured some of these burgers for a science project, and they concluded that the expected weight of such a burger is 0.251 pounds, but the standard deviation is 0.02 pounds. Afterwards, the 30 students celebrate by each ordering 1 burger. What is the probability that, among the 30 students, the average weight of a burger is less than the advertised weight of 1/4 pound?

Exercise 37.21. Customer arrivals. Consider customers who arrive at a checkout counter with an average rate of 8 per hour following a Poisson distribution. Find the probability that strictly more than 70 customers arrive during the eight-hour shift.

Exercise 37.22. Free throws. A particular basketball player successfully scores about 75% of his free throw attempts. Approximate the probability that he makes strictly more than 145 of 200 attempts.

Exercise 37.23. Jackpot. If a certain type of slot machine has only a probability of 0.0001 of yielding a jackpot on each game at a certain casino in Vegas, and the patrons play those slots 250,000 times during a given month, estimate the probability that the casino will have 30 or more jackpot rewards to payout during the month.

Exercise 37.24. Ice cream patrons. An ice cream shop estimates that its number of patrons per day is Poisson with mean 19. What is the estimated probability that they have 20 or more customers on a given day?

Exercise 37.25. Failing processors. In the early testing stages of processor manufacturing, 40% of the processors will fail in some way. If 500 processors are manufactured at this stage, what is a rough estimate for the probability that strictly less than 180 of the processors will fail in some way?

Exercise 37.26. Rotten berries. When buying a pack of strawberries at the farmer's market, there is a 0.10 chance that there will be a rotten berry in the pack. If a restaurant buys 300 packs of strawberries for their desserts, what is the approximate probability that fewer than 25 packs will contain a rotten berry?

Exercise 37.27. Airport security. If 6% of all passengers are screened with two rounds of security at the airport, and Southwest has 8 flights with 180 passengers each, what is the approximate probability that 80 or more of these Southwest passengers will receive this extra level of screening?

Exercise 37.28. Microwave popcorn. A bag of microwave popcorn comes with 200 kernels of corn. When microwaved for exactly 3 minutes, an individual kernel pops 90% of the time. What is the approximate probability that there are strictly less than 10 unpopped kernels in such a bag?

37.8.2 Extensions

Exercise 37.29. Burritos. Twenty-five students want burritos, but the dining hall only has one burrito maker. Each burrito takes an average of 72.5 seconds to cook, with standard deviation 3.2 seconds.

a. Estimate the probability that all twenty-five students can cook their burritos in half an hour or less, if we ignore the time in between the consecutive students.

b. Estimate the probability that all twenty-five students can cook their burritos in 2300 seconds or less, if we assume that there is an exact 20-second delay in between each pair of consecutive students. (Hint: There are 24 such delays between 25 students.)

c. Estimate the probability that all twenty-five students can cook their burritos in 2300 seconds or less, if we assume that there is a Normally distributed delay in between each pair of consecutive students, with average delay of 20 seconds between consecutive pairs of students, and standard deviation of 4 seconds.

Exercise 37.30. Online vs. traditional lecture. At a certain university, a large course has 1500 registered students; the course can be viewed online if a student does not feel like walking to the lecture hall. Each student decides each day (independently of the other students and independent of her/his own prior behavior) whether to attend the class in the lecture hall or just view the online version. The probability a student attends class in the lecture hall is only 0.30, so the registrar has quite a headache.

a. What is the expected number of students to attend class in the lecture hall on a particular day?

b. If the registrar assigns a lecture hall with 500 seats for the class, what is the probability that the lecture hall overflows (i.e., strictly more than 500 attend) on a particular day?

c. To save money, the registrar wants to use an even smaller lecture hall next semester, but the professor insists on having enough seats so the probability is 95% (or more) of no overflow on a given day. How many seats does the professor need in her lecture hall?

Exercise 37.31. Playing the lottery. A person plays a certain lottery game 10,000 times during his life. The chance of winning is 1/5000 during each attempt, and the attempts are independent. Approximate the probability that he wins 2 or more times during his life.

Chapter 38

Review of Named Continuous Random Variables

38.1 Summing-up: How To Tell Random Variables Apart

Your first step should be to identify whether X is discrete (how many?) or continuous (how long or how much?). Once you have decided that X is continuous, think about what values of X would be allowed.

The actual waiting time until your 1st customer is an Exponential random variable. The actual waiting time until your 3rd customer is a Gamma random variable. The customer could arrive at a time that is Uniform between 8:00 and 8:15. The percentage of customers who buy ice cream could be a Beta random variable. The number of customers during a long period of time is approximately Normally distributed.

The following are some **Pairs of Random Variables that are Related**:

Waiting until the 1st success:

- **Geometric**: X is the # of people you have to ask until you get your first "yes" answer (discrete)
- **Exponential**: X is the time you have to wait until you get your first "yes" answer (continuous)

Waiting until the 4th success:

- **Negative Binomial**: X is the # of people you have to ask until you get your 4th "yes" answer (discrete)
- **Gamma**: X is the time you have to wait until you get your 4th "yes" answer (continuous)

Summary of Named Continuous Random Variables

Name	Uniform	Exponential	Gamma	Beta	Normal
Density $f_X(x)$	$1/(b-a)$	$\lambda e^{-\lambda x}$	$\frac{\lambda e^{-\lambda x}(\lambda x)^{r-1}}{\Gamma(r)}$	$\frac{\Gamma(\alpha+\beta)x^{\alpha-1}(1-x)^{\beta-1}}{\Gamma(\alpha)\Gamma(\beta)}$	$\frac{e^{-(x-\mu)^2/(2\sigma^2)}}{\sqrt{2\pi\sigma^2}}$
Domain	$a \leq x \leq b$	$x \geq 0$	$x \geq 0$	$0 \leq x \leq 1$	$-\infty < x < \infty$
CDF $F_X(x)$	$(x-a)/(b-a)$	$1 - e^{-\lambda x}$;	$1 - e^{-\lambda x} \sum_{j=0}^{r-1} \frac{(\lambda x)^j}{j!}$, if $r \in \mathbb{N}$		convert to std. Normal $Z = (X - \mu_X)/\sigma_X$
$\mathbb{E}(X)$	$(a+b)/2$	$1/\lambda$	r/λ	$\alpha/(\alpha+\beta)$	μ
$\text{Var}(X)$	$(b-a)^2/12$	$1/\lambda^2$	r/λ^2	$\frac{\alpha\beta}{(\alpha+\beta)^2(\alpha+\beta+1)}$	σ^2
Param-eters	a, b are endpoints	λ is average # of successes per time unit	λ is average # of successes per time unit	usually some given constraints	$\mu =$ exp. value; $\sigma =$ st. dev.
What X is	location in an interval	wait time 'til 1st event (like continuous Geometric)	wait time 'til rth event (like continuous Neg. Binom.)	fraction, percentage, proportion,	practical random variable for quantities clustered around avg. value
When used	spread evenly	wait time until 1st event	wait time until rth event	Bayesian statistics	Central Limit Thm. and applications

Note: $\Gamma(r) = (r-1)!$ if $r \in \mathbb{N}$.

Remember, Exponential distributions have the memoryless property. This means that, for instance,

$$P(X > 20 \mid X > 8) = P(X > 12).$$

In practice, the question would read, given that nobody has arrived in the first 8 minutes, what is the probability nobody will arrive in the next 12 minutes? The first 8 minutes don't matter to us now. For Exponential, you can start from NOW. (Geometric distributions also have the memoryless property.)

38.2 Exercises

For Exercises 38.1 through 38.12, state which continuous or discrete distribution would be most appropriate and why you think so. The choices are:

Bernoulli	Hypergeometric	Exponential
Binomial	Poisson	Gamma
Geometric	Discrete Uniform	Beta
Negative Binomial	Continuous Uniform	Normal

Exercise 38.1. Let X be the diameter of a chocolate chip cookie selected at random if you know that the average diameter is 6 inches with a standard deviation of 0.4 inches and if you know that the distribution of cookies has a bell-shaped curve.

Exercise 38.2. Let X be the percent of flour which will be used up by the end of the day at the bakery.

Exercise 38.3. Let X indicate whether the next customer will want to buy a chocolate chip cookie.

Exercise 38.4. Let X be the length of time that the baker will have to wait until a customer wants to buy a chocolate chip cookie.

Exercise 38.5. Let X be the number of customers who will come into the store until a customer wants to buy a chocolate chip cookie.

Exercise 38.6. Let X be the number of customers, out of the next 20 who come in, who will want to buy a chocolate chip cookie.

Exercise 38.7. Let X be the number of customers who will want to buy a chocolate chip cookie in the next hour if, according to bakery records, an average of 3 customers per hour want to buy a chocolate chip cookie.

Exercise 38.8. Let X be the exact arrival time of the first customer if we know that exactly one customer arrived between 8:03 and 8:12 AM.

Exercise 38.9. Let X be the length of time that the baker will have to wait to bring out a new batch of dozen chocolate chip cookies if he just put out the first dozen and if on average 3 customers per hour want to buy a chocolate chip cookie.

Exercise 38.10. Let X be the position of the cookie on the tray (numbered 1 to 12) that the cookie you will be served has if all cookies are equally convenient to the waitress.

Exercise 38.11. Let X be the number of chocolate chip cookies which are brought to you if you ask the waitress to randomly select 4 cookies from a selection of 5 chocolate chip, 3 oatmeal, and 7 peanut butter.

Exercise 38.12. Let X be the number of cookies you will eat that until you find the 5th one that has more than 7 chocolate chips in it.

Exercise 38.13. **Dining hall dinners.** A total of 40,000 students on a university campus independently choose whether to go to the dining hall for dinner each day. Each student has dinner there, on a given day, with probability 0.84.

a. What is the distribution of the total number of visits by the students to the dining hall during a 7-day week?

b. Estimate the probability that 235,000 or fewer students go to the dining hall during a given week.

Exercise 38.14. **Survey.** Five hundred people participate in a 12-week survey to see how many times that they go to a religious service. Each person's attendance has a Binomial distribution with $n = 12$ and $p = 0.7$. Assume that the people behave independently with regard to attendance.

a. What is the actual distribution of the number of religious services attended altogether during the 12-week period?

b. Estimate the probability that the 500 people attended between 4170 and 4230 religious services altogether during the 12-week period.

Exercise 38.15. **Battery comparison.** It is known that the lifetime of SaveBucks batteries follows an Exponential distribution with a mean of 40 hours. The lifetime of LittleMoola batteries follows an Exponential distribution with a mean of 30 hours.

a. What is the probability a SaveBucks battery will last more than 48 hours?

b. What is the probability a LittleMoola battery will last more than 48 hours?

c. Given that a SaveBucks battery has lasted more than 48 hours, what is the probability it will last more than 96 hours? What property applies to this situation?

d. The batteries look identical, so if you have a battery you selected at random which has lasted more than 48 hours, what is the probability it is a SaveBucks battery? Which rule do you use to answer this question?

Exercise 38.16. Power outage. The power has gone out in your house, and you are using a small flashlight to read your probability book. You have three batteries available to you, all of the same brand, but you can't tell which brand in the dark. The flashlight uses only one battery at a time. Use the information about the two brands of batteries from Exercise 38.15 to answer the questions below.

a. If the 3 batteries are from SaveBucks, how long do you expect to have light from your flashlight? What is the standard deviation? What distribution are you using, and what are the parameters? What important assumption do you have to make about battery lives?

b. If the 3 batteries are from LittleMoola, how long do you expect to have light from your flashlight? What is the standard deviation? What distribution are you using and what are the parameters?

c. What is the probability you will need more than 3 batteries if you use the flashlight every evening for a week because you forgot to pay your electricity bill? (Assume there are 12 hours of darkness each night and 7 nights in the week.) What distribution are you using, and what are the parameters?

Exercise 38.17. Distribution of batteries sold. The SaveBucks battery company wants to wait until 60% of the batteries are sold at the corner convenience store before they ship more. After years of study, they have determined that the proportion of batteries which are sold after 6 weeks follows a Beta distribution with $\alpha = 2$ and $\beta = 3$.

a. What is the probability that more than 60% of the batteries will be sold at the end of the 6-week period?

b. What is the expected proportion of batteries which will be sold at the end of the 6-week period?

c. What is the standard deviation in the proportion of batteries which will be sold at the end of the 6-week period?

Exercise 38.18. Where is my engagement ring? While walking from the car into your dormitory, you dropped your engagement ring somewhere in the snow. The path is straight and 30 feet long. You are distraught because the density of its location seems to be constant along this 30-foot route.

a. What is the probability that the ring is within 12 feet of your car?

b. What is the probability that the ring is between 9 to 11 feet away from the car?

c. What is the expected distance and the standard deviation for the distance from the car where the ring fell?

Exercise 38.19. Lifeguard. A lifeguard has to jump in to save a swimmer in trouble at the lake at Possum Park an average of 2 times a week, according to a Poisson distribution.

a. What is the probability that he will save more than 25 swimmers in the 10-week season he works?

b. If you know he saved exactly 1 swimmer on his 8-hour shift today, what is the chance he did it in the last hour that he worked?

c. If there are 5 lifeguards working independently, and they each have the same average rate for saving swimmers, what is the probability that exactly three of them will save more than 25 swimmers (each) in the 10-week season?

d. The lifeguards receive a small raise if they save 10 swimmers. How long will the lifeguard expect to wait for his raise?

e. What is the probability our lifeguard have to wait at least 1 week after the start of the summer until he saves his first swimmer?

f. How many independent lifeguards will we expect to have to ask until we find our second lifeguard who had to wait at least 1 week after the start of the summer until he/she saved his/her first swimmer?

Exercise 38.20. Disease diagnosis. A patient has been diagnosed with a serious disease. The lifetimes of patients with this disease are approximately Exponential with expected value 10 years, after their initial diagnosis.

a. What is the probability that a patient who is diagnosed today will still be alive 15 years from now?

b. Given that a patient is still alive after 5 years, what is the probability the patient will still be alive 15 years after the initial diagnosis?

c. If 5 people are diagnosed with this disease today, what is the probability that all 5 of them will be alive 15 years from now?

Exercise 38.21. Nursing care. A nurse specializes in caring for patients diagnosed with the disease described in Exercise 38.20.

a. The nurse cares for one patient at a time since the care is given in the patient's home, and constant monitoring of the patient is needed. She begins caring for that patient as soon as the diagnosis is given, and she cares for the patient until death. Then she is reassigned to a new patient. What is the expected time until her 3rd patient dies?

b. The nurse wakes up and realizes that her patient's "call" button is lit. Unfortunately she doesn't know how long it has been lit. She knows that it wasn't lit 10 minutes ago, when she fell asleep. So it could have been pressed at anytime during the previous 10 minutes, and she is worried that her patient has been waiting too long. What is the probability that the patient pressed the call button in the first 30 seconds that the nurse was asleep?

c. Given that the call button was not pressed in the first 30 seconds, what is the probability it was pressed during the last 2 minutes?

Exercise 38.22. Apple juice. The fraction of juice which can be squeezed from an apple is a random variable with the following density:

$$f_X(x) = \begin{cases} k(1-x)^6 & \text{if } 0 \le x \le 1 \\ 0 & \text{elsewhere} \end{cases}$$

a. Find the value of k that makes this a valid probability density function.

b. What is the expected fraction of juice which can be squeezed from an apple?

c. What is the standard deviation in the fraction of juice which can be squeezed from an apple?

d. What is the probability that an apple will have more than half of the juice squeezed?

Exercise 38.23. Identify the distribution. If X has a mean 0.5 and a variance 0.25, find the parameters of the distribution (if possible) if X is:

a. Binomial

b. Exponential

c. Normal

d. Uniform

e. Geometric

Exercise 38.24. Identify the distribution. If X has a mean 3 and a variance 2.5, find the parameters of the distribution (if possible) if X is:

a. Binomial

b. Exponential

c. Normal

d. Uniform

e. Geometric

Exercise 38.25. Cable repairs. The cable technician guarantees he'll arrive sometime between 8 AM and noon, but he can't be any more specific than that.

a. What is the probability that he will come in between 9:30 and 10:45 AM?

b. You have a meeting at noon, and you're hoping that the cable technician will come before 11:30 AM so you'll have time to get to the meeting. If the cable technician has not come by 10 AM, what is the probability that he will come before 11:30 AM?

c. What is the probability that the next 4 times you need the cable technician to come out, he will come between 9:30 AM and 10:45 AM on at least 3 of those occasions?

Exercise 38.26. Cell phone usage. The amount of time you use on your cell phone each month is approximately Normally distributed with a mean of 623 minutes and a standard deviation of 24 minutes.

a. What is the probability that you will talk more than 700 minutes (when the extra charges apply) next month?

b. What range of talk-times represents the middle 50% of the times you will talk?

c. How many monthly bills do you expect to receive until you get the 6th one with extra charges (talk more than 700 minutes)?

Exercise 38.27. Student health center. The arrival times of students at the health center at a university is known to have a Poisson distribution with an average rate of 6 sick students arriving per hour.

a. What is the expected number of students who arrive in the next 30 minutes?

b. What is the expected time that will elapse between the 3rd and 4th arrivals? What is the standard deviation?

c. What is the expected time that will elapse between the 4th and 7th arrivals? What is the standard deviation?

d. What is the probability that there will be at least 15 minutes between each of the next 3 arrivals?

e. What is the probability there will be fewer than 3 visits to the health center during the next hour?

f. The health center is open from 9 AM to 5 PM. What is the probability that there will be exactly one hour (9–10, 10–11, etc.) with strictly fewer than 3 visits?

g. Yesterday, a temporary secretary was working the admissions desk. He kept records on how many students visited each hour, but not their exact times of arrival. If we know exactly one student visited the health center during the 2–3 PM hour, what is the probability the student arrived within 5 minutes of 1:30?

h. The fraction of students who leave the health center each day with a prescription for little blue Amoxycillin pills is modeled by the following probability density function:

$$f_X(x) = \begin{cases} k(x^2 - x^3) & \text{if } 0 \le x \le 1 \\ 0 & \text{elsewhere} \end{cases}$$

What is the expected fraction of students who will receive this prescription next week? (Hint: It might help to solve for k first.)

Exercise 38.28. Accident insurance. An insurance company expects that 10% of its moderate-risk drivers will be involved in an accident during the first 31 days of the year.

a. What is the average number of days that you would expect to wait for a moderate-risk driver, chosen at random, to be involved in an accident?

b. What portion of moderate-risk drivers are expected to be involved in an accident during the next quarter (90 days)?

Exercise 38.29. More accident insurance. For low-risk drivers, the waiting time for their first accident follows an Exponential distribution with a mean of 7 years. For high-risk drivers, the waiting time for their first accident follows an Exponential distribution with a mean of 2 years. Assuming that low-risk drivers and high-risk drivers are independent,

a. What is the probability that a low-risk driver and a high-risk driver will both have their first accident within the next 3 years?

b. What is the probability that either a low-risk driver or a high-risk driver will have an accident within the next 3 years?

c. Given that the low-risk driver will have an accident within the next 3 years, what is the probability that the low-risk driver will also have an accident within the next 3 years afterwards (i.e., years 4 through 6)?

d. Give an example of how a low-risk driver and a high-risk driver would NOT be independent. How reasonable is our assumption of independence? Why is the assumption of independence so important for our calculations?

Exercise 38.30. Renter's insurance. A renter's insurance policy is written to cover a loss, X, where X has a Uniform distribution with boundaries ranging from no loss up to a maximum of $2000.

a. What is the expected loss?

b. What is the variance in the loss?

c. If a deductible is set at $500, what is the probability that the loss would be above the deductible?

d. How much money will the insurance company expect to pay on a claim if the deductible is set at $500?

e. How much money will the renter expect to pay on a claim if the deductible is set at $500?

f. At what level must the deductible be set in order for the expected amount of money the insurance company pays on a claim to be only 10% of what it would be if there were no deductible?

Exercise 38.31. Stock investments. Three investments in high-tech stocks are thought to have returns which are Exponential random variables with means of 2, 3, and 4. Determine the probability that the maximum return from these 3 investments is more than 5. (Hint: Think about the opposite case.)

Exercise 38.32. **Computer lifetimes.** The lifetime of a computer component purchased at World Wide Computers is Exponentially distributed with a mean of 2 years.

a. What is the probability that the component fails in the first year?

b. What is the probability that the component fails in the second or third year (between the 1-year mark and the 3-year mark)?

c. What is the probability that the component fails after the third year?

Exercise 38.33. **Computer refunds.** Use the information in Exercise 38.32 about the computer components. If the component fails during the first year after purchase, the owner will refund the buyer's purchase price. If the component fails during the second or third year after purchase, he will refund 25% of the buyer's purchase price. Each component is purchased for $300.

a. How much does the owner expect to pay in refunds if a single component is purchased?

b. How much does the owner expect to pay in refunds if 50 components are purchased?

Exercise 38.34. **Energy-saving bulb.** The time an energy-saving light bulb will last without failing has an Exponential distribution with a median of 3 years. Calculate the probability that a light bulb, selected at random, will last at least 4 years. (Hint: Use the definition of median to help you get any parameters you may need.)

Part VII

Additional Topics

In this part, we discuss some more advanced topics that not all courses will have time to cover but are still important and interesting. Depending on the interests of the students and teachers, some of these topics (like moment generating functions) may be introduced earlier in the course. These topics regularly appear on the SOA/CAS P/1 exam.

By the end of this part of the book, you should be able to:

1. Calculate the variance of sums of random variables.
2. Calculate the covariance and correlation, and understand how these terms define the relationship between two random variables.
3. Calculate conditional expectations.
4. Use the Markov and Chebyshev inequalities to calculate limiting probabilities.
5. Calculate the PDF, CDF, joint PDF, and joint CDF for order statistics for independent, identically distributed, continuous random variables.
6. Understand what moment generating functions are and how they can be utilized.
7. Calculate various moments.
8. Transform random variables for calculations.

Math skills you will need: determinants, partial derivatives.

Additional resources: Calculators or a mathematical computation program may be used to assist in the calculations.

Chapter 39

Variance of Sums; Covariance; Correlation

William Feller was a probability theorist at Princeton University. One day he and his wife wanted to move a large table from one room of their large house to another, but, try as they might, they couldn't get it though the door. They pushed and pulled and tipped the table on its side and generally tried everything they could, but it just wouldn't go.

Eventually, Feller went back to his desk and worked out a mathematical proof that the table would never be able to pass through the door.

While he was doing this, his wife got the table through the door.

—*Professor Stewart's Hoard of Mathematical Treasures: Another Drawer from the Cabinet of Curiosities* by Ian Stewart (Basic Books, 2009)

I believe there is a direct correlation between love and laughter.

—Yakov Smirnoff

This morning, your mom made two different types of cookies, peanut butter and oatmeal raisin. She put them all in a big cookie jar in the kitchen. You and your brother both randomly (and hungrily!) grab cookies out of the jar, but your brother grabs first. Does the number of peanut butter cookies your brother grabs have an impact on the number of peanut butter cookies you later grab? What are the average and variance in the number of peanut butter cookies selected by each of you?

39.1 Introduction

Many times during this course, we have used the fact that the expected value of a sum of random variables is equal to the sum of the expected values of the random variables, i.e.,

$$\mathbb{E}(X_1 + \cdots + X_n) = \mathbb{E}(X_1) + \cdots + \mathbb{E}(X_n).$$

This nice fact holds *regardless of whether the X_j's are independent or dependent.* It works essentially because the sums or integrals used for the expected values are both linear, e.g., if X and Y are continuous:

$$\begin{aligned}\mathbb{E}(X+Y) &= \int_{-\infty}^{\infty}\int_{-\infty}^{\infty}(x+y)f_{X,Y}(x,y)\,dx\,dy \\ &= \int_{-\infty}^{\infty}\int_{-\infty}^{\infty} x f_{X,Y}(x,y)\,dx\,dy + \int_{-\infty}^{\infty}\int_{-\infty}^{\infty} y f_{X,Y}(x,y)\,dx\,dy \\ &= \mathbb{E}(X) + \mathbb{E}(Y) \qquad (39.1)\end{aligned}$$

Unfortunately, the analogous concept is more subtle for variances. *If the X_j's are independent*, then the variance of the sum of random variables is equal to the sum of the variances of the random variables. (Without the independence of the X_j's, this rule falls apart!) The reason that things do not work quite as nicely for variances is that $\mathrm{Var}(X) = \mathbb{E}(X^2) - (\mathbb{E}(X))^2$ does not have the same linearity properties used above, to show $\mathbb{E}(X+Y) = \mathbb{E}(X) + \mathbb{E}(Y)$.

39.2 Motivation for Covariance

If we dive deeper into $\mathrm{Var}(X_1 + \cdots + X_n)$, we recall from Section 12.2 that the variance of a random variable Y is

$$\mathrm{Var}(Y) = \mathbb{E}(Y^2) - (\mathbb{E}(Y))^2.$$

Using $Y = \sum_{i=1}^n X_i$, this gives

$$\mathrm{Var}\left(\sum_{i=1}^n X_i\right) = \mathbb{E}\left(\left(\sum_{i=1}^n X_i\right)^2\right) - \left(\mathbb{E}\left(\sum_{i=1}^n X_i\right)\right)^2. \qquad (39.2)$$

For the left term in (39.2), we just multiply to expand $(\sum_{i=1}^n X_i)^2$ as follows:

$$\begin{aligned}\left(\sum_{i=1}^n X_i\right)^2 &= \sum_{i=1}^n X_i \sum_{j=1}^n X_j \\ &= \sum_{i=1}^n\sum_{j=1}^n X_i X_j,\end{aligned}$$

but the expected value of the sum equals the sum of the expected values, so the left term in (39.2) is

$$\begin{aligned}\mathbb{E}\left(\left(\sum_{i=1}^n X_i\right)^2\right) &= \mathbb{E}\left(\sum_{i=1}^n\sum_{j=1}^n X_i X_j\right) \\ &= \sum_{i=1}^n\sum_{j=1}^n \mathbb{E}(X_i X_j).\end{aligned}$$

For the right term in (39.2), we first convert the expected value of the sum, i.e., $\mathbb{E}(\sum_{i=1}^{n} X_i)$, into the sum of expected values, i.e., $\sum_{i=1}^{n} \mathbb{E}(X_i)$, and then multiply to expand, so

$$\begin{aligned}\left(\mathbb{E}\left(\sum_{i=1}^{n} X_i\right)\right)^2 &= \left(\sum_{i=1}^{n} \mathbb{E}(X_i)\right)^2 \\ &= \sum_{i=1}^{n} \mathbb{E}(X_i) \sum_{j=1}^{n} \mathbb{E}(X_j) \\ &= \sum_{i=1}^{n} \sum_{j=1}^{n} \mathbb{E}(X_i)\mathbb{E}(X_j).\end{aligned}$$

Thus, equation (39.2) becomes

$$\operatorname{Var}\left(\sum_{i=1}^{n} X_i\right) = \sum_{i=1}^{n} \sum_{j=1}^{n} \mathbb{E}(X_i X_j) - \sum_{i=1}^{n} \sum_{j=1}^{n} \mathbb{E}(X_i)\mathbb{E}(X_j),$$

or equivalently, collecting analogous terms, we get

$$\operatorname{Var}\left(\sum_{i=1}^{n} X_i\right) = \sum_{i=1}^{n} \sum_{j=1}^{n} (\mathbb{E}(X_i X_j) - \mathbb{E}(X_i)\mathbb{E}(X_j)). \tag{39.3}$$

A key point from all of these manipulations with the sums and the squaring is that the terms $\mathbb{E}(X_i X_j) - \mathbb{E}(X_i)\mathbb{E}(X_j)$ play a fundamental role in the variance of the sum of random variables.

39.3 Properties of the Covariance

Due to the primary importance of $\mathbb{E}(X_i X_j) - \mathbb{E}(X_i)\mathbb{E}(X_j)$ above, we introduce the concept of the **covariance** of two random variables, which can be defined in two equivalent ways:

Definition 39.1. Covariance of two random variables

The covariance of two random variables X and Y is defined as

$$\operatorname{Cov}(X, Y) = \mathbb{E}((X - \mathbb{E}(X))(Y - \mathbb{E}(Y))).$$

Another form of the covariance is very commonly used, as we briefly explain:

$$\begin{aligned}\mathbb{E}((X - \mathbb{E}(X))(Y - \mathbb{E}(Y))) &= \mathbb{E}\left(XY - Y\mathbb{E}(X) - X\mathbb{E}(Y) + \mathbb{E}(X)\mathbb{E}(Y)\right) \\ &= \mathbb{E}(XY) - \mathbb{E}\left(Y\mathbb{E}(X)\right) \\ &\quad - \mathbb{E}\left(X\mathbb{E}(Y)\right) + \mathbb{E}\left(\mathbb{E}(X)\mathbb{E}(Y)\right)\end{aligned}$$

We note that $\mathbb{E}(X)$ is constant, so it can be pulled out of the term $\mathbb{E}\left(Y\mathbb{E}(X)\right) = \mathbb{E}(Y)\mathbb{E}(X)$. Similarly, $\mathbb{E}(Y)$ is a constant, so it can be pulled out of the term

$\mathbb{E}(X\mathbb{E}(Y)) = \mathbb{E}(X)\mathbb{E}(Y)$. Finally, $\mathbb{E}(X)$ and $\mathbb{E}(Y)$ are both constants, and thus $\mathbb{E}(\mathbb{E}(X)\mathbb{E}(Y)) = \mathbb{E}(X)\mathbb{E}(Y)$. So we obtain

$$\mathbb{E}((X - \mathbb{E}(X))(Y - \mathbb{E}(Y))) = \mathbb{E}(XY) - \mathbb{E}(Y)\mathbb{E}(X) - \mathbb{E}(X)\mathbb{E}(Y) + \mathbb{E}(X)\mathbb{E}(Y),$$

which simplifies to just

$$\mathbb{E}((X - \mathbb{E}(X))(Y - \mathbb{E}(Y))) = \mathbb{E}(XY) - \mathbb{E}(X)\mathbb{E}(Y).$$

Therefore, we have the following form, which we often utilize:

Theorem 39.2. Covariance of two random variables The covariance of two random variables X and Y can be expressed as

$$\text{Cov}(X, Y) = \mathbb{E}(XY) - \mathbb{E}(X)\mathbb{E}(Y).$$

The covariance gives some information about how far two random variables are spread from their expected values. Just like with the expected value or the variance, the covariance is only one number, so it can only provide a relatively small amount of information. Also, it is possible that many different pairs of X's and Y's could have exactly the same covariance. So the covariance does not tell the whole story about the relationship between two variables, but it does help. Very informally—just to build intuition—when a covariance between X and Y is positive, then the larger X turns out to be, the larger Y will tend to be, and vice versa. If the covariance is negative, then the larger that one of the random variables tends to be, the smaller the other will tend to be.

As an informal example, suppose X and Y are indicator random variables for the events that Alicia and Brent (respectively) get a peanut butter cookie from the same cookie jar (which contains several types of cookies). When Alicia gets a peanut butter cookie, then $X = 1$, and there is one less peanut butter cookie remaining for Brent, so Y becomes less likely to be 1 than it would be otherwise. Thus, X and Y have negative covariance. We elaborate on this specific example in Example 39.9.

Using the notation in equation (39.3), we conclude that the variance of the sum of random variables is equal to the sum of covariances of all pairs of the random variables:

Theorem 39.3. Variance of the sum of random variables (version 1)

If $X_1, \ldots, X_n$ are random variables (that are not necessarily independent), then

$$\text{Var}(X_1 + \cdots + X_n) = \sum_{i=1}^{n} \sum_{j=1}^{n} \text{Cov}(X_i, X_j).$$

Using $n = 1$, we see that the covariance of a random variable with itself is exactly equal to the variance of the random variable:

$$\text{Cov}(X, X) = \mathbb{E}((X - \mathbb{E}(X))^2) = \text{Var}(X),$$

which can also be recognized by writing

$$\begin{aligned}\mathrm{Cov}(X,X) &= \mathbb{E}(XX) - \mathbb{E}(X)\mathbb{E}(X)\\ &= \mathbb{E}(X^2) - (\mathbb{E}(X))^2\\ &= \mathrm{Var}(X).\end{aligned}$$

Corollary 39.4. Covariance of a random variable with itself

The covariance of X with itself is equal to the variance of X:

$$\mathrm{Cov}(X,X) = \mathrm{Var}(X).$$

Isolating all n of the terms of this flavor from Theorem 39.3, we get another way to express the variance of the sum of random variables:

Corollary 39.5. Variance of the sum of random variables (version 2)

If $X_1, \ldots, X_n$ are random variables (that are not necessarily independent), then

$$\mathrm{Var}(X_1 + \cdots + X_n) = \sum_{i=1}^{n} \mathrm{Var}(X_i) + \sum_{i=1}^{n}\sum_{j\neq i} \mathrm{Cov}(X_i, X_j).$$

We also note that the covariance is symmetric in terms of the two variables, i.e.,

$$\begin{aligned}\mathrm{Cov}(X,Y) &= \mathbb{E}(XY) - \mathbb{E}(X)\mathbb{E}(Y)\\ &= \mathbb{E}(YX) - \mathbb{E}(Y)\mathbb{E}(X)\\ &= \mathrm{Cov}(Y,X).\end{aligned}$$

Theorem 39.6. Covariance is symmetric

The covariance of X and Y equals the covariance of Y and X:

$$\mathrm{Cov}(X,Y) = \mathrm{Cov}(Y,X).$$

With this symmetry in mind, we see that each term $\mathrm{Cov}(X_i, X_j)$ in Corollary 39.5 will also appear as a $\mathrm{Cov}(X_j, X_i)$ term. So the following formulation makes calculations shorter to perform, since we are combining all of these duplicated pairs:

Corollary 39.7. Variance of the sum of random variables (version 3)

If $X_1, \ldots, X_n$ are random variables (that are not necessarily independent), then

$$\mathrm{Var}(X_1 + \cdots + X_n) = \sum_{i=1}^{n} \mathrm{Var}(X_i) + 2\sum\sum_{i<j} \mathrm{Cov}(X_i, X_j).$$

We have already observed that for any random variables X and Y and functions g and h, if X and Y are independent, then:

$$\mathbb{E}(g(X)h(Y)) = \mathbb{E}(g(X))\mathbb{E}(h(Y)).$$

(Recall: This was proved for discrete random variables in Section 12.2 and for continuous random variables in Section 29.4. The motivation for this is that independence of X and Y allows the joint mass $p_{X,Y}(x, y)$ to be factored into $p_X(x)p_Y(y)$, or in the continuous case, the joint density $f_{X,Y}(x, y)$ to be factored into $f_X(x)f_Y(y)$. So the double sum or double integral for $\mathbb{E}(g(X)h(Y))$ can be recomputed as the product of two sums or two integrals, i.e., $\mathbb{E}(g(X))\mathbb{E}(h(Y))$.)

As a result, using simply $g(X) = X$ and $h(Y) = Y$, we see that if X and Y are independent, then the covariance of X and Y is 0, i.e.,

$$\text{Cov}(X, Y) = \mathbb{E}(XY) - \mathbb{E}(X)\mathbb{E}(Y) = 0.$$

The covariance may or may not be 0 when X and Y are dependent; in fact, the covariance of dependent random variances is usually not 0. (For the interested reader, however, we give Examples 39.14 and 39.15, which show dependent random variables that nonetheless have 0 covariance.)

Theorem 39.8. Covariance of independent random variables is 0

The covariance of two *independent* random variables X and Y is 0:

$$\text{Cov}(X, Y) = 0.$$

This nice little fact, used in the context of Corollary 39.5 or Corollary 39.7, reinforces our understanding that, for *independent* random variables $X_1, \ldots, X_n$, the variance of the sum of the X_j's equals the sum of the variances of the X_j's:

$$\text{Var}(X_1 + \cdots + X_n) = \text{Var}(X_1) + \cdots + \text{Var}(X_n).$$

We put two more facts about covariance into Section 39.5: (1) the covariance of two sums equals two sums of covariances, and (2) constants can be factored out of covariances. (This is usually referred to as linearity.) First, however, we study some examples about covariance.

39.4 Examples of Covariance

Example 39.9. Consider an example in which Alicia grabs a cookie (without replacement) from a cookie jar that has 13 cookies, 5 of which are peanut butter. Let $X = 1$ if she gets a peanut butter cookie, and $X = 0$ otherwise, i.e., X is an indicator for whether she gets a peanut butter cookie. After Alicia picks, let Brent grab a cookie from the 12 that remain, and let $Y = 1$ if he gets a peanut butter cookie, or $Y = 0$ otherwise. We find several properties of X and Y.

Find the mass of X and the mass of Y:

When we focus on just one of the random variables at a time, the other random variable does not affect the situation. The mass of X is

$$p_X(1) = 5/13, \qquad \text{and} \qquad p_X(0) = 8/13.$$

The mass of Y is the same as the mass of X, although this might take a minute's thought to justify: Brent is equally likely to get any of the 13 cookies, so his chances of getting a peanut butter cookie must be 5/13. To see this rigorously, we can compute using the Bayes' rule, i.e., by conditioning on the value of X, i.e., by conditioning on whether Alicia received a peanut butter cookie:

$$\begin{aligned}
P(Y = 1) &= P(Y = 1 \text{ and } X = 0) + P(Y = 1 \text{ and } X = 1) \\
&= P(Y = 1 \mid X = 0)P(X = 0) + P(Y = 1 \mid X = 1)P(X = 1) \\
&= (5/12)(8/13) + (4/12)(5/13) \\
&= 5/13
\end{aligned}$$

Remember that indicator random variables are just Bernoulli random variables, i.e., just 1 or 0, depending on whether some event happens.

So Y has the same mass as X. In other words, Brent and Alicia are each equally likely to get a peanut butter cookie. On the other hand, when we take Alicia's situation into account first, there is an effect on the ability of Brent to get a cookie. So X and Y are dependent. When Alicia does not get a peanut butter cookie, then Brent is more likely to get one, because more peanut butter cookies remain, i.e.,

$$P(Y = 1 \mid X = 0) = 5/12.$$

When she does get one, then he is less likely to get one, because there are fewer peanut butter cookies remaining, i.e.,

$$P(Y = 1 \mid X = 1) = 4/12.$$

So

$$P(Y = 1 \mid X = 0) \neq P(Y = 1 \mid X = 1),$$

and these are both different from $P(Y = 1) = 5/13$. So Y is dependent on X.

Find the expected value of $X + Y$:

Even though X and Y are dependent, we can calculate the expected value of $X+Y$, the total number of peanut butter cookies eaten by Alicia and Brent altogether, by adding the expected values:

$$\mathbb{E}(X + Y) = \mathbb{E}(X) + \mathbb{E}(Y) = 5/13 + 5/13 = 10/13.$$

(Note that $\mathbb{E}(X) = 5/13$ since X is a Bernoulli random variable, and similarly, $\mathbb{E}(Y) = 5/13$ since Y is also a Bernoulli.)

Find the variance of $X + Y$:

Since X and Y are dependent, we cannot just get the variance of $X + Y$ by adding the variances of X and Y separately. We do still know that $\text{Var}(X) =$

$(5/13)(8/13)$ since X is Bernoulli, and similarly, $\text{Var}(Y) = (5/13)(8/13)$ since Y is Bernoulli.

We have $\text{Cov}(X, Y) = \mathbb{E}(XY) - \mathbb{E}(X)\mathbb{E}(Y)$. We emphasize that the product XY itself is a Bernoulli random variable because XY can only take on values 0 or 1. The only way to have $XY = 1$ is to have $X = 1$ and $Y = 1$ (otherwise, $X = 0$ or $Y = 0$ so $XY = 0$). In other words,

$$\begin{aligned}\mathbb{E}(XY) = 1P(&X = 1 \text{ and } Y = 1)\\ &+ 0P(X = 0 \text{ and } Y = 1)\\ &+ 0P(X = 1 \text{ and } Y = 0)\\ &+ 0P(X = 0 \text{ and } Y = 0)\end{aligned}$$

So we just have

$$\mathbb{E}(XY) = P(X = 1 \text{ and } Y = 1).$$

To get $X = 1$ and $Y = 1$, we need $X = 1$, and then conditioned on $X = 1$, we need $Y = 1$ too. If both Alicia and Brent want peanut butter cookies, Alicia needs to get one, and then (given that she got one), Brent will need one too. So

The product of Bernoullis is a Bernoulli too.

$$\begin{aligned}\mathbb{E}(XY) &= P(X = 1 \text{ and } Y = 1)\\ &= P(Y = 1 \mid X = 1)P(X = 1)\\ &= (4/12)(5/13).\end{aligned}$$

So

$$\text{Cov}(X, Y) = (4/12)(5/13) - (5/13)(5/13).$$

In summary,

$$\begin{aligned}\text{Var}(X + Y) &= \text{Var}(X) + \text{Var}(Y) + \text{Cov}(X, Y) + \text{Cov}(Y, X)\\ &= (2)(5/13)(8/13) + (2)((4/12)(5/13) - (5/13)(5/13))\\ &= 220/507\\ &= 0.4339\end{aligned}$$

Example 39.10. Consider an eight-hour work day. Huiping and Ravi each check their email just once per day. Let X and Y denote the time, respectively, until Huiping and Ravi each check their emails during the day. Assume that Huiping always checks email first, and that X and Y are Uniform on the region where $0 \le X \le Y \le 8$, so that the joint density is constant on the triangle in Figure 39.1. Since the joint density is constant on a triangle of area $(8)(8)(1/2) = 32$, then the joint density must be the reciprocal of the area, i.e.,

$$f_{X,Y}(x, y) = 1/32 \qquad \text{for } 0 \le x \le y \le 8,$$

and $f_{X,Y}(x, y) = 0$ otherwise.

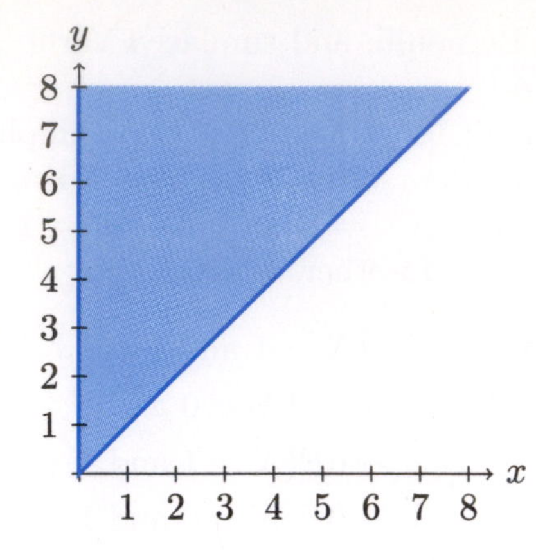

FIGURE 39.1: Region where X and Y have positive density.

Find the expected value and variance of $Y - X$, which is the time of the interval in between Huiping and Ravi checking their email.

The expected time $\mathbb{E}(X)$ until Huiping checks email is obtained by integrating $xf_{X,Y}(x,y)$ over the triangle. We can use integration with respect to y, for the outer integral; we can use integration over the x's in the range $0 \leq x\ leqy$, for the inner integral. This setup is shown in Figure 39.2.

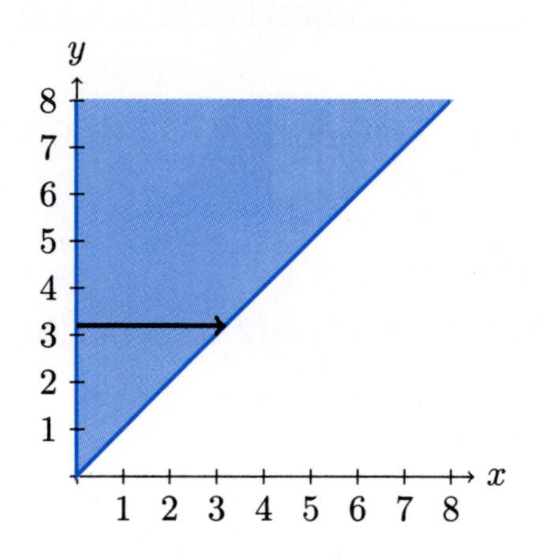

FIGURE 39.2: Fixed value of y (here, for example $y = 3.2$), and x ranging from 0 to y.

So the expected time until Huiping checks email is

$$\mathbb{E}(X) = \int_0^8 \int_0^y x/32\,dx\,dy = 8/3.$$

The expected time until Ravi checks email is

$$\mathbb{E}(Y) = \int_0^8 \int_0^y y/32\,dx\,dy = 16/3.$$

Thus

$$\mathbb{E}(Y - X) = \mathbb{E}(Y) - \mathbb{E}(X) = 16/3 - 8/3 = 8/3.$$

We cannot simply add the variances of X and Y, since X and Y are dependent. So we compute

$$\begin{aligned}\mathrm{Var}(Y - X) &= \mathbb{E}((Y - X)^2) - (\mathbb{E}(Y - X))^2 \\ &= \mathbb{E}((Y - X)^2) - (8/3)^2,\end{aligned}$$

and now we need to integrate $(y - x)^2 f_{X,Y}(x, y)$ over the triangle, to get the value of $\mathbb{E}((Y - X)^2)$. Since $f_{X,Y}(x, y) = 1/32$ on the triangle, we have

$$\begin{aligned}\mathbb{E}((Y - X)^2) &= \int_0^8 \int_0^y (y - x)^2/32 \, dx \, dy \\ &= 32/3.\end{aligned}$$

So, in summary,

$$\begin{aligned}\mathrm{Var}(Y - X) &= 32/3 - (8/3)^2 \\ &= 32/9.\end{aligned}$$

As an alternative way to find the variance of $Y - X$, we could use the covariance:

$$\begin{aligned}\mathrm{Var}(Y - X) &= \mathrm{Cov}(Y - X, Y - X) \\ &= \mathrm{Cov}(Y, Y) + (-1)\,\mathrm{Cov}(Y, X) \\ &\qquad + (-1)\,\mathrm{Cov}(X, Y) + (-1)(-1)\,\mathrm{Cov}(X, X) \\ &= \mathrm{Var}(X) + \mathrm{Var}(Y) - 2\,\mathrm{Cov}(X, Y).\end{aligned}$$

The variances of X and Y are:

$$\mathrm{Var}(X) = \int_0^8 \int_0^y x^2/32 \, dx \, dy - (8/3)^2 = 32/3 - (8/3)^2 = 32/9$$

and

$$\mathrm{Var}(Y) = \int_0^8 \int_0^y y^2/32 \, dx \, dy - (16/3)^2 = 32 - (16/3)^2 = 32/9,$$

and the covariance is

$$\begin{aligned}2\,\mathrm{Cov}(X, Y) &= 2\left(\int_0^8 \int_0^y xy/32 \, dx \, dy - (8/3)(16/3)\right) \\ &= 2(16 - (8/3)(16/3)) \\ &= 32/9.\end{aligned}$$

So we have verified that

$$\mathrm{Var}(Y - X) = 32/9 + 32/9 - 32/9 = 32/9.$$

Example 39.11. Consider a selection of 10 songs on a playlist, of which exactly 7 are rock songs. If two such songs are chosen, and repetition is not allowed, and all outcomes are equally likely, let X denote the number of songs that are rock songs.

We let X_1 and X_2 be indicator random variables that denote whether the first and second songs (respectively) are rock songs. Thus the expected number of rock songs chosen is

Remember that, for an indicator (i.e., a Bernoulli random variable) with probability of success p, and probability of failure $q = 1 - p$, the expected value is p and the variance is pq.

$$\mathbb{E}(X) = \mathbb{E}(X_1 + X_2) = \mathbb{E}(X_1) + \mathbb{E}(X_2) = 7/10 + 7/10 = 14/10 = 7/5 = 1.4.$$

Now consider the variance of X. We have

$$\text{Var}(X) = \text{Var}(X_1 + X_2) = \text{Var}(X_1) + \text{Var}(X_2) + 2\,\text{Cov}(X_1, X_2).$$

Each of the X_j's is a Bernoulli random variable, so

$$\begin{aligned}\text{Var}(X_j) &= P(X_j = 1)P(X_j = 0)\\ &= (\text{probability song is rock})(\text{probability song is not rock})\\ &= (7/10)(3/10)\\ &= 21/100.\end{aligned}$$

Also

$$\text{Cov}(X_1, X_2) = \mathbb{E}(X_1X_2) - \mathbb{E}(X_1)\mathbb{E}(X_2).$$

We know $\mathbb{E}(X_1) = 7/10$ and $\mathbb{E}(X_2) = 7/10$. Also X_1 and X_2 are Bernoullis, so X_1X_2 is Bernoulli too. Thus the expected value of X_1X_2 is just equal to the probability that X_1X_2 is equal to 1. So

$$\begin{aligned}\mathbb{E}(X_1X_2) &= P(X_1X_2 = 1)\\ &= P(X_1 = 1 \text{ and } X_2 = 1)\\ &= P(X_1 = 1)P(X_2 = 1 \mid X_1 = 1).\end{aligned}$$

As before, $P(X_1 = 1) = 7/10$. If $X_1 = 1$, then the second song is equally likely to be any of 9 remaining songs, of which 6 are rock songs. Thus

$$\begin{aligned}\mathbb{E}(X_1X_2) &= P(X_1 = 1)P(X_2 = 1 \mid X_1 = 1)\\ &= (7/10)(6/9).\end{aligned}$$

Putting this all together, we have

$$\begin{aligned}\text{Var}(X) &= \frac{21}{100} + \frac{21}{100} + 2\left(\left(\tfrac{7}{10}\right)\left(\tfrac{6}{9}\right) - \left(\tfrac{7}{10}\right)\left(\tfrac{7}{10}\right)\right)\\ &= 28/75\\ &= 0.37333.\end{aligned}$$

Notice that the scenarios in Examples 39.9 and 39.11 are specific cases of the Hypergeometric distribution. E.g., in Example 39.9, $X + Y$ has a Hypergeometric distribution, since we select $n = 2$ out of $N = 13$ items (all cookies); exactly $M = 5$ (the peanut butter cookies) are desirable. Thus, as we learned in the Hypergeometric chapter:

$$\mathbb{E}(X+Y) = n\frac{M}{N} = (2)\left(\frac{5}{13}\right) = \frac{10}{13},$$

and (as we will see below in the more general case, in Example 39.12) we have:

$$\text{Var}(X+Y) = n\frac{M}{N}\left(1-\frac{M}{N}\right)\frac{N-n}{N-1} = (2)\frac{5}{13}\left(1-\frac{5}{13}\right)\frac{13-2}{13-1} = \frac{220}{507}.$$

In Example 39.11, the random variable $X_1 + X_2$ is Hypergeometric, because there are $N = 10$ songs available altogether; $n = 2$ of the songs are chosen; and $M = 7$ of the songs are desirable.

Example 39.12. Recall from Chapter 19 that a Hypergeometric random variable is defined as follows: Consider N items altogether: M of the items are "desirable" and the other $N - M$ are "undesirable." We pick a collection of n items. Let X denote the number of "desirable" items that we get. Then X is a Hypergeometric random variable with parameters N, M, and n. The mass of X is

$$p_X(j) = P(X=j) = \frac{\binom{M}{j}\binom{N-M}{n-j}}{\binom{N}{n}}.$$

In Chapter 19, we already discussed the fact that a Hypergeometric random variable X can be decomposed as

$$X = X_1 + X_2 + \cdots + X_n,$$

where $X_j = 1$ if the jth item selected is desirable, and $X_j = 0$ otherwise. Thus $\mathbb{E}(X_j) = P(X_j = 1)$. There are N possible items that can be selected as the jth item, and M of these items are desirable. So $\mathbb{E}(X_j) = P(X_j = 1) = M/N$. Thus

$$\begin{aligned}\mathbb{E}(X) &= \mathbb{E}(X_1 + \cdots + X_n)\\ &= \mathbb{E}(X_1) + \cdots + \mathbb{E}(X_n)\\ &= M/N + \cdots + M/N\\ &= nM/N.\end{aligned}$$

In Remark 19.4 of Chapter 19, we computed the variance of a Hypergeometric random variable. Although the computation is very similar, it is perhaps helpful to recast this argument using the covariance. We have

$$\text{Var}(X) = \sum_{j=1}^{n}\text{Var}(X_j) + 2\sum\sum_{i<j}\text{Cov}(X_i, X_j).$$

Since each X_j is a Bernoulli, then

$$\begin{aligned}\mathrm{Var}(X_j) &= P(X_j = 1)P(X_j = 0)\\ &= \left(\frac{M}{N}\right)\left(1 - \frac{M}{N}\right).\end{aligned}$$

For $i < j$, we have

$$\mathrm{Cov}(X_i, X_j) = \mathbb{E}(X_iX_j) - \mathbb{E}(X_i)\mathbb{E}(X_j),$$

and we know $\mathbb{E}(X_i) = M/N$ and $\mathbb{E}(X_j) = M/N$. Also, X_i and X_j are each 0 or 1, so X_iX_j is 0 or 1. In other words, X_i and X_j are each Bernoulli, so X_iX_j is Bernoulli too. Thus

$$\begin{aligned}\mathbb{E}(X_iX_j) &= P(X_iX_j = 1)\\ &= P(X_i = 1 \text{ and } X_j = 1)\\ &= P(X_i = 1)P(X_j = 1 \mid X_i = 1).\end{aligned}$$

As before, $P(X_i = 1) = M/N$. If $X_i = 1$, then the jth item is equally likely to be any of $N - 1$ remaining items, of which $M - 1$ are desirable. Thus

$$\begin{aligned}\mathbb{E}(X_iX_j) &= P(X_i = 1)P(X_j = 1 \mid X_i = 1)\\ &= \left(\frac{M}{N}\right)\left(\frac{M-1}{N-1}\right).\end{aligned}$$

Putting this all together, we have

$$\mathrm{Var}(X) = \sum_{j=1}^{n}\left(\frac{M}{N}\right)\left(1 - \frac{M}{N}\right) + 2\sum\sum_{i<j}\left(\left(\frac{M}{N}\right)\left(\frac{M-1}{N-1}\right) - \left(\frac{M}{N}\right)\left(\frac{M}{N}\right)\right).$$

The n terms of the first type are all alike; none depend on j. The $\binom{n}{2} = (n)(n-1)/2$ terms of the second type are also alike; none of them depend on i or j. Thus

$$\mathrm{Var}(X) = n\left(\frac{M}{N}\right)\left(1 - \frac{M}{N}\right) + 2\frac{(n)(n-1)}{2}\left(\left(\frac{M}{N}\right)\left(\frac{M-1}{N-1}\right) - \left(\frac{M}{N}\right)\left(\frac{M}{N}\right)\right).$$

After a bit of simplification using the common denominator $N^2(N-1)$, this yields the result that the variance of a Hypergeometric random variable is

$$\mathrm{Var}(X) = \frac{nM(N-M)(N-n)}{N^2(N-1)} = n\frac{M}{N}\left(1 - \frac{M}{N}\right)\frac{N-n}{N-1}.$$

As motivation for the next two examples:

Remark 39.13. Comparing independence and covariance
If two random variables X and Y are independent, then $\mathrm{Cov}(X, Y) = 0$. The converse is not necessarily true. For instance, it is possible to have two random variables X and Y such that $\mathrm{Cov}(X, Y) = 0$ but X and Y are not independent.

Consider, for example, the following:

Example 39.14. Let X have mass $p_X(x) = P(X = x) = 1/5$ for $x = -2, -1, 0, 1, 2$. Define Y such that $Y = 1$ if $X = 0$, and $Y = 0$ otherwise.

In this example,

$$\mathbb{E}(X) = -2(1/5) - 1(1/5) + 0(1/5) + 1(1/5) + 2(1/5) = 0.$$

Also $XY = 0$ always, since either $X = 0$, or otherwise $Y = 0$. Thus $\mathbb{E}(XY) = 0$. So

$$\begin{aligned}\mathrm{Cov}(X, Y) &= \mathbb{E}(XY) - \mathbb{E}(X)\mathbb{E}(Y) \\ &= 0 - (0)\mathbb{E}(Y) \\ &= 0.\end{aligned}$$

On the other hand, X and Y are not independent, because the value of Y is completely dependent on the value of X.

The idea in the previous example can be generalized quite a bit:

Example 39.15. Let X be any random variable with expected value 0, and let Y be defined such that

$$Y = 0 \qquad \text{whenever } X \neq 0,$$

and let

$$Y \neq 0 \qquad \text{otherwise, e.g., let } Y = 13 \text{ when } X = 0.$$

Then the distribution of Y again depends on the distribution of X, but also X and Y again have covariance 0, since $XY = 0$ in all cases in this example, and thus

$$\begin{aligned}\mathrm{Cov}(X, Y) &= \mathbb{E}(XY) - \mathbb{E}(X)\mathbb{E}(Y) \\ &= 0 - (0)\mathbb{E}(Y) \\ &= 0.\end{aligned}$$

39.5 Linearity of the Covariance

Theorem 39.16. Covariance of two sums of random variables equals two sums of covariances

If $X_1, \ldots, X_n$ and $Y_1, \ldots, Y_m$ are random variables (not necessarily independent), and $a_1, \ldots, a_n$ and $b_1, \ldots, b_m$ are constants, then

$$\mathrm{Cov}(a_1X_1 + \cdots + a_nX_n,\ b_1Y_1 + \cdots + b_mY_m) = \sum_{i=1}^{n}\sum_{j=1}^{m} a_i b_j \,\mathrm{Cov}(X_i, Y_j).$$

To see this, we compute (in a similar fashion to the argument for sums of variances from Section 12.3),

$$\begin{aligned}
&\mathrm{Cov}(a_1X_1 + \cdots + a_nX_n,\ b_1Y_1 + \cdots + b_mY_m)\\
&\quad = \mathbb{E}\Big(\sum_{i=1}^{n} a_iX_i \sum_{j=1}^{m} b_jY_j\Big) - \mathbb{E}\Big(\sum_{i=1}^{n} a_iX_i\Big)\mathbb{E}\Big(\sum_{j=1}^{m} b_jY_j\Big)\\
&\quad = \mathbb{E}\Big(\sum_{i=1}^{n}\sum_{j=1}^{m} a_ib_jX_iY_j\Big) - \sum_{i=1}^{n} a_i\mathbb{E}(X_i) \sum_{j=1}^{m} b_j\mathbb{E}(Y_j)\\
&\quad = \sum_{i=1}^{n}\sum_{j=1}^{m} a_ib_j\mathbb{E}(X_iY_j) - \sum_{i=1}^{n}\sum_{j=1}^{m} a_ib_j\mathbb{E}(X_i)\mathbb{E}(Y_j)\\
&\quad = \sum_{i=1}^{n}\sum_{j=1}^{m} a_ib_j(\mathbb{E}(X_iY_j) - \mathbb{E}(X_i)\mathbb{E}(Y_j))\\
&\quad = \sum_{i=1}^{n}\sum_{j=1}^{m} a_ib_j \,\mathrm{Cov}(X_i, Y_j)
\end{aligned}$$

This result is often used in a context without any constants (i.e., all $a_i = 1$ and $b_j = 1$), because we often need to take the covariance of two sums of random variables. This happens, for instance, when we want to take the covariance of a sum of indicators with another sum of indicators. As another example, if n and m are large, and the X_i's are independent from each other, and the Y_j are independent from each other, but the X_i's and Y_j's have some dependencies, then $X_1 + \cdots + X_n$ is approximately Normal, and $Y_1 + \cdots + Y_m$ is approximately Normal too, and the covariance tells us how the two sums are related.

Corollary 39.17. Covariance of two sums of random variables equals two sums of covariances (version without constants)

If $X_1, \ldots, X_n$ and $Y_1, \ldots, Y_m$ are random variables (not necessarily independent), then

$$\mathrm{Cov}(X_1 + \cdots + X_n,\ Y_1 + \cdots + Y_m) = \sum_{i=1}^{n}\sum_{j=1}^{m} \mathrm{Cov}(X_i, Y_j).$$

When $m = 1$ and $n = 1$, and we write $a_1 = a$ and $b_1 = b$, we get another useful form:

Corollary 39.18. Constants can be factored out of covariances
If X and Y are random variables (not necessarily independent), and a and b are coefficients, then

$$\mathrm{Cov}(aX, bY) = ab\,\mathrm{Cov}(X, Y).$$

For instance, if X is the charge for an oil change, and $a = 1.07$, then aX is the charge for an oil change, with the 7% additional tax included. Similarly, if Y is the number of containers of oil that are needed, and b is the price per container, then bY is the price of the oil itself. Thus we can switch from $\mathrm{Cov}(aX, bY)$ to instead focusing on $ab\,\mathrm{Cov}(X, Y)$. This could be helpful, especially if the charge for an oil change is directly related to the number of containers of oil itself. The other things, like the tax and the charge per container of oil, can be factored out of the covariance equation, to simplify the situation and allow us to calculate $\mathrm{Cov}(X, Y)$ without the complications of tax or charge per container of oil.

39.6 Correlation

The correlation of two random variables X and Y is a one-number summary of the relationship between X and Y. The correlation is always between -1 and $+1$, as we will see below. A correlation near 1 indicates that Y basically increases as X increases. A correlation near -1 indicates that Y tends to decrease as X increases. A correlation near 0 indicates that Y is basically unaffected by the increase or decrease of X. These are all imprecise statements, as they must be, because the correlation is only a one-number way of stating how X and Y are related. The relationship between X and Y might be quite complicated, and one number (such as the correlation) cannot fully describe the relationship. So a rough approximation to the behavior of the correction is all that we can hope to quantify with such a one-number summary.

Example 39.19. If X is the number of hours that a student spends studying, and if Y is the score a student earns on an exam, then Y will (roughly speaking) increase as X increases. So the correlation will be a positive number and might even be somewhat close to 1, which indicates that more studying is strongly correlated to better performance on an exam. Of course, Y is capped (often at, say, $Y = 100$ for a perfect score), and X may only be capped by the possible studying time allowed before the exam takes place. So the correlation will probably not be exactly 1, but might be high, say, 0.723 or 0.881.

In the previous example, if X had been, for instance, time spent procrastinating, then X and Y would have had a negative correlation.

The correlation of X and Y is defined as follows:

Definition 39.20. Correlation of two random variables

The correlation of two random variables X and Y, usually written as ρ, is defined as

$$\rho(X,Y) = \frac{\mathrm{Cov}(X,Y)}{\sqrt{\mathrm{Var}(X)\,\mathrm{Var}(Y)}}.$$

Note: The variance is only zero when a random variable is constant. So, as long as X and Y are not constant, then the correlation between them is well-defined.

To see that the correlation of two random variables X and Y is always between -1 and $+1$, it is very helpful to write σ_X and σ_Y as the standard deviations of X and Y, and then to study how $\mathrm{Var}\left(\frac{X}{\sigma_X}+\frac{Y}{\sigma_Y}\right)$ and $\mathrm{Var}\left(\frac{X}{\sigma_X}-\frac{Y}{\sigma_Y}\right)$ behave. In fact, the correlation is like the covariance, but scaled by the sizes of the variances of the random variables.

We note that variances are always positive, so

$$\begin{aligned} 0 &\le \mathrm{Var}\left(\frac{X}{\sigma_X}+\frac{Y}{\sigma_Y}\right) \\ &= \mathrm{Var}\left(\frac{X}{\sigma_X}\right) + \mathrm{Var}\left(\frac{Y}{\sigma_Y}\right) + \mathrm{Cov}\left(\frac{X}{\sigma_X},\frac{Y}{\sigma_Y}\right) + \mathrm{Cov}\left(\frac{Y}{\sigma_Y},\frac{X}{\sigma_X}\right) \end{aligned}$$

but covariance is symmetric so the last two terms above are the same. Also, we can factor out the standard deviations to get $\mathrm{Var}\left(\frac{X}{\sigma_X}\right) = \frac{1}{\sigma_X^2}\mathrm{Var}(X) = 1$ and $\mathrm{Var}\left(\frac{Y}{\sigma_Y}\right) = \frac{1}{\sigma_Y^2}\mathrm{Var}(Y) = 1$. So we get $0 \le 2 + 2\,\mathrm{Cov}\left(\frac{X}{\sigma_X},\frac{Y}{\sigma_Y}\right)$, i.e.,

$$-1 \le \mathrm{Cov}\left(\frac{X}{\sigma_X},\frac{Y}{\sigma_Y}\right). \tag{39.4}$$

Similarly,

$$\begin{aligned} 0 &\le \mathrm{Var}\left(\frac{X}{\sigma_X}-\frac{Y}{\sigma_Y}\right) \\ &= \mathrm{Var}\left(\frac{X}{\sigma_X}\right) + \mathrm{Var}\left(\frac{Y}{\sigma_Y}\right) - \mathrm{Cov}\left(\frac{X}{\sigma_X},\frac{Y}{\sigma_Y}\right) - \mathrm{Cov}\left(\frac{Y}{\sigma_Y},\frac{X}{\sigma_X}\right) \end{aligned}$$

but again we simplify to get $0 \le 2 - 2\,\mathrm{Cov}\left(\frac{X}{\sigma_X},\frac{Y}{\sigma_Y}\right)$, i.e.,

$$1 \ge \mathrm{Cov}\left(\frac{X}{\sigma_X},\frac{Y}{\sigma_Y}\right). \tag{39.5}$$

Putting together (39.4) and (39.5) we get

$$-1 \le \mathrm{Cov}\left(\frac{X}{\sigma_X},\frac{Y}{\sigma_Y}\right) \le 1,$$

but $\frac{\text{Cov}(X,Y)}{\sigma_X \sigma_Y} = \rho(X,Y)$, so in other words,

$$-1 \le \rho(X,Y) \le 1.$$

Theorem 39.21. Correlation of two random variables is always between −1 and +1

The correlation $\rho(X,Y)$ of two random variables X and Y is always between -1 and $+1$:

$$-1 \le \rho(X,Y) \le +1.$$

Example 39.22. Let X be a Uniform random variable on the interval $[0,1]$, and let $Y = X^2$. Find the correlation between X and Y.

We see that

$$\mathbb{E}(X) = \int_0^1 (x)(1)\,dx = 1/2 \qquad \text{and} \qquad \mathbb{E}(X^2) = \int_0^1 (x^2)(1)\,dx = 1/3,$$

so

$$\text{Var}(X) = 1/3 - (1/2)^2 = 1/12.$$

Also

$$\mathbb{E}(Y) = \mathbb{E}(X^2) = 1/3 \qquad \text{and} \qquad \mathbb{E}(Y^2) = \mathbb{E}(X^4) = \int_0^1 (x^4)(1)\,dx = 1/5,$$

so

$$\text{Var}(Y) = 1/5 - (1/3)^2 = 4/45.$$

Also

$$\text{Cov}(X,Y) = \mathbb{E}(XY) - \mathbb{E}(X)\mathbb{E}(Y) = \mathbb{E}(X^3) - \mathbb{E}(X)\mathbb{E}(Y),$$

and we know $\mathbb{E}(X) = 1/2$ and $\mathbb{E}(Y) = 1/3$, and we see that $\mathbb{E}(X^3) = \int_0^1 x^3\,dx = 1/4$, so

$$\text{Cov}(X,Y) = 1/4 - (1/2)(1/3) = 1/12.$$

Thus X and Y are positively correlated, and the correlation between X and Y is

$$\rho(X,Y) = \frac{\text{Cov}(X,Y)}{\sqrt{\text{Var}(X)}\sqrt{\text{Var}(Y)}} = \frac{1/12}{\sqrt{1/12}\sqrt{4/45}} = \frac{\sqrt{15}}{4} = 0.968.$$

Example 39.23. Roll a die. Let X denote the value on the die. Let $Y = 1$ if the die is 4, 5, or 6, and let $Y = 0$ otherwise. Find the correlation between X and Y. Informally, when $Y = 1$, we know X must be larger (than when $Y = 0$). So we anticipate X and Y being strongly positively correlated.

We see that

$$\mathbb{E}(X) = \sum_{j=1}^{6}(j)(1/6) = 7/2, \qquad \text{and} \qquad \mathbb{E}(X^2) = \sum_{j=1}^{6}(j^2)(1/6) = 91/6,$$

so

$$\text{Var}(X) = 91/6 - (7/2)^2 = 35/12.$$

Also

$$\mathbb{E}(Y) = (3/6)(1) + (3/6)(0) = 1/2,$$

and

$$\mathbb{E}(Y^2) = (3/6)(1^2) + (3/6)(0^2) = 1/2,$$

so

$$\text{Var}(Y) = 1/2 - (1/2)^2 = 1/4.$$

Also

$$\text{Cov}(X, Y) = \mathbb{E}(XY) - \mathbb{E}(X)\mathbb{E}(Y),$$

and we know that

$$\begin{aligned}\mathbb{E}(XY) &= (1/6)(1)(0) + (1/6)(2)(0) + (1/6)(3)(0) \\ &\quad + (1/6)(4)(1) + (1/6)(5)(1) + (1/6)(6)(1) \\ &= 5/2.\end{aligned}$$

So we have

$$\text{Cov}(X, Y) = 5/2 - (7/2)(1/2) = 3/4.$$

Thus X and Y are positively correlated, and the correlation between X and Y is

$$\rho(X, Y) = \frac{\text{Cov}(X, Y)}{\sqrt{\text{Var}(X)}\sqrt{\text{Var}(Y)}} = \frac{3/4}{\sqrt{35/12}\sqrt{1/4}} = \frac{3\sqrt{105}}{35} = 0.878.$$

Example 39.24. Roll a die. Let X denote the value on the die. Let $Y = 1$ if the die is even, i.e., is 2, 4, or 6, and let $Y = 0$ otherwise. Find the correlation between X and Y. Informally, when $Y = 1$, we know X must be slightly larger (than when $Y = 0$), because for instance X is 2 not 1, or X is 4 not 3, or X is 6 not 5. So we anticipate X and Y being a little bit positively correlated.

The calculation goes just the same as above, with

$$\begin{aligned}\mathbb{E}(X) &= 7/2, \\ \text{Var}(X) &= 35/12, \\ \mathbb{E}(Y) &= 1/2, \\ \text{Var}(Y) &= 1/4, \\ \text{Cov}(X, Y) &= \mathbb{E}(XY) - \mathbb{E}(X)\mathbb{E}(Y),\end{aligned}$$

but in this case

$$\begin{aligned}\mathbb{E}(XY) &= (1/6)(1)(0) + (1/6)(2)(1) + (1/6)(3)(0) \\ &\quad + (1/6)(4)(1) + (1/6)(5)(0) + (1/6)(6)(1) \\ &= 2,\end{aligned}$$

so

$$\mathrm{Cov}(X, Y) = 2 - (7/2)(1/2) = 1/4.$$

Thus X and Y are positively correlated, and the correlation between X and Y is

$$\rho(X, Y) = \frac{\mathrm{Cov}(X, Y)}{\sqrt{\mathrm{Var}(X)}\sqrt{\mathrm{Var}(Y)}} = \frac{1/4}{\sqrt{35/12}\sqrt{1/4}} = \frac{\sqrt{105}}{35} = 0.293.$$

Thus, X and Y are positively correlated, but are not as strongly correlated as in the previous example.

Example 39.25. Roll a die. Let X denote the value on the die. Let $Y = 1$ if the die is odd, i.e., is 1, 3, or 5, and let $Y = 0$ otherwise. Find the correlation between X and Y. Informally, when $Y = 1$, we know X must be slightly smaller (than when $Y = 0$), because for instance X is 1 not 2, or X is 3 not 4, or X is 5 not 6. So we anticipate X and Y being a little bit negatively correlated. This is completely symmetric to Example 39.24.

Same as above, but now

$$\begin{aligned}\mathbb{E}(XY) &= (1/6)(1)(1) + (1/6)(2)(0) + (1/6)(3)(1) \\ &\quad + (1/6)(4)(0) + (1/6)(5)(1) + (1/6)(6)(0) \\ &= 3/2,\end{aligned}$$

so we have

$$\mathrm{Cov}(X, Y) = 3/2 - (7/2)(1/2) = -1/4.$$

Thus, X and Y are negatively correlated, and the correlation between X and Y is

$$\rho(X, Y) = \frac{\mathrm{Cov}(X, Y)}{\sqrt{\mathrm{Var}(X)}\sqrt{\mathrm{Var}(Y)}} = \frac{-1/4}{\sqrt{35/12}\sqrt{1/4}} = -\frac{\sqrt{105}}{35} = -0.293.$$

Example 39.26. Roll a die. Let X denote the value on the die. Let $Y = 1$ if the die is 1, 2, or 3, and let $Y = 0$ otherwise. Find the correlation between X and Y. Informally, when $Y = 1$, we know X must be smaller (than when $Y = 0$). So we anticipate X and Y being have a strong negative correlation. This is completely symmetric to Example 39.23.

Same as above, but now

$$\begin{aligned}\mathbb{E}(XY) &= (1/6)(1)(1) + (1/6)(2)(1) + (1/6)(3)(1)\\ &\quad + (1/6)(4)(0) + (1/6)(5)(0) + (1/6)(6)(0)\\ &= 1,\end{aligned}$$

so

$$\mathrm{Cov}(X,Y) = 1 - (7/2)(1/2) = -3/4.$$

Thus X and Y are negatively correlated, and the correlation between X and Y is

$$\rho(X,Y) = \frac{\mathrm{Cov}(X,Y)}{\sqrt{\mathrm{Var}(X)}\sqrt{\mathrm{Var}(Y)}} = \frac{-3/4}{\sqrt{35/12}\sqrt{1/4}} = -\frac{3\sqrt{105}}{35} = -0.878.$$

Example 39.27. Consider a random variable X that indicates if event A has occurred, and another random variable Y that indicates if event B has occurred.

In this example, if the occurrence of B makes A more likely to occur (i.e., if $P(A \mid B) > P(A)$), then X and Y have positive correlation.

If the occurrence of B makes A less likely to occur (i.e., if $P(A \mid B) < P(A)$), then X and Y have negative correlation.

If the occurrence of B does not affect the probability of A (i.e., if $P(A \mid B) = P(A)$), then X and Y have zero correlation. In fact, X and Y are independent in this last case.

To see all of this, we compute

$$\begin{aligned}\mathbb{E}(X) &= P(X=1) = P(A),\\ \mathrm{Var}(X) &= P(X=1)P(X=0) = P(A)P(A^c),\\ \mathbb{E}(Y) &= P(Y=1) = P(B),\\ \mathrm{Var}(Y) &= P(Y=1)P(Y=0) = P(B)P(B^c),\\ \mathbb{E}(XY) &= P(X=1 \text{ and } Y=1) = P(A\cap B),\end{aligned}$$

so

$$\begin{aligned}\rho(X,Y) &= \frac{\mathrm{Cov}(A,B)}{\sqrt{\mathrm{Var}(X)}\sqrt{\mathrm{Var}(Y)}}\\ &= \frac{P(A\cap B) - P(A)P(B)}{\sqrt{P(A)P(A^c)}\sqrt{P(B)P(B^c)}}.\end{aligned}$$

The denominator is always positive. Thus, the sign of the correlation of X and Y only depends on the sign of the numerator.

We see that the numerator is

$$P(A \cap B) - P(A)P(B),$$

which can be rewritten as

$$P(A \mid B)P(B) - P(A)P(B) = (P(A \mid B) - P(A))P(B).$$

Thus, if $P(A \mid B) > P(A)$, then $P(A \mid B) - P(A) > 0$, so the correlation of X and Y is positive.

If $P(A \mid B) < P(A)$, then $P(A \mid B) - P(A) < 0$, so the correlation of X and Y is negative.

If $P(A \mid B) = P(A)$, then $P(A \mid B) - P(A) = 0$, so the correlation of X and Y is zero, and A and B are independent in this case, so X and Y are independent too.

39.7 Exercises

39.7.1 Practice

Exercise 39.1. **Pizza for lunch.** Each day, Amy eats lunch at the cafeteria. She chooses pizza as her main dish with probability 40%, and her behavior each day is independent of all the other days. Let X denote the number of days she chooses pizza in a 10-day period. Let $Y = 10 - X$ denote the number of days in which Amy does not eat pizza.

a. Are X and Y dependent or independent?

b. Find the covariance $\mathrm{Cov}(X, Y)$ of X and Y.

c. Find the correlation $\rho(X, Y)$ of X and Y.

Exercise 39.2. Let X be Uniformly distributed on the interval $[0, 10]$, and let $Y = 10 - X$.

a. Find the covariance of X and Y.

b. Find the correlation of X and Y.

Exercise 39.3. **Client meeting.** An accountant must meet with one more client before he can go home. The amount of time X (in minutes) that he meets with the client is Uniformly distributed on the interval $[40, 60]$. The total length of time Y (also in minutes) that he must remain in the office is $1.3X + 10$.

a. Find the covariance of X and Y.

b. Find the correlation of X and Y.

Exercise 39.4. Broken crayons. You are babysitting for two children, Abby and Bill. They have a bucket of 30 crayons, 20 of which are unbroken, and the other 10 are broken. They each choose a crayon, without replacement. Let $X = 1$ if Abby gets an unbroken crayon, and $X = 0$ otherwise. Similarly, let $Y = 1$ if Bill gets an unbroken crayon, and $Y = 0$ otherwise.

a. Find the covariance of X and Y.

b. Find the correlation of X and Y.

39.7.2 Extensions

Find (a) the covariance, and (b) the correlation of the random variables X and Y defined in Exercises 39.5–39.15

Exercise 39.5. Let X be any random variable, and $Y = a - X$, where "a" is a constant.

Exercise 39.6. Let X be Uniformly distributed on the interval $[0, \pi]$, and let $Y = \cos X$.

Exercise 39.7. Sweet and sour. Henry and Sally each choose 1 candy from a bag of treats, without replacement. There are 20 sweet candies and 3 sour candies. Let $X = 1$ if Henry's candy is sweet, or $X = 0$ otherwise. Let $Y = 1$ if Sally's candy is sweet, or $Y = 0$ otherwise.

Exercise 39.8. Knitting sweaters. A grandmother is knitting wool sweaters to donate to a non-profit organization. She knits pink and blue ones for girls and boys, respectively. It takes 3 balls of yarn to knit one. She has promised to donate 10 sweaters, but she has forgotten how many of the recipients are boys and how many are girls. So the grandmother decides to roll a 6-sided die to determine how many blue sweaters to knit. Let X be the number of the die (and, thus, the number of blue sweaters that she knits), and let $Y = 10 - X$ be the number of pink sweaters that she knits.

Exercise 39.9. Let X and Y have a joint Uniform distribution on the triangle with corners at $(0, 2)$, $(2, 0)$, and the origin.

Exercise 39.10. Let X and Y be Uniformly distributed on the diamond-shaped region with corners located at $(1, 0)$, $(0, 1)$, $(-1, 0)$, and $(0, -1)$.

Exercise 39.11. Pizza pieces. Lloyd eats X pieces of pizza, where X is an integer-valued random variable that is equally likely to be any of the values 1 through 8, inclusive. He takes the extra pizza home to his family, but if he eats too much pizza, he buys an extra pizza to take home too. Let Y be the number of pieces that he takes home to his family. Then, if $1 \le X \le 4$, he does not need to buy any more pizza, so $Y = 8 - X$ is the leftover. On the other hand, if $5 \le X \le 8$, then $Y = 16 - X$, because he buys a second pizza in this case.

Exercise 39.12. Let X and Y be jointly distributed on the portion of the Cartesian plane between the curves $y = \sqrt{x}$ and $y = x^2$. Let the joint density of X, Y be $f_{X,Y}(x, y) = \frac{8y}{3x}$ in this regions, and $f_{X,Y}(x, y) = 0$ otherwise. (Notice that this is a density because $\int_0^1 \int_{x^2}^{\sqrt{x}} \frac{8y}{3x}\,dy\,dx = 0$.)

Exercise 39.13. Roll two dice. Let X denote the maximum value that appears, and let Y denote the minimum value that appears.

Exercise 39.14. Let X be Uniform on the interval $[0, 1]$, and let $Y = e^X$.

Exercise 39.15. Let X be an Exponentially distributed random variable with expected value $1/2$. Let $Y = X^2$.

Exercise 39.16. Consider X and Y such that the joint density $f_{X,Y}(x, y)$ of X and Y is Uniform on the square where $0 \leq X, Y \leq 1$. In other words,

$$f_{X,Y}(x, y) = 1 \qquad \text{if } 0 \leq x \leq 1 \text{ and } 0 \leq y \leq 1,$$

and $f_{X,Y}(x, y) = 0$ otherwise.

a. Are X and Y dependent or independent?

b. Find the covariance $\text{Cov}(X^2, X + Y)$ of X^2 and $X + Y$.

c. Find the correlation $\text{Cov}(X^2, X + Y)$ of X^2 and $X + Y$.

Exercise 39.17. Roll two 4-sided dice (*not 6-sided dice*). Let X be the minimum value, and let Y be the maximum value. Find the covariance of X and Y.

39.7.3 Advanced

Exercise 39.18. Let X and Y correspond to the horizontal and vertical coordinates in the triangle with corners at $(2, 0)$, $(0, 2)$, and the origin. Let $f_{X,Y}(x, y) = \frac{15}{28}(xy^2 + y)$ for (x, y) inside the triangle, and $f_{X,Y}(x, y) = 0$ otherwise.

a. Find the covariance of X and Y.

b. Find the correlation of X and Y.

Exercise 39.19. If X and Y have a constant joint density on the triangle where $0 \leq y \leq x \leq 1$, compute $\text{Cov}(X, Y)$.

Exercise 39.20. Draw 5 cards, without replacement, from a standard deck of cards. Let X be the number of hearts selected. Find the variance of X.

Exercise 39.21. Suppose that Sasha picks an integer X at random between 1 and 6 (inclusive). Then Ravi picks a different integer Y at random, from the 5 integers that remain. Assume that all $(6)(5) = 30$ choices are equally likely. Find the correlation $\rho(X, Y)$ of X and Y.

Exercise 39.22. A total of $3n$ bears are in a bucket: n are red, n are yellow, and n are blue. A child begins grabbing the bears at random, with all selections equally likely. The bears are selected "without replacement," i.e., she never puts the bears back after she grabs them. Find the variance of the number of bears she grabs until the first red bear appears.

Exercise 39.23. Roll 10 differently colored, 6-sided dice. Make a list of all $\binom{10}{3} = 120$ triples using the 10 values that appear. Let X be the total number of such triples for which the three values in the triple agree. Find $\mathrm{Var}(X)$.

Chapter 40

Conditional Expectation

Statistically the probability of any one of us being here is so small that you would think the mere fact of existence would keep us all in a contented dazzlement of surprise. We are alive against the stupendous odds of genetics, infinitely outnumbered by all the alternates who might, except for luck, be in our places.

—*The Lives of a Cell: Notes of a Biology Watcher* by Lewis Thomas (Viking, 1974)

If a woman is waiting on her spouse, how does her arrival time affect the expected time until he arrives?

40.1 Introduction

The idea of conditional expectation of one random variable (say, X), given the value of another random variable (say, $Y = y$), is to use the conditional mass or the conditional density to calculate the expected value of one random variable (X) when the value of the other is known (Y). In other words, we know some information ahead of time about one of our random variables that may affect the mass or density—and, thus, the probabilities and expected value—of the other random variable.

Definition 40.1. Conditional Expectation

In the continuous case, the conditional expectation of X, given $Y = y$, is

$$\mathbb{E}(X \mid Y = y) = \int_{-\infty}^{\infty} x f_{X|Y}(x \mid y)\, dx.$$

In the discrete case, the conditional expectation of X, given $Y = y$, is

$$\mathbb{E}(X \mid Y = y) = \sum_{x} x p_{X|Y}(x \mid y).$$

40.2 Examples

As an example, returning to the situation in Example 39.23, we have the following:

Example 40.2. Roll a die. Let X denote the value on the die. Let $Y = 1$ if the die is 4, 5, or 6, and let $Y = 0$ otherwise.

Given $Y = 1$, what is the expected value of X?

When $Y = 1$ is known, then X is equally likely to be any of the values 4, 5, or 6, and thus $\mathbb{E}(X \mid Y = 1) = \frac{1}{3}(4) + \frac{1}{3}(5) + \frac{1}{3}(6) = 5$.

On the other hand, when $Y = 0$ is known, then X is equally likely to be any of the values 1, 2, or 3, and thus $\mathbb{E}(X \mid Y = 0) = \frac{1}{3}(1) + \frac{1}{3}(2) + \frac{1}{3}(3) = 2$.

As another example, returning to the situation in Example 39.24, we have the following:

Example 40.3. Roll a die. Let X denote the value on the die. Let $Y = 1$ if the die is even, i.e., is 2, 4, or 6, and let $Y = 0$ otherwise. Given $Y = 1$, what is the expected value of X?

When $Y = 1$ is known, then X is equally likely to be any of the values 2, 4, or 6, and thus $\mathbb{E}(X \mid Y = 1) = \frac{1}{3}(2) + \frac{1}{3}(4) + \frac{1}{3}(6) = 4$.

On the other hand, when $Y = 0$ is known, then X is equally likely to be any of the values 1, 3, or 5, and thus $\mathbb{E}(X \mid Y = 0) = \frac{1}{3}(1) + \frac{1}{3}(3) + \frac{1}{3}(5) = 3$.

Now we turn our attention to conditional expectation for pairs of continuous random variables. As an example:

Example 40.4. Consider the scenario in Example 33.2. David works at a customer call center. He talks to customers on the telephone. The length (in hours) of each conversation has density $f_X(x) = 3e^{-3x}$ for $x > 0$ and $f_X(x) = 0$ otherwise. The lengths of calls are independent. As soon as one conversation is finished, he hangs up the phone, and immediately picks up the phone again to start another call (i.e., there are no gaps in between the calls).

Thus, if David conducts 2 phone calls in a row, the total amount of time he spends on the telephone is $X_1 + X_2$, where the X_j's are independent, and each X_j has the Exponential density above.

We let X (also called X_1) denote the time of the first call, so X has the density above. We let Y (also called $X_1 + X_2$) denote the total time of call one plus call two, combined, so as in Example 33.2, Y is a Gamma random variable with $n = 2$ and $\lambda = 3$, so Y has density $f_Y(y) = 9ye^{-3y}$ for $y > 0$, and $f_Y(y) = 0$ otherwise.

Given that the first two calls take 5/6 of an hour total, i.e., given that $Y = 5/6$, find the expected length of the first call. We compute

$$\mathbb{E}(X \mid Y = 5/6) = \int_{-\infty}^{\infty} x f_{X|Y}(x \mid 5/6)\, dx = \int_{-\infty}^{\infty} x \frac{f_{X,Y}(x, 5/6)}{f_Y(5/6)}\, dx.$$

We see that

$$f_Y(5/6) = 9(5/6)e^{-(3)(5/6)} = (15/2)e^{-5/2}.$$

Also

$$f_{X,Y}(x, y) = f_{Y|X}(y \mid x) f_X(x).$$

Once X is known, then $Y > X$. For any $y > X$, we have $P(Y < y \mid X = x) = P(Y - x < y - x \mid X = x) = P(X_2 < y - x)$. Differentiating with respect to y gives $f_{Y|X}(y \mid x) = f_{X_2}(y - x)$. (Less formally, this says that, once the length of the first call is known, the remaining length until the end of both calls is just equal to the length of the second call.) So $f_{Y|X}(y \mid x) = 3e^{-(3)(y-x)}$ for $y - x > 0$, i.e., for $y > x$. Therefore, for $0 < x < 5/6$, we have

$$f_{X,Y}(x, 5/6) = 3e^{(-3)((5/6)-x)} 3e^{-3x} = 9e^{-5/2}$$

Now we compute

$$\begin{aligned}
\mathbb{E}(X \mid Y = 5/6) &= \int_{-\infty}^{\infty} x \frac{f_{X,Y}(x, 5/6)}{f_Y(5/6)}\, dx \\
&= \int_0^{5/6} x \frac{9e^{-5/2}}{(15/2)e^{-5/2}}\, dx \\
&= \int_0^{5/6} (x)(6/5)\, dx \\
&= 5/12
\end{aligned}$$

So given that the two calls last $5/6$ of an hour altogether, then the first call is expected to last $5/12$ of an hour.

This should make sense intuitively, because when only the total length of two calls is known, we have no reason to believe that one call will be longer than the other, so the expected length of the first call plus the expected length of the second call must equal the total length of the two calls, i.e., $5/6$. So the conditional expected length of each call must be $5/12$. (Each call is expected to last half of the total time.)

Now we focus on computing the expected value of one random variable by first conditioning, and then taking an expected value. Here, we want to compute $\mathbb{E}(X)$. We first suppose that $Y = y$ and we use integration to compute $\mathbb{E}(X \mid Y = y)$. Then we integrate over all possible y's, weighting each value by $f_Y(y)$, to get the expected value of X, regardless of what Y is. This is shown below; at the end, no Y's or y's remain:

$$\begin{aligned}\mathbb{E}(X) &= \int_{-\infty}^{\infty}\int_{-\infty}^{\infty} x f_{X,Y}(x,y)\,dx\,dy \\ &= \int_{-\infty}^{\infty}\int_{-\infty}^{\infty} x f_{X|Y}(x \mid y) f_Y(y)\,dx\,dy \\ &= \int_{-\infty}^{\infty}\int_{-\infty}^{\infty} x f_{X|Y}(x \mid y)\,dx\, f_Y(y)\,dy \\ &= \int_{-\infty}^{\infty} \mathbb{E}(X \mid Y = y)\, f_Y(y)\,dy\end{aligned}$$

In the last line above, note that only "y" remains; no "x" is still in the picture. Thus, we are basically computing the expected value of a function that only depends on Y, i.e., $\mathbb{E}(X \mid Y = y)$. So the last line above is equal to $\mathbb{E}(\,\mathbb{E}(X \mid Y = y)\,)$, where the outer integral is of course taken with regard to Y, and the inner integral is with regard to X. This is sometimes written as $\mathbb{E}(\,\mathbb{E}(X \mid Y)\,)$. So the outer expected value refers to possible Y, while the inner expected value refers to possible X. So we have

$$\begin{aligned}\mathbb{E}(X) &= \int_{-\infty}^{\infty} \mathbb{E}(X \mid Y = y)\, f_Y(y)\,dy \\ &= \mathbb{E}(\,\mathbb{E}(X \mid Y = y)\,) \\ &= \mathbb{E}(\,\mathbb{E}(X \mid Y)\,).\end{aligned}$$

Similarly, if X and Y are discrete, we have

$$\begin{aligned}\mathbb{E}(X) &= \sum_y \sum_x x p_{X,Y}(x,y) \\ &= \sum_y \sum_x x p_{X|Y}(x \mid y) p_Y(y) \\ &= \sum_y \mathbb{E}(X \mid Y = y)\, p_Y(y)\end{aligned}$$

As before, in the last line, only "y" remains; no "x" is still in the picture. So again we are computing the expected value of $\mathbb{E}(X \mid Y = y)$, which only depends on Y. Thus, the last line is equal to $\mathbb{E}(\,\mathbb{E}(X \mid Y = y)\,)$, and again the outer summation is taken with regard to Y, while the inner summation is taken over all possible X. This is also sometimes written as $\mathbb{E}(\,\mathbb{E}(X \mid Y)\,)$, where once again the outer expected value is with regard to Y, and the inner expected value is with regard to X. So we have

$$\begin{aligned}\mathbb{E}(X) &= \sum_y \mathbb{E}(X \mid Y = y)\, p_Y(y) \\ &= \mathbb{E}(\,\mathbb{E}(X \mid Y = y)\,) \\ &= \mathbb{E}(\,\mathbb{E}(X \mid Y)\,).\end{aligned}$$

Example 40.5. If X and Y are continuous random variables that are *independent*, then

$$\mathbb{E}(X \mid Y = y) = \mathbb{E}(X).$$

To see this, we just write

$$\mathbb{E}(X \mid Y = y) = \int_{-\infty}^{\infty} x f_{X|Y}(x \mid y)\, dx,$$

but $f_{X|Y}(x \mid y) = f_X(x)$ since X and Y are independent, so

$$\mathbb{E}(X \mid Y = y) = \int_{-\infty}^{\infty} x f_X(x)\, dx = \mathbb{E}(X).$$

Of course the arguments in the previous example and the following example only work when X and Y are independent. This argument would fail if X and Y are dependent.

Example 40.6. If X and Y are discrete random variables that are *independent*, then

$$\mathbb{E}(X \mid Y = y) = \mathbb{E}(X).$$

To see this, we just write

$$\mathbb{E}(X \mid Y = y) = \sum_x x p_{X|Y}(x \mid y),$$

but $p_{X|Y}(x \mid y) = p_X(x)$ since X and Y are independent, so

$$\mathbb{E}(X \mid Y = y) = \sum_x x p_X(x) = \mathbb{E}(X).$$

Theorem 40.7. Conditional Expected Values of Independent Random Variables If X and Y are independent random variables—either continuous or discrete—then

$$\mathbb{E}(X \mid Y = y) = \mathbb{E}(X).$$

Example 40.8. Suppose that Alice and Bob each eat one cookie per day, on five consecutive days, so that they eat a total of ten cookies altogether. They both love peanut butter cookies, but the cookie selection in each of their dormitories is random. Let $X_1, \ldots, X_5$ be indicators for Alice's five cookies, and let $X_6, \ldots, X_{10}$ be indicators for Bob's five cookies, where $X_j = 1$ if the jth cookie is peanut butter, and $X_j = 0$ otherwise. Assume that the X_j's are all independent Bernoulli random variables, each with parameter $p = 0.40$, because there is always a fresh supply of cookies (in particular, Alice and Bob's choices each day do not affect each other).

Write $X = X_1 + \cdots + X_5$ for the total number of peanut butter cookies that Alice gets to eat over a five day period. Write $Y = X_1 + \cdots + X_{10}$ for the total number of peanut butter cookies that the couple gets to eat. Find the expected number of cookies that Alice got to eat, given that the couple got to eat $Y = 7$ peanut butter cookies altogether, i.e., find $\mathbb{E}(X \mid Y = 7)$.

Given $Y = 7$, we claim that X has conditional distribution as a Hypergeometric random variable with parameters $M = 5$ and $N = 10$ and $n = 7$. Using what we already know about the Hypergeometric distribution, this yields

$$\mathbb{E}(X \mid Y = 7) = nM/N = (7)(5)/10 = 3.5.$$

This makes sense because given $Y = 7$, there are 7 peanut butter cookies that the couple gets, so the number of cookies that Alice gets is Hypergeometric, i.e., it is 5 out of these 7 peanut butters and 3 non-peanut-butters. Each of the 7 peanut butter cookies is equally likely to belong to Alice or Bob, so Alice expects to get 3.5 peanut butter cookies, and Bob expects to get 3.5 peanut butter cookies.

In case the reader is unconvinced by the claim that the conditional distribution of X is Hypergeometric (given $Y = 7$), we can instead compute the conditional mass $P(X|Y = 7)$ before we can get to the conditional expected value.

$$p_{X|Y}(x \mid 7) = \frac{p_{X,Y}(x, 7)}{p_Y(7)} = \frac{p_X(x)p_{Y|X}(7 \mid x)}{p_Y(7)}$$

Since X is the number of peanut butter cookies Alice ate over five days, then X is Binomial with parameters $n = 5$ and $p = 0.40$, so

$$p_X(x) = \binom{5}{x}(0.4)^x(0.6)^{5-x}.$$

Since Y is the total number of peanut butter cookies eaten by the couple over five days, then Y is Binomial with parameters $n = 10$ and $p = 0.40$, so

$$p_Y(7) = \binom{10}{7}(0.4)^7(0.6)^3.$$

Now we need the conditional probability for $P_{Y|X}(7|x)$, i.e., for the probability that the couple has eaten 7 peanut butter cookies total over the five days, given that Alice ate x of them. Basically, the piece we are missing is the number of peanut butter cookies that Bob ate. The only way that they eat 7 peanut butter cookies altogether—given that Alice ate x of them—is for Bob to eat $7 - x$ of them. The number of peanut butter cookies that Bob eats is Binomial with $n = 5$ and $p = 0.4$. So

$$p_{Y|X}(7 \mid x) = P(Y = 7 \mid X = x) = \binom{5}{7-x}(0.4)^{7-x}(0.6)^{5-(7-x)}.$$

Thus, putting all the pieces together for the probability $p_{X|Y}(x \mid 7)$,

$$p_{X|Y}(x \mid 7) = \frac{\binom{5}{x}(0.4)^x(0.6)^{5-x}\binom{5}{7-x}(0.4)^{7-x}(0.6)^{5-(7-x)}}{\binom{10}{7}(0.4)^7(0.6)^3} = \frac{\binom{5}{x}\binom{5}{7-x}}{\binom{10}{7}}.$$

This verifies the claim that X has a Hypergeometric distribution, given that $Y = 7$.

Example 40.9. Suppose that each person who logs onto an online retailer's website on Black Friday is expected to spend \$27.50. Exactly 120 people are surveyed to see how much is spent by the people within this group. Each behaves independently of the others, and each visits the online retailer with probability 0.90. How much money do we expect the group of 120 people to spend at the online retailer?

We let Y denote the number of the people within the group of 120 who decide to visit the online retailer. Thus Y is Binomial with parameters $n = 120$ and $p = 0.90$. Once $Y = y$ is given, we know that exactly y people visit the store, so the money spent by the group is $X_1 + \cdots + X_y$, so the expected money spent by the group given $Y = y$ is

$$\mathbb{E}(X_1 + \cdots + X_y \mid y) = \overbrace{27.50 + \cdots + 27.50}^{y} = (y)(27.50).$$

Thus, the expected money spent by the group, averaged over all possible values of $Y = y$, is

$$\mathbb{E}\left(\sum_{j=1}^{Y} X_j\right) = \mathbb{E}((Y)(27.50)) = (120)(0.90)(27.50) = 2970.$$

This can also be written as

$$\begin{aligned}
\mathbb{E}\Big(\sum_{j=1}^{Y} X_j\Big) &= \sum_{y=0}^{120} \mathbb{E}\Big(\sum_{j=1}^{Y} X_j \mid Y = y\Big) p_Y(y) \\
&= \sum_{y=0}^{120} \mathbb{E}\Big(\sum_{j=1}^{y} X_j\Big) p_Y(y) \\
&= \sum_{y=0}^{120} (27.50)(y) p_Y(y) \\
&= 27.50 \sum_{y=0}^{120} (y) p_Y(y) \\
&= 27.50\, \mathbb{E}(Y) \\
&= (27.50)(120)(0.9) \\
&= 2970.
\end{aligned}$$

Example 40.10. Example 40.9 is a specific case of a more general phenomenon: Consider a nonnegative, integer valued random variable Y, and a sequence of identically distributed random variables $X_1, X_2, X_3, \ldots$. Also assume Y and all of the X_j's are independent. Now suppose that we want to find the expected value of the first Y of the X_j's. In other words, we want to find $\mathbb{E}(X_1 + \cdots + X_Y)$. We emphasize that there are a random number of X_j's that we are summing. Once we know that value of Y, the problem is easy:

$$\begin{aligned}
\mathbb{E}(X_1 + \cdots + X_Y \mid Y = y) &= \mathbb{E}(X_1 + \cdots + X_y \mid Y = y) \\
&= \mathbb{E}(X_1 + \cdots + X_y) \qquad \text{since } X_j\text{'s \& } Y \text{ are indep.} \\
&= \mathbb{E}(X_1) + \cdots + \mathbb{E}(X_y) \\
&= y\mathbb{E}(X_1) \quad \text{since } X_j\text{'s are identically distributed.}
\end{aligned}$$

Thus (using the outer $\mathbb{E}$ for Y and the inner $\mathbb{E}$ for X), we have

$$\begin{aligned}
\mathbb{E}(\mathbb{E}(X_1 + \cdots + X_Y \mid Y)) &= \sum_{y=0}^{\infty} \mathbb{E}(X_1 + \cdots + X_Y \mid Y = y) P(Y = y) \\
&= \sum_{y=0}^{\infty} y\mathbb{E}(X_1) P(Y = y) \\
&= \mathbb{E}(X_1) \sum_{y=0}^{\infty} y P(Y = y) \\
&= \mathbb{E}(X_1)\mathbb{E}(Y).
\end{aligned}$$

So the expected value of $X_1 + \cdots + X_Y$ is just equal to the expected value of one of the X_j's, multiplied by the expected value of Y, i.e., by the number of random variables that we expect to add.

Theorem 40.11. If Y is a nonnegative, integer valued random variable and if $X_1, X_2, X_3, \ldots$ is a sequence of identically distributed random variables, which are independent of each other and independent of Y, then (using the outer $\mathbb{E}$ for Y and the inner $\mathbb{E}$ for X)

$$\mathbb{E}(\mathbb{E}(X_1 + \cdots + X_Y \mid Y)) = \mathbb{E}(X_1)\mathbb{E}(Y).$$

Example 40.12. Suppose that the number of shoppers coming into a store follows a Poisson distribution with an expected value of 10 per hour. Suppose also that each shopper is—independent of all other shoppers—equally likely to be a man or a woman.

Given that exactly 7 men are shopping, how many women do we expect are shopping?

Let X and Y denote the number of men and women shoppers, respectively, and let $N = X + Y$ be the total number of shoppers, which is known to be a Poisson random variable with mean 10.

We see that

$$\begin{aligned} p_{X,Y}(x,y) &= P(X = x \text{ and } Y = y) \\ &= P(X = x \text{ and } Y = y \mid N = x+y)P(N = x+y) \\ &= \frac{(x+y)!}{x!y!}(1/2)^x(1/2)^y \frac{e^{-10}10^{x+y}}{(x+y)!} \\ &= \left(\frac{e^{-5}(1/2)^x 10^x}{x!}\right)\left(\frac{e^{-5}(1/2)^y 10^y}{y!}\right) \\ &= \left(\frac{e^{-5}5^x}{x!}\right)\left(\frac{e^{-5}5^y}{y!}\right) \end{aligned}$$

So we have factored the joint mass into "x stuff" and "y stuff," each of which is a mass, and thus X and Y must be independent. We see from each of their masses that X and Y are independent Poissons, each with expected value 5.

So, X is Poisson with expected value 5 and Y is Poisson with expected value 5. Also X and Y are independent. Thus, *regardless of how many men are shopping* (e.g., regardless of the fact that exactly 7 men are shopping), the number of female shoppers is unaffected by the value of X. Thus, even when given that 7 men are shopping, the expected number of women shopping is still 5.

The phenomenon in Example 40.12 can be generalized.

Theorem 40.13. Consider a set of N objects, where N is a Poisson random variable with mean λ. Suppose that, independent of N, and independent of the other objects, each object is of exactly 1 of j possible different types, as follows: it is of type k with probability p_k, where (of course) $p_1 + \cdots + p_j = 1$. Now let X_k denote the number of the N objects that are of type k, so that $X_1 + \cdots + X_j = N$ (since all of the objects have some type). Then, in general, the random variables $X_1, \ldots, X_j$'s each turn out to be Poisson random variables, which also turn out to be *independent*, and $\mathbb{E}(X_j) = p_j\lambda$.

For comparison, in Example 40.12, there was a Poisson number N of shoppers, each of which was one of $j = 2$ types, that is, a man (type 1) with probability 1/2 or a woman (type 2) with probability 1/2. In general, the p_k probabilities do not have to be the same.

40.3 Exercises

40.3.1 Practice

Exercise 40.1. Let X and Y have a joint uniform distribution on the triangle with corners at $(0, 2)$, $(2, 0)$, and the origin. Find $\mathbb{E}(Y \mid X = 1/2)$.

Exercise 40.2. **Errors on a page.** Suppose that the number of errors per page of a book has a Poisson distribution with parameter $\lambda = 0.12$. Also suppose that the expected number of pages in a randomly selected new book is 400. Find the mean number of errors in such a book.

Exercise 40.3. **Arrival times.** Consider a man and a woman who arrive at a certain location; whoever arrives first will wait for the other to arrive. If X and Y denote (respectively) the arrival times of the man and the woman after noon, in minutes, assume that X and Y are independent and each Uniformly distributed on $[0, 60]$. (In other words, the man and woman arrive independently, at Uniform times between noon and 1 PM.)

If the woman arrives at 12:35 PM, find the expected time spent waiting, i.e., find

$$\mathbb{E}(\,|X - Y| \mid Y = 35\,).$$

Hint: Since we know that $Y = 35$, we may as well just substitute "35" for Y, so that we only have X's in our lives. So we just need to find

$$\mathbb{E}(\,|X - 35|\,).$$

Exercise 40.4. **Female heights.** As in Exercise 35.24 and Example 36.10, assume that the height (in inches) of an American female is Normal with expected value $\mu_1 = 64$ and standard deviation $\sigma_1 = 2.5$. Also assume that the height of an American male is Normal with expected value $\mu_2 = 69$ and standard deviation $\sigma_2 = 3.0$.

A man and a woman are chosen at random. The woman's height is measured, and she is found to be exactly 68.2 inches tall. How much taller do we expect the man to be (as compared to this particular woman)?

Exercise 40.5. Annual precipitation. Let X be the amount of annual snowfall in a region in North America. Assume that, given the value Y of annual precipitation in the region, we know X is Normally distributed with $\mathbb{E}(X \mid Y) = 0.3Y$ and $\sigma_X = 0.1Y$. Given that the annual precipitation value Y in a student's hometown is 36.3 inches, what is the expected amount of snowfall?

Exercise 40.6. Mac and cheese. While Juanita waits for her Mac and Cheese to boil, she works on her homework for 1/2 of the time. She has been waiting for 3 minutes already (so she has already done 1.5 minutes of homework). She doesn't know how much more time is needed. She estimates that the remaining time (in minutes) until the water boils is an Exponential random variable Y with density $f_Y(y) = \frac{1}{4}e^{-y/4}$ for $y > 0$, and $f_Y(y) = 0$ otherwise. Find the total length of time that she expects to work on her homework while waiting for the water to boil.

Exercise 40.7. Flowers. Sally and David each pick 10 flowers from the case without paying attention to what type of flowers they are picking. There are a large quantity of flowers available, 20% of which are roses. Let X be the number of roses that Sally picks, and let Y be the number of roses that the couple picks altogether. Find the number of roses that we expect Sally to pick if the total number picked is $Y = 12$.

Exercise 40.8. Selling cookies. In a particular girl scout troop, each girl sells an average of 30 boxes of cookies. Let Y be the number of girls in the troop, and let X be the number of boxes of cookies sold. Find $\mathbb{E}(X \mid Y = y)$.

Exercise 40.9. Candy machine. There are 20 pieces of candy in a machine: 5 Hershey kisses, 6 Kit Kats, 4 Snickers, and 5 Paydays. The machine randomly selects one of the remaining candies when someone makes a purchase, and the candies are not replenished after the purchase (i.e., purchases are made without replacement). If X Kit Kats are contained in a purchase of Y candies, find $\mathbb{E}(X \mid Y = 4)$.

Exercise 40.10. Wake up times. Suppose that, on a random Sunday, your roommate wakes up at some time Uniformly distributed between 12 noon and 3 PM. Suppose that you awake at 2 PM that day. If your roommate is not yet awake when you get up at 2 PM, when do you expect him to wake up?

Exercise 40.11. Doctor's appointments. A doctor has two consecutive appointments with patients. The duration of each appointment is Exponential with expected value 20 minutes. The durations of the two appointments are assumed to be independent. Given that the total duration of the two appointments turns out to be 38 minutes altogether, how long do we expect the first appointment to be?

Exercise 40.12. **First aid.** A nurse has 10 minutes to administer first aid to various soldiers in the middle of a battle. The time (in minutes) devoted to the first soldier is Y, and the time (in minutes) devoted to the second soldier is X. Assume that (X, Y) is Uniformly distributed on the triangle where $X \geq 0$, $Y \geq 0$, and $X + Y \leq 10$. If we know that $Y = 3$, i.e., that she spends exactly 3 minutes with the first soldier, how much time do we expect that she will spend with the second soldier?

Exercise 40.13. Let X_1 and X_2 be independent exponential random variables, each with mean 1. Let $Y = X_1 + X_2$. Find $\mathbb{E}(X_1 \mid Y = 3)$.

40.3.2 Extensions

Exercise 40.14. Roll two 6-sided dice. Let X denote the minimum value that appears, and let Y denote the maximum value that appears.

a. Find $\mathbb{E}(Y \mid X = 3)$.

b. Find $\mathbb{E}(X + Y \mid X = 3)$.

Exercise 40.15. A child rolls a pair of dice, one of which is blue and one of which is red.

a. Given that the sum of the dice is 8, find the probability that the red die shows the value 4.

b. Now the child rolls a pair of dice that look the same (i.e., which are not painted). Given that the sum of the dice is 8, find the probability that both of the dice simultaneous show the value 4.

c. Are your answers the same or different in the two parts above? Why?

Exercise 40.16. Two 6-sided dice are rolled. Let X be the minimum of the two values, and let Y be the absolute value of the difference of the two values. If D_1, D_2 are the two values on the two dice, then

$$X = \min(D_1, D_2),$$

and

$$Y = |D_1 - D_2|.$$

If $Y = 4$, what is the expected value of X?

Exercise 40.17. **Guitar songs.** Helen and Joe play guitar together every day at lunchtime. The number of songs that they play on a given day has a Poisson distribution, with an average of 5 songs per day. Regardless of how many songs they will play that day, Helen and Joe always flip a coin at the start of each song, to decide who will play the solo on that song. If we know that Joe plays exactly 4 solos on a given day, then how many solos do we expect that Helen will play on that same day?

Exercise 40.18. **Black Jack.** Let Y denote the value of the dealer's card that can be seen by all the players, in a game of Black Jack. Let X be a Bernoulli random variable that indicates whether the dealer must stay (i.e., not take another card). Given $Y = 10$, find the expected value of X. Hint: In Black Jack, if the dealer has a total of 17 or greater in her hand, then she must stay; if her total is 16 or less, she will draw. For the purposes of this question, the Ace has value 11, and each face card (Jack, Queen, or King) has the value 10.

Exercise 40.19. Let X and Y denote, respectively, the x- and y-coordinates of the location of a random chosen point that is Uniformly distributed in a region. Given $Y = 1$, find the expected value of X when the region is:

a. a circle of radius 2, centered at the origin

b. a semicircle corresponding to the right-hand portion of the circle from part a.

Exercise 40.20. Sandra rolls two 10-sided dice. Let Y be the sum of the two dice, and let X be the value on the first die that she rolls. Given $Y = 15$, what is the expected value of X?

Exercise 40.21. **Date wait.** Suppose that Harrison comes to pick up Rosita for a date. Rosita will be ready at a time that is Uniformly distributed between 7 PM and 7:10 PM. If Harrison arrives at 7:07 PM, how long does he expect to have to wait until Rosita is ready? (If she is already ready when he arrives, then his waiting time is 0, since a waiting time cannot be negative.)

Exercise 40.22. A point is chosen Uniformly inside a circle of radius 2, centered at the origin. If we are given that the point lands exactly on the x-axis, find the expected distance of the point to the origin.

40.3.3 Advanced

Exercise 40.23. Let X and Y correspond to the horizontal and vertical coordinates in the triangle with corners at $(2, 0)$, $(0, 2)$, and the origin. Let $f_{X,Y}(x, y) = \frac{15}{28}(xy^2 + y)$ for (x, y) inside the triangle, and $f_{X,Y}(x, y) = 0$ otherwise. Find $\mathbb{E}(X \mid Y = 1.5)$.

Exercise 40.24. Show that the generalization of Example 40.12, given in a box at the end of the chapter, is true.

Chapter 41

Markov and Chebyshev Inequalities

The worst form of inequality is to try to make unequal things equal.
—Aristotle

If we know the average speed of drivers on a highway, what kind of bounds can we give on the probability that a randomly selected driver is speeding? How about bounds on the probability that a randomly chosen driver is driving within 5 miles of the average speed?

41.1 Introduction

One subdiscipline within probability theory is the study of probability inequalities. These inequalities are used to put bounds on how large or small a probability can be, under certain conditions. Some bounds are stronger than others. Some inequalities have many conditions, and other inequalities are quite simple. Many different types of inequalities are used in limiting situations, i.e., what happens to a limit of a sequence of random variables and/or their associated probabilities. As with many topics covered during this course, the study of probability inequalities is very rich and deep. We could spend many chapters on probability inequalities, but here we only scratch the surface.

41.2 Markov Inequality

The Markov inequality gives bounds on how large (the absolute value of) a random variable can be.

Theorem 41.1. Markov Inequality

If X is any random variable and $a > 0$, then

$$P(|X| \geq a) \leq \frac{\mathbb{E}(\ |X|\ )}{a}.$$

To see that the Markov inequality is true, we just consider whether $|X| \geq a$. More precisely, we let $Y = 1$ if $|X| \geq a$, and $Y = 0$ otherwise. Then Y is an indicator, so

$$\begin{aligned} P(|X| \geq a) &= P(Y = 1) \\ &= \mathbb{E}(Y). \end{aligned}$$

The novel step is that Y is always less than $|X|/a$:

- If $Y = 1$, then $|X| \geq a$, so we have

$$Y = 1 \leq |X|/a.$$

- If $Y = 0$, then

$$Y = 0 \leq |X|/a$$

  because $|X|$ and a are both greater than or equal to 0.

Since $Y \leq |X|/a$ always, then

$$\mathbb{E}(Y) \leq \mathbb{E}(|X|/a) = \mathbb{E}(|X|)/a.$$

Putting this together with the series of equations above, we get the Markov inequality:

$$P(|X| \geq a) \leq \frac{\mathbb{E}(\ |X|\ )}{a}.$$

As a straightforward extension of the original Markov inequality, we notice that, if X is always nonnegative (i.e., if $X \geq 0$ always), then $|X| = X$. So, in such a case, the Markov inequality works for not just $|X|$ but also for X too. So the second version of the Markov inequality gives bounds on how large a nonnegative random variable can be.

Corollary 41.2. Markov Inequality (version 2)

If X is a nonnegative random variable and $a > 0$, then

$$P(X \geq a) \leq \frac{\mathbb{E}(X)}{a}.$$

Notice that the Markov inequality does not require us to know anything about the distribution of the random variable X, except that we need to know the expected value. We do not even need to know if X is discrete or continuous.

Example 41.3. Consider a class of students for which the class average is 60%.

Find a bound for the probability that a randomly chosen student's score is 72% or higher. We let X denote a randomly chosen student's score. Using the Markov inequality, we have

$$P(X \geq 72) \leq \frac{\mathbb{E}(X)}{72} = \frac{60}{72} = 5/6.$$

Notice, in this example, that the distribution of X is not even given! All that is required is that we know the expected value of X.

Example 41.4. The average salary of actuaries in a certain specialty is known to be \$83,000 per year.

If an actuary in that specialty is chosen at random, we can let X denote the actuary's salary. By the Markov inequality, the probability that the actuary makes \$90,000 or more can be bounded as follows:

$$P(X \geq 90{,}000) \leq \frac{\mathbb{E}(X)}{90{,}000} = \frac{83{,}000}{90{,}000} = 83/90 = 0.92.$$

Example 41.5. On a certain highway, the speed limit is 55 miles per hour, but most drivers are not driving so fast. The average speed on the highway is 49 miles per hour.

If X denotes a randomly chosen driver's speed, then the probability that such a person is driving faster than the speed limit is

$$P(X \geq 55) \leq \frac{\mathbb{E}(X)}{55} = \frac{49}{55} = 0.89.$$

One important thing to emphasize is that, in none of these examples do we get to actually calculate a probability directly. We only get upper bounds on probabilities. The advantage of using the Markov inequality, however, is that we do not need to know anything about the distribution of a random variable X except for the expected value. Thus, the Markov inequality can be applied in lots and lots of situations. We do not need to know much to use it, but on the other hand, it tells us relatively little, as compared to a situation in which we can build an inequality based on having more information about the random variable's actual distribution.

One unfortunate thing about the Markov inequality is that we do not necessarily get sharp bounds (i.e., precise bounds) for $P(|X| \geq a)$. Another unfortunate thing about the Markov inequality, however, is that it does not help us to calculate values $P(|X| \geq a)$ when a is smaller than the expected value of $|X|$ (and thus $\mathbb{E}(|X|)/a$ is bigger than 1). If we try to apply the Markov inequality when $a > \mathbb{E}(|X|)$, then the results that we get are not too interesting, because we will have

$$P(|X| \geq a) \leq \frac{\mathbb{E}(|X|)}{a}$$

but if $\frac{\mathbb{E}(|X|)}{a} > 1$, this is not very informative, because we can automatically write the tighter bound:

$$P(|X| \geq a) \leq 1,$$

which is true trivially, just because all probabilities of all events are bounded above by 1. So we do not gain any additional information in such a case. We give one such *useless example*.

Example 41.6. Return to the scenario of Example 41.3, in which a class of students has a class average is 60%.

Find a bound for the probability that a randomly chosen student's score is 55% or higher. We let X denote a randomly chosen student's score. Using the Markov inequality, we have

$$P(X \geq 55) \leq \frac{\mathbb{E}(X)}{55} = \frac{60}{55} = 1.09.$$

This is completely uninteresting, because we could just as well have written

$$P(X \geq 55) \leq 1,$$

and we have still learned nothing new at all. So the Markov inequality is not useful to us in such a situation.

41.3 Chebyshev Inequality

The Chebyshev inequality gives a different type of information about a random variable. As with the Markov inequality, in order to apply the Chebyshev inequality, we do not need to know the distribution of the random variable being considered. We just need to know the expected value of the random variable, as well as the variance, which automatically tells us the standard deviation too. Since we must provide two facts about the random variable here (both the average value and the variance), then we should expect to get a more informative bound, as compared to the Markov inequality (which only requires use of the average value).

Theorem 41.7. Chebyshev Inequality

If X is any random variable and k is any positive number, then

$$P(|X - \mathbb{E}(X)| \geq k) \leq \frac{\mathrm{Var}(X)}{k^2}.$$

To see that the Chebyshev inequality is true, we just apply the Markov inequality, using $(X - \mathbb{E}(X))^2$ as the random variable, and using $a = k^2$. Then the Markov inequality yields

$$P((X - \mathbb{E}(X))^2 \geq k^2) \leq \frac{\mathbb{E}((X - \mathbb{E}(X))^2)}{k^2}.$$

On the left hand side, $(X - \mathbb{E}(X))^2 \geq k^2$ if and only if $|X - \mathbb{E}(X)| \geq k$, so we can rewrite the left hand side as $P(|X - \mathbb{E}(X)| \geq k)$. On the right hand side, $\mathbb{E}((X - \mathbb{E}(X))^2) = \mathrm{Var}(X)$. So we get

$$P(|X - \mathbb{E}(X)| \geq k) \leq \frac{\mathrm{Var}(X)}{k^2},$$

which is exactly the statement of the Chebyshev inequality.

We can get a second version of the Chebyshev inequality if we just use "$k = a\sigma_X$" in the version above, where σ_X is the standard deviation of X. This yields

$$P(|X - \mathbb{E}(X)| \geq a\sigma_X) \leq \frac{\mathrm{Var}(X)}{(a\sigma_X)^2},$$

for any $a > 0$. Since $\mathrm{Var}(X)/\sigma_X^2 = 1$, this simplifies to

$$P(|X - \mathbb{E}(X)| \geq a\sigma_X) \leq \frac{1}{a^2}.$$

The choice of letter that we use is arbitrary, so we write the second version of the Chebyshev inequality using k's instead of a's:

Corollary 41.8. Chebyshev Inequality (version 2)

If X is any random variable and k is any positive number, then

$$P(|X - \mathbb{E}(X)| \geq k\sigma_X) \leq 1/k^2.$$

If we read the second version of the Chebyshev inequality literally, it means that a random variable is more than k standard deviations away from its expected value at most $1/k^2$ of the time. This is a way of quantifying the fact that a random variable is "relatively close" to its expected value "most of the time." Chebyshev gives bounds that quantify both "how close" and "how much of the time."

We could also take complements on both sides of this equation, and rewrite the equation

$$P(|X - \mathbb{E}(X)| \geq k\sigma_X) \leq 1/k^2.$$

as the following:

$$1 - P(|X - \mathbb{E}(X)| \geq k\sigma_X) \geq 1 - \frac{1}{k^2},$$

or equivalently,

$$P(|X - \mathbb{E}(X)| \leq k\sigma_X) \geq \frac{k^2 - 1}{k^2}.$$

This says that a random variable is within k standard deviations from its expected value at least $\frac{k^2-1}{k^2}$ of the time.

Corollary 41.9. Chebyshev Inequality (version 3)

If X is any random variable and k is any positive number, then

$$P(|X - \mathbb{E}(X)| \leq k\sigma_X) \geq \frac{k^2 - 1}{k^2}.$$

Example 41.10. Return to the scenario of Example 41.3, in which a class of students has a class average is 60%. Also suppose it is known that the standard deviation of the class's scores is 10.

Find a bound for the probability that a randomly chosen student's score is between 45% and 75%. We let X denote a randomly chosen student's score. Using the Chebyshev inequality, since the standard deviation is $\sigma_X = 10$, we use $k = 1.5$, and then we have

$$\begin{aligned} P(45 \leq X \leq 75) &= P(|X - 60| \leq 15) \\ &= P(|X - 60| \leq (1.5)(10)) \\ &\geq \frac{(1.5)^2 - 1}{(1.5)^2} \\ &= 0.5556. \end{aligned}$$

As with the Markov inequality, the Chebyshev inequality does not require that we know that distribution of the scores at all. We just need to know the expected value and the variance or the standard deviation. Also, as with the Markov inequality, we do not get a precise answer, but rather, we only get some bounds on the desired probability.

Example 41.11. Return to the scenario of Example 41.4, in which the average salary of actuaries in a certain specialty is known to be \$83,000 per year. Also assume that the standard deviation is \$20,000 per year.

Let X be the salary of a randomly chosen actuary in the profession under consideration. The probability that such an actuary makes more than \$123,000 or less than \$43,000 is

$$\begin{aligned} P(X \geq 123{,}000 \text{ or } X \leq 43{,}000) &= P(|X - 83{,}000| \geq 40{,}000) \\ &= P(|X - 83{,}000| \geq (2)(20{,}000)) \\ &\leq 1/2^2 \\ &= 1/4. \end{aligned}$$

Another way to express this is to say that the probability a randomly chosen actuary's salary is between \$43,000 and \$123,000 is

$$\begin{aligned} P(43{,}000 \leq X \leq 123{,}000) &= P(|X - 83{,}000| \leq 40{,}000) \\ &= P(|X - 83{,}000| \leq (2)(20{,}000)) \\ &\geq \frac{2^2 - 1}{2^2} \\ &= 3/4. \end{aligned}$$

We do not have sufficient information to tell the exact probability that a randomly chosen actuary's salary is between \$43,000 and \$123,000, but the bounds given here are better than nothing.

Example 41.12. Return to the scenario of Example 41.5, in which, on a certain highway, the speed limit is 55 miles per hour, but most drivers are not driving so fast. The average speed on the highway is 49 miles per hour. Also assume that the variance is 144.

If X denotes a randomly chosen driver's speed, then the variance of X is 144, so the standard deviation of X is 12. The probability that a randomly selected driver is travelling between 29 and 69 miles per hour is

$$\begin{aligned} P(29 \leq X \leq 69) &= P(|X - 49| \leq 20) \\ &= P(|X - 49| \leq (20/12)(12)) \\ &\geq \frac{(20/12)^2 - 1}{(20/12)^2} \\ &= 16/25 \\ &= 0.64 \end{aligned}$$

We already noted, after Example 41.10, that the Chebyshev inequality does not require us to know the distribution of the random variable, but it does require us to know the expected value and to know the variance, or equivalently, to know the standard deviation. We only get bounds on the probabilities from the Chebyshev inequality. Also, we do not get any information about the fraction of the time that a random variable is less than (or is more than) k standard deviations of the expected value, for $k < \sigma_X$ in version 1, or equivalently for $k < 1$ in versions 2 or 3. The Chebyshev inequality is just not helpful in such situations. In version 1, the right hand side will be $\mathrm{Var}(X)/k^2 > 1$ in such a case, which is not helpful. In version 2, the right hand side will be $1/k^2 > 1$, which is not helpful. In version 3, the right hand side will be $\frac{k^2-1}{k^2} < 0$, which again is not helpful. So these extreme cases do not yield any useful information from the Chebyshev inequality.

41.4 Exercises

41.4.1 Practice

Exercise 41.1. **Waiting for a bus.** While waiting for the bus on a snowy morning, the expected waiting time (including unusual delays for snow!), is 12 minutes.

a. Use the Markov inequality to find an upper bound on the probability that the bus takes 15 minutes or more to arrive.

b. In the same scenario as above, assume that the *standard deviation* of the waiting time is 3. Use the Chebyshev inequality to find a bound on the probability that the bus takes between 6 to 18 minutes to arrive.

Exercise 41.2. **Students in class.** The average number of students in a class at a certain university is 31.

a. Use the Markov inequality to find an upper bound on the probability that a class selected at random will have 40 or more students.

b. In the scenario above, now assume that the *variance* of the class size is 64. Use the Chebyshev inequality to find a bound on the probability that a class selected at random will have between 17 and 45 people.

Exercise 41.3. **Basketball shots.** A basketball player has improved his scoring ability. During a game, he can be expected to make 12 shots.

a. Give a bound on the probability that he makes at least 16 shots.

b. If the number of shots successfully scored by the same basketball player is known to have a standard deviation of 2.5, find a bound on the probability that he successfully makes between 7 and 17 shots.

Exercise 41.4. Guitar players. The expected number of guitar players in a classroom is 8. Given an upper bound on the probability that there are 11 or more guitar players in a classroom.

Exercise 41.5. Chicken feathers. A certain type of chicken's wing has 138 feathers on average, with standard deviation 5. Find a bound on the probability that the chicken has between 120 and 156 feathers.

Exercise 41.6. Sneezing. Henry caught a cold recently and therefore he has been sneezing a lot. His expected waiting time between sneezes is 45 seconds. The standard deviation of the waiting time between his sneezes is 8 seconds. Find a bound on the probability that the time between two consecutive sneezes is between 30 and 60 seconds.

Exercise 41.7. Sleepy dog. The expected time for a student's dog to fall asleep is 12 minutes. Give an upper bound on the time that it takes the dog more than 20 minutes to fall asleep.

Exercise 41.8. Music library. In a student's music library on his mp3 player, the expected length of a randomly chosen song is 3.2 minutes.

a. Find an upper bound on the probability that a randomly chosen song is at least 5 minutes long.

b. If the standard deviation in the song lengths is known to be 0.8 minutes, find a bound on the probability that the song is between 2.9 and 3.5 minutes long.

Exercise 41.9. Snowy winter. The average amount of snow in a student's hometown during the winter months is 10 inches.

a. Find a bound on the probability that the snowfall in a given winter will exceed 16 inches.

b. If the standard deviation of snowfall is 3.25 inches, find a bound on the probability that there is between 6 and 14 inches of snow.

Exercise 41.10. Soccer goals. A talented soccer player is expected to score 21 goals in a given soccer season.

a. Find an upper bound on the probability that he scores 30 goals or more during the season.

b. If he has a standard deviation of 6.5 goals during the season, give a bound on the probability that he scores between 8 and 34 goals during the season.

Exercise 41.11. Flight time. The flight time of a plane flight from Denver to New York City is has an average flight time of 3 hours and 16 minutes, and standard deviation is 30 minutes.

a. Give a bound on the probability that such a flight takes 6 hours or longer.

b. Give a bound on the probability that the announced arrival time (3 hours, 16 minutes) and the actual arrival time differ by 1 hour or more.

Exercise 41.12. Cold weather. The average temperature in December is 27°.

a. Find a bound on the probability of the temperature falling outside the range −50° to 50°.

b. If the standard deviation of the temperature is 7°, then what is the probability that the temperature falls outside the range of 14° to 40°?

Exercise 41.13. Final exam. Let X be the number of problems on the final exam. The professor puts, on average, 30 problems on each exam. The standard deviation of the number of problems he puts on a final exam is 2. Given a bound for the likelihood that the final exam contains between 25 and 35 problems.

Exercise 41.14. Video download. A video can be downloaded from the web and moved to a student's mobile device in 12 minutes, on average. If the standard deviation of the time required is 2 minutes, then give a bound on the probability that 9–15 minutes are needed.

Exercise 41.15. Eating habits. In a study on eating habits, a particular participant averages 750 cm^3 of food during per meal.

a. It is extraordinarily rare for this participant to eat more than 1000 cm^3 of food at once. Find a bound on the probability of such an event.

b. If the standard deviation of a meal size is 100 cm^3, the find a bound on the event that the meal is either too much food, i.e., more than 1000 cm^3, or an insufficient amount of food, namely, less than 500 cm^3.

Exercise 41.16. Long jumps. An athletic director is recording long-jump scores for a group of students. The expected value of each of their long jumps is 7 meters, and the standard deviation is 0.2 meters. If he chooses a student at random, find a bound on the probability that the student's long-jump is between 6.7 and 7.3 meters.

Exercise 41.17. Sled runs. An energetic student can manage to get 9 sled runs down a hill, on average, within a 30 minute time period.

a. Find an upper bound on the probability that the student achieves 12 or more runs during a one hour period.

b. If the standard deviation of the number of runs that they can manage in a 30 minute period is 2, then give a bound on the probability that the student gets between 6 and 12 sled runs during a 30 minute period.

Exercise 41.18. String theory. Suppose that the expected number of students who take String Theory is 23, with variance 169. Suppose that the assigned classroom can hold only 40 students. Also, 6 students is the minimum enrollment, i.e., if 5 or less people sign up, the class will be combined with another. What is a bound on the probability that the class is held (i.e., has a sufficient number of students) and can be taught in the intended classroom (i.e., does not have too many students)?

Exercise 41.19. Fair coin. Trying to figure out if a coin is fair, a student flips it one thousand times (!!!). Give an upper bound on the probability that he gets 700 or more heads, assuming that the coin was actually a fair coin.

Exercise 41.20. Failing lights. On a strand of lights, the expected number of lights that fail to work correctly is 20.

a. What is an upper bound on the probability that 30 or more lights fail to work?

b. If the standard deviation is 5, what is a bound on the probability that between 12 and 28 lights fail to work correctly?

Exercise 41.21. Buying gifts. This holiday season, assume that a person will, on average, spend $673 on gifts, with a standard deviation of $79. Use a Chebyshev inequality to find a bound on the probability that a randomly selected person spends between $568 and $778.

Exercise 41.22. Accountant age. Assume that the average age of tax accountants at a certain CPA firm is 44.

a. Find a bound for the probability that a randomly chosen employee of the firm is 48 or older.

b. If the standard deviation of the age of a randomly chosen employee is 5 years, then find a bound for the probability that a randomly chosen employee's age is between 36 and 52.

41.4.2 Extensions

Exercise 41.23. With the same assumptions as in Exercise 41.2, three classes are independently selected at random.

a. Let A denote the event that all three of the classes have 40 or more students (i.e., 40 or more in each class). Find a bound on the probability of A. Hint: Separate A into three independent events, find a bound on the probability of each, and then think about how to appropriately combine your bounds.

b. In the scenario above, let B denote the event that all three classes selected at random will have between 20 and 42 people (i.e., 20 to 42 people in each class). Find a bound on the probability of B. Hint: Again, separate B into three independent events, find a bound on each, and then recombine appropriately.

Chapter 42

Order Statistics

Good order is the foundation of all great things.
—Edmund Burke

Order is the shape upon which beauty depends.
—Pearl S. Buck

When five students gather and compare their grades, what is the probability that the highest grade exceeds 97%? What is the probability that at least one student failed the exam? What is the probability that three or more students earned a "B" grade or higher?

42.1 Introduction

The concept of order statistics is usually used with continuous random variables, because the idea of determining the rank of the random variables (i.e., which is smallest, which is largest, which is second-smallest, etc.) is key to the concept of order statistics. Any "ties" between two random variables—i.e., any two random variables that happen to equal exactly the same value—complicate the issue. With continuous random variables, the probability of any ties is 0, so the use of continuous random variables allows us to safely ignore the possibility of ties. It is usually impossible to remove this complication with discrete random variables, so we do not discuss order statistics of discrete random variables in this text at all.

Usually we speak of the "order statistics" among independent, identically distributed random variables. So if $X_1, X_2, \ldots, X_n$ are the random variables under consideration, then the X_j's are independent, and all of their densities $f_j(x)$ are the same function (or, equivalently, all of their cumulative distribution functions $F_j(x)$ are the same function). This is not always the case when working with order statistics (i.e., we could talk about order statistics of random variables that are either not independent or not identically distributed), but the independent, identically distributed situation is the most common scenario for studying order statistics.

In this text, when speaking about order statistics, we always assume that the random variables under study are independent, identically distributed, continuous random variables.

42.2 Examples

Example 42.1. Consider the waiting times X_1 and X_2 (in minutes) until Samuel hears from two of his friends, Mary and Josephine, respectively. He doesn't know which one will call first. Both of the X_j's are Exponential with expected value 5. The X_j's are also independent. Then the first order statistic is $X_{(1)} = \min\{X_1, X_2\}$, i.e., the minimum of the two waiting times. The second order statistic is $X_{(2)} = \max\{X_1, X_2\}$, i.e., the maximum of the two waiting times.

In this case, $X_{(1)}$ is an Exponential random variable. To see this, for any $a > 0$,

$$\begin{aligned} P(X_{(1)} > a) &= P(X_1 > a \text{ and } X_2 > a) \\ &= P(X_1 > a)P(X_2 > a) \\ &= e^{-a/5}e^{-a/5} \\ &= e^{-2a/5} \end{aligned}$$

Thus $X_{(1)}$ is Exponential with $\mathbb{E}(X_{(1)}) = 5/2$. We emphasize that the parameter of $X_{(1)}$ is different that the parameters of X_1 and X_2. With this in mind, we can treat $X_{(1)}$ in just the same way that we would treat any other Exponential random variable with expected value 5/2. For instance, we know immediately that the variance of the time until he hears from his first friend is $(5/2)^2 = 25/4 = 6.25$ minutes. Also, the probability he hears from the first friend within the first four minutes is $P(X_{(1)}) = 1 - e^{-2(4)/5} = 0.798$.

On the other hand, we emphasize that $X_{(2)}$ is not an Exponential random variable. For instance, the CDF of $X_{(2)}$ does not have the form of the CDF of an Exponential random variable. For $a > 0$:

$$\begin{aligned} P(X_{(2)} \le a) &= P(X_1 \le a \text{ and } X_2 \le a) \\ &= P(X_1 \le a)P(X_2 \le a) \\ &= (1 - e^{-a/5})(1 - e^{-a/5}). \end{aligned}$$

Incidentally, we can also find the densities of $X_{(1)}$ and $X_{(2)}$. Since $X_{(1)}$ is Exponential with expected value $5/2$, then no calculation is necessary. We know

$$f_{X_{(1)}}(x_1) = \frac{2}{5}e^{-2x_1/5} \qquad \text{for } x_1 > 0,$$

and $f_{X_{(1)}}(x_1) = 0$ otherwise.

To find the density of $X_{(2)}$, we just differentiate the CDF

$$F_{X_{(2)}}(x_2) = P(X_{(2)} \leq x_2) = (1 - e^{-x_2/5})(1 - e^{-x_2/5})$$

with respect to x_2, and we get

$$f_{X_{(2)}}(x_2) = \frac{2}{5}e^{-x_2/5}(1 - e^{-x_2/5}) \qquad \text{for } x_2 > 0,$$

and $f_{X_{(2)}}(x_2) = 0$ otherwise.

Example 42.2. Consider the heights X_1, X_2, X_3 of three randomly chosen people, Alicia, Bernadette, and Charlotte, respectively. Suppose that the X_j's are independent and identically distributed, e.g., if their heights are measured in inches then perhaps the expected value of each height is 63.5 and the standard deviation is 2.5.

In this example, the first order statistic, $X_{(1)}$, is the minimum of the three people's heights. The second order statistic is $X_{(2)}$, the height of the middle person (who is neither shortest nor tallest). Finally, the third order statistic is $X_{(3)}$, the maximum of the three people's heights. So, for example, if

$$X_1 = 63.076$$
$$X_2 = 62.849$$
$$X_3 = 63.870$$

then the first, second, and third order statistics are, respectively:

$$X_{(1)} = 62.849$$
$$X_{(2)} = 63.076$$
$$X_{(3)} = 63.870$$

Definition 42.3. Order Statistics

If we put $X_1, X_2, \ldots, X_n$ in order, so that $X_{(1)}$ is the smallest of these random variables, and $X_{(2)}$ is the second-smallest of these random variables, and in general, $X_{(j)}$ is the jth-smallest of these random variables, and thus $X_{(n)}$ is the largest of these random variables, then we say that $X_{(1)}, X_{(2)}, \ldots, X_{(n)}$ are the 1st, 2nd, ..., nth order statistics.

Example 42.4. For example, if a driver experiences the following ten waiting times at ten red lights on a trip,

$$X_1 = 29.186,\quad X_2 = 26.169,\quad X_3 = 3.422,\quad X_4 = 12.995,\quad X_5 = 3.127,$$
$$X_6 = 1.360,\quad X_7 = 29.617,\quad X_8 = 10.420,\quad X_9 = 9.484,\quad X_{10} = 18.837.$$

then we could put the waiting times in order, to get the order statistics.

The first order statistic is the minimum, written as $X_{(1)} = 1.360$. The second order statistic is the second-smallest value, written as $X_{(2)} = 3.127$. The tenth order statistic is the maximum value, written as $X_{(10)} = 29.617$. So the 1st, 2nd, 3rd, ..., 10th order statistics are, respectively:

$$X_{(1)}=1.360,\quad X_{(2)}=3.127,\quad X_{(3)}=3.422,\quad X_{(4)}=9.484,\quad X_{(5)}=10.420,$$
$$X_{(6)}=12.995,\quad X_{(7)}=18.837,\quad X_{(8)}=26.169,\quad X_{(9)}=29.186,\quad X_{(10)}=29.617.$$

Example 42.5. The joint density of the order statistics can only be nonzero in the region where, for instance, $X_{(1)} < X_{(2)} < X_{(3)}$. If we consider the joint density $f_{X_{(1)},X_{(2)},X_{(3)}}(2.7, 5.22, 3.9)$ of the first, second, and third order statistics of three random variables X_1, X_2, X_3 that are each defined in the interval $[0, 10]$, then we know that the joint density will be zero at this point, because we cannot have the second order statistic $X_{(2)} = 5.22$ and a smaller third order statistic $X_{(3)} = 3.9$. On the other hand, $f_{X_{(1)},X_{(2)},X_{(3)}}(2.7, 3.9, 5.22)$ will be positive because the order statistics are in ascending order!

42.3 Joint Density and Joint CDF of Order Statistics

Throughout this section, we assume that the random variables $X_1, \ldots, X_n$ under consideration are independent and identically distributed.

Example 42.6. As in Example 42.1, consider the waiting times X_1 and X_2 (in minutes) until Samuel hears from two of his friends, Mary and Josephine, respectively. He doesn't know which one will call first. Both of the X_j's are Exponential with expected value 5. The X_j's are also independent. Then the first order statistic is $X_{(1)} = \min\{X_1, X_2\}$, i.e., the minimum of the two waiting times. The second order statistic is $X_{(2)} = \max\{X_1, X_2\}$, i.e., the maximum of the two waiting times.

Consider any a, b with $0 < a < b$. In order to calculate the joint density of $(X_{(1)}, X_{(2)})$ evaluated at (a, b), we first calculate the joint CDF:

$$F_{X_{(1)},X_{(2)}}(a, b) = P(X_{(1)} \le a \text{ and } X_{(2)} \le b)$$

To have $X_{(1)} \le a$ and $X_{(2)} \le b$, we need either:

- Both X_1 and X_2 are smaller than a, or
- Only $X_2 < a$, and $a < X_1 < b$, or
- Only $X_1 < a$, and $a < X_2 < b$.

Thus

$$\begin{aligned} F_{X_{(1)},X_{(2)}}(a, b) &= P(X_1 \le a \text{ and } X_2 \le a) \\ &\quad + P(a < X_1 \le b \text{ and } X_2 \le a) \\ &\quad + P(X_1 \le a \text{ and } a < X_2 \le b) \\ &= (1 - e^{-a/5})(1 - e^{-a/5}) \\ &\quad + (e^{-a/5} - e^{-b/5})(1 - e^{-a/5}) \\ &\quad + (1 - e^{-a/5})(e^{-a/5} - e^{-b/5}) \\ &= (1 - e^{-a/5})(1 - e^{-a/5}) + 2(e^{-a/5} - e^{-b/5})(1 - e^{-a/5}) \end{aligned}$$

Thus, for $0 < x_1 < x_2$,

$$F_{X_{(1)},X_{(2)}}(x_1, x_2) = (1 - e^{-x_1/5})(1 - e^{-x_1/5}) + 2(e^{-x_1/5} - e^{-x_2/5})(1 - e^{-x_1/5}).$$

Now we differentiate with respect to x_1 and x_2 to get the joint density of $(X_{(1)}, X_{(2)})$. It does not matter which order we choose to differentiate, so we first differentiate with respect to x_2, so that the entire first term (which depends only on x_1) is removed. We get

$$\frac{\partial}{\partial x_2} F_{X_{(1)},X_{(2)}}(x_1, x_2) = 2\left(\frac{1}{5}e^{-x_2/5}\right)(1 - e^{-x_1/5})$$

and then

$$f_{X_{(1)},X_{(2)}}(x_1, x_2) = \frac{\partial}{\partial x_1}\frac{\partial}{\partial x_2} F_{X_{(1)},X_{(2)}}(x_1, x_2) = 2\left(\frac{1}{5}e^{-x_2/5}\right)\left(\frac{1}{5}e^{-x_1/5}\right).$$

Thus, if we write

$$f_X(x) = \frac{1}{5}e^{-x/5},$$

which is the common distribution of both X_1 and X_2, then we see that the joint density of $X_{(1)}$ and $X_{(2)}$ is

$$f_{X_{(1)},X_{(2)}}(x_1, x_2) = 2f_X(x_1)f_X(x_2) \qquad \text{for } 0 < x_1 < x_2,$$

and $f_{X_{(1)},X_{(2)}}(x_1, x_2) = 0$ otherwise.

Now we emphasize that this concept works much more generally.

Example 42.7. Consider n random variables $X_1, \ldots, X_n$ that are independent and all have the same distribution.

The joint density of $X_1, \ldots, X_n$ can be factored into a product of the densities of the X_j's, since the X_j's are independent.

$$f_{X_1,X_2,\ldots,X_n}(x_1, x_2, \ldots, x_n) = f_{X_1}(x_1) f_{X_2}(x_2) \cdots f_{X_n}(x_n).$$

Of course, not only are the X_j's all independent, but they also all have the same distribution, which we can just write as $f(x)$. Thus, the joint density of $X_1, \ldots, X_n$ becomes

$$f_{X_1,X_2,\ldots,X_n}(x_1, x_2, \ldots, x_n) = f(x_1) f(x_2) \cdots f(x_n).$$

Thus, if we integrate all X_j's over all real numbers, we have

$$\int_{-\infty}^{\infty} \int_{-\infty}^{\infty} \cdots \int_{-\infty}^{\infty} f(x_1) f(x_2) \cdots f(x_n) \, dx_1 \, dx_2 \, \cdots \, dx_n = 1.$$

The joint density of the order statistics $X_{(1)}, \ldots, X_{(n)}$ can also be written as a product, but we must insist now that the x_j's are in ascending order. *The main point driving this idea is that there are $n!$ equally likely ways that the underlying $X_1, \ldots, X_n$ can be placed in order.* So by multiplying by $n!$, the integral of the joint density of the order statistics will be 1. Therefore, the joint density of $X_{(1)}, \ldots, X_{(n)}$ is

$$f_{X_{(1)},\ldots,X_{(n)}}(x_1, \ldots, x_n) = n! f(x_1) \cdots f(x_n) \qquad \text{for } x_1 < \cdots < x_n,$$

and $f_{X_{(1)},\ldots,X_{(n)}}(x_1, \ldots, x_n) = 0$ otherwise.

Theorem 42.8. Joint Density of Order Statistics

If $X_1, \ldots, X_n$ are independent, continuous random variables that each have density $f_X(x)$, then the order statistics $X_{(1)}, X_{(2)}, \ldots, X_{(n)}$ have joint density

$$f_{X_{(1)},\ldots,X_{(n)}}(x_1, \ldots, x_n) = n! f(x_1) \cdots f(x_n) \qquad \text{for } x_1 < \cdots < x_n,$$

and $f_{X_{(1)},\ldots,X_{(n)}}(x_1, \ldots, x_n) = 0$ otherwise.

Now we can easily recompute some of the previous problems from earlier chapters.

Example 42.9. As in Exercise 31.29, let X_1, X_2, X_3 be independent continuous random variables, each Uniformly distributed in the interval $[0, 10]$. As an example, X_1, X_2, X_3 might be the arrival times of three people to a bus stop between 8:50 AM and 9:00 AM, when they are not wearing wrist watches, so they arrive Uniformly during a 10-minute interval. Let Y denote the middle of the three values, i.e., the arrival time of the 2nd person.

In the language of order statistics, Y is always the second-smallest of X_1, X_2, X_3, so we have $Y = X_{(2)}$. The joint density of $X_{(1)}, X_{(2)}, X_{(3)}$ is

$$\begin{aligned} f_{X_{(1)},X_{(2)},X_{(3)}}(x_1, x_2, x_3) &= 3! f(x_1) f(x_2) f(x_3) \qquad \text{for } x_1 < x_2 < x_3, \\ &= 6\Big(\frac{1}{10}\Big)\Big(\frac{1}{10}\Big)\Big(\frac{1}{10}\Big) \qquad \text{for } 0 \le x_1 < x_2 < x_3 \le 10, \\ &= \frac{6}{1000} \qquad \text{for } 0 \le x_1 < x_2 < x_3 \le 10, \end{aligned}$$

and $f_{X_{(1)},X_{(2)},X_{(3)}}(x_1, x_2, x_3) = 0$ otherwise.

Thus, the density of $X_{(2)}$ can be found by integrating $X_{(1)}$ and $X_{(3)}$ out of the picture. If we integrate the joint density $f_{X_{(1)},X_{(2)},X_{(3)}}(x_1, x_2, x_3)$ over all $X_{(1)}$'s that are less than $X_{(2)}$, and integrate over all $X_{(3)}$'s that are greater than $X_{(2)}$, then we get the density $f_{X_{(2)}}(x_2)$ of the middle value, $Y = X_{(2)}$. So the density of $X_{(2)}$ is

$$f_{X_{(2)}}(x_2) = \int_{x_2}^{10} \int_0^{x_2} \frac{6}{1000}\, dx_1\, dx_3 = \frac{6}{1000} x_2(10 - x_2) \qquad \text{for } 0 \le x_2 \le 10,$$

and $f_{X_{(2)}}(x_2) = 0$ otherwise. We rewrite this result in the following way:

$$f_{X_{(2)}}(x_2) = 6\Big(\frac{1}{10}\Big)\Big(\frac{x_2}{10}\Big)\Big(\frac{10 - x_2}{10}\Big) \qquad \text{for } 0 \le x_2 \le 10,$$

and $f_{X_{(2)}}(x_2) = 0$ otherwise. If we write $f_X(x) = 1/10$ as the density of each of the X_j's, and we write $F_X(x) = \frac{x}{10}$ as the cumulative distribution function of each of the X_j, and $1 - F_X(x) = 1 - \frac{x}{10}$, then we notice that $f_{X_{(2)}}(x_2)$ has the really nice form

$$f_{X_{(2)}}(x_2) = 3! f_X(x_2) F_X(x_2)(1 - F_X(x_2)) \qquad \text{for } 0 \le x_2 \le 10,$$

and $f_{X_{(2)}}(x_2) = 0$ otherwise. This might not look very general, but it is actually a very special case of the following nice, general result:

Example 42.10. Consider independent continuous random variables X_1, X_2, $\dots$, X_n that each have density $f_X(x)$ and cumulative distribution function $F_X(x)$. Then the density of the jth order statistic, $f_{X_j}(x_j)$, is exactly

$$f_{X_{(j)}}(x_j) = \binom{n}{j-1, 1, n-j} f_X(x_j)(F_X(x_j))^{j-1}(1 - F_X(x_j))^{n-j},$$

where

$$\binom{n}{j-1, 1, n-j} = \frac{n!}{(j-1)!1!(n-j)!}.$$

In particular, $\binom{n}{j-1,1,n-j} = \frac{n!}{(j-1)!1!(n-j)!}$ is exactly the number of ways to pick exactly 1 of the n variables $X_1, \dots, X_n$ to be the jth order statistic (i.e., to be the jth smallest of the collection $X_1, \dots, X_n$), while also choosing $j-1$ of the variables to be smaller than X_j, and finally letting the other $n-j$ variables be larger than X_j.

Theorem 42.11. Density of a Particular Order Statistic

If $X_1, X_2, \dots, X_n$ are independent, continuous random variables that each have density $f_X(x)$, then the density of the jth order statistic, $f_{X_j}(x_j)$, is exactly

$$f_{X_{(j)}}(x_j) = \binom{n}{j-1, 1, n-j} f_X(x_j)(F_X(x_j))^{j-1}(1 - F_X(x_j))^{n-j},$$

where

$$\binom{n}{j-1, 1, n-j} = \frac{n!}{(j-1)!1!(n-j)!}.$$

is exactly the number of ways to pick exactly 1 of the n variables $X_1, \dots, X_n$ to be the jth order statistic (i.e., to be the jth smallest of the collection $X_1, \dots, X_n$), while also choosing $j-1$ of the variables to be smaller than X_j, and finally letting the other $n-j$ variables be larger than X_j.

In the special case discussed in Example 42.9, we have

$$n = 3 \qquad \text{and} \qquad j = 2,$$

as well as

$$f_X(x) = 1/10 \qquad \text{and} \qquad F_X(x) = \frac{x}{10} \qquad \text{and} \qquad 1 - F_X(x) = 1 - \frac{x}{10}.$$

Therefore, we are able to quickly verify the earlier result, from Example 42.9. For $0 \le x_2 \le 10$,

$$\begin{aligned} f_{X_{(2)}}(x_2) &= \binom{3}{1,1,1} f_X(x_2)F_X(x_2)(1 - F_X(x_2)) \\ &= 6\Big(\frac{1}{10}\Big)\Big(\frac{x_2}{10}\Big)\Big(\frac{10 - x_2}{10}\Big). \end{aligned}$$

Otherwise, $f_{X_{(2)}}(x_2) = 0$.

Example 42.12. As in Example 42.1, consider the waiting times X_1 and X_2 (in minutes) until Samuel hears from two of his friends, Mary and Josephine, respectively. He doesn't know which one will call first. Both of the X_j's are Exponential with expected value 5. The X_j's are also independent.

Using what we learned in Example 42.10, we can verify the calculation of the densities of $X_{(1)}$ and $X_{(2)}$ from Example 42.1. We have

$$f_{X_{(1)}}(x_1) = \binom{2}{0,1,1}\frac{1}{5}e^{-x_1/5}(1-e^{-x_1/5})^0(e^{-x_1/5})^1 = \frac{2}{5}e^{-2x_1/5}.$$

Also

$$f_{X_{(2)}}(x_2) = \binom{2}{1,1,0}\frac{1}{5}e^{-x_2/5}(1-e^{-x_2/5})^1(e^{-x_2/5})^0 = \frac{2}{5}e^{-x_2/5}(1-e^{-x_2/5}).$$

This agrees with the calculations of the densities made in Example 42.1.

Example 42.13. Consider seven students whose grades are independent and are each Uniform on the interval from 82 to 98. Let $X_1, \ldots, X_7$ denote their grades. Rank the student's scores in order, and denote the order statistics as, respectively, $X_{(1)}, \ldots, X_{(7)}$.

Since each X_j is Uniform on the interval from 82 to 98, then the density of each X_j is $\frac{1}{98-82}$ on the interval $82 \le x \le 98$. Also, the cumulative distribution function of each X_j is $F_X(x) = \frac{x-82}{98-82}$ for $82 \le x \le 98$. Thus, applying the general model from Example 42.10, we see that the fifth-smallest student's score, $X_{(5)}$, has the following density, for $82 \le x_5 \le 98$:

$$f_{X_{(5)}}(x_5) = \binom{7}{4,1,2}\left(\frac{1}{98-82}\right)\left(\frac{x_5-82}{98-82}\right)^{5-1}\left(1-\frac{x_5-82}{98-82}\right)^{7-5}.$$

Otherwise, $f_{X_{(5)}}(x_5) = 0$.

This can be simplified to the following:

$$f_{X_{(5)}}(x_5) = 105\left(\frac{1}{16}\right)\left(\frac{x_5-82}{16}\right)^4\left(\frac{98-x_5}{16}\right)^2 \qquad \text{for } 82 \le x_5 \le 98,$$

and $f_{X_{(5)}}(x_5) = 0$ otherwise.

Example 42.14. Now we generalize the formula for the density of the jth order statistic of n independent random variables $X_1, \ldots, X_n$ that are each Uniform on the interval from a to b. As usual, we rank the random variables in order, and we denote the order statistics as, respectively, $X_{(1)}, \ldots, X_{(n)}$.

Since each X_j is Uniform on the interval from a to b, then the density of each X_j is $\frac{1}{b-a}$ on the interval $a \leq x \leq b$. Also, the cumulative distribution function of each X_j is $F_X(x) = \frac{x-a}{b-a}$ for $a \leq x \leq b$. Thus, using the model from Example 42.10,

Theorem 42.15. The Order Statistics of Independent, Identically Distributed Uniform Random Variables
If $X_1, X_2, \ldots, X_n$ are independent and Uniformly distributed on the interval $[a, b]$, then the density of the jth order statistic, i.e., the density of the jth smallest of the n Uniform random variables, is

$$f_{X_{(j)}}(x_j) = \binom{n}{j-1, 1, n-j} \left(\frac{1}{b-a}\right) \left(\frac{x_j - a}{b-a}\right)^{j-1} \left(1 - \frac{x_j - a}{b-a}\right)^{n-j},$$

for $a \leq x_j \leq b$, and $f_{X_{(j)}}(x_j) = 0$ otherwise.

In particular, we notice that the jth order statistic, in a collection of independent random variables that have identical Uniform distributions, is not a Uniform random variable itself.

Corollary 42.16. The Order Statistics of Independent, Identically Distributed Uniform Random Variables
If $X_1, X_2, \ldots, X_n$ are independent and Uniformly distributed on the interval $[a, b]$, then none of the order statistics $X_{(1)}, X_{(2)}, \ldots, X_{(n)}$ are Uniformly distributed.

This observation makes sense intuitively as well. For instance, if we look at the smallest random variable, i.e., the first order statistic, $X_{(1)}$, from a collection of 10 Uniform random variables distributed on the interval $[0, 2]$, we know that $X_{(1)}$ should have a density that is more heavily weighted toward the "0" end of $[0, 2]$ (i.e., toward the smaller end of the possibilities), and only very lightly weighted on the "2" end of the interval $[0, 2]$. It would be relatively unlikely, for instance, that the smallest of 10 random variables is at-or-above 1.72, because that would mean that the other 9 random variables (which must be larger than $X_{(1)}$) are forced to be larger than $X_{(1)}$ too, and, in particular, are forced to be larger than 1.72. It would be much more likely for $X_{(1)}$ to be, say, between 0 and 0.3, i.e., somewhere closer to the lower end of the spectrum of possibilities, because this puts less limitations on the ranges of possible values of the other nine random variables.

Example 42.17. Consider 10 people who are each waiting for an email to arrive, and the waiting times are independent Exponential random variables, each with average 30 minutes.

In this example, the waiting times can be referred to as $X_1, \ldots, X_n$, and we are given $\mathbb{E}(X_j) = 30$ for each j, so each X_j has parameter $\lambda = 1/30$. Thus the density of each X_j is

$$f_{X_j}(x) = \frac{1}{30}e^{-x/30} \qquad \text{for } x > 0,$$

and $f_{X_j}(x) = 0$ otherwise. Also, each X_j has cumulative distribution function

$$F_{X_j}(x) = 1 - e^{-x/30} \qquad \text{for } x > 0,$$

and $F_{X_j}(x) = 0$ otherwise. Thus, the density of the jth order statistic, i.e., the density of the time of the jth email to arrive, is

$$f_{X_{(j)}}(x_j) = \binom{10}{j-1, 1, 10-j}\frac{1}{30}e^{-x_j/30}\left(1 - e^{-x_j/30}\right)^{j-1}\left(e^{-x_j/30}\right)^{10-j}.$$

Nothing in this equation is special about 10 emails arriving altogether, or about the particular parameter $\lambda = 1/30$. We can generalize the previous example—which also generalizes Examples 42.1 and 42.12—as follows:

Example 42.18. Consider n Exponential random variables (usually viewed as waiting times) that are independent and have the same parameter, say, λ.

In this example, we refer to the Exponential random variables as $X_1, \ldots, X_n$, and we are told that all of the X_j's have the same parameter λ. Then the density of each X_j is

$$f_{X_j}(x) = \lambda e^{-\lambda x} \qquad \text{for } x > 0,$$

and $f_{X_j}(x) = 0$ otherwise. Also, each X_j has cumulative distribution function

$$F_{X_j}(x) = 1 - e^{-\lambda x} \qquad \text{for } x > 0,$$

and $F_{X_j}(x) = 0$ otherwise. Thus,

Theorem 42.19. The Order Statistics of Independent, Identically Distributed Exponential Random Variables

If $X_1, X_2, \ldots, X_n$ are independent and Exponentially distributed with parameter λ (i.e., with expected value $1/\lambda$), then the density of the jth order statistic, i.e., the density of the jth smallest of the n Exponential random variables, is

$$f_{X_{(j)}}(x_j) = \binom{n}{j-1, 1, n-j}\lambda e^{-\lambda x_j}\left(1 - e^{-\lambda x_j}\right)^{j-1}\left(e^{-\lambda x_j}\right)^{n-j} \qquad \text{for } x_j > 0,$$

and $f_{X_{(j)}}(x_j) = 0$ otherwise.

In particular, we notice that the jth order statistic, in a collection of independent random variables that have identical Exponential distributions, is not an Exponential random variable itself, with one exception: The first order statistic—the minimum of the collection of independent Exponential random variables—is an Exponentially distributed random variable.

Corollary 42.20. The Order Statistics of Independent, Identically Distributed Exponential Random Variables
If $X_1, X_2, \ldots, X_n$ are independent and Exponentially distributed with expected value $1/\lambda$, then the first order statistic $X_{(1)}$ is the only one of the order statistics order statistics $X_{(1)}, X_{(2)}, \ldots, X_{(n)}$ that is also Exponentially distributed.
The density of the first order statistic is

$$f_{X_{(1)}}(x_1) = \binom{n}{0, 1, n-1} \lambda e^{-\lambda x_1} \left(e^{-\lambda x_1}\right)^{n-1} = n\lambda e^{-n\lambda x_1} \qquad \text{for } x_1 > 0,$$

and $f_{X_{(1)}}(x_1) = 0$ otherwise. So $X_{(1)}$ is Exponentially distributed with parameter $n\lambda$, i.e., with expected value $\mathbb{E}(X_{(1)}) = \frac{1}{n\lambda}$.

Finally, we conclude with a few more examples.

Example 42.21. As in Example 22.5, consider the following: The lifetime of a music player (before it permanently fails) is a random variable X with density

$$f_X(x) = \frac{1}{3} e^{-x/3} \qquad \text{for } x > 0,$$

and $f_X(x) = 0$ otherwise.

If seven friends have these music players, and $X_1, \ldots, X_7$ denote the lifetimes of the 7 devices, then the 5th order statistic, $X_{(5)}$, is the time until the 5th one out of 7 permanently fails. For $x_5 > 0$, the 5th order statistic has density

$$\begin{aligned} f_{X_{(5)}}(x_5) &= \binom{7}{4, 1, 2} (1/3) e^{-x_5/3} \left(1 - e^{-x_5/3}\right)^{5-1} \left(e^{-x_5/3}\right)^{7-5} \\ &= (105)(1/3) e^{-x_5/3} \left(1 - e^{-x_5/3}\right)^{4} \left(e^{-x_5/3}\right)^{2}; \end{aligned}$$

otherwise, $f_{X_{(5)}}(x_5) = 0$.

Example 42.22. As in Example 26.6, consider the following: Each time a pitcher delivers a fastball, the speed is distributed between 90 and 100 miles per hour, with density 1/10. Assume that the speeds of pitches are independent. Suppose that the pitcher throws n such fastballs.

By the results of Example 42.14, applied to Uniform random variables on the interval [90, 100], we know that the density of the jth order statistic, i.e., the jth slowest of the fastballs is

$$f_{X_{(j)}}(x_j) = \frac{n!}{(j-1)!1!(n-j)!}\Big(\frac{1}{10}\Big)\Big(\frac{x_j-90}{10}\Big)^{j-1}\Big(1-\frac{x_j-90}{10}\Big)^{n-j},$$

for $90 \le x_j \le 100$, and $f_{X_{(j)}}(x_j) = 0$ otherwise.

If the pitcher throws 5 fastballs, and we want to compute the probability that 2 or more of them are below 93 miles per hour, it suffices to prove that the 2nd order statistic (i.e., the 2nd-slowest of the 5 fastballs) is less than 93 miles per hour.

The density for $j = 2$, i.e., for the 2nd order statistic, i.e., for the 2nd-slowest of the $n = 5$ fastballs, is, for $90 \le x_2 \le 100$,

$$\begin{aligned} f_{X_{(2)}}(x_2) &= \frac{5!}{(2-1)!1!(5-2)!}\Big(\frac{1}{10}\Big)\Big(\frac{x_2-90}{10}\Big)^{2-1}\Big(1-\frac{x_2-90}{10}\Big)^{5-2} \\ &= 20\Big(\frac{1}{10}\Big)\Big(\frac{x_2-90}{10}\Big)\Big(1-\frac{x_2-90}{10}\Big)^{3}. \end{aligned}$$

Otherwise, $f_{X_{(j)}}(x_2) = 0$.

So the probability that the 2nd-slowest of the fastballs is less than 93 miles per hour is

$$\int_{90}^{93} f_{X_{(2)}}(x)\,dx = \int_{90}^{93} 20\Big(\frac{1}{10}\Big)\Big(\frac{x-90}{10}\Big)\Big(1-\frac{x-90}{10}\Big)^{3}dx = \frac{23589}{50000} = 0.47178.$$

Example 42.23. As in Example 42.14, consider independent random variables $X_1, \ldots, X_n$ that are each Uniform on the interval from a to b. Rank the random variables in order, and denote the order statistics as, respectively, $X_{(1)}, \ldots, X_{(n)}$.

We showed in Example 42.14 that the jth-smallest order statistic, $X_{(j)}$, has the following density:

$$f_{X_{(j)}}(x_j) = \binom{n}{j-1, 1, n-j}\Big(\frac{1}{b-a}\Big)\Big(\frac{x_j-a}{b-a}\Big)^{j-1}\Big(1-\frac{x_j-a}{b-a}\Big)^{n-j},$$

for $a \le x_j \le b$, and $f_{X_{(j)}}(x_j) = 0$ otherwise.

In particular, if $a = 0$ and $b = 1$, i.e., if $X_1, \ldots, X_n$ are each Uniform on $[0, 1]$, then the jth order statistic has density

$$\begin{aligned} f_{X_{(j)}}(x) &= \frac{n!}{(j-1)!1!(n-j)!}x^{j-1}(1-x)^{n-j} \qquad \text{for } 0 \le x \le 1, \\ &= j\binom{n}{j}x^{j-1}(1-x)^{n-j} \qquad \text{for } 0 \le x \le 1, \end{aligned}$$

and $f_{X_{(j)}}(x) = 0$ otherwise. Thus, the expected value of the jth order statistic, $X_{(j)}$, is

$$\begin{aligned}\mathbb{E}(X_{(j)}) &= \int_0^1 x f_{X_{(j)}}(x)\,dx \\ &= \int_0^1 j\binom{n}{j} x^j (1-x)^{n-j}\,dx \\ &= j\binom{n}{j}\int_0^1 x^j(1-x)^{n-j}\,dx\end{aligned}$$

Also

$$\int_0^1 x^j(1-x)^{n-j}\,dx = \frac{(n-j)!j!}{(n+1)!}$$

(the inquisitive reader may want to consider why—the reasoning is connected to the density of a Beta random variable), and thus we conclude

$$\mathbb{E}(X_{(j)}) = \frac{j}{n+1}.$$

Example 42.24. In general (i.e., when a and b are not necessarily 0 and 1), then if $X_1, \ldots, X_n$ are independent and Uniform on $[a, b]$, then we can normalize that X_j's so that they are distributed on the interval $[0, 1]$. To do this, we define $Y_j = \frac{X_j - a}{b-a}$, and then each Y_j is Uniform on $[0, 1]$. So the jth order statistic $Y_{(j)}$ of $Y_1, \ldots, Y_n$ has expected value $\mathbb{E}(Y_{(j)}) = \frac{j}{n+1}$, by the results of the previous example. Equivalently, $\mathbb{E}\left(\frac{X_{(j)}-a}{b-a}\right) = \frac{j}{n+1}$, so the expected value of the jth order statistic $X_{(j)}$ is

$$\mathbb{E}(X_{(j)}) = \frac{j(b-a)}{n+1} + a.$$

42.4 Exercises

42.4.1 Practice

Exercise 42.1. Dancers. In the middle of a complicated dance routine, five dancers are located across the width of the stage. Let $X_1, \ldots, X_5$ denote their positions across the stage. Assume that the X_j's are independent and that each X_j is Uniformly distributed on $[0, 100]$, where "0" denotes the far left side of the stage, and "100" denotes the far right side of the stage, and the measurements are given in feet.

a. Find the density of the person located closest to the left-hand side of the stage, i.e., find $f_Y(y)$ for $Y = \min(X_1, \ldots, X_5)$. (Hint: Y is just the 1st order statistic, i.e., $Y = X_{(1)}$.)

b. Find the probability that nobody is located within 20 feet of the left-hand side of the stage.

c. Find the probability that at least two people are located within 30 feet of the left-hand side of the stage. (Hint: You can use the 2nd order statistic for this part.)

Exercise 42.2. Throwing darts. Alfredo, Barbara, and Cathy throw darts at a dartboard of radius 9 inches. Let X_1, X_2, X_3 denote the distance of Alfredo, Barbara, and Cathy's darts (respectively) from the center of the board. Thus, $X_{(1)}$ (the first order statistic) is the minimum distance from the center to any of the darts; $X_{(3)}$ is the distance to the farthest dart; and $X_{(2)}$ is the distance to the middle of the three darts with respect to the origin.

a. Find the density of $X_{(2)}$.

b. Check that $f_{X_{(2)}}(x_2)$ is a density, i.e., check that $\int_0^9 f_{X_{(2)}}(x_2)\,dx_2 = 1$.

Exercise 42.3. Phone calls. Suppose that three male students are making telephone calls to their girlfriends. Let X_1, X_2, X_3 denote the times of their phone calls. Suppose that the X_j's are independent Exponential random variables, each with expected value 20 minutes.

a. Find the density of $X_{(1)}$, the first order statistic. (Easy check: $X_{(1)}$ is the minimum of three independent Exponentials, so we remember that $X_{(1)}$ is Exponential too.)

b. Find the probability that none of the calls are less than 12 minutes long, i.e., they all exceed 12 minutes.

c. Find the probability that two or more of the calls are less than 23 minutes each.

Exercise 42.4. Let X_1, X_2, X_3 be three independent random variables that are Uniformly distributed on the interval $[0, 20]$. Find the probability that the minimum of the three random variables is between 12 and 15.

Exercise 42.5. Standing in a line. Five people stand along a street waiting for a bus to arrive. The street is 9 yards away from a building. The distance (in yards) of each person from the street has density $f_X(x) = \frac{\sqrt{x}}{18}$, for $0 \le x \le 9$. These distances are independent. Find the density of the greatest of the five distances from the street.

Exercise 42.6. Student presentations. In four sections of a course, running (independently) in parallel, there are four students giving presentations that are each Exponential in length, with expected value of 10 minutes each. How time much do we expect to be needed until all four of the presentations are completed?

Exercise 42.7. Soda cans. The weight of a can of soda is Uniformly distributed, between 238 and 242 grams. Find the density of:

a. the minimum of the three weights among three cans of soda;

b. the maximum of the three weights among three cans of soda;

c. the 2nd-largest of the three weights among three cans of soda.

Exercise 42.8. Pizza delivery. Suppose that the delivery time for a pizza delivery is Uniformly distributed between 15 to 20 minutes. Now suppose that 4 people from a dormitory independently order pizza from 4 different drivers (so the times for delivery are independent). What is the density of:

a. the minimum time until a pizza arrives?

b. the maximum time until a pizza arrives?

c. the time until the second pizza?

d. the time until the third pizza?

Exercise 42.9. Homework times. The time in which a student starts his homework each night—during a seven day week—is Uniformly distributed between 6 PM and 8 PM. Let X and Y denote, respectively, the earliest and latest times that he starts his homework that week.

a. Find the density of X.

b. Find the density of Y.

c. Find the expected value of X.

d. Find the expected value of Y.

Exercise 42.10. Marble drops. On a Rube Goldberg machine, three marbles are dropped from a ledge, and they land at a location which is Uniformly distributed between 0 to 12 inches away from a wall. What is the expected location of the marble that lands farthest from the wall?

Exercise 42.11. Student arrivals. A class of 40 students is supposed to arrive at 8:30 AM, but in practice they each arrive at a time Uniformly distributed between 8:20 AM and 8:35 AM, and their arrival times are independent.

a. What is the expected arrival time of the last student to arrive?

b. What is the probability that 38 or more of the students have arrived by 8:30 AM?

Exercise 42.12. Shot put throws. Three athletes, competing in the shot put event at a track and field competition, throw the ball (called the "shot") as far as they are able. The athletes have relatively similar strengths, so assume that the distances they throw the ball are independently, Uniformly distributed random variables on the interval $[16.5, 19.5]$, measured in meters. What is the probability that at least two of the three throws exceed 17 meters?

42.4.2 Extensions

Exercise 42.13. Violinist's hands. When an extremely talented violin player is soloing, the position of her hand along the violin's neck seems to be Uniform on a 9-inch interval (the total length is about 13 inches from nut to bridge; but not all of the string is used). The violinist is photographed 6 times during a performance. Assume that the location of her hand is independently placed in each of the photos.

a. Find the jth order statistic of the location of her hand ($1 \leq j \leq 6$).

b. Find the probability that her hand is within 3 inches of the nut during at least 2 of the 6 photographs.

Exercise 42.14. Let X_1, X_2, X_3, X_4 be independent random variables, each uniformly distributed on the interval $[0, 10]$. Let $X_{(1)} = \min\{X_1, X_2, X_3, X_4\}$ be the first order statistic. Find the probability that $X_{(1)}$ is smaller than 3, i.e., find $P(X_{(1)} \leq 3)$.

Chapter 43

Moment Generating Functions

A generating function is a clothesline on which we hang up a sequence of numbers for display.

—*generatingfunctionology* by Herbert S. Wilf (A. K. Peters, 2006)

How can we systematically calculate the moments of a random variable?

43.1 A Brief Introduction to Generating Functions

The study of generating functions is a topic in mathematics that has applications and meanings in many topics, including discrete mathematics, complex analysis, probability theory, asymptotics, algorithm analysis, etc. Generating functions are also used in applied mathematics and in other disciplines (e.g., computer science) as well.

Neil Sloane's *On-Line Encyclopedia of Integer Sequences*, `http://oeis.org/`, contains 250,000 integer sequences that might be of interest. We can also think about non-integer sequences as well.

The whole idea of generating functions simply goes back to Maclaurin series, a fundamental topic in calculus. Remember that a Maclaurin series is just a Taylor series expansion of a function $g(t)$ around the point $t = 0$. In other words, a Maclaurin series is just a way to write a function $g(t)$ as an infinite series, in which the jth term is the jth derivative of g, evaluated at $t = 0$, divided by $j!$.

$$g(t) = \sum_{j=0}^{\infty} \frac{g^{(j)}(0)}{j!} t^j.$$

The idea of Maclaurin series can perhaps appear more friendly upon first encounter if we consider them from Wilf's point of view, as a "clothesline on which we hang up a sequence of numbers for display." If the numbers $\frac{g^{(j)}(0)}{j!}$ are of some interest to us (e.g., because they form a sequence we want to study), then we can view a Maclaurin series as a nice way to display these numbers. We just put $\frac{g^{(j)}(0)}{j!}$ as a coefficient of t^j, and then we sum up all of the terms; anytime we need $\frac{g^{(j)}(0)}{j!}$, it will be sitting in front of t^j, ready to be retrieved. The clothesline looks like this:

$$g(t) = \frac{g(0)}{0!}t^0 + \frac{g'(0)}{1!}t^1 + \frac{g''(0)}{2!}t^2 + \frac{g'''(0)}{3!}t^3 + \frac{g^{(4)}(0)}{4!}t^4 + \frac{g^{(5)}(0)}{5!}t^5 + \frac{g^{(6)}(0)}{6!}t^6 + \cdots$$

At this point, one might wonder what to do if we do not have a function in mind with interesting derivatives. Can we start from a sequence of numbers and build such a function? Yes, we can start with a sequence of numbers and construct—from the sequence—a new function $g(t)$ that has interesting coefficients in its Maclaurin series, so we want to display them on a clothesline. One motivation for doing this is that a series representation is often a much more compact way to store a sequence of integers, especially when no general, succinct form of the numbers themselves are available. To build such functions ourselves, we start with an interesting sequence of numbers $a_0, a_1, a_2, \ldots$, and we can just put them onto such a clothesline and add things up, and see what function we get as a result. We can define g by writing

$$\begin{aligned} g(t) &= \sum_{j=0}^{\infty} \frac{a_j}{j!} t^j \\ &= a_0 + a_1 t + \frac{a_2}{2!}t^2 + \frac{a_3}{3!}t^3 + \frac{a_4}{4!}t^4 + \frac{a_5}{5!}t^5 + \frac{a_6}{6!}t^6 + \frac{a_7}{7!}t^7 + \cdots \end{aligned}$$

Taking j derivatives, we get $a_j = g^{(j)}(0)$ from this function g that we created! The reason behind $a_j = g^{(j)}(0)$ should be clear after taking a few derivatives and trying it, but nonetheless, we put the reasoning in the Appendix to this chapter; see Section 43.5.

Definition 43.1. Generating Function

If $a_0, a_1, a_2, \ldots$ is an interesting sequence of numbers from which we want to build a generating function $g(t)$, we write

$$\begin{aligned} g(t) &= \sum_{j=0}^{\infty} \frac{a_j}{j!} t^j \\ &= a_0 + a_1 t + \frac{a_2}{2!}t^2 + \frac{a_3}{3!}t^3 + \frac{a_4}{4!}t^4 + \frac{a_5}{5!}t^5 + \frac{a_6}{6!}t^6 + \frac{a_7}{7!}t^7 + \cdots \end{aligned}$$

and then we get

$$a_j = g^{(j)}(0)$$

from this function g that we created!

There are several kinds of generating functions that are interesting to study, e.g., ordinary generating functions, Exponential generating functions, multivariate generating functions, etc. (See, for instance, *Analytic Combinatorics* by P. Flajolet and R. Sedgewick [3] for a comprehensive introduction to generating functions.) In this chapter, we build moment generating functions.

43.2 Moment Generating Functions

The **moment generating function** associated with random variable X, is defined by defining a_j as

$$a_j = \mathbb{E}(X^j),$$

the jth moment of a random variable. Thus, the moment generating function $g(t)$ or $M_X(t)$ associated with X should have the property that

$$M_X(t) = \sum_{j=0}^{\infty} \frac{a_j}{j!} t^j = \sum_{j=0}^{\infty} \frac{\mathbb{E}(X^j)}{j!} t^j.$$

(We use $g(t)$ and $M_X(t)$ interchangeably in the rest of the discussion, since moment generating functions are the only kind of generating function we handle in this book.) The equation above lets us wrap all moments $\mathbb{E}(X^j)$ of a random variables into a compact form. They are divided by $j!$ and then conveniently stored as the coefficients of $M_X(t)$. Since the t has nothing to do with X, i.e., the t is a constant with regard to X, we can pull the t from each term into the expected value, and we get

$$M_X(t) = \sum_{j=0}^{\infty} \frac{\mathbb{E}((tX)^j)}{j!}.$$

The sum of expected values is equal to the expected value of the sum, so this implies

$$M_X(t) = \mathbb{E}\left(\sum_{j=0}^{\infty} \frac{(tX)^j}{j!} \right).$$

Finally, we note that, regardless of the value of X, we always have

$$\sum_{j=0}^{\infty} \frac{(tX)^j}{j!} = e^{tX},$$

so we conclude that the moment generating function associated with X is

$$M_X(t) = \mathbb{E}(e^{tX}).$$

Definition 43.2. Moment generating function The moment generating function $M_X(t)$ of a random variable X is equal to the expected value of e^{tX}:

$$M_X(t) = \sum_{j=0}^{\infty} \frac{\mathbb{E}(X^j)}{j!} t^j = \mathbb{E}\left(\sum_{j=0}^{\infty} \frac{(tX)^j}{j!} \right) = \mathbb{E}(e^{tX}).$$

We know that, as discussed above, for all generating functions of this form, $a_j = g^{(j)}(0)$. In this case, $a_j = \mathbb{E}(X^j)$ and $M_X(t) = \mathbb{E}(e^{tX})$. Thus, if we compute $\mathbb{E}(e^{tX})$, and we then take j derivatives and substitute in $t = 0$, we will get the jth moment of X. This is sometimes referred to as "moment pumping" because it just causes moment after moment of the random variable to be pumped out. One new moment appears each time we take a derivative.

(We emphasize that t is just used as a variable (not a random variable), which has nothing to do with X. It is, in a sense, a dummy variable or a placeholder, that allows the scheme of the generating functions to work properly.)

The formula $M_X(t) = \mathbb{E}(e^{tX})$ gives us the usual, straightforward way to calculate the moment generating function of a random variable. We are just calculating the expected value of the function e^{tX}, and we have lots and lots of experience with calculating the expected value of a function of a random variable. If X is discrete, then

$$M_X(t) = \mathbb{E}(e^{tX}) = \sum_x e^{tx} p_X(x).$$

If X is continuous, then

$$M_X(t) = \mathbb{E}(e^{tX}) = \int_{-\infty}^{\infty} e^{tx} f(x)\, dx.$$

Definition 43.3. Moment generating function: discrete and continuous cases

If X is discrete, the moment generating function of X is

$$\mathbb{E}(e^{tX}) = \sum_x e^{tx} p_X(x).$$

If X is continuous, the moment generating function of X is

$$\mathbb{E}(e^{tX}) = \int_{-\infty}^{\infty} e^{tx} f(x)\, dx.$$

At this point, one might wonder what is so great about moment generating function. For instance, if we want the 2nd moment of a random variable X, we can just compute

$$\mathbb{E}(X^2) = \sum_x x^2 p_X(x),$$

if X is discrete. Alternatively, if X is continuous, we have

$$\mathbb{E}(X^2) = \int_{-\infty}^{\infty} x^2 f(x)\, dx.$$

The trouble, however, is that if we want another moment of X, (say, for instance, the 3rd moment of X), **we have to just turn around and do the whole computation again** (using, for instance, x^3 instead of x^2). So the advantage of using moment generating functions is that, if we compute

$M_X(t) = \mathbb{E}(e^{tX})$ at the start, then we can get *any moment* of X by simply taking derivatives of $M_X(t)$ and then evaluating at $t = 0$. To get the jth moment of X, we just take j derivatives and then evaluate at $t = 0$. Taking derivatives is generally much, much easier than taking sums or evaluating integrals, so the method of moment generating functions has some advantages over the direct calculation of moments. Based on our experience from calculus, we are already good at taking derivatives, so the method of moment generating functions should be appealing.

43.3 Moment Generating Function: Discrete Case

Example 43.4. Consider a Binomial random variable X with parameters n and p. Then the mass of X is

$$p_X(x) = \binom{n}{x} p^x (1-p)^{n-x} \qquad \text{for } x = 0, 1, 2, \ldots, n,$$

and $p_X(x) = 0$ otherwise. So the moment generating function of X is

$$\begin{aligned} M_X(t) &= \mathbb{E}(e^{tX}) \\ &= \sum_{x=0}^{n} e^{tx} \binom{n}{x} p^x (1-p)^{n-x} \\ &= \sum_{x=0}^{n} \binom{n}{x} (e^t p)^x (1-p)^{n-x} \\ &= (e^t p + 1 - p)^n, \end{aligned}$$

where the last line follows from the Binomial theorem, i.e., from

$$\sum_{j=0}^{n} \binom{n}{j} a^j b^{n-j} = (a+b)^n.$$

So the moment generating function of a Binomial random variable X with parameters n and p is

$$M_X(t) = \mathbb{E}(e^{tX}) = (e^t p + 1 - p)^n.$$

Thus, taking one derivative with respect to t, we see that the first moment of X is

$$\begin{aligned} \mathbb{E}(X) = g'(0) &= \frac{d}{dt}(e^t p + 1 - p)^n \bigg|_{t=0} \\ &= (n)(e^t p + 1 - p)^{n-1}(e^t p)\big|_{t=0} \\ &= (n)(p + 1 - p)^{n-1}(p) \\ &= np, \end{aligned}$$

and the second moment of X is

$$\begin{aligned}
\mathbb{E}(X^2) &= g''(0) \\
&= \frac{d^2}{dt^2}(e^t p + 1 - p)^n \bigg|_{t=0} \\
&= \frac{d}{dt}(n)(e^t p + 1 - p)^{n-1}(e^t p) \bigg|_{t=0} \\
&= \Big((n)(n-1)(e^t p + 1 - p)^{n-2}(e^t p)^2 + (n)(e^t p + 1 - p)^{n-1}(e^t p)\Big)\Big|_{t=0} \\
&= (n)(n-1)(p + 1 - p)^{n-2}(p)^2 + (n)(p + 1 - p)^{n-1}(p) \\
&= (n)(n-1)(p)^2 + np.
\end{aligned}$$

So the variance of X is

$$\begin{aligned}
\mathrm{Var}(X) &= \mathbb{E}(X^2) - (\mathbb{E}(X))^2 \\
&= (n)(n-1)(p)^2 + np - (np)^2 \\
&= (n)(-1)(p)^2 + np \\
&= np(1-p).
\end{aligned}$$

All of this agrees with what we computed earlier in the course about Binomial random variables. This technique also enables us to readily calculate $\mathbb{E}(X^j)$ for higher values of j.

Example 43.5. If X is a Poisson random variable with parameter λ, the mass of X is

$$p_X(x) = \frac{e^{-\lambda}\lambda^x}{x!} \qquad \text{for } x = 0, 1, 2, \ldots,$$

and $p_X(x) = 0$ otherwise. So the moment generating function of X is

$$\begin{aligned}
M_X(t) &= \mathbb{E}(e^{tX}) \\
&= \sum_{x=0}^{\infty} e^{tx} \frac{e^{-\lambda}\lambda^x}{x!} \\
&= e^{-\lambda} \sum_{x=0}^{\infty} \frac{(e^t \lambda)^x}{x!} \\
&= e^{-\lambda} e^{(e^t \lambda)} \\
&= e^{((e^t - 1)\lambda)}.
\end{aligned}$$

So the moment generating function of a Poisson random variable X with parameter λ is

$$M_X(t) = \mathbb{E}(e^{tX}) = e^{((e^t - 1)\lambda)}.$$

Thus, the first moment of X is

$$\begin{aligned}\mathbb{E}(X) &= g'(0)\\ &= \frac{d}{dt}e^{((e^t-1)\lambda)}\Big|_{t=0}\\ &= e^{((e^t-1)\lambda)}e^t\lambda\Big|_{t=0}\\ &= e^{(0\lambda)}\lambda\\ &= \lambda,\end{aligned}$$

and the second moment of X is

$$\begin{aligned}\mathbb{E}(X^2) &= g''(0)\\ &= \frac{d^2}{dt^2}e^{((e^t-1)\lambda)}\Big|_{t=0}\\ &= \frac{d}{dt}e^{((e^t-1)\lambda)}e^t\lambda\Big|_{t=0}\\ &= \left(e^{((e^t-1)\lambda)}(e^t\lambda)^2 + e^{((e^t-1)\lambda)}e^t\lambda\right)\Big|_{t=0}\\ &= \left(e^{(0\lambda)}\lambda^2 + e^{(0\lambda)}\lambda\right)\\ &= \lambda^2 + \lambda.\end{aligned}$$

So the variance of X is

$$\begin{aligned}\mathrm{Var}(X) &= \mathbb{E}(X^2) - (\mathbb{E}(X))^2\\ &= \lambda^2 + \lambda - (\lambda)^2\\ &= \lambda.\end{aligned}$$

All of this agrees with what we computed earlier in the course about Poisson random variables. This technique also enables us to readily calculate $\mathbb{E}(X^j)$ for higher values of j.

43.4 Moment Generating Function: Continuous Case

Example 43.6. If X is a Uniform random variable on the interval $[a, b]$, the density of X is

$$f_X(x) = \frac{1}{b-a} \qquad \text{for } a \leq x \leq b,$$

and $f_X(x) = 0$ otherwise. So the moment generating function of X is

$$\begin{aligned} M_X(t) &= \mathbb{E}(e^{tX}) \\ &= \int_{-\infty}^{\infty} e^{tx} f_X(x)\,dx \\ &= \int_a^b e^{tx}\frac{1}{b-a}\,dx \\ &= \frac{1}{b-a}\frac{e^{tx}}{t}\bigg|_{x=a}^{b} \\ &= \frac{e^{tb}-e^{ta}}{(b-a)(t)}. \end{aligned}$$

So the moment generating function of a random variable X that is Uniform on $[a, b]$ is

$$M_X(t) = \mathbb{E}(e^{tX}) = \frac{e^{tb}-e^{ta}}{(b-a)(t)}.$$

We use the series expansion of e^{tb} and e^{ta} because the computation of the derivative with quotient rule (especially for the second moment) becomes quite intricate and much more tedious. The first moment of X is

$$\begin{aligned} \mathbb{E}(X) &= g'(0) \\ &= \frac{d}{dt}\frac{e^{tb}-e^{ta}}{(b-a)(t)}\bigg|_{t=0} \\ &= \frac{1}{b-a}\frac{d}{dt}\Big(\frac{1}{t} + \sum_{j=1}^{\infty}\frac{(tb)^j}{j!t} - \frac{1}{t} - \sum_{j=1}^{\infty}\frac{(ta)^j}{j!t}\Big)\bigg|_{t=0} \\ &\qquad \text{since } e^{tb} = \textstyle\sum_{j=0}^{\infty}\frac{(tb)^j}{j!} \text{ and } e^{ta} = \sum_{j=0}^{\infty}\frac{(ta)^j}{j!} \\ &= \frac{1}{b-a}\frac{d}{dt}\Big(\sum_{j=1}^{\infty}\frac{t^{j-1}b^j}{j!} - \sum_{j=1}^{\infty}\frac{t^{j-1}a^j}{j!}\Big)\bigg|_{t=0} \\ &= \frac{1}{b-a}\Big(\sum_{j=1}^{\infty}\frac{(j-1)t^{j-2}b^j}{j!} - \sum_{j=1}^{\infty}\frac{(j-1)t^{j-2}a^j}{j!}\Big)\bigg|_{t=0} \\ &= \frac{1}{b-a}\Big(\frac{b^2}{2!} - \frac{a^2}{2!}\Big) \qquad \text{only the } j=2 \text{ term remains when } t=0 \\ &= \frac{(b+a)(b-a)}{(b-a)(2)} \\ &= \frac{a+b}{2}, \end{aligned}$$

and the second moment of X is

$$\begin{aligned}
\mathbb{E}(X^2) &= g''(0) \\
&= \frac{d^2}{dt^2}\frac{e^{tb}-e^{ta}}{(b-a)(t)}\bigg|_{t=0} \\
&= \frac{1}{b-a}\frac{d}{dt}\Big(\sum_{j=1}^{\infty}\frac{(j-1)t^{j-2}b^j}{j!}-\sum_{j=1}^{\infty}\frac{(j-1)t^{j-2}a^j}{j!}\Big)\bigg|_{t=0} \\
&= \frac{1}{b-a}\Big(\sum_{j=1}^{\infty}\frac{(j-1)(j-2)t^{j-3}b^j}{j!}-\sum_{j=1}^{\infty}\frac{(j-1)(j-2)t^{j-3}a^j}{j!}\Big)\bigg|_{t=0} \\
&= \frac{1}{b-a}\Big(\frac{(2)(1)b^3}{3!}-\frac{(2)(1)a^3}{3!}\Big) \quad \text{only the } j=3 \text{ term remains, as } t=0 \\
&= \frac{b^3-a^3}{(b-a)(3)} \\
&= \frac{(b-a)(a^2+ab+b^2)}{(b-a)(3)} \\
&= \frac{a^2+ab+b^2}{3}.
\end{aligned}$$

So the variance of X is

$$\begin{aligned}
\mathrm{Var}(X) &= \mathbb{E}(X^2)-(\mathbb{E}(X))^2 \\
&= \frac{a^2+ab+b^2}{3}-\Big(\frac{a+b}{2}\Big)^2 \\
&= \frac{(b-a)^2}{12}.
\end{aligned}$$

All of this agrees with what we computed earlier in the course about Uniform random variables. This technique also enables us to readily calculate $\mathbb{E}(X^j)$ for higher values of j.

Example 43.7. If X is a standard Normal random variable, i.e., with $\mathbb{E}(X)=0$ and $\mathrm{Var}(X)=1$, then (using exp for the Exponential function)

$$f_X(x)=\frac{1}{\sqrt{2\pi}}\exp(-x^2/2) \qquad \text{for all } x.$$

So the moment generating function of X is

$$\begin{aligned} M_X(t) &= \mathbb{E}(e^{tX}) \\ &= \int_{-\infty}^{\infty} e^{tx} f_X(x)\,dx \\ &= \int_{-\infty}^{\infty} e^{tx} \frac{1}{\sqrt{2\pi}} \exp(-x^2/2)\,dx \\ &= \int_{-\infty}^{\infty} \frac{1}{\sqrt{2\pi}} \exp\left(tx - \frac{x^2}{2}\right) dx. \end{aligned}$$

By adding and subtracting t^2 in the numerator, we see that $tx - x^2/2$ can be written as the difference of two squares:

$$tx - \frac{x^2}{2} = \frac{2tx - x^2}{2} = \frac{t^2 - (t^2 - 2tx + x^2)}{2} = \frac{t^2 - (x-t)^2}{2}.$$

Writing things in this way will help us to change variables in the computation above. Returning to our computation,

$$M_X(t) = \int_{-\infty}^{\infty} \frac{1}{\sqrt{2\pi}} \exp\left(\frac{t^2 - (x-t)^2}{2}\right) dx = e^{t^2/2} \int_{-\infty}^{\infty} \frac{1}{\sqrt{2\pi}} \exp\left(\frac{-(x-t)^2}{2}\right) dx.$$

Now we substitute $u = x - t$ and $du = dx$, so that the equation becomes

$$M_X(t) = e^{t^2/2} \int_{-\infty}^{\infty} \frac{1}{\sqrt{2\pi}} \exp\left(\frac{-u^2}{2}\right) du = e^{t^2/2},$$

where the last line follows since $\frac{1}{\sqrt{2\pi}} \exp(\frac{-u^2}{2})$ is exactly the density of a standard Normal random variable, and we get "1" when we integrate it over all u's.

So the moment generating function of a standard Normal random variable X is

$$M_X(t) = \mathbb{E}(e^{tX}) = e^{t^2/2}.$$

We can easily check now that the first moment of X is 0 as follows:

$$\mathbb{E}(X) = g'(0) = \left.\frac{d}{dt} e^{t^2/2}\right|_{t=0} = \left. te^{t^2/2}\right|_{t=0} = 0,$$

and the second moment of X is

$$\mathbb{E}(X^2) = g''(0) = \left.\frac{d^2}{dt^2} e^{t^2/2}\right|_{t=0} = \left.\frac{d}{dt} te^{t^2/2}\right|_{t=0} = \left. t^2 e^{t^2/2} + e^{t^2/2}\right|_{t=0} = 0 + 1 = 1.$$

So the variance of X is

$$\text{Var}(X) = \mathbb{E}(X^2) - (\mathbb{E}(X))^2 = 1 - 0^2 = 1.$$

All of this agrees with what we already knew about X since X is a standard Normal random variable. This technique also enables us to readily calculate $\mathbb{E}(X^j)$ for higher values of j.

Note: The 3rd and 4th moments of a random variable are called the *skewness* and the *kurtosis*, respectively. The skewness measures how "skewed" the distribution is, i.e., whether the density or mass is heavier on one side than the other. The kurtosis measures whether the concentration of the density or mass is tall and narrow, or short and wide.

Example 43.8. If Y is a Normal random variable with parameters $\mathbb{E}(Y) = \mu$ and $\text{Var}(Y) = \sigma^2$, then we note that $\frac{Y-\mu}{\sigma}$ is a standard Normal random variable. Thus, writing $X = \frac{Y-\mu}{\sigma}$, and using the facts about standard Normal random variables from the previous example, we see that the moment generating function of $\frac{Y-\mu}{\sigma}$ is

$$\mathbb{E}(e^{t(Y-\mu)/\sigma}) = \mathbb{E}(e^{tX}) = e^{t^2/2}$$

but we also note that

$$\mathbb{E}(e^{t(Y-\mu)/\sigma}) = \mathbb{E}(e^{(tY/\sigma)-(t\mu/\sigma)}) = e^{-t\mu/\sigma}\mathbb{E}(e^{tY/\sigma})$$

Putting these together, we have

$$e^{-t\mu/\sigma}\mathbb{E}(e^{tY/\sigma}) = e^{t^2/2}$$

or, multiplying on both sides by $e^{t\mu/\sigma}$,

$$\mathbb{E}(e^{tY/\sigma}) = \exp\left(\frac{t^2}{2} + \frac{t\mu}{\sigma}\right)$$

Temporarily replacing t/σ by v and t by σv, this yields

$$\mathbb{E}(e^{vY}) = \exp\left(\frac{(\sigma v)^2}{2} + v\mu\right)$$

and thus the moment generating function of a Normal random variable Y with $\mathbb{E}(Y) = \mu$ and $\text{Var}(Y) = \sigma^2$ is

$$M_Y(v) = \mathbb{E}(e^{vY}) = \exp\left(\frac{(\sigma v)^2}{2} + v\mu\right).$$

We can easily check now that the first moment of Y is μ as follows:

$$\begin{aligned}
\mathbb{E}(Y) &= g'(0) \\
&= \frac{d}{dt}\exp\left(\frac{(\sigma t)^2}{2} + t\mu\right)\bigg|_{t=0} \\
&= \left(\frac{2(\sigma t)(\sigma)}{2} + \mu\right)\exp\left(\frac{(\sigma t)^2}{2} + t\mu\right)\bigg|_{t=0} \\
&= \mu,
\end{aligned}$$

and the second moment of Y is

$$\begin{aligned}
\mathbb{E}(Y^2) &= g''(0) \\
&= \frac{d^2}{dt^2} \exp\Big(\frac{(\sigma t)^2}{2} + t\mu\Big)\Big|_{t=0} \\
&= \frac{d}{dt}\Big(\frac{2(\sigma t)(\sigma)}{2} + \mu\Big) \exp\Big(\frac{(\sigma t)^2}{2} + t\mu\Big)\Big|_{t=0} \\
&= \Big(\Big(\frac{2(\sigma t)(\sigma)}{2} + \mu\Big)^2 \exp\Big(\frac{(\sigma t)^2}{2} + t\mu\Big) + \sigma^2 \exp\Big(\frac{(\sigma t)^2}{2} + t\mu\Big)\Big)\Big|_{t=0} \\
&= \mu^2 + \sigma^2.
\end{aligned}$$

So the variance of Y is

$$\mathrm{Var}(Y) = \mathbb{E}(Y^2) - (\mathbb{E}(Y))^2 = \mu^2 + \sigma^2 - (\mu)^2 = \sigma^2.$$

All of this agrees with what we already knew about Y. This technique also enables us to readily calculate $\mathbb{E}(Y^j)$ for higher values of j.

43.5 Appendix: Building a Generating Function

In this chapter, we claim that if we start with a sequence $a_0, a_1, a_2, \ldots$ of numbers that are interesting to us, and if we if define g by writing

$$\begin{aligned}
g(t) &= \sum_{j=0}^{\infty} \frac{a_j}{j!} t^j \\
&= a_0 + a_1 t + \frac{a_2}{2!} t^2 + \frac{a_3}{3!} t^3 + \frac{a_4}{4!} t^4 + \frac{a_5}{5!} t^5 + \frac{a_6}{6!} t^6 + \frac{a_7}{7!} t^7 + \cdots
\end{aligned}$$

then we get $a_j = g^{(j)}(0)$ from this function g that we created! Now we explain the reasoning behind this phenomenon.

If we evaluate at $t = 0$, we get $g(0)$ on the left hand side, and on the right hand side, we get a_0; all of the other terms on the right-hand-side are just 0; thus

$$g(0) = a_0.$$

If we take one derivative with respect to t, we get

$$g'(t) = a_1 + \frac{a_2}{2!} 2t + \frac{a_3}{3!} 3t^2 + \frac{a_4}{4!} 4t^3 + \frac{a_5}{5!} 5t^4 + \frac{a_6}{6!} 6t^5 + \cdots$$

and then substitute in $t = 0$, we conclude

$$g'(0) = a_1.$$

If we take two derivatives with respect to t, we get

$$g''(t) = \frac{a_2}{2!}(2)(1) + \frac{a_3}{3!}(3)(2)t + \frac{a_4}{4!}(4)(3)t^2 + \frac{a_5}{5!}(5)(4)t^3 + \frac{a_6}{6!}(6)(5)t^4 + \cdots$$

and then substitute in $t = 0$, we conclude

$$g''(0) = a_2.$$

If we take three derivatives with respect to t, we get

$$g'''(t) = \frac{a_3}{3!}(3)(2)(1) + \frac{a_4}{4!}(4)(3)(2)t + \frac{a_5}{5!}(5)(4)(3)t^2 + \frac{a_6}{6!}(6)(5)(4)t^3 + \cdots$$

and then substitute in $t = 0$, we conclude

$$g'''(0) = a_3.$$

More generally, if we take j derivatives with respect to t, we get

$$g^{(j)}(t) = \frac{a_j}{j!}j! + \frac{a_{j+1}}{(j+1)!}(j+1)^{\underline{j}}t + \frac{a_{j+2}}{(j+2)!}(j+2)^{\underline{j}}t^2 + \cdots$$

and then substitute in $t = 0$, we conclude

$$g^{(j)}(0) = a_j.$$

Thus, the jth derivative of $g(t)$, evaluated at $t = 0$, is exactly a_j, as desired.

43.6 Exercises

Exercise 43.1. If X is a Geometric random variable with parameter p (in other words,

$$P(X = x) = p_X(x) = (1-p)^{x-1}p \qquad \text{for } x = 1, 2, 3, \ldots,$$

and $p_X(x) = 0$ otherwise), then find the moment generating function of X.

Exercise 43.2. Use the moment generating function from Exercise 43.1 to verify that, if X is Geometric with parameter p, then $\mathbb{E}(X) = 1/p$.

Exercise 43.3. Now use the moment generating function from Exercise 43.1 to verify that, if X is Geometric with parameter p, then $\mathbb{E}(X^2) = (2-p)/p^2$.

Exercise 43.4. Verify that if X is Geometric with parameter p, then $\text{Var}(X) = q/p^2$.

Exercise 43.5. If X is a Negative Binomial random variable with parameters p and r, then find the moment generating function of X. (Hint: Use the result from Exercise 43.1 as an aid.)

Exercise 43.6. Use the moment generating function from Exercise 43.5 to verify that, if X is Negative Binomial with parameters p, r, then $\mathbb{E}(X) = r/p$.

Exercise 43.7. Now use the moment generating function from Exercise 43.5 to verify that, if X is Negative Binomial with parameters p, r, then $\mathbb{E}(X^2) = r(r+1-p)/p^2$.

Exercise 43.8. Verify that if X is Negative Binomial with parameters p, r, then $\text{Var}(X) = rq/p^2$.

Exercise 43.9. If X is an Exponential random variable with parameter λ, then find the moment generating function of X. (Assume, for this problem, that we are using t in the range where $t < \lambda$.)

Exercise 43.10. Use the moment generating function from Exercise 43.9 to verify that, if X is Exponential with parameter λ, then $\mathbb{E}(X) = 1/\lambda$.

Exercise 43.11. Now use the moment generating function from Exercise 43.9 to verify that, if X is Exponential with parameter λ, then $\mathbb{E}(X^2) = 2/\lambda^2$.

Exercise 43.12. Verify that if X is Exponential with mean $1/\lambda$, then $\text{Var}(X) = 1/\lambda^2$.

Exercise 43.13. If X is a Gamma random variable with parameters λ and r, then find the moment generating function of X. (Hint: Use the result from Exercise 43.9 as an aid. Again, for this problem, that we are using t in the range where $t < \lambda$.)

Exercise 43.14. Use the moment generating function from Exercise 43.13 to verify that, if X is Gamma with parameters λ, r, then $\mathbb{E}(X) = r/\lambda$.

Exercise 43.15. Now use the moment generating function from Exercise 43.13 to verify that, if X is Gamma with parameters λ, r, then $\mathbb{E}(X^2) = r(r+1)/\lambda^2$.

Exercise 43.16. Verify that if X is a Gamma random variable with parameters λ, r, then $\text{Var}(X) = r/\lambda^2$.

Chapter 44

Transformations of One or Two Random Variables

Math is the only place where truth and beauty mean the same thing.
—Danica McKellar

"Obvious" is the most dangerous word in mathematics.
—E. T. Bell

In a toy store, there are balls of many sizes available in a big bin. You randomly select a ball from the bin. How does the size of the radius affect the volume of the ball?

44.1 The Distribution of a Function of One Random Variable

Example 44.1. Choose a ball at random from a large bin at the toy store. Suppose that the radius (in inches) is Uniformly distributed on the interval $[0, 15]$. What is the expected value of the volume of the ball?

We recall that a spherical ball of radius r has volume $\frac{4}{3}\pi r^3$. So if X denotes the radius of the ball, and Y denotes the volume of the ball, then

$$Y = \frac{4}{3}\pi X^3.$$

Unfortunately, Y is *not linearly related* to X (i.e., we do *not* have $Y = aX + b$). So we compute directly

$$\begin{aligned}
\mathbb{E}(Y) &= \mathbb{E}\left(\frac{4}{3}\pi X^3\right) \\
&= \int_0^{15} \frac{4}{3}\pi x^3 \frac{1}{15}\,dx \\
&= \frac{4\pi}{45}\int_0^{15} x^3\,dx \\
&= \frac{4\pi}{45}\frac{15^4}{4} \\
&= 1125\pi \\
&= 3534.291736
\end{aligned}$$

So the expected volume of the ball is 3534.291736 cubic inches.

A different way to approach the problem above is to compute the density of Y. Since $0 \le X \le 15$, then $0 \le \frac{4}{3}\pi X^3 \le \frac{4}{3}\pi 15^3 = 4500\pi$. Again, we emphasize that Y is *not Uniform* on the interval $[0, 4500\pi]$, but we do know that Y cannot be outside this interval. So the density of Y is $f_Y(y) = 0$ for y outside the interval. For Y inside this interval, we have

$$\begin{aligned}
F_Y(y) &= P(Y \le y) \\
&= P\left(\frac{4}{3}\pi X^3 \le y\right) \\
&= P\left(X \le \left(\frac{3y}{4\pi}\right)^{1/3}\right) \\
&= \frac{(\frac{3y}{4\pi})^{1/3} - 0}{15 - 0} \qquad \text{since } X \text{ is Uniformly distributed on } [0, 15] \\
&= \frac{(\frac{3y}{4\pi})^{1/3}}{15}
\end{aligned}$$

Now taking a derivative with respect to Y, we get the density of Y:

$$\begin{aligned}
f_Y(y) &= \frac{d}{dy}F_Y(y) \\
&= \frac{d}{dy}\frac{(\frac{3y}{4\pi})^{1/3}}{15} \\
&= \frac{(1/3)(\frac{3y}{4\pi})^{-2/3}\frac{3}{4\pi}}{15} \\
&= \frac{(\frac{3y}{4\pi})^{-2/3}}{60\pi}
\end{aligned}$$

With the density of Y in hand, we can compute probabilities that involve Y, or we can compute $\mathbb{E}(Y)$, or $\mathrm{Var}(Y)$, etc. We compute

$$\begin{aligned}
\mathbb{E}(Y) &= \int_0^{4500\pi} y f_Y(y)\,dy \\
&= \int_0^{4500\pi} y \frac{\left(\frac{3y}{4\pi}\right)^{-2/3}}{60\pi}\,dy \\
&= \frac{\left(\frac{3}{4\pi}\right)^{-2/3}}{60\pi} \int_0^{4500\pi} y^{1/3}\,dy \\
&= \frac{\left(\frac{3}{4\pi}\right)^{-2/3}}{60\pi} \frac{(4500\pi)^{4/3}}{4/3} \\
&= 1125\pi
\end{aligned}$$

which agrees with the value of $\mathbb{E}(Y)$ we had computed above. The advantage of this method is that we now have the density of Y, so we compute other things if we want to, like other probabilities involving Y, or the variance of Y, etc.

Remark 44.2. How to find the CDF and density of a function of a continuous random variable

If X is a continuous random variable and g is a function, then $Y = g(X)$ is a function of a continuous random variable. Often we know the distribution (e.g., either the density or the CDF) of X but not the distribution of (e.g., either the density or CDF) of Y.

To find the CDF of Y, follow these basic steps:

1. Determine where Y is defined. In other words, find where $f_Y(y) = 0$ and where $f_Y(y) > 0$.
2. Focus on y in the region where Y is defined. Find $F_Y(y) = P(Y \leq y)$ by using the inverse function g^{-1} if one exists, or use a similar technique if g does not have a unique inverse.
3. Differentiate $F_Y(y)$ with respect to y to get the density $f_Y(y)$ of Y.
4. If desired, check to see that $f_Y(y)$ is really a density, i.e., that integrating $f_Y(y)$ over all possible y's gives a result of 1.

Step 1 of Remark 44.2 can be trickier than it sounds.

Example 44.3. Let X be a Uniform random variable on the interval $[-15, 15]$. Let $Y = X^2$. First, we find the density of Y.

The range where X is located is

$$-15 \leq X \leq 15,$$

and thus

$$0 \leq X^2 \leq 15^2 = 225,$$

or in other words,

$$0 \leq Y \leq 225.$$

So the density $f_Y(y)$ of Y is 0 for y outside the interval $[0, 225]$. Notice that, in particular, even though X can be negative, that Y can never be negative, because Y is a "square" of a quantity.

Now consider y in the interval $[0, 225]$. We want to compute

$$P(Y \leq y) = P(X^2 \leq y)$$

or equivalently

$$P(Y \leq y) = P(-\sqrt{y} \leq X \leq \sqrt{y})$$

This line is perhaps unexpected; note that $f(x) = x^2$ does not have a unique inverse, but we have used something like the inverse of f to find the correct range for X.

Since X is Uniformly distributed, it follows that

$$F_Y(y) = P(Y \leq y) = \frac{\sqrt{y} - (-\sqrt{y})}{30} = \frac{2\sqrt{y}}{30}.$$

Differentiating with respect to Y yields

$$f_Y(y) = \frac{y^{-1/2}}{30} \qquad \text{for } 0 \leq y \leq 225,$$

and $f_Y(y) = 0$ otherwise.

We can perform an easy check to see that this is plausible for the density of Y (i.e., that we did not make any mistakes). Of course we could get an integral of "1" and still have a subtle mistake somewhere, but if the integral is exactly 1, then we are relatively sure that we performed the computation correctly. We compute

$$\int_0^{225} \frac{y^{-1/2}}{30}\, dy = \frac{y^{1/2}}{(1/2)(30)}\bigg|_{y=0}^{225} = \frac{(225)^{1/2}}{(1/2)(30)} = \frac{15}{15} = 1$$

With the density of Y in hand, we can compute probabilities; for instance

$$P(Y \leq 64) = \int_0^{64} \frac{y^{-1/2}}{30}\, dy = \frac{y^{1/2}}{(1/2)(30)}\bigg|_{y=0}^{64} = \frac{(64)^{1/2}}{(1/2)(30)} = \frac{8}{15}.$$

We can also compute the expected value of Y:

$$\begin{aligned}
\mathbb{E}(Y) &= \int_0^{225} (y)\left(\frac{y^{-1/2}}{30}\right) dy \\
&= \int_0^{225} \frac{y^{1/2}}{30}\, dy \\
&= \frac{y^{3/2}}{(3/2)(30)}\bigg|_{y=0}^{225} \\
&= \frac{(225)^{3/2}}{(3/2)(30)} \\
&= 75.
\end{aligned}$$

Example 44.4. Let X be a Continuous Uniform $(0, 1)$ random variable.

Let $Y = X^3$. First we find the density of Y. Since $0 < X < 1$, then also $0 < Y < 1$ too. So $f_Y(y) = 0$ for y outside the interval $(0, 1)$. Now consider y inside the interval $(0, 1)$, i.e., consider $0 < y < 1$. We have

$$\begin{aligned}
F_Y(y) &= P(Y \leq y) \\
&= P(X^3 \leq y) \\
&= P(X \leq y^{1/3}) \\
&= \frac{y^{1/3} - 0}{1 - 0} \\
&= y^{1/3}
\end{aligned}$$

So we can differentiate with respect to y to get the density of Y:

$$\begin{aligned}
f_Y(y) &= \frac{d}{dy} F_Y(y) \\
&= \frac{d}{dy} y^{1/3} \\
&= (1/3)y^{-2/3}
\end{aligned}$$

If in doubt, we can test

$$\int_0^1 (1/3)y^{-2/3}\, dy = y^{1/3}\Big|_{y=0}^{1} = 1,$$

so $f_Y(y)$ is a density.

Now we find the expected value of Y:

$$\begin{aligned}
\mathbb{E}(Y) &= \int_0^1 (y)(1/3)y^{-2/3}\,dy \\
&= (1/3)\int_0^1 y^{1/3}\,dy \\
&= (1/3)\left.\frac{y^{4/3}}{4/3}\right|_{y=0}^{1} \\
&= (1/3)\frac{1}{4/3} \\
&= 1/4
\end{aligned}$$

44.2 The Distributions of Functions of Two Random Variables

In the previous section, when one random variable Y is a function of random variable X, i.e., $Y = f(X)$, we investigated how to determine the density of Y from the density of X. In the present section, we push this concept one step further. If U, V are random variables that are each functions of another pair of random variables, X and Y, then we study how to get the joint density $f_{U,V}(u, v)$ of U and V from the joint density $f_{X,Y}(x, y)$ of X and Y. The procedure is relatively straightforward every time, but solving several examples will help improve our understanding.

First consider two random variables X and Y for which the joint density $f_{X,Y}(x, y)$ is known.

Write U as a function of X and Y, i.e.,

$$U = g(X, Y)$$

and write V as a function of X and Y, i.e.,

$$V = h(X, Y).$$

(The functions g and h should already be given, so the above two equations do not require any work! Some examples of functions g and h are given in the examples later in this section.)

Compute four partial derivatives of g and h, with respect to x and y:

$$\frac{\partial}{\partial x} g(x, y), \qquad \frac{\partial}{\partial y} g(x, y), \qquad \frac{\partial}{\partial x} h(x, y), \qquad \frac{\partial}{\partial y} h(x, y)$$

(These partial derivatives are often very straightforward to compute.)

Write the joint density of U and V (we will need to make some adjustments afterwards, so this notation will look strange mathematically, but we remember that x and y depend on u and v):

$$f_{U,V}(u,v) = \frac{f_{X,Y}(x,y)}{\left|\frac{\partial}{\partial x}g(x,y)\frac{\partial}{\partial y}h(x,y) - \frac{\partial}{\partial y}g(x,y)\frac{\partial}{\partial x}h(x,y)\right|}$$

Notice that the left-hand side has u's and v's, but the right hand side has x's and y's (because, as we noted, the x's and y's depend on u and v). So we still need to make a substitution. Before we can use such a result, we will have to solve for X and Y in terms of U and V, and substitute appropriate combinations of u's and v's for every x and y.

Example 44.5. Consider random variables U, V, X, Y such that

$$U = 1 - X - Y$$

and

$$V = X - Y.$$

Suppose that the joint density $f_{X,Y}(x,y)$ of X and Y is known. Find the joint density $f_{U,V}(u,v)$ of U and V.

Notice that, in this example, the new random variables are just linear combinations of the old random variables in this example. (In Example 44.8 we will see a non-linear example.)

We write

$$U = g(X,Y) = 1 - X - Y$$

and

$$V = h(X,Y) = X - Y.$$

Then we compute four partial derivatives of g and h, with respect to x and y:

$$\frac{\partial}{\partial x}g(x,y) = -1, \qquad \frac{\partial}{\partial y}g(x,y) = -1, \qquad \frac{\partial}{\partial x}h(x,y) = 1, \qquad \frac{\partial}{\partial y}h(x,y) = -1$$

So we get

$$\begin{aligned} f_{U,V}(u,v) &= \frac{f_{X,Y}(x,y)}{\left|\frac{\partial}{\partial x}g(x,y)\frac{\partial}{\partial y}h(x,y) - \frac{\partial}{\partial y}g(x,y)\frac{\partial}{\partial x}h(x,y)\right|} \\ &= \frac{f_{X,Y}(x,y)}{|(-1)(-1) - (1)(-1)|} \\ &= \frac{1}{2}f_{X,Y}(x,y) \end{aligned}$$

Finally, we need to substitute u's and v's instead of x's and y's on the right hand side of the equation. This means that—when possible (as in this problem)—we can just solve for the values of x and y in the equations analogous to the definitions of U and V:

$$U = 1 - X - Y$$

and

$$V = X - Y.$$

For instance, in this case, if we add U and V, we get $U + V = 1 - 2Y$, so $Y = \frac{1-U-V}{2}$. If we subtract V from U, we get $U - V = 1 - 2X$, and so $X = \frac{1-U+V}{2}$.

So we conclude that

$$f_{U,V}(u,v) = \frac{1}{2} f_{X,Y}\left(\frac{1-u+v}{2}, \frac{1-u-v}{2}\right).$$

This is a very general result because we did not even need to specify the actual joint density $f_{X,Y}(x,y)$ of X and Y; rather, this solution works regardless of what the joint density is. A second question, however, arises: For which u's and v's is the joint density $f_{X,Y}\left(\frac{1-u+v}{2}, \frac{1-u-v}{2}\right)$ going to be strictly positive? This involves some understanding of the two-dimensional geometry of the problem. So we consider some specific cases below.

Example 44.6. Consider independent random variables X and Y that are each Uniform on the interval $[0,1]$. Suppose that

$$U = 1 - X - Y$$

and

$$V = X - Y.$$

Find the joint density $f_{U,V}(u,v)$ of U and V.

We have already concluded, in Example 44.5 above, that

$$f_{U,V}(u,v) = \frac{1}{2} f_{X,Y}\left(\frac{1-u+v}{2}, \frac{1-u-v}{2}\right).$$

Since X and Y are independent and Uniform on the square where $0 \le X \le 1$ and $0 \le Y \le 1$, then $f_{X,Y}(x,y) = 1$ on this square, and thus the equation

$$f_{U,V}(u,v) = \frac{1}{2} f_{X,Y}\left(\frac{1-u+v}{2}, \frac{1-u-v}{2}\right)$$

reduces to just

$$f_{U,V}(u,v) = \frac{1}{2},$$

but the second question (about the geometry of the problem, which we mentioned above) still remains: Where is the joint density of U and V defined? (Where is the joint density of U and V strictly positive?)

To answer this question, we can step around the four sides of the square where X and Y are defined, and just find the analogous curves for U and V.

Side 1: When $Y = 0$, then $Y = \frac{1-U-V}{2}$ becomes $V = 1 - U$.

Side 2: When $X = 1$, then $X = \frac{1-U+V}{2}$ becomes $V = 1 + U$.

Side 3: When $Y = 1$, then $Y = \frac{1-U-V}{2}$ becomes $V = -1 - U$.

Side 4: When $X = 0$, then $X = \frac{1-U+V}{2}$ becomes $V = U - 1$.

In the plane for U and V, the four pieces fit together at the places indicated in Figure 44.1.

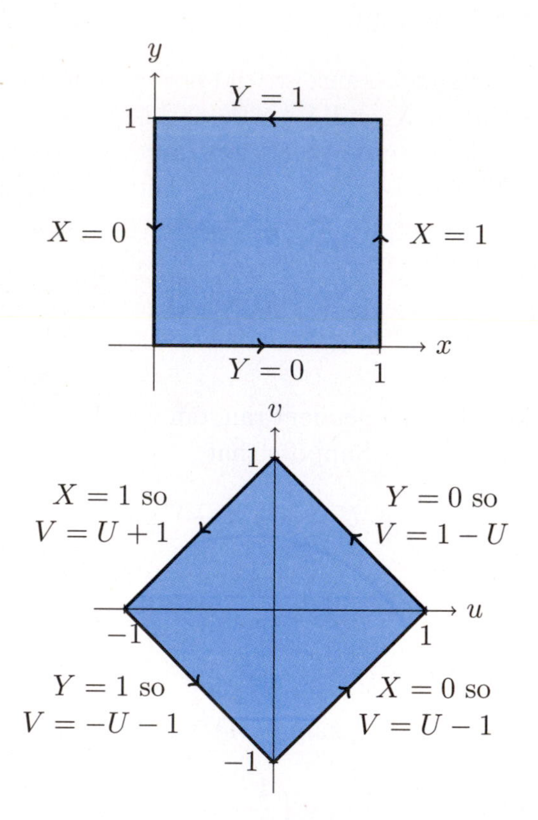

FIGURE 44.1: Transforming the square $0 \leq X \leq 1$, $0 \leq Y \leq 1$ into the U, V plane.

So we see that the pair of random variables (U,V) have a joint density that is Uniform. In fact, $f_{U,V}(u, v) = 1/2$, on a diamond-shaped region of the U, V plane. This diamond is a square tilted on its side, and the diamond has sides of length $\sqrt{2}$, so the diamond has area 2. Any time that a function is constant on a region, then the integral over that region is just the constant times the area.

Thus, in this case, if we integrate $f_{U,V}(u,v) = 1/2$ over the whole diamond, we get $(2)(1/2) = 1$. This check reassures us that the joint density seems to have been calculated correctly, without mistakes.

Example 44.7. Consider independent, standard Normal random variables X and Y. Suppose that

$$U = 1 - X - Y$$

and

$$V = X - Y.$$

Find the joint density $f_{U,V}(u,v)$ of U and V.

Again, we have already concluded, in Example 44.5, that

$$f_{U,V}(u,v) = \frac{1}{2} f_{X,Y}\left(\frac{1-u+v}{2}, \frac{1-u-v}{2}\right).$$

Since X and Y are independent and standard Normal, then the joint density of X and Y is

$$f_{X,Y}(x,y) = \frac{1}{\sqrt{2\pi}} e^{-x^2/2} \frac{1}{\sqrt{2\pi}} e^{-y^2/2}.$$

Thus, the equation

$$f_{U,V}(u,v) = \frac{1}{2} f_{X,Y}\left(\frac{1-u+v}{2}, \frac{1-u-v}{2}\right)$$

reduces to

$$\begin{aligned}
f_{U,V}(u,v) &= \frac{1}{2}\frac{1}{\sqrt{2\pi}} \exp\left(-\left(\frac{1-u+v}{2}\right)^2 /2\right) \frac{1}{\sqrt{2\pi}} \exp\left(-\left(\frac{1-u-v}{2}\right)^2 /2\right) \\
&= \frac{1}{2}\frac{1}{\sqrt{2\pi}}\frac{1}{\sqrt{2\pi}} \exp\left(-\frac{\left(\frac{1-u+v}{2}\right)^2}{2} - \frac{\left(\frac{1-u-v}{2}\right)^2}{2}\right) \\
&= \frac{1}{2}\frac{1}{\sqrt{2\pi}}\frac{1}{\sqrt{2\pi}} \exp\left(-(u-1)^2/4 - v^2/4\right) \\
&= \frac{1}{\sqrt{(2\pi)(2)}} e^{-(u-1)^2/4} \frac{1}{\sqrt{(2\pi)(2)}} e^{-v^2/4}
\end{aligned}$$

Thus U and V are *independent* Normal random variables. In fact, $\mathbb{E}(U) = 1$ and $\text{Var}(U) = 2$, and also $\mathbb{E}(V) = 0$ and $\text{Var}(V) = 2$; we can read these facts directly from the setup in the last line of the equation above. In other words, we can tell that U and V are independent because $f_{U,V}(u,v)$ factors into a product for which the first part has only u's and the second part has only v's.

To put the power of the methods above into their proper context, we point out that without these methods, we still could also have determined

$$\mathbb{E}(U) = \mathbb{E}(1 - X - Y) = 1$$

and

$$\text{Var}(U) = \text{Var}(1 - X - Y) = \text{Var}(X) + \text{Var}(Y) = 2$$

and

$$\mathbb{E}(V) = \mathbb{E}(X - Y) = 0$$

and

$$\text{Var}(V) = \text{Var}(X - Y) = \text{Var}(X) + \text{Var}(Y) = 2$$

by simply using the equations $U = 1 - X - Y$ and $V = X - Y$ and by using the knowledge that X, Y are independent. On the other hand, we emphasize that we *could not have determined*, a priori, that U and V are independent. To know that U and V are independent, this calculation of the joint density of U and V was absolutely necessary.

One more quick comment about the domain where the joint density of the pairs of random variables is defined: Since the joint density of X and Y is defined throughout the X, Y plane, then also the joint density of U and V is defined throughout the U, V plane too.

Example 44.8. Consider random variables U, V, X, Y such that

$$U = XY$$

and

$$V = X/Y.$$

Suppose that the joint density $f_{X,Y}(x, y)$ of X and Y is known. Find the joint density $f_{U,V}(u, v)$ of U and V.

Notice that the new random variables are not just linear combinations of the old random variables in this example.

We write

$$U = g(X, Y) = XY$$

and

$$V = h(X, Y) = X/Y.$$

Then we compute four partial derivatives of g and h, with respect to x and y:

$$\frac{\partial}{\partial x} g(x, y) = y, \quad \frac{\partial}{\partial y} g(x, y) = x, \quad \frac{\partial}{\partial x} h(x, y) = 1/y, \quad \frac{\partial}{\partial y} h(x, y) = -xy^{-2}$$

So we get

$$f_{U,V}(u,v) = \frac{f_{X,Y}(x,y)}{\left|\frac{\partial}{\partial x}g(x,y)\frac{\partial}{\partial y}h(x,y) - \frac{\partial}{\partial y}g(x,y)\frac{\partial}{\partial x}h(x,y)\right|}$$
$$= \frac{f_{X,Y}(x,y)}{|(y)(-xy^{-2}) - (x)(1/y)|}$$
$$= \frac{y}{2x}f_{X,Y}(x,y)$$

Finally, we need to substitute u's and v's instead of x's and y's on the right hand side of the equation. The fraction $\frac{y}{2x}$ is just $\frac{1}{2v}$. Also $uv = x^2$, so $\sqrt{uv} = x$; and $u/v = y^2$, so $\sqrt{u/v} = y$.

So we conclude that

$$f_{U,V}(u,v) = \frac{1}{2v}f_{X,Y}(\sqrt{uv}, \sqrt{u/v}).$$

Again, this is very general, so we will consider a specific case below.

Example 44.9. Consider independent random variables X and Y that are each Uniform on the interval $[0,1]$. Suppose that

$$U = XY$$

and

$$V = X/Y.$$

Find the joint density $f_{U,V}(u,v)$ of U and V.

As before, we first refer back to the general situation, discussed in Example 44.8 above. This establishes the fact that

$$f_{U,V}(u,v) = \frac{1}{2v}f_{X,Y}(\sqrt{uv}, \sqrt{u/v}).$$

Since X and Y are independent and Uniform on the square where $0 \leq X \leq 1$ and $0 \leq Y \leq 1$, then $f_{X,Y}(x,y) = 1$ on this square, and thus the equation

$$f_{U,V}(u,v) = \frac{1}{2v}f_{X,Y}(\sqrt{uv}, \sqrt{u/v}).$$

reduces to just

$$f_{U,V}(u,v) = \frac{1}{2v}.$$

but, again, one question remains: Where is the joint density of U and V defined?

To answer this question, we can step around the four sides of the square where X and Y are defined, and just find the analogous curves for U and V.

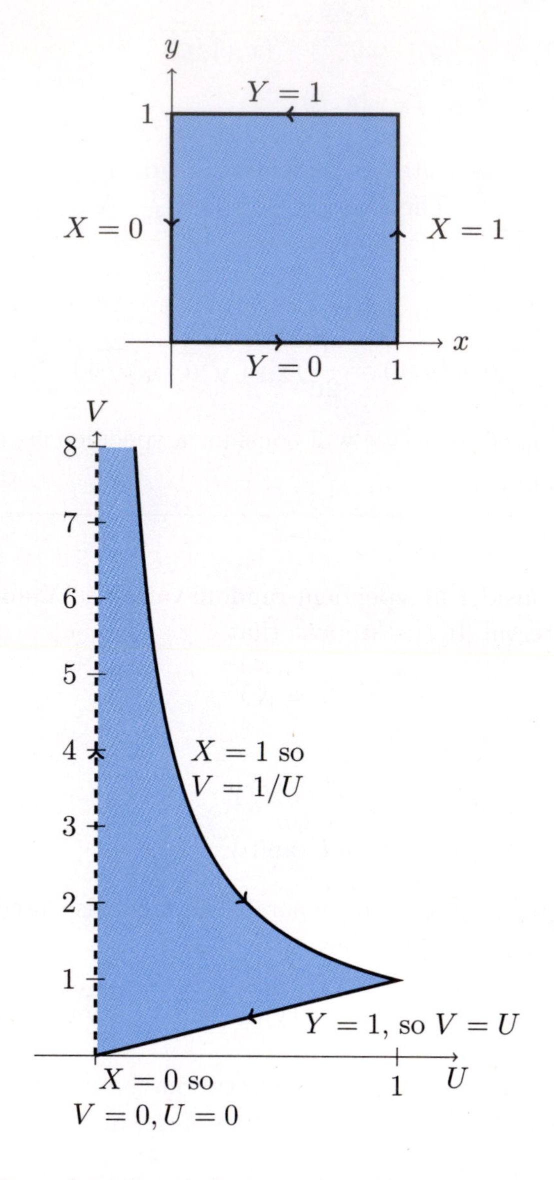

FIGURE 44.2: Transforming the square $0 \leq X \leq 1$, $0 \leq Y \leq 1$ into the U, V plane.

Side 1: When $Y = 0$, then $U = XY = 0$ and $V = X/Y$ is undefined. So for fixed $Y = c$ where c is small and near 0, we see that $U = Xc$, and $V = X/c$, so $V = U/c^2$ (e.g., for $Y = 1/100$, we have $V = 10{,}000\,U$). So in the U, V plane, the lines near $Y = 0$ are arbitrarily close to the line $U = 0$ and $V > 0$ (shown as a dashed line in Figure 44.2).

Side 2: When $X = 1$, then $U = XY = Y$ and $V = X/Y = 1/Y$, so $V = 1/U$.

Side 3: When $Y = 1$, then $U = XY = X$ and $V = X/Y = X$, so $V = U$.

Side 4: When $X = 0$, then $U = XY = 0$ and $V = X/Y = 0$.

In the plane for U and V, the four pieces fit together at the places indicated in Figure 44.2. The side $X = 0$ gets completely compressed into the origin.

In this example, the joint density of U and V is not Uniform. We can still perform a check to make sure that we actually have a density. We integrate $f_{U,V}(u, v) = 1/(2v)$ over all u's and v's depicted in Figure 44.2, and we check that we get 1. To see this, we compute

$$\begin{aligned} \int_0^1 \int_u^{1/u} \frac{1}{2v}\,dv\,du &= \frac{1}{2}\int_0^1 \ln(v)\big|_{v=u}^{1/u}\,du \\ &= \frac{1}{2}\int_0^1 (\ln(1/u) - \ln(u))\,du \\ &= \int_0^1 -\ln(u)\,du, \end{aligned}$$

and using integration by parts, we conclude

$$\int_0^1 -\ln(u)\,du = (u - u\ln(u))\big|_{u=0}^{1} = 1.$$

So this joint density integrates to 1 over the region where it is defined, which helps verify that everything was calculated correctly. We strongly encourage the use of such pictures and double-checks of the transformed joint density.

44.3 Exercises

44.3.1 Practice

Exercise 44.1. Bottle volume. A certain type of cylindrical bottle always has height 14 cm. During the manufacturing process however, the radius of the bottom is Uniformly distributed between 2.3 cm and 2.7 cm. (Whatever radius is chosen for the bottom is automatically given to the rest of the cylinder too. Once the bottom is fixed, the whole can is manufactured that way.) What is the probability that the bottle has a volume of less than 275 cm^3?

Exercise 44.2. Gift box. A gift is chosen at random, with the price Uniformly distributed between \$5 and \$12. Each gift needs (additionally) to be placed in a box that costs \$2. What is the expected total price of the gift and its box?

Exercise 44.3. Fabric size. A customer at the fabric store buys fabric that is 40 inches wide, i.e., 1.11 yards wide. The length is cut by the employee at the store. When she is asked to cut 1 yard of fabric, the actual length is Uniformly distributed between 0.87 and 1.05 yards. What is the probability that the entire piece of fabric has area 1 square yard or larger?

Exercise 44.4. Chess board. You want to buy a square chess board from a local artists' collective. Since each chessboard is uniquely handcrafted and you didn't bring your ruler, you are not sure of the exact dimensions. Let X be the length of the side of one of the boards, and assume that X is Uniformly distributed between 12 and 18 inches.

a. What is the expected area of the board that you buy?

b. What is the standard deviation of the area of the board?

Exercise 44.5. Summer temperature. Suppose that the temperature X in summer, given in Kelvin, is Normally distributed with mean 300K and standard deviation 10K. Let Y be the same temperature, given in degrees Fahrenheit. So

$$Y = \frac{9}{5}(X - 273) + 32.$$

Find the probability that the temperature is higher than 90 degrees Fahrenheit.

Exercise 44.6. Phone bill. Lucas calls his girlfriend Margaret every day on his cell phone. The time X, in hours, that they talk in a day is Uniformly distributed between 0 and 2. Lucas pays \$2 per hour that he is on the phone, plus a flat rate of \$5 per day. So $Y = 2X + 5$ is the size of his bill per day.

a. Find the probability that he spends more than \$6 on a given day for his cell phone service.

b. Find the expected amount that he spends on a given day for his cell phone service.

c. Find the standard deviation of the amount that he spends on a given day for his cell phone service.

Exercise 44.7. Cookie dough. If the amount of cookie dough, X, used in a cookie is Uniformly distributed between 1.7 and 2.6 tablespoons, then the height Y of the cookie is

$$Y = \frac{3}{10}X^{1/3}.$$

Find $P(Y > 0.4)$.

Exercise 44.8. **Cheeseburger price.** Sally owns a burger restaurant. The cost of a pound of beef is $3.75. Each cheeseburger contains an amount of beef that is Uniformly distributed between 0.23 and 0.27 pounds. Sally includes a slice of cheese on each burger, for 35 cents, and the bun itself costs 33 cents. Burgers are listed on the dollar menu. Assuming that the only costs of making a burger are the beef, cheese, and bun, what is the probability that the costs exceed the price of the burger?

44.3.2 Extensions

Exercise 44.9. Let X be a random variable that is Uniformly distributed on the interval $[0, \pi/2]$. Find the expected value of $Y = \sin X$.

Exercise 44.10. Generalize Example 44.4 as follows:

Let X be a Uniform random variable on the interval $(0, 1)$, i.e., X is Uniformly distributed with $0 < X < 1$. Let $Y = X^n$ where n is any positive integer. Find $\mathbb{E}(Y)$.

Exercise 44.11. Consider an Exponential random variable X with parameter $\lambda > 0$.

Is it always true that, if a and b are positive constants, then $Y = aX + b$ is an Exponential random variable too?

If your answer is "yes," then give a justification (e.g., give an argument in favor).

If your answer is "no," then give a very concrete counterexample (e.g., for at least one specific a and b of your choice, show that $Y = aX + b$ is not Exponential).

44.3.3 Advanced

Exercise 44.12. Consider the scenario in which

$$U = g(X, Y) = X^2$$

and

$$V = h(X, Y) = X + Y.$$

Suppose the joint density $f_{X,Y}(x, y)$ of X and Y is Uniform on the square where $0 \le X, Y \le 1$. In other words,

$$f_{X,Y}(x, y) = 1 \qquad \text{if } 0 \le x \le 1 \text{ and } 0 \le y \le 1,$$

and $f_{X,Y}(x, y) = 0$ otherwise.

a. Find

$$\frac{\partial}{\partial x} g(x, y), \qquad \frac{\partial}{\partial y} g(x, y), \qquad \frac{\partial}{\partial x} h(x, y), \qquad \frac{\partial}{\partial y} h(x, y).$$

b. Now write the joint density of U and V:

$$f_{U,V}(u,v) = \frac{f_{X,Y}(x,y)}{\left|\frac{\partial}{\partial x}g(x,y)\frac{\partial}{\partial y}h(x,y) - \frac{\partial}{\partial y}g(x,y)\frac{\partial}{\partial x}h(x,y)\right|}$$

using the values derived above, and using the fact that $f_{X,Y}(x,y) = 1$ in the region where X, Y is defined. Make sure that, in your expression of $f_{U,V}(u,v)$, you convert all x's and y's to u's and v's in an appropriate way.

Exercise 44.13. Returning to Exercise 44.12, we still need to know the region where U and V can occur in the U, V plane, in other words, we need to identify the region where the density $f_{U,V}(u,v)$ is relevant.

a. Find equations for the curves in the U, V plane that correspond to the four lines in the X, Y plane around the box $0 \leq X \leq 1$ and $0 \leq Y \leq 1$.

b. Draw the region in the U, V plane where the joint density $f_{U,V}(u,v)$ is of interest.

Exercise 44.14. Check your calculations in Exercises 44.12 and 44.13 by making sure that you get the usual "1" when you integrate the joint density $f_{U,V}(u,v)$ from Exercise 44.12, over *all values* u and v in the picture from Exercise 44.13.

Exercise 44.15. Are the variables U and V from the Exercises 44.12, 44.13, and 44.14 dependent or independent? Justify your answer.

Chapter 45

Review Questions for All Chapters

Exercise 45.1. Rolling dice. A player rolls a huge bag of 200 dice. Approximate the probability that 40 or more 1's appear.

Exercise 45.2. Dice game. In a certain game, a player wants to roll five dice and get as many 1's as possible. Here is the scheme:

Round 1: The player rolls all five of the dice, and notices how many 1's appear.

Round 2: The player sets the 1's aside and only rolls the dice which did not show a 1 the first time.

Round 3: The player sets the 1's aside from rounds 1 and 2 and only rolls the dice which still did not yet show a 1.

After three rounds, the player is tired so she stops. How many 1's does she expect to have at this point?

(Hint: Consider one Bernoulli for each of the dice, so $\mathbb{E}(X_1 + \cdots + X_5) = \mathbb{E}(X_1) + \cdots + \mathbb{E}(X_5)$.)

Exercise 45.3. Let X and Y be independent continuous random variables that are each Uniformly distributed in the interval $[0, 30]$. Let Z denote the larger of X and Y; in other words, $Z = \max(X, Y)$.

a. Find the cumulative distribution function $F_Z(a) = P(Z \leq a)$ of the random variable Z.

b. Find the density $f_Z(z)$ of Z.

c. Find the expected value $\mathbb{E}(Z)$ of Z.

Exercise 45.4. Cold drinks. There are many kinds of drinks in a large cooler in the cafeteria. Fifteen percent of them are decaffeinated. After class (five days a week), Alice always needs a drink. She is always running to her next class, and the cafeteria is busy, so never has time to pick a specific one. She likes caffeine, so she will be "happy" if she gets caffeine four or more times during the five day week. What is the probability that Alice is happy this week?

Exercise 45.5. Snowfall. The amount of snow during the winter in a certain town follows a Normal distribution with mean 15 inches and standard deviation 4 inches. What is the probability that the snow is more than 7 inches during the winter?

Exercise 45.6. Texas Hold 'Em. Suppose one is dealing a hand of Texas Hold 'Em poker at a standard 9-man table (each of the 9 people receives 2 cards, and there are 5 additional community cards placed face-up in front of the dealer). What is the probability that, after these 23 cards have been dealt, all 4 of the aces have been dealt?

Exercise 45.7. Tennis players. There are 50 people in a tennis club. In their professional lives, 25 of the people are accordion players, 15 are basketball players, and 10 are carrot farmers. (People tend to register for the tennis club in groups since they are friends.) The club is selecting 10 people at random to go on a tennis tour around the world.

a. What is the probability that exactly 5 carrot farmers are picked?

b. What is the expected number of accordion players to be picked?

Exercise 45.8. Waiting for a bus. A passenger is sitting at the mall waiting for the bus to arrive. The expected waiting time is 30 minutes, i.e., half an hour. Let X be waiting time (in minutes) until the bus arrives. What is the probability that the waiting time is more than 20 minutes?

Exercise 45.9. Laundry room. While doing laundry, it seems that each student spends between 3 to 15 minutes in the laundry room per week, and the distribution for each student is assumed to be Uniform on this interval.

a. How long is each student expected to be in the laundry room?

b. What is the variance of the time a student spends in the laundry room?

c. Given that a student has already been in the laundry room for 7 minutes, how long is the student's expected total stay?

d. During a seven day week, in a dorm with 1000 students who each do laundry once, how many minutes do we expect the students to spend altogether in the laundry room (i.e., what is the expected value of the sum of times, with the sum over all 1000 students).

e. What is the approximate probability that, among the 1000 students, the average time spent in the laundry room (per student) exceeds 9.1 minutes during a week? Assume that the students' times are independent.

Exercise 45.10. Flight time. The flight time of a plane flight from Denver to New York City is Normally distributed, with average flight time of 3 hours and 16 minutes, and standard deviation of 30 minutes. What is the probability that such a flight takes 4 hours or longer?

Exercise 45.11. Checkmate. When presented with a certain position in a chess game, and asked whether a checkmate is possible from that position, only 29% of people will understand the method to obtain a checkmate. If people are surveyed until finding the first person who sees the potential to make a checkmate from such a position, how many people do we expect to need to survey, until such an intelligent player is found?

Exercise 45.12. Presents. In an affluent suburban home, a certain child expects to get 35 presents for Christmas. Give an upper bound on the probability that they receive 40 or more presents.

Exercise 45.13. Chalk length. The length of a randomly selected piece of chalk is Uniformly distributed between 0 to 1.5 inches. Suppose that there are 10 pieces of chalk in a classroom. How many pieces of chalk do we expect to be an inch or longer?

Exercise 45.14. Waiting for a bus. Joe finds that, when he waits for a bus, his waiting time is Exponential, with an average waiting time of 6 minutes. Assume that he waits for one bus in the morning and again for one bus in the evening.

a. What is the probability that he spends 15 minutes or less at the bus stop altogether during the day?

b. What is his expected waiting time at the bus stop (mornings and evenings included) during a 20 day period?

c. What is the approximate probability that he spends 200 minutes or less at the bus stop during a 20 day period (again, mornings and evenings included)?

Exercise 45.15. Artichokes. People shopping at the grocery store are interviewed to see whether they enjoy artichokes. Only 11% of people like artichokes.

a. How many people does the interviewer expect to meet until finding the 25th person who likes artichokes?

b. What is the variance of the number of people he meets, to find this 25th person who likes artichokes?

Exercise 45.16. Prison escapes. There are 10,000 prison inmates in a certain state. Independently of each other, and independent of their behavior on previous days, assume that, on a given day, a prisoner has probability $p = 0.000001$ of escaping.

a. What is the probability, on a given day, that none of the prisoners escapes?

b. What is the probability that the state has n consecutive days without any escapes?

c. Let X be the number of days until the next escape occurs. What distribution does X have?

d. How many days does it take until we are at least 99% sure that at least one prisoner has escaped in the state?

e. Focus attention on a specific criminal. What is the probability that he escapes during a 50-year period?

f. If 10,000 prisoners are imprisoned, each for 50 years, how many prisoners do we expect to escape?

g. There is a very intelligent criminal named Scar-Foot. Instead of having probability of 0.000001 of escaping each day, he has probability 0.0001 of escaping each day. What is the probability that Scar-Foot is able to escape during a 50-year period?

Exercise 45.17. Typing monkeys. A group of scientists is testing the learning ability of monkeys. They want to see if monkeys have any language preferences. They acquire a monkey that has lived its entire life in Moscow, and they put the monkey in front of a Russian keyboard, which contains 33 characters.

a. They want to see if the monkey can accurately type the entire first line of a famous poem written by Alexander Pushkin:

Я помню чудное мгновенье

The poem has 21 characters in this first line (the scientists do not take spaces into account at all). What is the probability that the monkey types all 21 characters, in order, correctly? (He just types the keys independently, at random.)

b. Next, the scientists put the monkey in front of a 26-character English keyboard and see if he can type the phrase in English. It happens to also take 21 characters when written in English: "A MAGIC MOMENT I REMEMBER" (again, they do not take spaces into account). What is the probability that the monkey types all 21 characters, in order, correctly?

c. What is the ratio of these two probabilities, i.e., how much more likely is the monkey to type the 21 character phrase correctly in English versus in Russian? Give the ratio of the two probabilities that were calculated in the previous two parts of this problem.

Exercise 45.18. Cheap boots. A company has a strange policy: It sells inexpensive boots, but it only ships one boot at a time, and it does not tell the buyer whether he will receive a left or right boot (hence, the need to sell the boots for an inexpensive price, to still attract buyers). How many boots should a person buy, to be at least 95% sure of ending up with a complete pair?

Exercise 45.19. Dead pixels. A high-quality computer company sells monitors which rarely have a dead pixel. The number of dead pixels per monitor has a Poisson distribution with average 0.05 per monitor. If the company sells 30 monitors, and the distribution of pixels is independent from monitor to monitor, what is the distribution of the total number of dead pixels found on all 30 monitors altogether?

Exercise 45.20. Candies. In a large bag of 40 Starburst candies, there are 8 orange, 9 yellow, 12 red, and 11 pink. You only like orange and red. If you take 8 from the bag, what is the probability that at least 5 out of the 8 are ones you like (i.e., there are 5 or more reds and oranges altogether)?

Exercise 45.21. Study time. Let X be the time (in hours) that Stephen spends studying on one particular day during "dead week" (the nickname for the week before final exams). Then X has density $\frac{1}{4}e^{-x/4}$. Let Y be the number of hours that Stephen spends studying for all 7 days during "dead week." Assume that the time spent studying on distinct days is independent.

a. Find $\mathbb{E}(Y)$.

b. Find $\text{Var}(Y)$.

Exercise 45.22. Walk time. If the walking time to the dining hall is Exponential with mean 8 minutes, what is the probability that it takes 15 or more minutes to get there?

Exercise 45.23. Solitaire. A student wins at solitaire about 17% of the time. How many games does the student expect to play until winning her 3rd game?

Exercise 45.24. Exam scores. On a 50-question Geography exam, the average score is 25.5 out of 50. The standard deviation of the score is 8. Find a bound on the probability that a randomly selected student's score is greater than 42 or less than 9.

Exercise 45.25. Couples. Consider n pairs of husbands and wives, sitting randomly in a row of $2n$ chairs. What is the probability that each person is sitting beside her/his spouse, and also no two women are adjacent, and no two men are adjacent (i.e., the sexes are alternating)?

Exercise 45.26. Flower arranging. Each student in a flower arranging class gets to take home a house plant. Each student has a variety of plants to choose from, and the selections are made independently (in particular, there is an ample number of each kind of plant). There is a 20% chance that, when a student selects her/his plant, the choice will be a peperomia. If there are 60 students in the class, what is the probability that exactly 14 of the students will take home a peperomia?

Exercise 45.27. Garden mole. My garden is a square plot of land that measures 10 feet by 10 feet, and it has a mole living somewhere inside. The location of the mole is Uniformly distributed in the garden. I have a small tomato area that is a 2 feet by 2 feet square region. What is the probability that the mole is found in the tomato patch?

Exercise 45.28. Evil scientist. An evil scientist has devised an experiment that takes one hour to run to completion. It has a $1/12$ chance of causing the world to explode. Otherwise, it fails and the evil scientist immediately resets the experiment and tries again! (Her attempts are independent and each have probability $1/12$ of succeeding.) Knowing that a nearby hero will find the evil scientist and foil her plan after 8 hours have passed, what is the probability that the evil scientist will succeed before the hero catches her, i.e., that the evil scientist will have a success within her first 8 attempts?

Exercise 45.29. Exam questions. On a probability exam, each of the n questions are created independently of the others. Each question has a ten percent chance of involving a Geometric random variable.

a. How many questions on the exam are expected to have a Geometric random variable involved?

b. What is the variance of the number of questions that have a Geometric random variable?

Exercise 45.30. Take-home exam. Consider a student who starts a take-home exam at midnight and expects to take 8 hours to finish the exam. If the student has 34 hours remaining until the exam must be completed, give a bound on the probability that the student is able to complete the exam within the 34 hours until it is due for submission.

Exercise 45.31. Pillows. A girl buys pillows to decorate her apartment living room. There are square and circular pillows in stock at the store where she is shopping. There are a total of 400 pillows available:

- 50 are square, with down feathers;
- 250 are square, with cotton fluff;
- 20 are circular, with down feathers;
- 80 are circular, with cotton fluff.

What is the probability that, if she grabs a square pillow at random, it is stuffed with cotton fluff?

Exercise 45.32. Auditions. A string quartet is being assembled. Each player that auditions is only likely to be successful at the audition with probability 27%. How many musicians are expected to audition until 4 of them are selected for the quartet?

Exercise 45.33. Basketball shooting. A basketball player has been developing his skill during the season, and he now scores 75% of his free throws successfully. If he shoots 200 free throws this season, how many points should he be expected to contribute? (Each free throw is worth 1 point.)

Exercise 45.34. Online dating. Approximately 1 in 4 relationships begin online. Suppose that we interview couples until we find 10 couples who met online.

a. What is the expected number of couples we will need to interview?

b. What is the variance of the number of couples we will need to interview?

Exercise 45.35. Biased coin. Consider a sequence of independent flips of a biased coin. The coin is a Head or Tail with probabilities p or $q := 1 - p$, respectively.

Let X be the smallest j such that the jth and $(j+1)$st flips are both Heads.

Let Y be the smallest j such that the jth flip is a Head and the $(j+1)st$ flip is a Tail.

Let Z be the smallest j such that the jth flip is a Tail and the $(j+1)$st flip is a Head.

a. Find $P(X < Y)$.

b. Find $P(X < Z)$.

Exercise 45.36. Deck of cards. A deck is thoroughly shuffled, and 5 cards are chosen, without replacement. What is the probability that the selection of 5 cards contains at least one card from each of the 4 suits?

Exercise 45.37. Let X, Y be independent exponential(1) random variables.

a. Compute the conditional probability that $X < \ln 2$, given $X - Y \geq 0$. In other words, find

$$P\,(X < \ln 2 \mid X - Y \geq 0)\,.$$

b. Compute the conditional probability that $X < \ln 2$, but now given $X - Y$ is exactly 0. In other words, find

$$P\,(X < \ln 2 \mid X - Y = 0)\,.$$

Answers to Exercises

Chapter 1

1.1 Answers will vary. E.g., (1) $(0.25, 0.5)$;
(2) $\{(x, y) \mid 0.2 < x < 0.5,\ 0.36 < y < 0.42\}$; (3) $\{(x, y) \mid x^2 + y^2 \leq 1\}$;
1.3 Answers will vary. E.g., (1) Chris grabs lemon-lime, lemon-lime, orange;
(2) $\{(x_1, \ldots, x_4, \text{orange}) \mid$ each of $x_1, \ldots, x_4$ is lemon-lime or fruit punch$\}$;
(3) $\{(x_1, \ldots, x_j, \text{orange}) \mid 0 \leq j \leq 12;$ the x_j's are lemon-lime or fruit punch, with ≤ 6 of each$\}$;
1.5 $\{(x_1, x_2, \ldots, x_{75}) \mid x_1 + x_2 + \cdots + x_{75} \leq 400\}$;
1.7 a. 6; $S = \{(rrbb), (rbrb), (rbbr), (bbrr), (brbr), (brrb)\}$; b. $2^6 = 64$;
1.9 a. $\{(x_1, x_2, x_3) \mid x_j \in \mathbb{R}^{>0}\}$; b. $\{(x_1, y_1, x_2, y_2, x_3, y_3) \mid x_j, y_k \in \mathbb{R}^{>0}\}$;
c. $\{(x_1, y_1, x_2, y_2, x_3, y_3) \mid x_j, y_k \in \mathbb{R}^{>0}; x_1 < x_2 < x_3; y_1 > y_2 > y_3\}$;
1.11 $S = \{(x, y) \mid x, y \in \mathbb{R}^{\geq 0};\ x + y \leq 2\}$;
1.13 28;
1.15

j	3	4	5	6	7	8	9	10	11	12	13	14	15	16	17	18
$\lvert A_j \rvert$	1	3	6	10	15	21	25	27	27	25	21	15	10	6	3	1

Chapter 2

2.1 a. $330/27333 = 0.0121$; b. $9153/27333 = 0.3349$; c. No. The table gives a partition of the songs, i.e., each song is only included in one genre and is therefore only counted once. The genres are disjoint subsets of the sample space; d. $18180/27333 = 0.6651$;
2.3 0.83; **2.5** a. 13/30; b. 17/30; c. Answers will vary. E.g., Partition the shoes into those that are black and brown and those that are another color;
2.7 a. $10.222, 24.799, 5.000, 63.1, 30$ inches of rain; b. $\{x \mid x \geq 0\}$; c. Answers will vary. E.g., partition the sample space into those values which are less than the average annual rainfall and those which are greater than or equal to the average; d. Each possible outcome is in exactly one part of the partition.
2.9 a. 5/12; b. 3/4; c. 1/2; **2.11** $(1/101)^7$;
2.13 a. 0.01; b. 0.1; c. 0.11; d. 0.89; **2.15** a. 2/3; b. 1/6;
2.17 $P\{(P)\} = 1/3$; $P\{(G, P), (Y, P)\} = 1/3$; $P\{(G, P), (G, Y, P)\} = 1/3$; $P\{(Y, G, P), (G, Y, P)\} = 1/3$; $P\{(P), (Y, P)\} = 1/2$;
2.19 920; **2.21** a. 2^7; b. 21; c. $\binom{7}{j}$; d. $P(A_j) = \binom{7}{j}(1/2)^7$; **2.23** 0.47;
2.27 a. If A and B are disjoint, then equality is true. Otherwise, A and B will have an intersection, so $P(A \cup B) = P(A) + P(B) - P(A \cap B) < P(A) + P(B)$;

b. Yes. $P(A\cup B\cup C) = P(A)+P(B\setminus A)+P(C\setminus(A\cup B)) \le P(A)+P(B)+P(C)$.
2.29 $15/216 = 5/72 = 0.0694$; 2.31 $P(A_0) = 1/6$, $P(A_1) = 2/3$, $P(A_2) = 1/6$;
2.33 $P(B_k) = (k/6)^3$

Chapter 3

3.1 0.622; 3.3 58/125; 3.5 a. 63/64; b. 27/64; 3.7 Yes. 3.9 0.7273 or 8/11;
3.11 Yes. E.g., $P(A) = 0.7$, $P(B) = 0.4$, $P(A \cap B) = 0.28$, $P(A \cup B) = 0.82$

Chapter 4

4.1 $P(B \mid C) = p^2$, $P(C \mid B) = p^2$, $P(A \mid B) = 0$,
$P(B \mid A) = 0$, $P(A \mid C) = (1-p)^3$, $P(C \mid A) = p^5$;
4.3 $P(B \mid A) = 330/9153 = 110/3051 = 0.0361$,
$P(J \mid A) = 537/9153 = 179/3051 = 0.0587$,
$P(R \mid A) = 8286/9153 = 2762/3051 = 0.9053$;
4.5 a. 5/9; b. 2/5; c. 2/4; 4.7 47/62; 4.9 2/5; 4.11 2/3; 4.13 2/3;
4.15 $\binom{n}{j}/(2^n - 1)$

Chapter 5

5.1 0.1633; 5.3 0.6; 5.5 0.765; 5.7 a. 0.5326; b. 0.7176; c. 0.6372;
5.9 a. 3/5; b. 2/5; 5.11 a. 2/15; b. 0.4333; 5.13 1/3; 5.15 5/12;
5.17 a. 0.71; b. 0.29; c. 48/71; d. 1; e. 25/29; f. 25/39

Chapter 6

6.1 $0.03 < P(B) < 0.27 < P(D) \le 1$; 6.3 0.19; 6.5 a.1/5; b. 1/2; 6.7 0.48

Chapter 7

7.1 discrete; $X \in \{0, 1, 2, 3, 4, 5\}$;
7.3 X is continuous; $X \in [0, 4]$;
Y is continuous; $Y \in [0, 100]$;
Z is discrete; $Z \in \mathbb{Z}^{\geq 0}$;
7.5 X, Y, Z are all discrete and take values in $\mathbb{Z}^{\geq 0}$;
7.7 X, Y are both discrete and have possible values $\{0, 1, \ldots, 10\}$;
7.9 X, Y are both continuous and have possible values $(0, \infty)$;
7.11 X is discrete; $X \in \mathbb{N}$;
Y is continuous; $Y \in (0, \infty)$;
7.13 X is discrete; $X \in \{1, 2, \ldots, 10\}$;
Y is continuous; for instance, $Y \in (0, 0.52)$ if the half-gallon of milk could hold up to 0.52 gallons;
Z is discrete; $Z \in 1, 2, 3, 4$;
7.15 X is discrete; Answers will vary depending on the age of students in the course, e.g., $X \in \{3000, \ldots, 44350\}$;
Y is discrete; $Y \in \mathbb{N}$;
7.17 a. 10/17; b. 80/221; c. 11/221

Chapter 8

8.1 a. $S = \{0, 1, 2, 3, 4\}$; b. $P(X = x) = \binom{4}{x}(7/10)^x(3/10)^{4-x}$ for $0 \le x \le 4$;

c. $$F_X(x) = \begin{cases} 0 & \text{for } x < 0, \\ 0.0081 & \text{for } 0 \le x < 1, \\ 0.0837 & \text{for } 1 \le x < 2, \\ 0.3483 & \text{for } 2 \le x < 3, \\ 0.7599 & \text{for } 3 \le x < 4, \\ 1 & \text{for } x \ge 4 \end{cases};$$

8.3 Mass:

x	-1	1	2	3
$P(X = x)$	0.6651	0.0121	0.0196	0.3032

;

CDF: $$F_X(x) = \begin{cases} 0 & \text{for } x < -1, \\ 0.6651 & \text{for } -1 < x < 1, \\ 0.6772 & \text{for } 1 \le x < 2, \\ 0.6968 & \text{for } 2 \le x < 3, \\ 1 & \text{for } x \ge 3 \end{cases};$$

8.5 a. $p_X(x) = (0.9)^{x-1}(0.1)$;

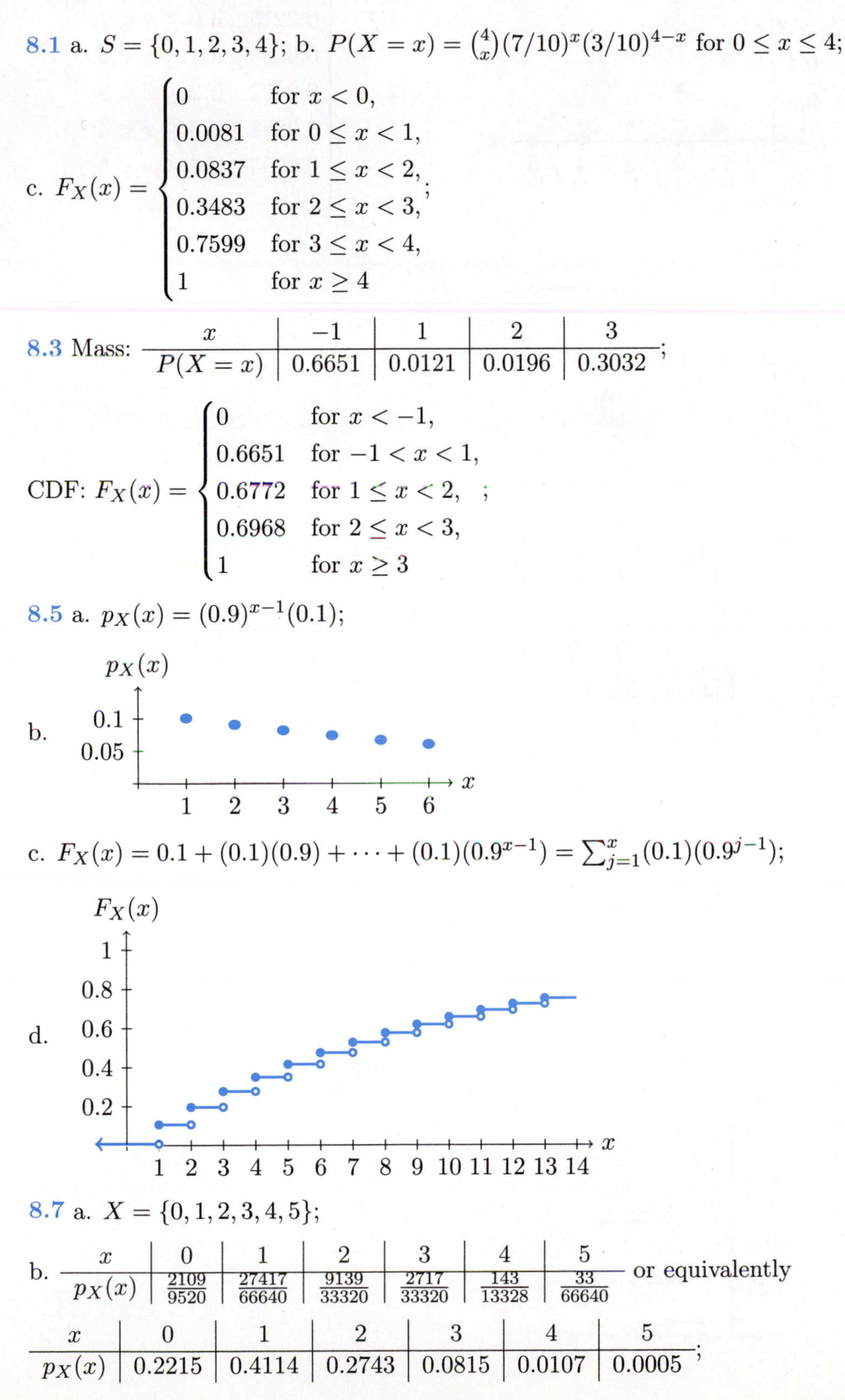

b.

c. $F_X(x) = 0.1 + (0.1)(0.9) + \cdots + (0.1)(0.9^{x-1}) = \sum_{j=1}^{x}(0.1)(0.9^{j-1})$;

d.

8.7 a. $X = \{0, 1, 2, 3, 4, 5\}$;

b.

x	0	1	2	3	4	5
$p_X(x)$	$\frac{2109}{9520}$	$\frac{27417}{66640}$	$\frac{9139}{33320}$	$\frac{2717}{33320}$	$\frac{143}{13328}$	$\frac{33}{66640}$

or equivalently

x	0	1	2	3	4	5
$p_X(x)$	0.2215	0.4114	0.2743	0.0815	0.0107	0.0005

;

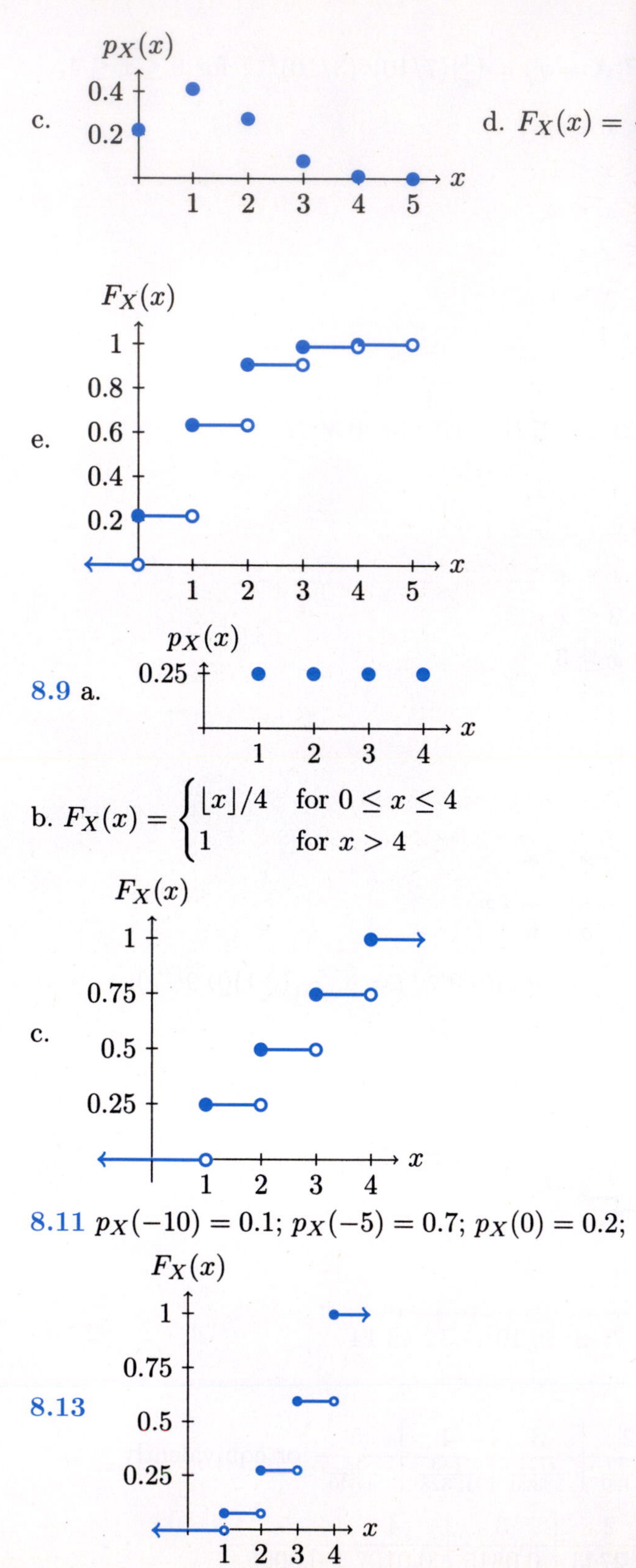

d. $F_X(x) = \begin{cases} 0 & \text{if } x < 0; \\ 0.2215 & \text{if } 0 \leq x < 1; \\ 0.6330 & \text{if } 1 \leq x < 2; \\ 0.9072 & \text{if } 2 \leq x < 3; \\ 0.9888 & \text{if } 3 \leq x < 4; \\ 0.9995 & \text{if } 4 \leq x < 5; \\ 1 & \text{if } 5 \leq x; \end{cases}$

b. $F_X(x) = \begin{cases} \lfloor x \rfloor / 4 & \text{for } 0 \leq x \leq 4 \\ 1 & \text{for } x > 4 \end{cases}$

8.11 $p_X(-10) = 0.1$; $p_X(-5) = 0.7$; $p_X(0) = 0.2$;

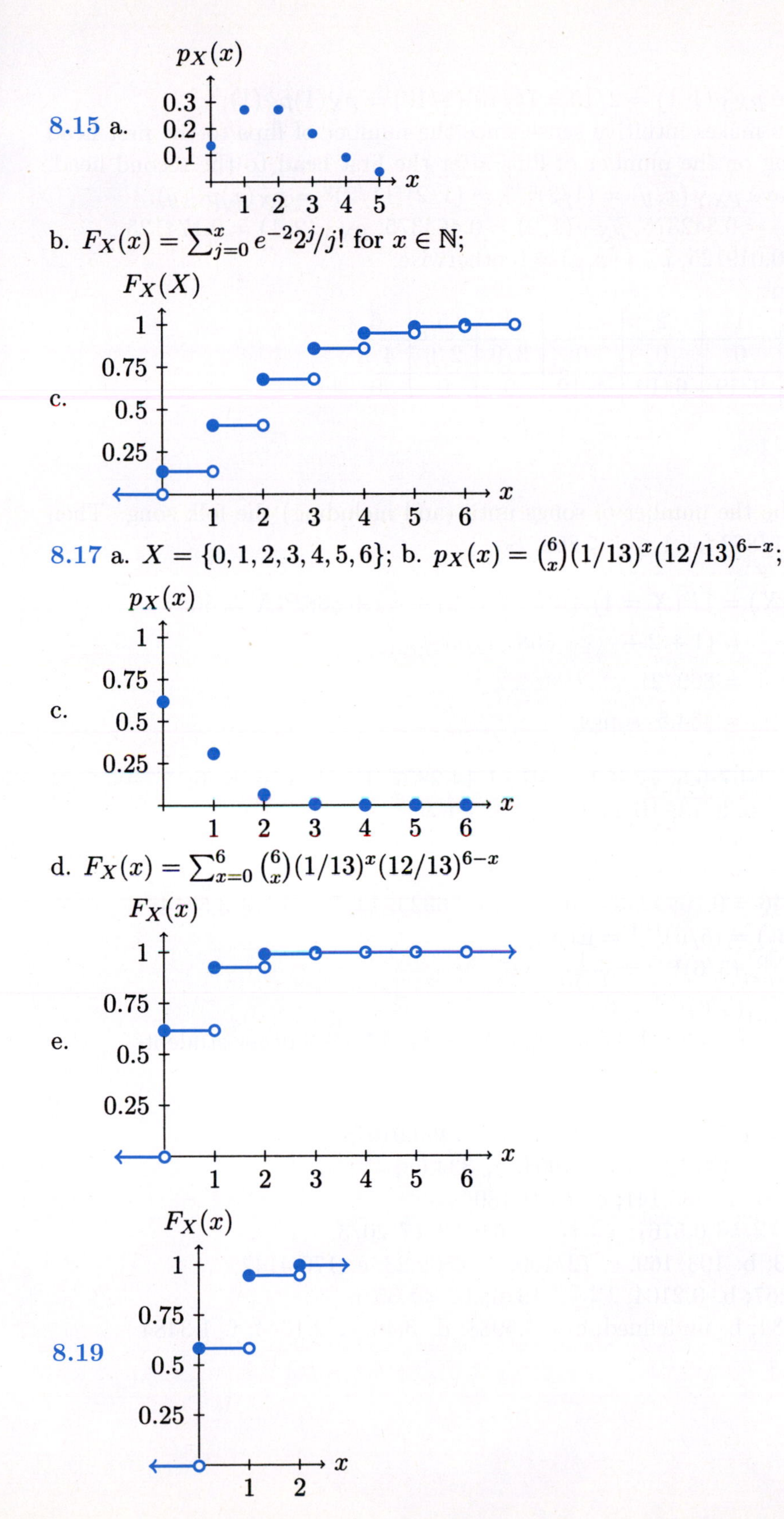

8.15 a.

b. $F_X(x) = \sum_{j=0}^{x} e^{-2}2^j/j!$ for $x \in \mathbb{N}$;

c.

8.17 a. $X = \{0, 1, 2, 3, 4, 5, 6\}$; b. $p_X(x) = \binom{6}{x}(1/13)^x(12/13)^{6-x}$;

c.

d. $F_X(x) = \sum_{x=0}^{6} \binom{6}{x}(1/13)^x(12/13)^{6-x}$

e.

8.19

Chapter 9

9.1 Yes, since $p_{X,Y}(1,1) = 2/10 = (4/10)(5/10) = p_X(1)p_Y(1)$;
9.3 Yes. This makes intuitive sense since the number of flips to the first head has no bearing on the number of flips after the first head to the second head. Indeed, we have $p_{X,Y}(x,y) = (1/2)^{x+y} = (1/2)^x(1/2)^y = p_X(x)p_Y(y)$.
9.5 $p_{X,Y}(0,3) = 0.342375$, $p_{X,Y}(1,2) = 0.464375$, $p_{X,Y}(2,1) = 0.174125$, $p_{X,Y}(3,0) = 0.019125$, $p_{X,Y}(x,y) = 0$ otherwise;
9.7 Dependent;
9.9

$Y\backslash X$	1	2	3	4	5	6
0	0	0	0	3/9	2/9	4/9
1	9/19	6/19	4/19	0	0	0

Chapter 10

10.1 0.76;
10.3 Let X be the number of songs until (and including) the folk song. Then $P(X = j) = 1/868$ for $1 \leq j \leq 868$. So

$$\begin{aligned}\mathbb{E}(X) &= 1P(X=1) + 2P(X=2) + \cdots + 868P(X=868) \\ &= (1+2+\cdots+868)(1/868) \\ &= 869/2 \\ &= 434.5 \quad \text{songs.}\end{aligned}$$

10.5 \$45,200; 10.7 5.5; 10.9 1.5; 10.11 14.2857; 10.13 a. 6; b. 6;
10.15 a. 6/13; b. 2/13; 10.17 Claw; 10.19 7

Chapter 11

11.1 $0.3 + 0.46 = 0.76$; 11.3 150.5; 11.5 7.6923; 11.7 6; 11.9 3.5714;
11.11 a. $P(A_j) = (5/6)^{j-1} = \mathbb{E}(X_j)$
so $\mathbb{E}(X) = \sum_{j=1}^{\infty}(5/6)^{j-1} = \frac{1}{1-5/6} = 6$;
b. $\mathbb{E}(Y) = \sum_{j=1}^{\infty}(5/6)^{j-1} = 6$;
11.13 15/8; 11.15 3.5; 11.17 a. 1391; b. 2781; 11.19 9 other students

Chapter 12

12.1 $\text{Var}(X) = p(1-p)$; 12.3 0.96; 12.5 7499.9167;
12.7 a. 15.36; b. 396.12; c. 160.1904; d. 334.68;
12.9 a. 2781; b. 15,465,141; c. 7,731,180;
12.11 −\$38; 12.13 0.5761; 12.15 55/64; 12.17 20/3;
12.19 a. 6/13; b. 108/169; c. 72/169; d. \$509.23; e. 170.4142;
12.21 a. 0.4267; b. 0.2104; 12.23 19/6; 12.25 35/6;
12.27 a. −3.34; b. undefined; c. −0.5988; d. 3.46; e. 2.1644; f. 1.3484

Chapter 13

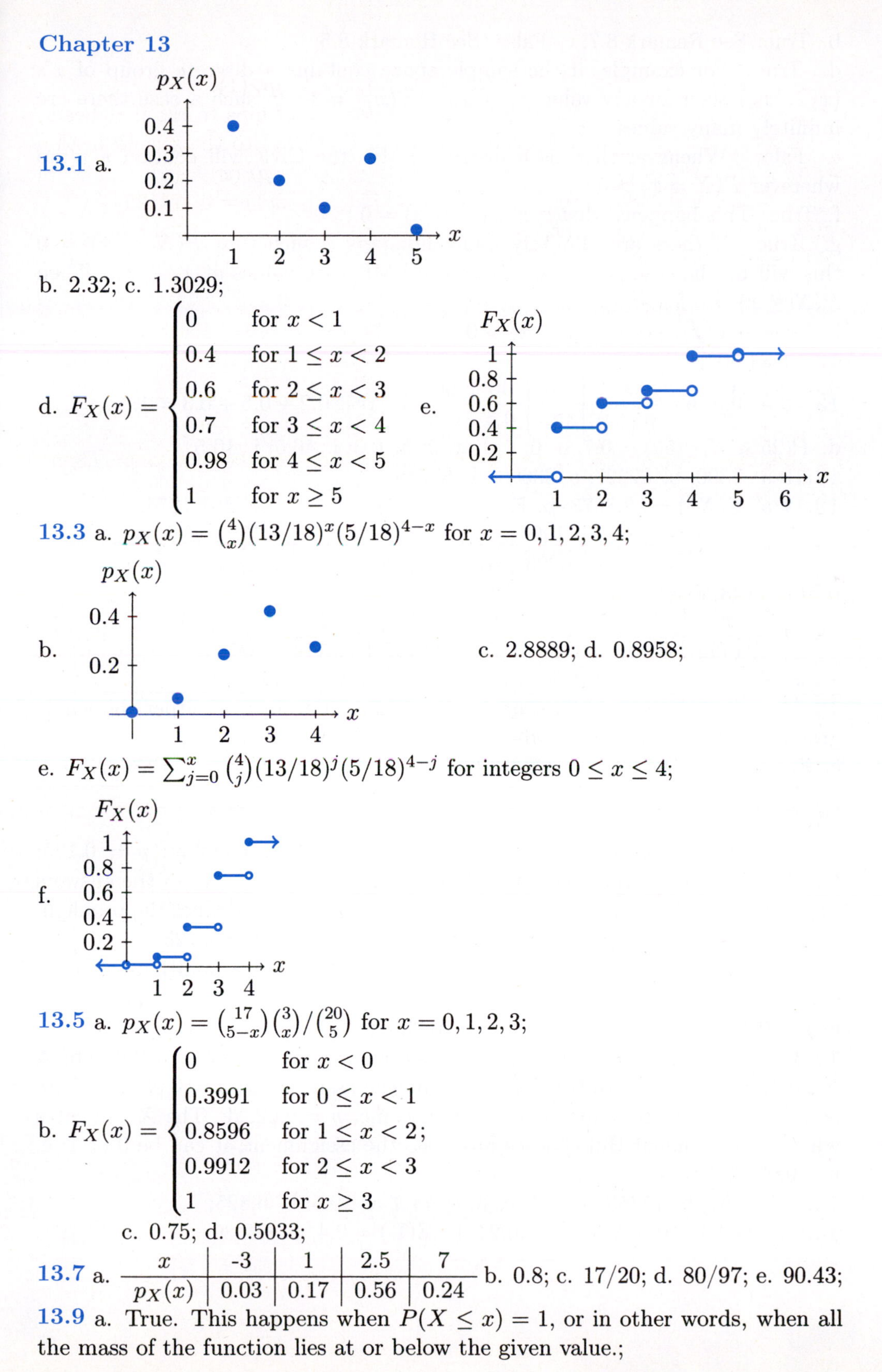

13.1 a. 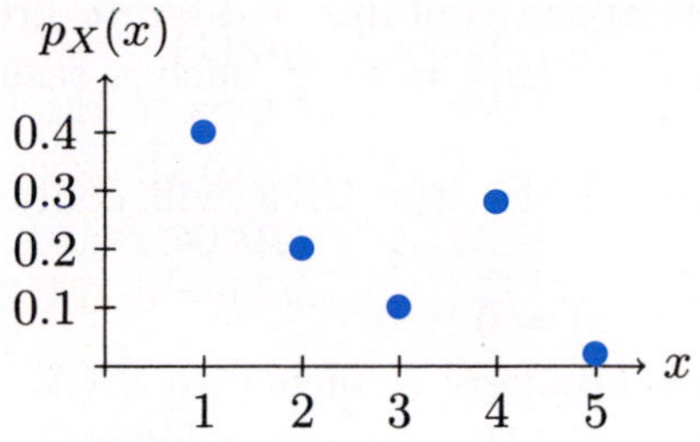

b. 2.32; c. 1.3029;

d. $F_X(x) = \begin{cases} 0 & \text{for } x < 1 \\ 0.4 & \text{for } 1 \leq x < 2 \\ 0.6 & \text{for } 2 \leq x < 3 \\ 0.7 & \text{for } 3 \leq x < 4 \\ 0.98 & \text{for } 4 \leq x < 5 \\ 1 & \text{for } x \geq 5 \end{cases}$ e. 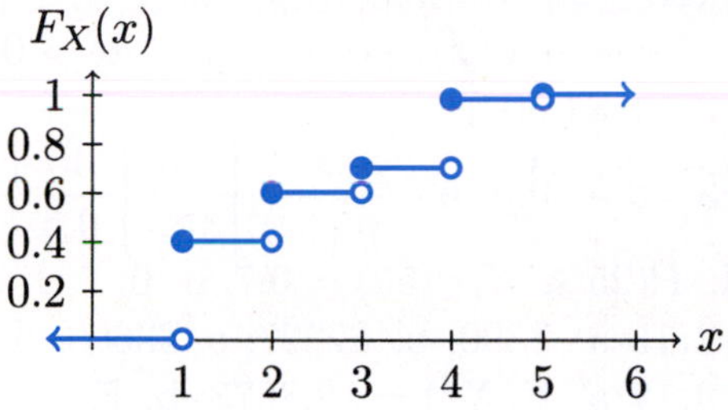

13.3 a. $p_X(x) = \binom{4}{x}(13/18)^x(5/18)^{4-x}$ for $x = 0, 1, 2, 3, 4$;

b. 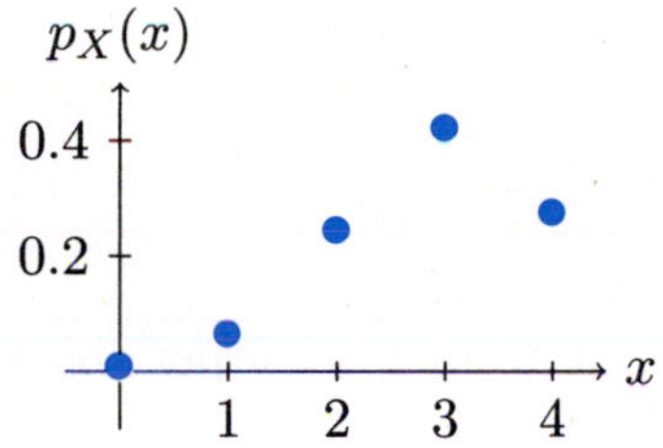

c. 2.8889; d. 0.8958;

e. $F_X(x) = \sum_{j=0}^{x} \binom{4}{j}(13/18)^j(5/18)^{4-j}$ for integers $0 \leq x \leq 4$;

f. 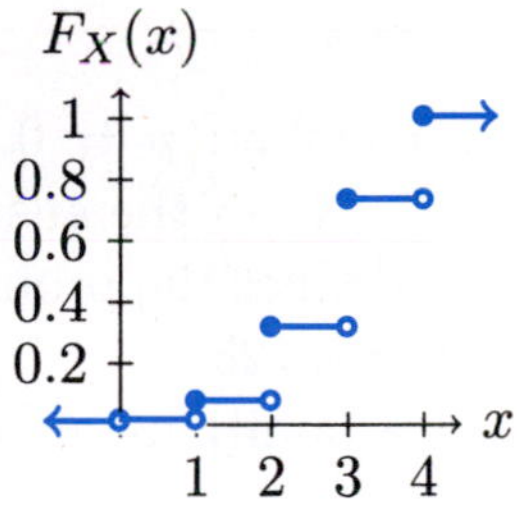

13.5 a. $p_X(x) = \binom{17}{5-x}\binom{3}{x}/\binom{20}{5}$ for $x = 0, 1, 2, 3$;

b. $F_X(x) = \begin{cases} 0 & \text{for } x < 0 \\ 0.3991 & \text{for } 0 \leq x < 1 \\ 0.8596 & \text{for } 1 \leq x < 2; \\ 0.9912 & \text{for } 2 \leq x < 3 \\ 1 & \text{for } x \geq 3 \end{cases}$

c. 0.75; d. 0.5033;

13.7 a.

x	-3	1	2.5	7
$p_X(x)$	0.03	0.17	0.56	0.24

b. 0.8; c. 17/20; d. 80/97; e. 90.43;

13.9 a. True. This happens when $P(X \leq x) = 1$, or in other words, when all the mass of the function lies at or below the given value.;

b. True. See Remark 8.7; c. False. See Remark 8.5;
d. True. For example, if the sample space contains a discrete group of x's $(x_1 \ldots x_n)$ then for any value $x_j > x_n$, $F_X(x_j) = 1$. In such a case there are infinitely many values $> x_n$.;
e. False. Whenever there is a discrete PMF, the CDF will contain a jump wherever $P(X = x) > 0$.;
f. True. This happens whenever $P(X \leq x) = 0$.;
g. True. If there are infinitely many numbers x such that $P(X \leq x) = 0$ this will be the case. Consider a discrete PMF with values at $x_1 \ldots x_n$. Then $P(X \leq x) = 0$ for any $x < x_1$.
h. True. $F_X(x) > 0$ if $P(X \leq x) > 0$;
13.11 $4.50;
13.13 a. 0.5; b.

x	30	45	60
$p_X(x)$	0.2	0.5	0.3

c. Yes. $0.2 + 0.5 + 0.3 = 1$.;
d. $P(25 < X < 50) = 0.7$; e. 0; f. 1; g. 0; h. 0.3; i. 46.5; j. 10.5;
13.15 a. $300; b. $2227; c. $3600; d. $7715;
13.17 a. $\mathbb{E}(X^2) = 12.8673$; b. $\mathbb{E}(-12X + 10) = -27.44$; c. 451.1376;
13.19 a.

x	50	100	150	200	250	270	290
$p_X(x)$	7/28	6/28	5/28	4/28	3/28	2/28	1/28

b.$145.7143; c. $78.5779;
13.21

x	0	1	2	3	4
$p_X(x)$	0.92	0.034	0.020	0.0145	0.0113

13.23 Average price = $78.75; Standard deviation = $21;
13.25 a. 0.0157; b. No. Passengers often come in groups, so either the whole group would be there or the whole group would miss the flight.;
c. $\mathbb{E}(X) = \$15,592.50$; $\sigma_X = \$294.2257$; d. $\mathbb{E}(X) = \$192.50$; $\sigma_X = \$32.69$

Chapter 14

14.1 a. A "success" occurs if the call is from a family member; $p = 0.125$; b. A "failure" occurs if a non-family member calls; $q = 0.875$; Note: the answers to a and b may be switched, depending on how the student defines the problem.
c. There is a single trial with a success/failure outcome, $p = 0.125$;
d. Random variable X indicates whether the call is from a family member; it can be 0 or 1;
e. $p = 0.125$; f. $p = 0.125$; g. 0.125; h. 0.3307;
14.3 a. A "success" is when a student has done the assignment; $p = 0.02$.; b. A "failure" is when the student has not done the assignment; $q = 0.98$; c. There is a single trial with a success/failure outcome, $p = 0.02$; d. The X indicates whether the student Hui chooses has done the assignment; it can be 0 or 1; e. $p = 0.02$; f. 0.02;
14.5 a. 7/55; b. 15/55 = 3/11; c. 3/7; 14.7 a. 19.5; b. 6.825;
14.9 a. $\mathbb{E}(X) = 0.4$; $\text{Var}(X) = 0.24$; b. $\mathbb{E}(Y) = 0.4$; c. $12.80

Chapter 15

15.1 a. A "success" is when a Skittle is purple; $p = 0.2$;
b. a "failure" is when a Skittle is not purple, $p = 0.8$;
c. Random variable X gives the total number of purple Skittles found. It can take on any integer value from 0 to 25;
d. There are $n = 25$ independent trials (because it was a random sample), with equal probability of success($p = 0.2$) on each trial, and we are counting up the number of successes (# of purple Skittles);
e. 0.196; f. 0.9726; g. 5; h. 2; i. 32.5 cents; j. 3;
15.3 a. Random variable X gives the number of babies who are not born by C-section. It can take an integer value from 0 to 9;
b. There are $n = 9$ independent trials (because it was a random sample), with equal probability of success($p = 0.8$), and we are counting up the number of successes (# of babies not born by C-section);
c. 0.1762; d. 0.2587; e. 2; f. 7.2; g. 1.44;

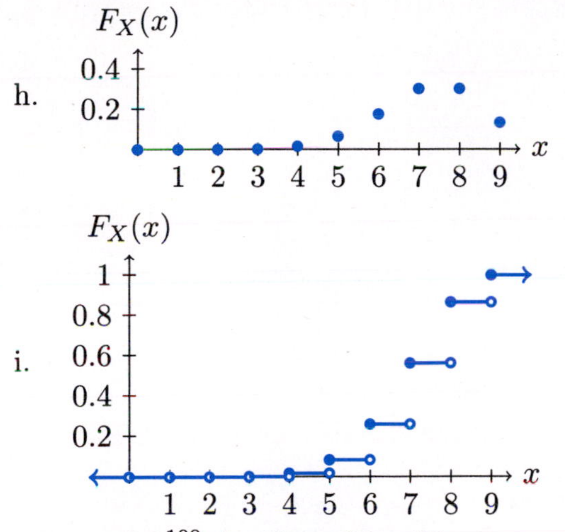

15.5 a. 0.8^{100}; b. 20; c. Yes. The expected profit is \$650; d. 14 boxes;
15.7 9 times **15.9** a. 0.2503; b. 0.7759; c. 0.9274; d. 0.7625;
15.11 a. 0.992; b. 0.936; c. 0.9285; d. 0.9995; e. 0.7273;
15.13 3.2726×10^{-11}; **15.15** a. -4; b. 194.4; c. 0.01229;
15.17 a. 0.161; b. 1.4; c. 1.302;
15.19 a. 1.75; b. No. Since there is no replacement, they are not independent Bernoulli trials;
15.21 School B, because $\mathbb{E}(X_A) = 222.8 < 243.81 = \mathbb{E}(X_B)$;
15.23 a. 270; b. $\sum_{j=45}^{50} \binom{50}{j}(7/10)^j(3/10)^{50-j} = 0.0007$;
15.25 $\lim_{n->\infty} P(X_n > 0) = 1$. Yes because as n grows very large, the probability (0.4^n) that no student would eat cereal for breakfast approaches 0.

Chapter 16

16.1 a. The variable X is the number of Skittles that come down the line until the inspector finds the first one that is striped; here, X can be 1,2,....

b. The trials are independent, each with equal probability of success ($p = 0.05$). The trials are performed until the first success of finding a striped Skittle.

c. 0.0315; d. 0.4312; e. 0.6634; f. 0.6983; g. 20; h. 19.4936;

16.3 a. The variable X is the number of parents you have to ask until you find one who did not have a baby by C-section; the X can be 1,2,. . . .

b. The trials are independent, each with equal probability of success ($p = 0.8$). The trials are performed until the first success of finding parent of a baby not born by C-section.

c. 0.0016; d. 0.008; e. 0.0079; f. 1.25; g. 0.3125;

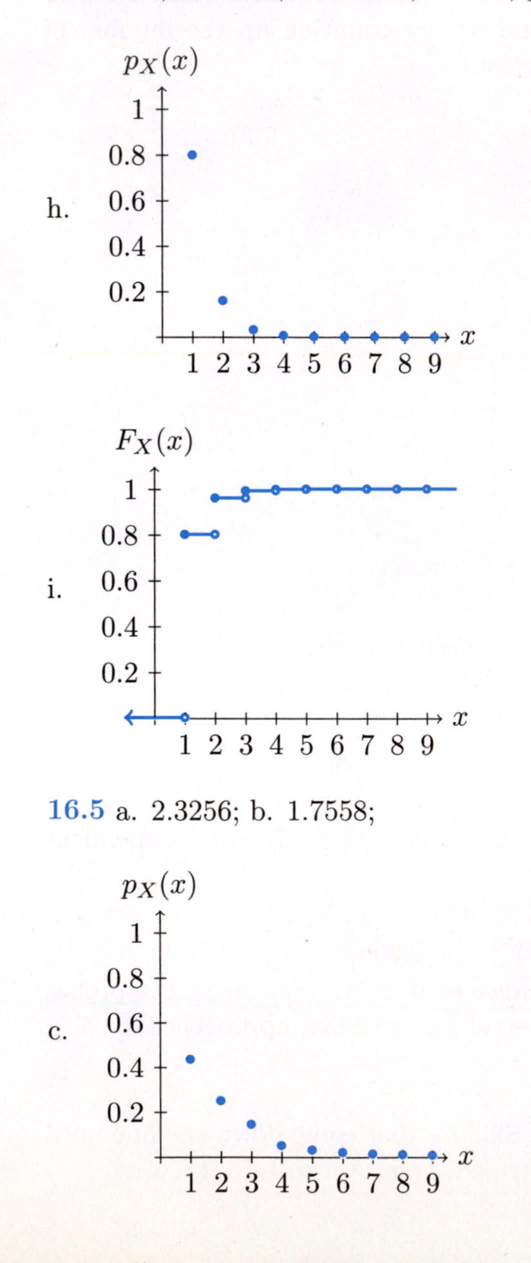

16.5 a. 2.3256; b. 1.7558;

d. 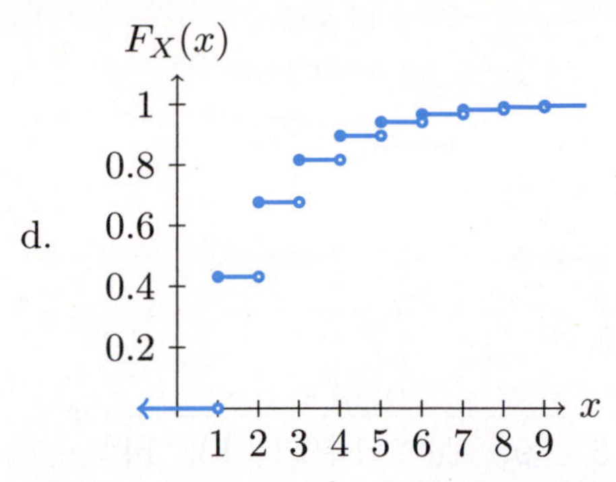

16.7 a. 13 times; b. 0.7865; c. 0.2367; Binomial; $n = 6$, $p = 0.7865$;

16.9 a. 3.5714; b. 17.8571 minutes; c. 0.0374;

16.11 a. 0.065498; b. 0.19988; c. 5;

16.13 a. 2.5 games; b. 3.75 games; c. 0.216;

16.15 a. $\mathbb{E}(X) = 14.2857$; b. $\text{Var}(X) = 189.7959$; c. $P(Y = y) = q^{x-3}p$;

16.17 $P(X > n) = (7/8)^n$;

16.19 There is only one possible way for $X = x$ if X Geometric, but there are $\binom{n}{x}$ ways for X to equal x if X Binomial. Geometric variables can have an infinite number of values, whereas Binomials can only take values between 0 and n. So if X Geometric then $P(X > x|X > y) = q^x/q^y = q^{x-y}$, but if X Binomial then $P(X > x|X > y) = \sum_{j=x+1}^{n} \binom{n}{j} p^j q^{n-j} / \sum_{j=y+1}^{n} \binom{n}{j} p^j q^{n-j}$, which clearly does not have the memoryless property.

Chapter 17

17.1 a. The variable X is the number of Skittles until the 3rd striped Skittle is found; this X can take values 3, 4, 5, ...

b. We must count the number of trials until we get the 3rd success, which is the 3rd striped Skittle found. Each Skittle is independent and each trial has the same probability of success ($p = 0.05$).

c. 0.0089; d. 0.0012; e. 0.9978; f. 60; g. 33.7639;

17.3 a. The variable X is the number of parents you have to ask until you find the 7th whose baby was not born by C-section. The X can be 7,8,....

b. We must count the number of trials until we get the rth success ($r = 7$), each set of parents is independent, and each trial has the same probability of success ($p = 0.8$).

c. 0.1409; d. 0.2618; e. 8.75; f. 2.1875;

g. 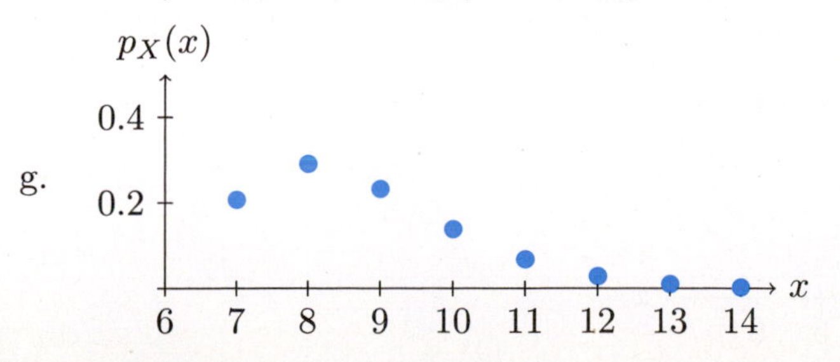

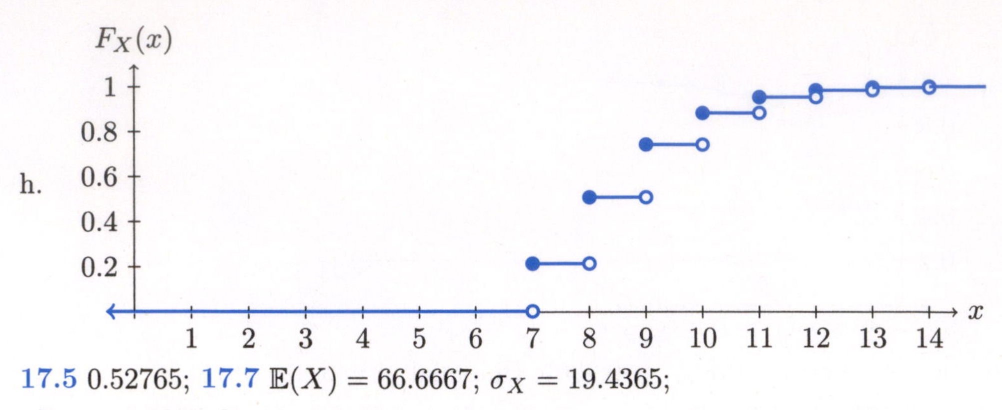

17.5 0.52765; **17.7** $\mathbb{E}(X) = 66.6667$; $\sigma_X = 19.4365$;

17.9 a. 0.3907; b. 60 minutes;

17.11 a. 0.6405; b. 4/27; Geometric; The number of tries to the next success does not depend on how many tries it took to get the previous successes.

Chapter 18

18.1 a. The X is a nonnegative integer valued random variable. We are given a rate (marriages per day) and we are asked to count up the number of marriages in a particular time frame (5 minutes). The average rate is $\lambda = 35/36 = 0.9722$ marriages/5 minutes.

b. $e^{-280}280^{300}/300! = 0.0115$; c. 35/3; d. 3.4157; e. 0.3413;

f. 0.0451; Binomial(24,0.3413);

18.3 a. 0.064998 for a child; 6.3524×10^{-23} for an adult;

b. There is a an average rate and we must count the number of times an adult/child laughs in a certain period of time (hour).;

c. 0.2094; d. 0.8823;

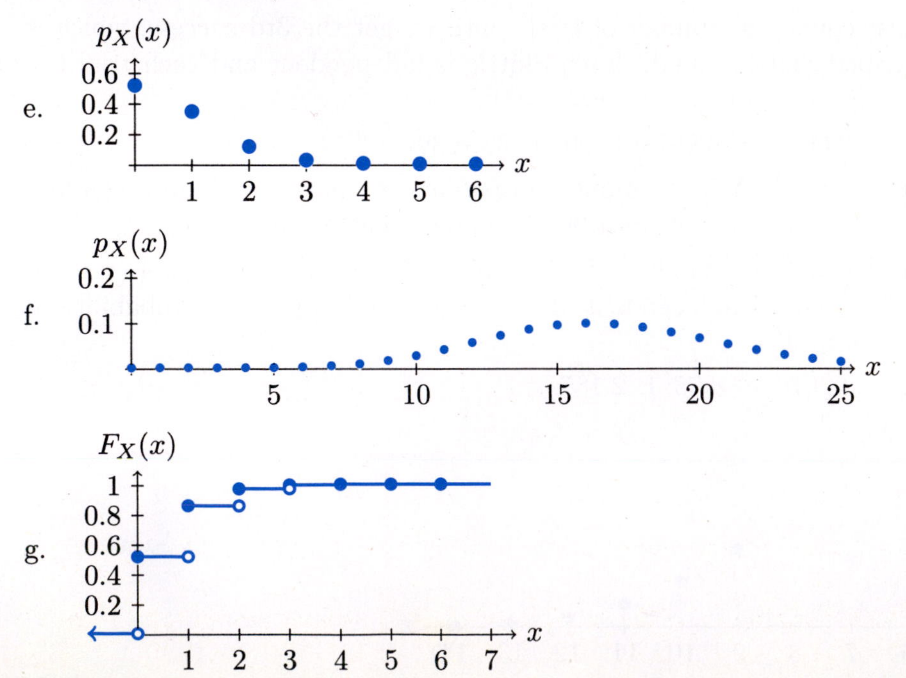

h.

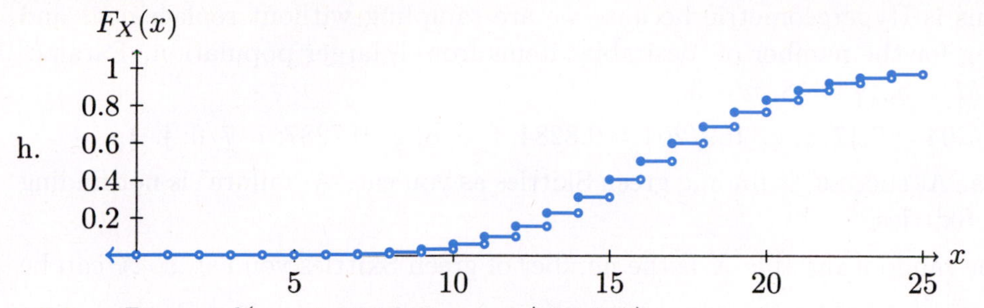

18.5 a. Binomial($n = 1{,}000{,}000, p = 1/729{,}000$);
b. $P(X = 3) = \binom{1{,}000{,}000}{3}(1/729{,}000)^3(728{,}999/729{,}000)^{999{,}997}$;
c. 1.3717; d. Poisson($\lambda = 1000000/729000 = 1.3717$); e. $P(X = 3) \approx 0.1091$;
18.7 $P(X \geq 1) \approx 0.86466$;

18.9 a.

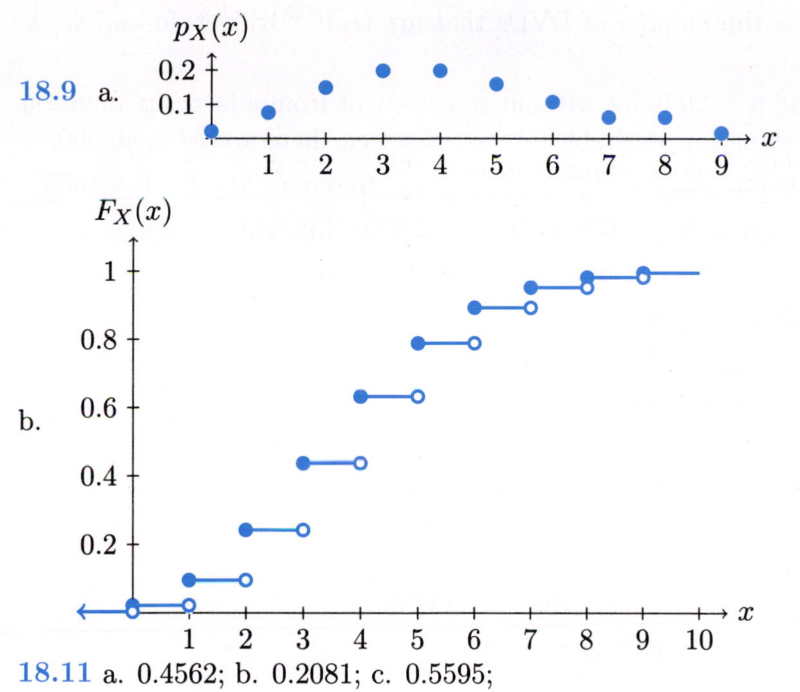

b.

18.11 a. 0.4562; b. 0.2081; c. 0.5595;
18.13 a. 0.1804; b. 0.3233; c. 0.5581; d. \$5; e. \$3.54;
18.15 a. 0.1606; b. 0.000003;
c. 0.0362; the 35 defects can occur any time in the 168 hours (there does not have to be a certain amount each day).
18.17 a. 0.1126; b. 8.8811; Geometric; 18.19 a. 4; b. 0.1563;
18.21 a. 42; b. 0.04388;
18.23 a. $P(X = 8) = \binom{60{,}000}{8}(1/10{,}000)^8(9999/10{,}000)^{59{,}992}$;
b. $P(X = 8) \approx e^{-6}6^8/8!$; c. 0.1033; 18.25 Yes

Chapter 19

19.1 a. A "success" is a bag of cookies. A "failure" is a bag of potato chips or pretzels.
b. The random variable X is the number of bags of cookies you get, so X can be 0,1,2 or 3.

c. This is Hypergeometric because we are sampling without replacement and looking for the number of "desirable" items from a larger population. Parameters: $M = 5$, $N = 18$, $n = 3$.

d. $35/204 = 0.1716$; e. $169/204 = 0.8284$; f. $5/6$; g. 0.7287; i. $7/6$; j. 1;

19.3 a. A "success" is finding green Skittles as you eat. A "failure" is not finding green Skittles.

b. The random variable X is the number of green Skittles you eat, so X can be $0, 1, \ldots, 9$.

c. We are sampling $n = 10$ items without replacement from a larger population of $N = 45$ items, looking for "desirable" items, of which there are $M = 9$.

d. 0.3133; e. 3.6024×10^{-8}; f. 2;

19.5 a. Here, X is the number of DVDs that are truly "Bridget Jones," so X can be $0,1,\ldots,20$.

b. We are sampling $n = 20$ items without replacement from a larger population of $N = 50{,}000$, looking for "desirable" items, of which there are $M = 40{,}000$.

c. 16; d. $P(X = 18) = \binom{40{,}000}{18}\binom{10{,}000}{2}/\binom{50{,}000}{20}$; e. Binomial$(20, 0.8)$; f. 0.1369;

19.7 a. $120/1771 = 0.0678$; b. $150/1771 = 0.0847$; c. $135/161 = 0.8385$;

d. $2/191 = 0.0105$; e. 1.3043; f. 0.8187; g. 0.004591; Binomial;

19.9 $10/19 = 0.5263$

19.11 a. $116/329 = 0.3526$; b. 2.4; 19.13 $P(X = 0) = 0.0725$; $\mathbb{E}(X) = 2$;

19.15 a. $20/39 = 0.5128$; b. $28/55 = 0.5091$;

c. No. Part b is based upon a given outcome on the first night, so the probability would not be the same. A conditional probability is not the same as a non-conditional probability, unless the events are independent. But here the events are dependent.

19.17 a. $56/19 = 2.9474$; b. $1232/1083 = 1.13758$;

c. $770/12597 = 0.0611$; d. $770/849 = 0.9069$;

19.19 a. 0.01008; b. $13/4$

Chapter 20

20.1 a. All 5 outcomes are equally likely ($p = 1/5$), and you are only selecting one at random.

b. $1/5$;

c. The random variable X is the number of the color that you pick, so X can be 1, 2, 3, 4, or 5.

d. $4/5$;

e.

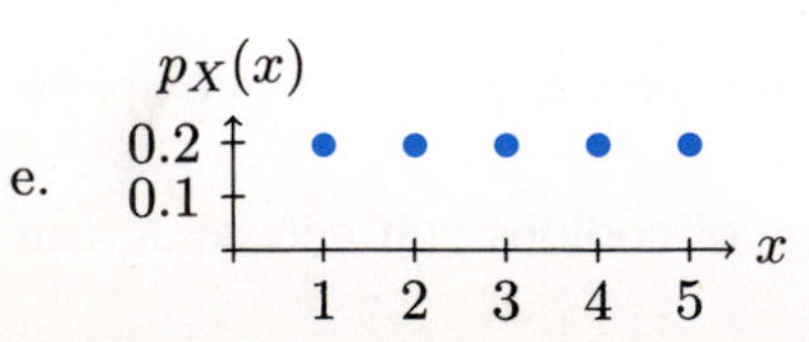

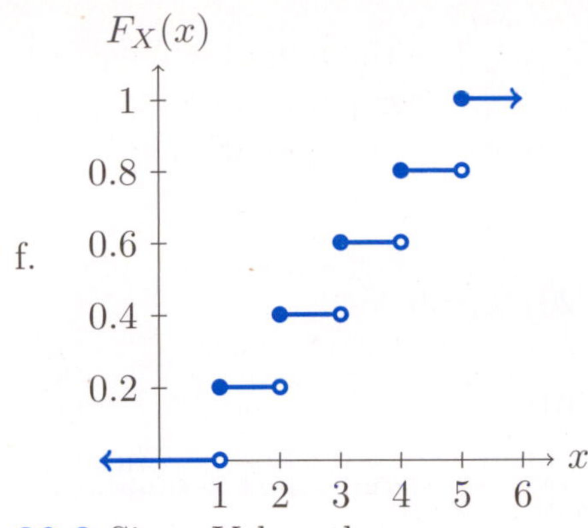

20.3 Since Y has the same mass as $X+1$, it follows that $\mathbb{E}(Y) = \mathbb{E}(X) + 1$, and $\text{Var}(Y) = \text{Var}(X)$.

Chapter 21

21.1 Binomial; $n = 50$, counting up total number of broken cones, each with $p = 0.12$;
21.3 Poisson; there is a rate $\lambda = 2/\text{minute}$ and a set interval (1 hour);
21.5 Hypergeometric; sampling without replacement, population and sample sizes, number of successes in population;
21.7 Discrete Uniform; equal probability of success for each outcome;
21.9 Binomial (or Poisson approximation); $n = 10000$, counting up total number of undercooked cones, each with $p = 0.00005$; since n is large and p is small, the approximation with $\lambda = 0.5$ is appropriate.
21.11 a. 0.2; b. Binomial(10, 0.2); 0.6242; c. Geometric(0.2); 0.1024; d. Negative Binomial(0.2, 3); 15; e. Binomial (7, 0.6242); 0.1698; f. Discrete Uniform(5); 1/5;
21.13 a. Hypergeometric ($M = 10$, $N = 15$, $n = 3$); $24/91 = 0.2637$; b. Binomial(3, 2/3); $8/27 = 0.2963$; c. Geometric(2/3); $2/27 = 0.0741$; d. Bernoulli(2/3); 2/3; e. Bernoulli(2/3); 2/3;
21.15 a. Geometric(0.08); 0.0164; b. Negative Binomial(0.08, 4); 50; c. Binomial(150, 0.08); 12; d. Bernoulli(0.08); 0.92;
21.17 a. Poisson(30); 0.1755; b. Geometric(0.1755); 5.699; c. Negative Binomial(0.1755, 4); 0.0504;
21.19 a. Hypergeometric($M = 10, N = 45, n = 5$); 0.7343; b. Binomial approximation to the Hypergeometric(5, 1/36); 0.1314; c. Geometric(0.1314); 7.6113;
21.21 a. Discrete Uniform(7); 1/7; b. Geometric(1/7); 7; c. Binomial(20,0.25); 0.1897; d. Poisson approximation to the Binomial (0.365); 0.0005;
21.23 Hypergeometric($M = 5, N = 18, n = 3$); a. $\mathbb{E}(X) = 15/18$; $\sigma_X = 0.7287$; b. 0.1716; c. Binomial approximation to the Hypergeometric(3, 5/18); 0.1886

Chapter 22

22.1 a. 4060; b. 24,360;
22.3 a. $840/34650 = 4/165 = 0.0242$; b. $210/34650 = 1/165 = 0.0061$;
c. $8/55$; d. $6/11$;
22.5 a. 0.3087; b. 0.03087;
22.7 a. $29/52 = 0.5577$; b. $3/52 = 0.0577$ c. $20/52 = 0.3846$;
22.9 0.00003; **22.11** 1/11; **22.13** 0.0738;
22.15 a. $57/616 = 0.0925$; b. $39/1496 = 0.0261$;
c. No. These are only the cases where all boys or all girls are chosen. There are several cases where some boys and some girls are chosen.
22.17 0.0002; **22.19** a. $64/425 = 0.1506$; b. $3/4$;
22.21 $p_X(0) = 44/120 = 11/30$; $p_X(1) = 45/120 = 3/8$; $p_X(2) = 20/120 = 1/6$;
$p_X(3) = 10/120 = 1/12$; $p_X(4) = 0; p_X(5) = 1/120$
22.23 7.037; **22.25** a. 0.6644; b. 0.0055; **22.27** 17/45;
22.29 a. 1/64; b. 0.0757; c. 0.0002; d. 0.0023;
22.31 a. 35; b. $15/35 = 3/7$; c. $5/35 = 1/7$; d. 210;
22.33 a. 3,628,800 ways to arrange seating;
b. 181,440 ways to arrange seating with the family sitting together;
22.35 1/19; **22.37** a. 0.0319; b. 0.049; c. 0.1976;
22.39 The probability of not sitting next to each other is $\frac{n-3}{n-1}$ for $n \geq 3$, or 0 for $n = 2$.
22.41 2/15; **22.43** 2/3; **22.45** $n!n!/(2n-1)!$ **22.47** 0.1

Chapter 24

24.1 a. continuous; b. discrete; c. continuous; d. discrete;

24.3

	X is discrete	X is continuous
$P(X \geq 2)$	???	0.3
$P(X < 2)$	???	0.7
$P(X \leq 2)$	0.7	0.7
$P(X = 2)$	???	0

24.5 a. $k = 30$; b. $53/512 = 0.1035$;

24.7 a. 4/5; b. $F_X(x) = \begin{cases} 0 & \text{for } x < 2; \\ (x-2)/5 & \text{for } 2 \leq x \leq 7; \\ 1 & \text{for } x > 7; \end{cases}$

c. 3/5; d. 0; e. 3/10; f. 0;

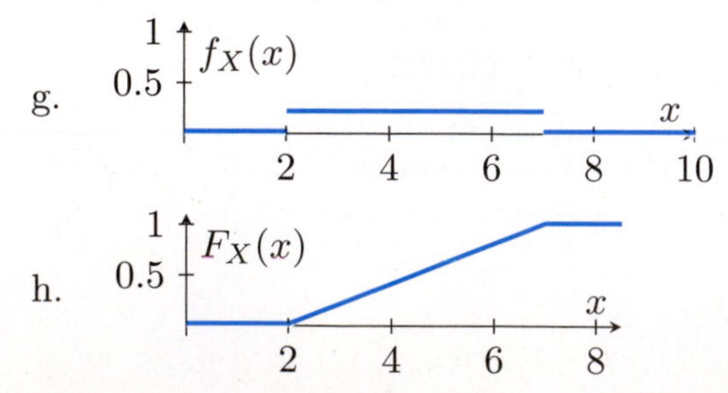

24.9 a. $F_X(x) = \begin{cases} 0 & \text{for } x \le 4; \\ (x-4)/6 & \text{for } 4 < x < 10; \\ 1 & \text{for } x \ge 10; \end{cases}$

b.

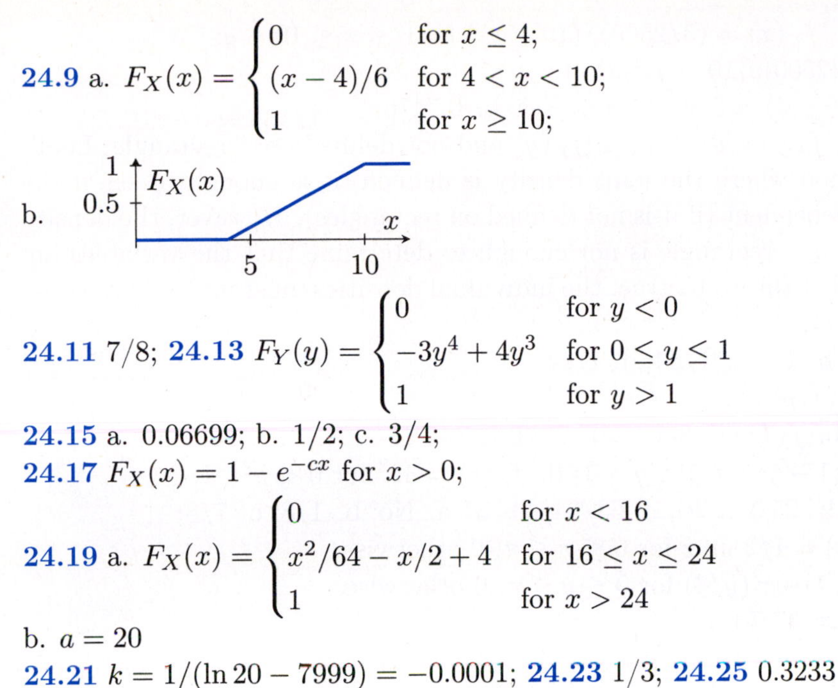

24.11 7/8; **24.13** $F_Y(y) = \begin{cases} 0 & \text{for } y < 0 \\ -3y^4 + 4y^3 & \text{for } 0 \le y \le 1 \\ 1 & \text{for } y > 1 \end{cases}$

24.15 a. 0.06699; b. 1/2; c. 3/4;
24.17 $F_X(x) = 1 - e^{-cx}$ for $x > 0$;

24.19 a. $F_X(x) = \begin{cases} 0 & \text{for } x < 16 \\ x^2/64 - x/2 + 4 & \text{for } 16 \le x \le 24 \\ 1 & \text{for } x > 24 \end{cases}$

b. $a = 20$
24.21 $k = 1/(\ln 20 - 7999) = -0.0001$; **24.23** 1/3; **24.25** 0.3233
24.27 Yes. An example of the CDFs of two such random variables is given below.

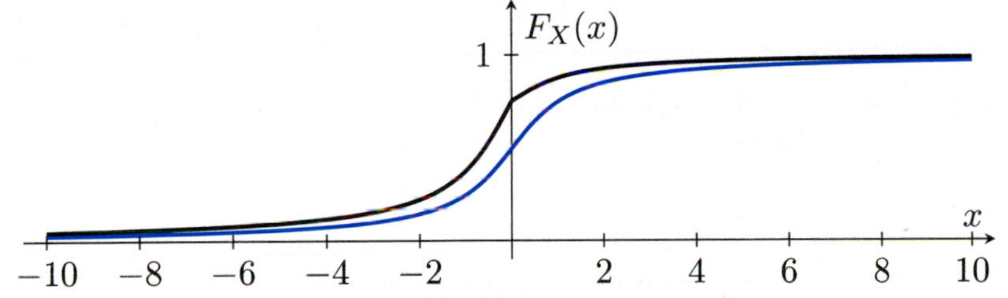

Chapter 25

25.1 5/9; **25.3** 7/9; **25.5** 10/27; **25.7** 0.0252;
25.9 a. 1/64; b. 7/64; d. 49/64; e. (1+7+7+49)/64=1;

25.11 $F_{X,Y}(x, y) = \begin{cases} 0 & \text{for } x < 0,\ y < 0 \\ x^2y/2 + xy^2/2 & \text{for } 0 \le x, y \le 1 \\ 1 & \text{for } x, y \ge 1 \end{cases}$

25.13 0.9997; **25.15** 1/4;

25.17 $F_W(w) = \begin{cases} 0 & \text{for } w \le 0 \\ 1 - e^{-3w} - e^{-5w} + e^{-8w} & \text{for } w > 0 \end{cases}$

25.19 0.06699; **25.21** 9/25 = 0.36; **25.23** 1/16

Chapter 26

26.1 a. Yes. $f_{X,Y}(x, y)$ is defined on a rectangle and can be factored into $f_X(x)f_Y(y)$.
b. $f_X(x) = (2/9)(3 - x)$ for $0 \le x \le 3$; c. $f_Y(y) = (1/2)(2 - y)$ for $0 \le y \le 2$;
26.3 a. $f_X(x) = 3x^2$ for $0 \le x \le 1$; b. $f_Y(y) = 1/(3y \ln 2)$ for $1/2 \le y \le 4$;

26.5 a. No; b. $f_X(x) = (3/2500)x(10-x)^2$ for $0 \le x \le 10-y$;
c. $f_Y(y) = (3/2500)y(10-y)^2$ for $0 \le y \le 10-x$;
26.7 $P(X+Y \le 4) = 2/9$; 26.9 a. Yes; b. 0.7476;
c. 0.7476; No; $f_{X,Y}(x,y) \ne f_X(x)f_Y(y)$ and not defined on a rectangle; Looking at the region where the joint density is defined, it is enough to see if the variables are dependent (if it is not defined on rectangles). However, the density being defined on a rectangle is not enough to determine that the variables are independent. For this to be true, the individual densities must multiply to equal the joint density.
26.11 a. Yes; b. 1/8 c. 1/64; d. 1/84;
26.13 a. $f_{X,Y}(x,y) = 1$; b. 7/8;
26.15 a. No; b. $f_X(x) = 3(1-x)^2$ for $0 < x < 1$;
c. $f_Y(y) = 6y(1-y)$ for $0 < y < 1$; d. $f_Z(z) = 3z^2$ for $0 < z < 1$;
26.17 a. Yes; b. 25/64; 26.19 63/64; 26.21 a. No; b. 1/8; c. 7/8;
26.23 a. $f_X(x) = 1/2 \sin x$ for $0 \le x \le \pi$, 0 otherwise;
b. $f_Y(y) = (1/4)\sec^2(y/4)$ for $0 \le y \le \pi$, 0 otherwise;
c. 0.2071; 26.25 17/81

Chapter 27

27.1 a. $f_Y(y) = (5/256)y(4-y)^3$ for $0 \le y \le 4$;
b. $f_{X|Y}(x|y) = 3x^2/(4-y)^3$ for $0 \le x \le 4, 0 \le y \le 4, x+y \le 4$;
27.3 a. $f_Y(y) = 1/10 + (3/40)y$ for $0 \le y \le 4$;
b. $f_{X|Y}(x|y) = (3/2)(x^2+y)/(4+3y)$ for $0 \le x \le 2, 0 \le y \le 4$;
27.5 a. $f_Y(y) = 5e^{-5y}$ for $0 < y$; b. $f_{X|Y}(x|y) = e^{-x+2y}$ for $0 < y < x/2$;
27.7 a. $f_Y(y) = (6/7)(1/3 + y + y^2)$ for $0 \le y \le 1$;
b. $f_{X|Y}(x|y) = 3(x+y)^2/(1+3y+3y^2)$ for $0 \le x, y \le 1$;
27.9 a. $f_Y(y) = (1/450)(30-y)$ for $0 \le y \le 30$;
b. $f_{X|Y}(x|y) = 1/(30-y)$ for $0 \le y, 0 \le x, x+y \le 30$;
27.11 a. $f_{X|Y}(x|y) = 1/(3-y)$for $0 < x < 3$; b. 1/2; c. 2/5;
27.13 a. $f_{X|Y}(x|y) = 2e^{-2x}$; b. 0.1353; c. 0.7534;
27.15 a. $f_{X|Y}(x|y) = (2/9)(3-x)$; b. 8/9; c. 1/4; 27.17 2/35

Chapter 28

28.1 $\mathbb{E}(X) = 20.3333$; 28.3 $\mathbb{E}(X) = 3$; 28.5 $\mathbb{E}(X) = 7.5$; 28.7 $\mathbb{E}(X) = 3.1378$;
28.9 $\mathbb{E}(X) = 1$; 28.11 a. $\mathbb{E}(X) = 1/2$; b. $\mathbb{E}(Y) = 1/7$;
28.13 a. $\mathbb{E}(X) = 1$; b. $\mathbb{E}(Y) = 2/3$; 28.15 $\mathbb{E}(X) = 1.25$;
28.17 a. $\mathbb{E}(X) = 0$; b. $\mathbb{E}(Y) = -1/9$; 28.19 $\mathbb{E}(Y) = 10/3$; 28.21 $\mathbb{E}(X) = 6/5$

Chapter 29

29.1 $\text{Var}(X) = 100/3$; 29.3 $\text{Var}(X) = 25/12$; 29.5 $\text{Var}(X) = 50/9$;
29.7 $\text{Var}(X) = 1/4$; 29.9 a. $\mathbb{E}(X+Y) = 2$; b. $\text{Var}(X) = 1/2$; 29.11 0.2704;
29.13 $\mathbb{E}(X^2+Y^3) = 2.3$; 29.15 a. $\mathbb{E}(X+Y) = 5$; b. $\text{Var}(X+Y) = 13/3$;
29.17 9; 29.19 9; 29.21 549; 29.23 $27.55; 29.25 22/49;
29.27 a. $\mathbb{E}(X) = 35/9$; b. $\text{Var}(X) = 5.7099$; 29.29 0.1945;
29.31 a. 162; b. $\mathbb{E}(X^n) = n!3^n$

Chapter 30

30.1 a. $F_X(x) = 2x$ for $0 \le x \le 1/2$; b. 0.4; c. 0.1; d. 0; e. 1/4; f. 0.1443;

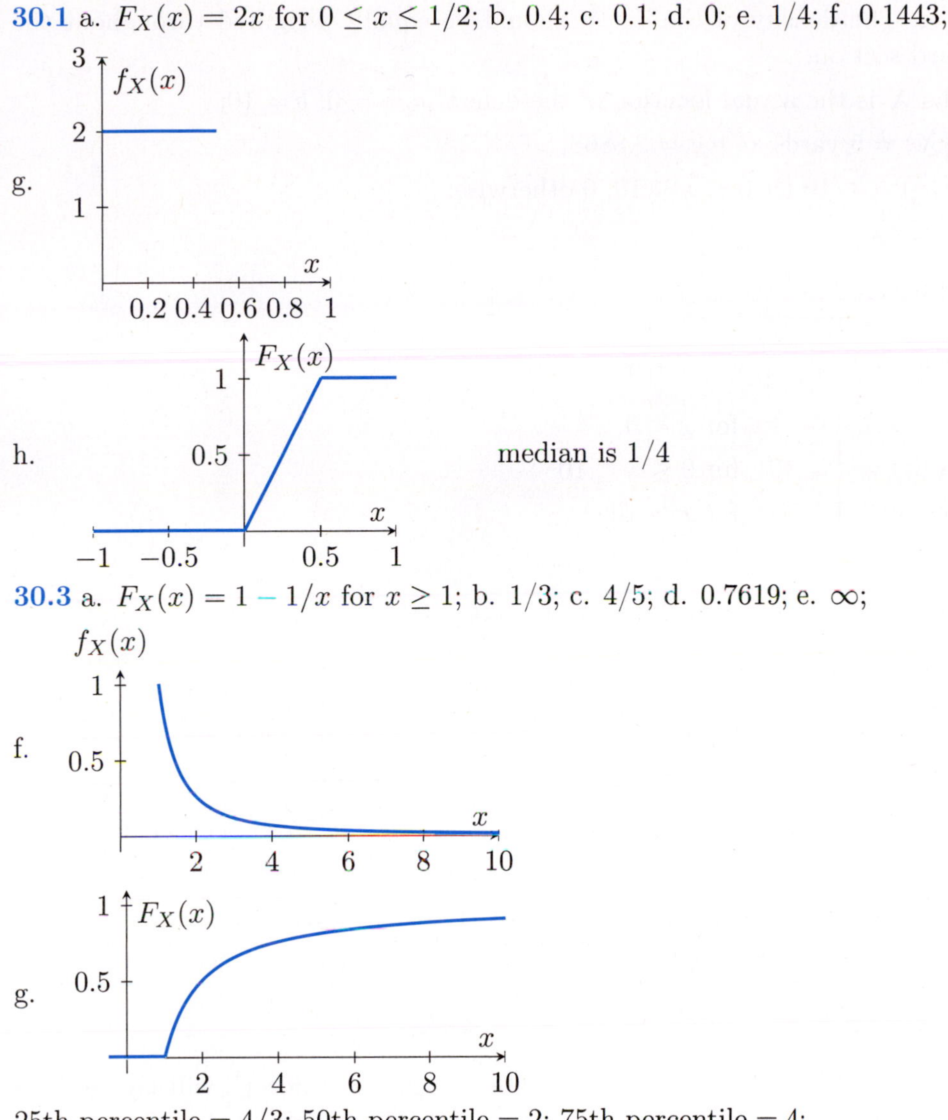

30.3 a. $F_X(x) = 1 - 1/x$ for $x \ge 1$; b. 1/3; c. 4/5; d. 0.7619; e. ∞;

25th percentile = 4/3; 50th percentile = 2; 75th percentile = 4;

30.5 $k = 4/5$; **30.7** a. False; $0 \le F_X(x) \le 1$ for all x; b. True;

c. False; $0 \le p_X(x) \le 1$ for all x;

30.9 a. False; the derivative of $F_X(x)$ is $f_X(x)$; b. True; c. True;

d. False; the reverse is true.; e. True;

f. False; the reverse is true since $F_X(x)$ is the Cumulative Distribution Function which is the area under the density.

30.11 a. False; $0 \le F_X(x)$ for all x since $F_X(x)$ is a probability, and probabilities are always nonnegative;

b. False; $0 \le f_X(x)$ for all x since densities are integrated to get probabilities, and probabilities are always nonnegative;

c. False; $0 \le p_X(x)$ for all x since $p_X(x)$ is a probability, and probabilities are always nonnegative.

Chapter 31

31.1 a. The density of the defect should be equally weighted throughout the 10-yard section.

b. The X is the actual location of the defect. c. $a = 0$, $b = 10$;

d. $\mathbb{E}(X) = 5$ yards; e. $\sigma_X = 2.8868$;

f. $f_X(x) = 1/10$ for $0 \le x \le 10$, 0 otherwise;

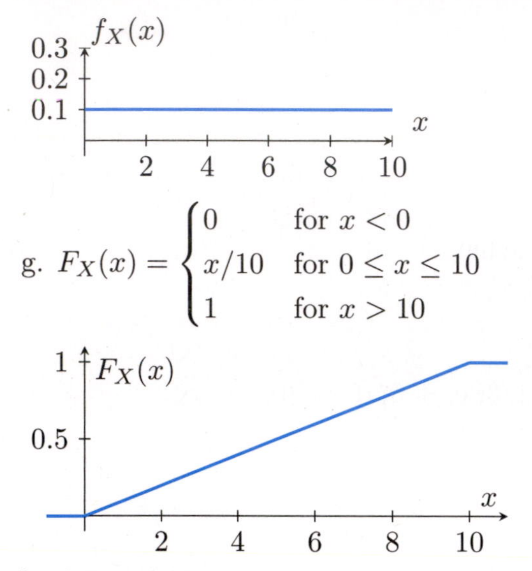

g. $F_X(x) = \begin{cases} 0 & \text{for } x < 0 \\ x/10 & \text{for } 0 \le x \le 10 \\ 1 & \text{for } x > 10 \end{cases}$

h. 0.2; i. 0.29; j. 0.4;

31.3 a. 1/4; b. 80; 31.5 4/15; 31.7 1/33; 31.9 a. \$46.75; b. \$18.0422; 31.11 a. 4.4444; b. 8.025; c. 7.7778; d. 22.84; 31.13 0.2637; 31.15 2; 31.17 $\mathbb{E}(\min(X, Y, Z)) = 2.25$; 31.19 \$0.21; 31.21 0.075; 31.23 0.19; 31.25 0.6434; 31.27 $\mathbb{E}(X) = 1$; 31.29 $F_Y(y) = 3y^2/100 - y^3/500$

Chapter 32

32.1 a. We know the average daily rate of eggs the chickens will lay, and we waiting for the first egg to be laid (one event). We are measuring the farmer's wait in minutes (continuous). b. The X is the amount of time (in minutes) until the first egg is laid. c. $\lambda = 0.0125$/minute; d. $\mathbb{E}(X) = 80$ minutes; e. $\mathrm{Var}(X) = 80$ minutes;

f. $f_X(x) = 0.0125e^{-0.0125x}$, $x > 0$

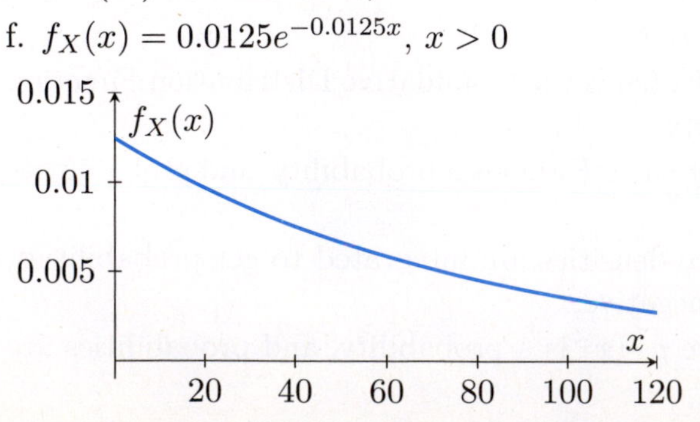

g. $F_X(x) = 1 - e^{-x/200}$, $x > 0$

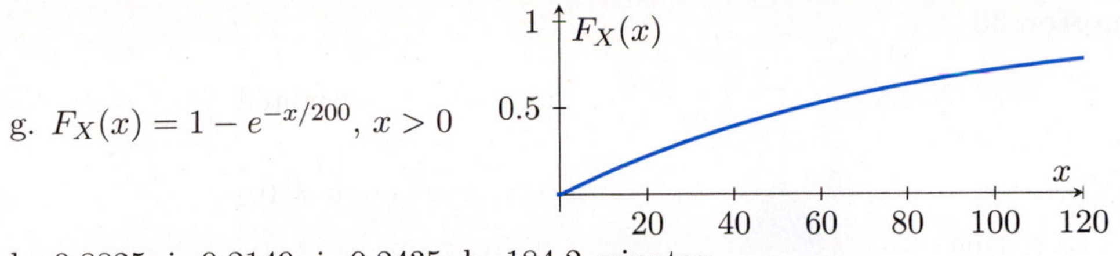

h. 0.8825; i. 0.2149; j. 0.2435; k. 184.2 minutes;
32.3 a. 0.3679; b. 0.4724; c. 0.4420; d. 0.3912; e. 0.1386 hours;
32.5 0.8047 hours; **32.7** a. 0.2835; b. 0.1353; c. 0.3679; **32.9** \$1461.62;
32.11 a. 0.5666; b. 3.0787; **32.13** $\ln 2/\lambda$; **32.15** 0.9965;
32.17 a. $p_Y(y) = e^{-\lambda y} - e^{-\lambda(y+1)}$;
b. Geometric, where we count number of losses (do not count the win), and probability of a win on a trial is $p = 1-e^{-\lambda}$, and probability of a loss is $q = e^{-\lambda}$.
32.19 a. $\frac{2a}{\lambda^2} + \frac{b}{\lambda} + c$, if X is exponential; b. $\frac{b^2}{\lambda^2} + \frac{8ab}{\lambda^3} + \frac{20a^2}{\lambda^4}$

Chapter 33

33.1 a. We know the average egg-laying rate, and we are measuring the wait until the 6th egg. Waiting for just 1 egg would be an Exponential variable.
b. The random variable X is the amount of time (in minutes) until the 6th egg is laid.
c. $\lambda = 0.0125$, $r = 6$; d. 480 minutes; e. 195.9592;
f. $f_X(x) = \frac{0.0125^6}{5!} x^5 e^{-0.0125x}$, $x > 0$;

g. $F_X(x) = 1 - e^{-0.0125x} \sum_{j=0}^{5} \frac{(0.0125x)^j}{j!}$, $x > 0$

h. $e^{-360\times 0.0125} \sum_{j=0}^{5} (360 \times 0.0125)^j/j! = 0.7029$;
i. $e^{-720\times 0.0125} \sum_{j=0}^{5} (720 \times 0.0125)^j/j! - e^{-780\times 0.0125} \sum_{j=0}^{5} (780 \times 0.0125)^j/j! = 0.0385$;
33.3 a. 315 minutes; b. 119.0588 minutes; **33.5** a. 5 hours; b. 1.5811 hours;
33.7 21 minutes; **33.9** 0.6916; **33.11** 50 minutes; **33.13** 0.4232; **33.15** 0.2381;
33.17 0.406

Chapter 34

34.1 a. 3/7; b. 0.17496;
34.3 a. It gives the proportion of horses; the density is $kx^2(1-x)^2$.
b. $\alpha = 3$, $\beta = 3$; c. $k = 30$; d. $\mathbb{E}(X) = 1/2$

Chapter 35

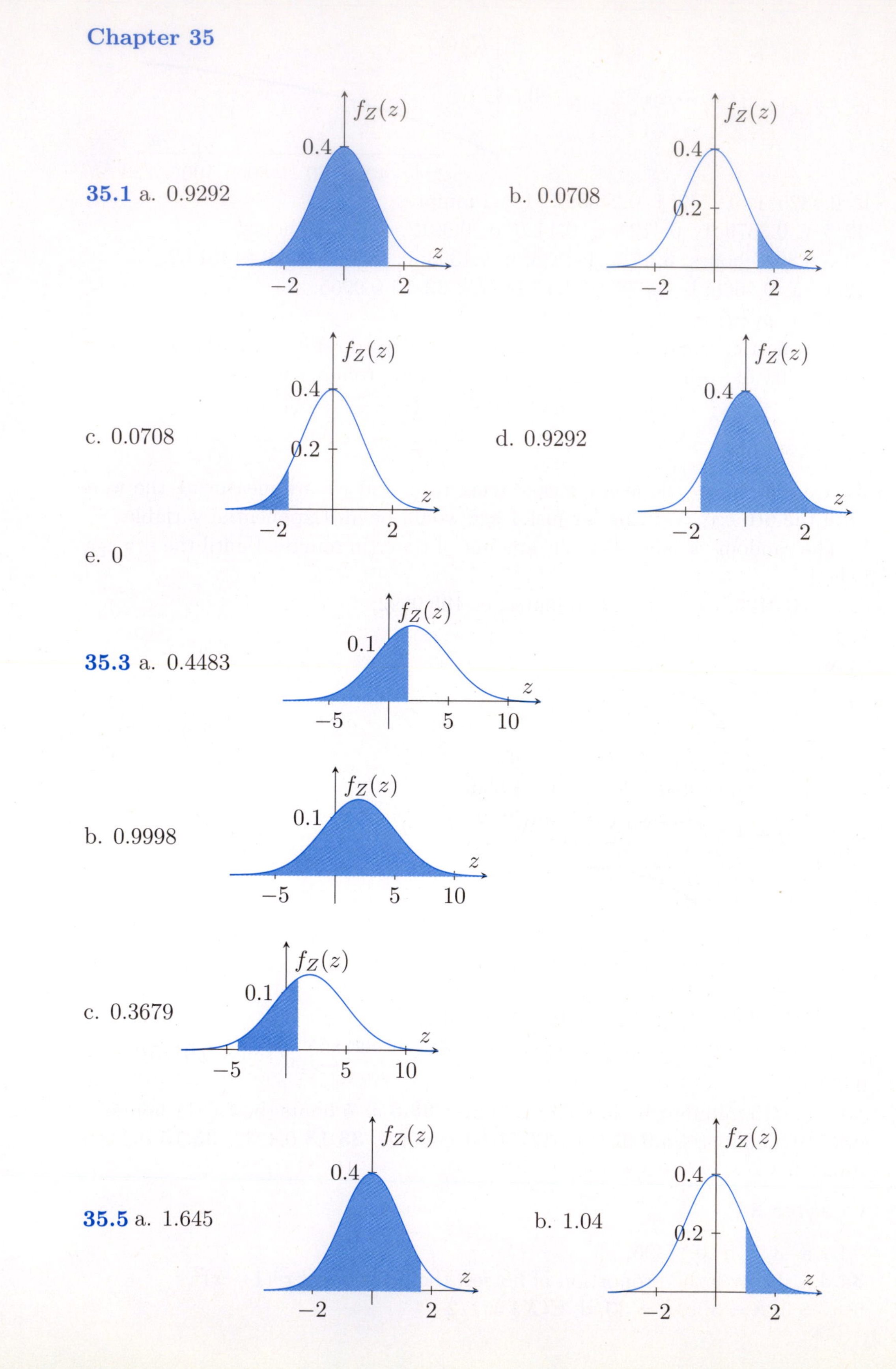

35.1 a. 0.9292

b. 0.0708

c. 0.0708

d. 0.9292

e. 0

35.3 a. 0.4483

b. 0.9998

c. 0.3679

35.5 a. 1.645

b. 1.04

c. 0.935

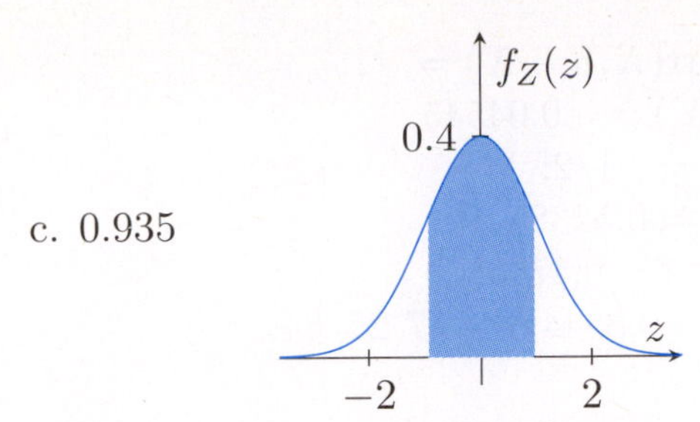

35.7 a. 4.31; b. 4.43; c. −2.935 and 6.935;
35.9 $a = -0.48$; **35.11** 0.2389; **35.13** a. 1.464 or less; b. 0.0475; c. 0.095;
35.15 0.9418; **35.17** 0.0548; **35.19** 0.6826; **35.21** 0.0548;
35.23 a. 0.0062; b. 0.3944; c. 0; d. 16.256 ounces;
e. $(15.832, 16.168)$; f. $(16.392, \infty)$

Chapter 36

36.1 0.3015; **36.3** 0.8888; **36.5** 0.1685; **36.7** 0.2810; **36.9** 0.3409; **36.11** 0.3778;
36.13 0.7372; **36.15** 0.5

Chapter 37

37.1 0.1685; **37.3** 0.1271; **37.5** 0.0384; **37.7** 0.1251; **37.9** 0.881; **37.11** 0.8098;
37.13 0.3936; **37.15** 0.6228; **37.17** 0.3156; **37.19** 0.7372; **37.21** 0.2090;
37.23 0.1841; **37.25** 0.0307; **37.27** 0.7794; **37.29** a. 0.2177; b. 0.6808; c. 0.6179

Chapter 38

38.1 Normal; **38.3** Bernoulli; **38.5** Geometric; **38.7** Poisson; **38.9** Gamma;
38.11 Hypergeometric; **38.13** a. Binomial($n = 280{,}000, p = 0.84$); b. 0.1515;
38.15 a. 0.3012; b. 0.2019; c. 0.3012; d. 0.5987;
38.17 a. 0.1792; b. 2/5; c. 1/5; **38.19** a. 0.1122; b. 1/8; c. 0.01113;
d. 5 weeks; e. 0.1353; f. 14.78 people;
38.21 a. 30 years; b. 0.05; c. 0.2105;
38.23 a. $p = 1/2$, $n = 1$; b. $\lambda = 2$; c. $\mu = 1/2 = \sigma$;
d. $a = (1 - \sqrt{3})/2$, $b = (1 + \sqrt{3})/2$; e. not possible;
38.25 a. 0.3125; b. 0.75; c. 0.0935;
38.27 a. 3 students; b. 10 minutes; $\sigma(\mathrm{X}) = 10$ minutes;
c. 30 minutes; $\sigma(\mathrm{X}) = 17.3205$ minutes;
d. 0.0111; e. 0.06197; f. 0.3168; g. 1/6; h. 3/5;
38.29 a. 0.2708; b. 0.8546; c. 0.3486;
d. A high-risk driver may cause an accident with a low-risk driver, in which case they both have an accident, so the assumption of independence is not very realistic. This assumption allows us to calculate the probability that both types of drivers have an accident by multiplying the individual probabilities.
38.31 0.4688; **38.33** a. \$146.80; b. \$7340

Chapter 39

39.1 a. Dependent; b. $\mathrm{Cov}(X, Y) = -2.4$; c. $\rho(X, Y) = -1$;
39.3 a. $\mathrm{Cov}(X, 1.3X - 10) = 43.3333$; b. $\mathrm{Corr}(X, 1.3X - 10) = 1$;

39.5 a. $\text{Cov}(X, a - X) = -Var(X)$; b. $\text{Corr}(X, a - X) = -1$;
39.7 a. $\text{Cov}(X, Y) = -0.0052$; b. $\text{Corr}(X, Y) = -0.04545$;
39.9 a. $\text{Cov}(X, Y) = -1/9$; b. $\text{Corr}(X, Y) = -1/2$;
39.11 a. $\text{Cov}(X, Y) = 2.75$; b. $\text{Corr}(X, Y) = 0.5238$;
39.13 a. $\text{Cov}(X, Y) = 0.9452$; b. $\text{Corr}(X, Y) = 0.4795$;
39.15 a. $\text{Cov}(X, Y) = 1/2$; b. $\text{Corr}(X, Y) = 0.8944$; **39.17** 25/64

Chapter 40

40.1 3/4; **40.3** 15.42; **40.5** 10.89; **40.7** 6; **40.9** 1.2; **40.11** 19 minutes;
40.13 1.5; **40.15** a. 1/5; b. 1/5;
c. Same; Given that the sum of the dice is 8, there is only one possible combination of rolls where one die shows the value 4. This is the case in which both dice simultaneously show 4, since if one die shows 4, the other die must show $8 - 4 = 4$ as well. Thus painting the dice makes no difference to the probability.
40.17 2.5 songs; **40.19** a. 0; b. $\sqrt{3}/2$; **40.21** 9/20; **40.23** $3/11 = 0.2727$

Chapter 41

41.1 a. 4/5; b. $P(6 \le X \le 18) \ge 3/4$; **41.3** a. 3/4; b. 3/4;
41.5 $P(120 \le X \le 156) \ge 0.9228$; **41.7** 3/5;
41.9 a. 5/8; b. $P(6 \le X \le 14) \ge 0.3398$;
41.11 a. $P(X \ge 6) \le 0.5444$; b. $P(|X - 196| \ge 60) \le 1/4$;
41.13 $P(|X - 30| \le 5) \ge 0.84$;
41.15 a. $P(X \ge 1000) \le 0.75$; b. $P(|X - 750| \ge 250) \le 0.16$;
41.17 a. 3/4; b. $P(|X - 9| \le 3) \ge 0.5556$; **41.19** 5/7;
41.21 $P(|X - 673| \le 105) \ge 0.434$;
41.23 a. $P(A) \le 0.4655$; b. $P(B) \ge 0.1045$

Chapter 42

42.1 a. $f_Y(y) = 1/20\,(1 - y/100)^4$ for $0 \le y \le 100$, 0 otherwise;
b. 0.3277; c. 0.4718;
42.3 a. $f_{X_{(1)}}(x_1) = 3/20\,e^{-3x_1/20}$ for $x_1 \ge 0$;
b. 0.1653; c. 0.7627;
42.5 $f_{X_{(5)}}(x_5) = 5(x_5^{3/2}/27)^4(\sqrt{x_5}/18)$ for $0 \le x_5 \le 9$, 0 otherwise;
42.7 a. $f_{X_{(1)}}(x_1) = 3/4((242 - x_1)/4)^2$ for $238 \le x_1 \le 242$, 0 otherwise;
b. $f_{X_{(3)}}(x_3) = 3/4((x_3 - 238)/4)^2$ for $238 \le x_3 \le 242$, 0 otherwise;
c. $f_{X_{(2)}}(x_2) = 6x_2(4 - x_2)/64$ for $238 \le x_2 \le 242$, 0 otherwise;
42.9 a. $f_X(x) = 7/2((8 - x)/2)^6$ for $6 \le X \le 8$, 0 otherwise;
b. $f_Y(y) = 7/2((y - 6)/2)^6$ for $6 \le Y \le 8$, 0 otherwise;
c. 6:15 PM; d. 7:45 PM; **42.11** a. 8:34:38 AM; b. 0.00002;
42.13 a. $P(X_{(j)} = x) = \binom{6}{j-1,1,6-j}(x/9)^{j-1}(1/9)(1 - x/9)^{6-j}$ for $1 \le j \le 6$;
b. 0.6488

Chapter 43

43.1 $M_X(t) = pe^t/(1 - e^t(1 - p))$; **43.5** $M_X(t) = (pe^t/(1 - e^t(1 - p)))^r$

Chapter 44

44.1 0.5013; 44.3 0.8333; 44.5 0.3015; 44.7 0.2551; 44.9 $2/\pi$;
44.13 a. $y = 0 \to v = \sqrt{u}$; $y = 1 \to v = 1 + \sqrt{u}$;
$x = 0 \to u = 0$; $x = 1 \to u = 1$;

b. 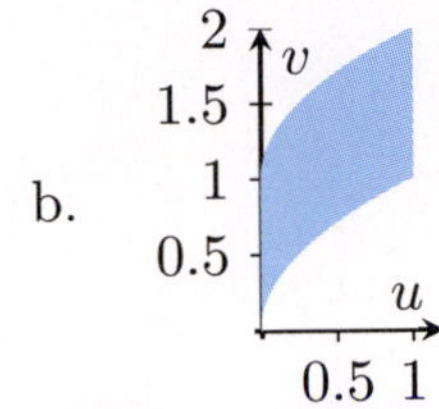

44.15 U and V are independent, since $f_{U,V}(u, v)$ can be factored into $f_U(u) = 1/(2\sqrt{u})$ and $f_V(v) = 1$, which are both bonafide densities in the defined regions $0 \le u \le 1$ and $\sqrt{u} \le v \le 1 + \sqrt{u}$.

Chapter 45

45.1 0.1210;

45.3 a. $F_Z(z) = \begin{cases} 0 & \text{for } x < 0 \\ (z/30)^2 & \text{for } 0 \le x \le 30; \\ 1 & \text{for } x > 30 \end{cases}$

b. $f_Z(z) = 2z/30^2$ for $0 \le z \le 30$, 0 otherwise; c. $\mathbb{E}(X) = 20$;
45.5 0.9772; 45.7 a. 0.0161; b. 5;
45.9 a. 9 minutes; b. 12 minutes; c. 11 minutes; d. 9000 minutes; e. 0.1814;
45.11 3.45; 45.13 10/3; 45.15 a. 227.2727; b. 1838.843;
45.17 a. 1.2918×10^{-32}; b. 1.93×10^{-30}; c. 149.4; 45.19 Poisson($\lambda = 1.5$);
45.21 a. 28 hours; b. 112; 45.23 17.6471; 45.25 $2n!/(2n)! = (n-1)!/(2n-1)!$;
45.27 0.04; 45.29 a. $0.1n$; b. $0.09n$; 45.31 5/6; 45.33 150

Bibliography

[1] Patrick Billingsley. *Probability and Measure.* Wiley, Anniversary edition, 2012.

[2] Rick Durrett. *Probability: Theory and Examples.* Cambridge, 4th edition, 2010.

[3] Philippe Flajolet and Robert Sedgewick. *Analytic Combinatorics.* Cambridge, 2009.

[4] Jim Pitman. *Probability.* Springer, 1993.

[5] Sheldon Ross. *A First Course in Probability.* Prentice Hall, 9th edition, 2014.

Index

Summary of Named Discrete Random Variables

Name	Mass	Expected value	Variance	Parameters	What X is	When used
Bernoulli	$p_X(1) = p$ $p_X(0) = q$	p	pq	$p =$ prob. succ./trial	0 or 1 (no or yes)	1 success or failure
Binomial	$\binom{n}{x} p^x q^{n-x}$	np	npq	$n = \#$ trials; $p =$ prob. succ./trial	$0, 1, 2, \ldots, n$ (successes)	successes in n trials
Geometric	$q^{x-1}p$	$1/p$	q/p^2	$p =$ prob. succ./trial	$1, 2, 3, \ldots$ (trials)	# trials to 1st succ.
Negative Binomial	$\binom{x-1}{r-1} q^{x-r} p^r$	r/p	qr/p^2	$p =$ prob. succ./trial; $r = \#$ of succ. needed	$r, r+1, \ldots$ (trials)	# trials to rth succ.
Poisson	$e^{-\lambda}\lambda^x/x!$	λ	λ	λ rate	$0, 1, 2, 3, \ldots$ (events)	# events in period
Hyper-geometric	$\frac{\binom{M}{x}\binom{N-M}{n-x}}{\binom{N}{n}}$	$n\frac{M}{N}\left(1 - \frac{M}{N}\right)\frac{N-n}{N-1}$	nM/N	M good, $N - M$ bad; n selected	$1, 2, \ldots, M$	# of good selected
Discrete Uniform	$1/N$	$(N+1)/2$	$(N^2 - 1)/12$	N outcomes	$1, 2, \ldots, N$	equally likely

Summary of Named Continuous Random Variables

Name	Uniform	Exponential	Gamma	Beta	Normal
Density $f_X(x)$	$1/(b-a)$	$\lambda e^{-\lambda x}$	$\frac{\lambda e^{-\lambda x}(\lambda x)^{r-1}}{\Gamma(r)}$	$\frac{\Gamma(\alpha+\beta)x^{\alpha-1}(1-x)^{\beta-1}}{\Gamma(\alpha)\Gamma(\beta)}$	$\frac{e^{-(x-\mu)^2/(2\sigma^2)}}{\sqrt{2\pi\sigma^2}}$
Domain	$a \le x \le b$	$x \ge 0$	$x \ge 0$	$0 \le x \le 1$	$-\infty < x < \infty$
CDF $F_X(x)$	$(x-a)/(b-a)$	$1 - e^{-\lambda x}$;	$1 - e^{-\lambda x} \sum_{j=0}^{r-1} \frac{(\lambda x)^j}{j!}$, if $r \in \mathbb{N}$		convert to std. Normal $Z = (X - \mu_X)/\sigma_X$
$\mathbb{E}(X)$	$(a+b)/2$	$1/\lambda$	r/λ	$\alpha/(\alpha+\beta)$	μ
$\mathrm{Var}(X)$	$(b-a)^2/12$	$1/\lambda^2$	r/λ^2	$\frac{\alpha\beta}{(\alpha+\beta)^2(\alpha+\beta+1)}$	σ^2
Param-eters	a, b are endpoints	λ is average # of successes per time unit	λ is average # of successes per time unit	usually some given constraints	μ = exp. value; σ = st. dev.
What X is	location in an interval	wait time 'til 1st event (like continuous Geometric)	wait time 'til rth event (like continuous Neg. Binom.)	fraction, percentage, proportion,	practical random variable for quantities clustered around avg. value
When used	spread evenly	wait time until 1st event	wait time until rth event	Bayesian statistics	Central Limit Thm. and applications

Note: $\Gamma(r) = (r-1)!$ if $r \in \mathbb{N}$.

Standard Normal Table

z	_._0	_._1	_._2	_._3	_._4	_._5	_._6	_._7	_._8	_._9
0.0	0.5000	0.5040	0.5080	0.5120	0.5160	0.5199	0.5239	0.5279	0.5319	0.5359
0.1	0.5398	0.5438	0.5478	0.5517	0.5557	0.5596	0.5636	0.5675	0.5714	0.5753
0.2	0.5793	0.5832	0.5871	0.5910	0.5948	0.5987	0.6026	0.6064	0.6103	0.6141
0.3	0.6179	0.6217	0.6255	0.6293	0.6331	0.6368	0.6406	0.6443	0.6480	0.6517
0.4	0.6554	0.6591	0.6628	0.6664	0.6700	0.6736	0.6772	0.6808	0.6844	0.6879
0.5	0.6915	0.6950	0.6985	0.7019	0.7054	0.7088	0.7123	0.7157	0.7190	0.7224
0.6	0.7257	0.7291	0.7324	0.7357	0.7389	0.7422	0.7454	0.7486	0.7517	0.7549
0.7	0.7580	0.7611	0.7642	0.7673	0.7704	0.7734	0.7764	0.7794	0.7823	0.7852
0.8	0.7881	0.7910	0.7939	0.7967	0.7995	0.8023	0.8051	0.8078	0.8106	0.8133
0.9	0.8159	0.8186	0.8212	0.8238	0.8264	0.8289	0.8315	0.8340	0.8365	0.8389
1.0	0.8413	0.8438	0.8461	0.8485	0.8508	0.8531	0.8554	0.8577	0.8599	0.8621
1.1	0.8643	0.8665	0.8686	0.8708	0.8729	0.8749	0.8770	0.8790	0.8810	0.8830
1.2	0.8849	0.8869	0.8888	0.8907	0.8925	0.8944	0.8962	0.8980	0.8997	0.9015
1.3	0.9032	0.9049	0.9066	0.9082	0.9099	0.9115	0.9131	0.9147	0.9162	0.9177
1.4	0.9192	0.9207	0.9222	0.9236	0.9251	0.9265	0.9279	0.9292	0.9306	0.9319
1.5	0.9332	0.9345	0.9357	0.9370	0.9382	0.9394	0.9406	0.9418	0.9429	0.9441
1.6	0.9452	0.9463	0.9474	0.9484	0.9495	0.9505	0.9515	0.9525	0.9535	0.9545
1.7	0.9554	0.9564	0.9573	0.9582	0.9591	0.9599	0.9608	0.9616	0.9625	0.9633
1.8	0.9641	0.9649	0.9656	0.9664	0.9671	0.9678	0.9686	0.9693	0.9699	0.9706
1.9	0.9713	0.9719	0.9726	0.9732	0.9738	0.9744	0.9750	0.9756	0.9761	0.9767
2.0	0.9772	0.9778	0.9783	0.9788	0.9793	0.9798	0.9803	0.9808	0.9812	0.9817
2.1	0.9821	0.9826	0.9830	0.9834	0.9838	0.9842	0.9846	0.9850	0.9854	0.9857
2.2	0.9861	0.9864	0.9868	0.9871	0.9875	0.9878	0.9881	0.9884	0.9887	0.9890
2.3	0.9893	0.9896	0.9898	0.9901	0.9904	0.9906	0.9909	0.9911	0.9913	0.9916
2.4	0.9918	0.9920	0.9922	0.9925	0.9927	0.9929	0.9931	0.9932	0.9934	0.9936
2.5	0.9938	0.9940	0.9941	0.9943	0.9945	0.9946	0.9948	0.9949	0.9951	0.9952
2.6	0.9953	0.9955	0.9956	0.9957	0.9959	0.9960	0.9961	0.9962	0.9963	0.9964
2.7	0.9965	0.9966	0.9967	0.9968	0.9969	0.9970	0.9971	0.9972	0.9973	0.9974
2.8	0.9974	0.9975	0.9976	0.9977	0.9977	0.9978	0.9979	0.9979	0.9980	0.9981
2.9	0.9981	0.9982	0.9982	0.9983	0.9984	0.9984	0.9985	0.9985	0.9986	0.9986
3.0	0.9987	0.9987	0.9987	0.9988	0.9988	0.9989	0.9989	0.9989	0.9990	0.9990
3.1	0.9990	0.9991	0.9991	0.9991	0.9992	0.9992	0.9992	0.9992	0.9993	0.9993
3.2	0.9993	0.9993	0.9994	0.9994	0.9994	0.9994	0.9994	0.9995	0.9995	0.9995
3.3	0.9995	0.9995	0.9995	0.9996	0.9996	0.9996	0.9996	0.9996	0.9996	0.9997
3.4	0.9997	0.9997	0.9997	0.9997	0.9997	0.9997	0.9997	0.9997	0.9997	0.9998
3.5	0.9998	0.9998	0.9998	0.9998	0.9998	0.9998	0.9998	0.9998	0.9998	0.9998
3.6	0.9998	0.9998	0.9999	0.9999	0.9999	0.9999	0.9999	0.9999	0.9999	0.9999

For a standard Normal random variable Z, these are the values of the cumulative distribution function $F_Z(z) = P(Z \leq z)$, also corresponding to the area under the density f_Z to the left of z.

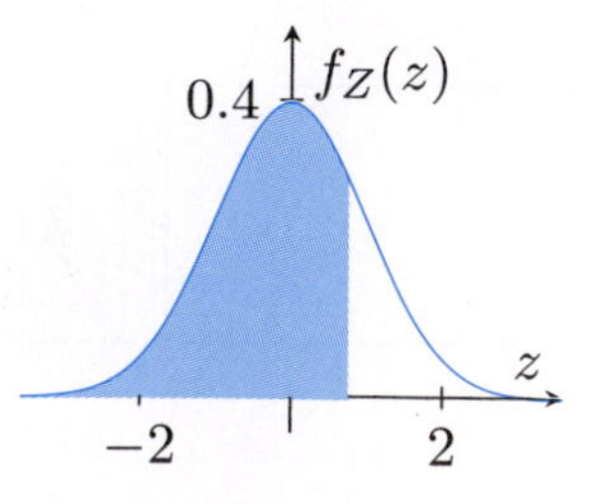

For example, in this graph, the area under the curve is $P(Z \leq 0.75) = 0.7734$.